T0212335

Lecture Notes in Artificial Intelligence 11107

Subseries of Lecture Notes in Computer Science

More information about this series at http://www.springer.com/series/1244

Petr Sojka · Aleš Horák
Ivan Kopeček · Karel Pala (Eds.)

Text, Speech, and Dialogue

21st International Conference, TSD 2018
Brno, Czech Republic, September 11–14, 2018
Proceedings

 Springer

Editors
Petr Sojka
Faculty of Informatics
Masaryk University
Brno, Czech Republic

Ivan Kopeček
Faculty of Informatics
Masaryk University
Brno, Czech Republic

Aleš Horák
Faculty of Informatics
Masaryk University
Brno, Czech Republic

Karel Pala
Faculty of Informatics
Masaryk University
Brno, Czech Republic

ISSN 0302-9743   ISSN 1611-3349   (electronic)
Lecture Notes in Artificial Intelligence
ISBN 978-3-030-00793-5   ISBN 978-3-030-00794-2   (eBook)
https://doi.org/10.1007/978-3-030-00794-2

Library of Congress Control Number: 2018954548

LNCS Sublibrary: SL7 – Artificial Intelligence

This Springer imprint is published by the registered company Springer Nature Switzerland AG
The registered company address is: Gewerbestrasse 11, 6330 Cham, Switzerland

Preface

The annual Text, Speech and Dialogue Conference (TSD), which originated in 1998, has entered its third decade. In the course of this time, thousands of authors from all over the world have contributed to the proceedings. TSD constitutes a recognized platform for the presentation and discussion of state-of-the-art technology and recent achievements in the field of natural language processing (NLP). It has become an interdisciplinary forum, interweaving the themes of speech technology and language processing. The conference attracts researchers not only from Central and Eastern Europe but also from other parts of the world. Indeed, one of its goals has always been to bring together NLP researchers with different interests from different parts of the world and to promote their mutual cooperation.

One of the declared goals of the conference has always been, as its title says, twofold: not only to deal with language processing and dialogue systems as such, but also to stimulate dialogue between researchers in the two areas of NLP, i.e., between text and speech people. In our view, the TSD Conference was successful in this respect in 2018 again. We had the pleasure of welcoming three prominent invited speakers this year: Kenneth Ward Church presented a keynote with a proposal of an organizing framework for deep nets titled "Minsky, Chomsky & Deep Nets"; Piek Vossen presented the Pepper robot in "Leolani: A Reference Machine with a Theory of Mind for Social Communication"; and Isabel Trancoso reported on "Speech Analytics for Medical Applications".

This volume contains the proceedings of the 21st TSD Conference, held in Brno, Czech Republic, in September 2018. In the review process, 53 papers were accepted out of 110 submitted papers, leading to an acceptance rate of 48%.

We would like to thank all the authors for the efforts they put into their submissions and the members of the Program Committee and reviewers who did a wonderful job selecting the best papers. We are also grateful to the invited speakers for their contributions. Their talks provide insight into important current issues, applications, and techniques related to the conference topics.

Special thanks go to the members of the Local Organizing Committee for their tireless effort in organizing the conference.

We hope that the readers will benefit from the results of this event and disseminate the ideas of the TSD Conference all over the world. Enjoy the proceedings!

July 2018

Aleš Horák
Ivan Kopeček
Karel Pala
Petr Sojka

Organization

TSD 2018 was organized by the Faculty of Informatics, Masaryk University, in cooperation with the Faculty of Applied Sciences, University of West Bohemia in Plzeň. The conference webpage is located at http://www.tsdconference.org/tsd2018/.

Program Committee

Elmar Nöth (General Chair), Germany
Rodrigo Agerri, Spain
Eneko Agirre, Spain
Vladimir Benko, Slovakia
Archna Bhatia, USA
Jan Černocký, Czech Republic
Simon Dobrisek, Slovenia
Kamil Ekstein, Czech Republic
Karina Evgrafova, Russia
Yevhen Fedorov, Ukraine
Volker Fischer, Germany
Darja Fiser, Slovenia
Eleni Galiotou, Greece
Björn Gambäck, Norway
Radovan Garabík, Slovakia
Alexander Gelbukh, Mexico
Louise Guthrie, UK
Tino Haderlein, Germany
Jan Hajič, Czech Republic
Eva Hajičová, Czech Republic
Yannis Haralambous, France
Hynek Hermansky, USA
Jaroslava Hlaváčová, Czech Republic
Aleš Horák, Czech Republic
Eduard Hovy, USA
Denis Jouvet, France
Maria Khokhlova, Russia
Aidar Khusainov, Russia
Daniil Kocharov, Russia
Miloslav Konopík, Czech Republic
Ivan Kopeček, Czech Republic
Valia Kordoni, Germany

Evgeny Kotelnikov, Russia
Pavel Král, Czech Republic
Siegfried Kunzmann, Germany
Nikola Ljubešić, Croatia
Natalija Loukachevitch, Russia
Bernardo Magnini, Italy
Oleksandr Marchenko, Ukraine
Václav Matoušek, Czech Republic
France Mihelić, Slovenia
Roman Mouček, Czech Republic
Agnieszka Mykowiecka, Poland
Hermann Ney, Germany
Karel Oliva, Czech Republic
Juan Rafael Orozco-Arroyave, Colombia
Karel Pala, Czech Republic
Nikola Pavesić, Slovenia
Maciej Piasecki, Poland
Josef Psutka, Czech Republic
James Pustejovsky, USA
German Rigau, Spain
Marko Robnik Šikonja, Slovenia
Leon Rothkrantz, The Netherlands
Anna Rumshinsky, USA
Milan Rusko, Slovakia
Pavel Rychlý, Czech Republic
Mykola Sazhok, Ukraine
Pavel Skrelin, Russia
Pavel Smrž, Czech Republic
Petr Sojka, Czech Republic
Stefan Steidl, Germany
Georg Stemmer, Germany
Vitomir Štruc, Slovenia

Marko Tadić, Croatia
Tamas Varadi, Hungary
Zygmunt Vetulani, Poland
Aleksander Wawer, Poland
Pascal Wiggers, The Netherlands

Yorick Wilks, UK
Marcin Wołinski, Poland
Alina Wróblewska, Poland
Victor Zakharov, Russia
Jerneja Žganec Gros, Slovenia

Additional Reviewers

Ladislav Lenc
Marton Makrai
Malgorzata Marciniak
Montse Maritxalar
Jiří Martinek
Elizaveta Mironyuk

Arantza Otegi
Bálint Sass
Tadej Skvorc
Jan Stas
Ivor Uhliarik

Organizing Committee

Aleš Horák (Co-chair), Ivan Kopeček, Karel Pala (Co-chair), Adam Rambousek (Web System), Pavel Rychlý, Petr Sojka (Proceedings)

Sponsors and Support

The TSD conference is regularly supported by International Speech Communication Association (ISCA). We would like to express our thanks to the Lexical Computing Ltd. and IBM Česká republika, spol. s r. o. for their kind sponsoring contribution to TSD 2018.

Contents

Speech

Dialogue

Invited Papers

Minsky, Chomsky and Deep Nets

Kenneth Ward Church[✉]

Baidu, Sunnyvale, CA, USA
KennethChurch@baidu.com

Abstract. When Minsky and Chomsky were at Harvard in the 1950s, they started out their careers questioning a number of machine learning methods that have since regained popularity. Minsky's *Perceptrons* was a reaction to neural nets and Chomsky's *Syntactic Structures* was a reaction to ngram language models. Many of their objections are being ignored and forgotten (perhaps for good reasons, and perhaps not). While their arguments may sound negative, I believe there is a more constructive way to think about their efforts; they were both attempting to organize computational tasks into larger frameworks such as what is now known as the Chomsky Hierarchy and algorithmic complexity. Section 5 will propose an organizing framework for deep nets. Deep nets are probably not the solution to all the world's problems. They don't do the impossible (solve the halting problem), and they probably aren't great at many tasks such as sorting large vectors and multiplying large matrices. In practice, deep nets have produced extremely exciting results in vision and speech, though other tasks may be more challenging for deep nets.

Keywords: Minsky · Chomsky · Deep nets · Perceptrons

1 A Pendulum Swung Too Far

There is considerable excitement over deep nets, and for good reasons. More and more people are attending more and more conferences on Machine Learning. Deep nets have produced substantial progress on a number of benchmarks, especially in vision and speech.

This progress is changing the world in all kinds of ways. Face recognition and speech recognition are everywhere. Voice-powered search is replacing typing.[1] Cameras are everywhere as well, especially in China. While the West finds it creepy to live in a world with millions of cameras,[2] the people that I talk to in China believe that cameras reduce crime and make people feel safer [1]. The big commercial opportunity for face recognition is likely to be electronic

---

[1] http://money.cnn.com/2017/05/31/technology/mary-meeker-internet-trends/index.html.

[2] http://www.dailymail.co.uk/news/article-4918342/China-installs-20-million-AI-equipped-street-cameras.html.

© Springer Nature Switzerland AG 2018
P. Sojka et al. (Eds.): TSD 2018, LNAI 11107, pp. 3–14, 2018.
https://doi.org/10.1007/978-3-030-00794-2_1

payments, but there are many smaller opportunities for face recognition. Many people use face recognition to unlock their phones. My company, Baidu, uses face recognition to unlock the doors to the building. After I link my face to my badge, I don't need to bring my badge to get into the building; all I need is my face. Unlike American Express products,[3] it is hard to leave home without my face.

What came before deep nets? When Minsky and Chomsky were at Harvard in the 1950s, they started out their careers questioning a number of machine learning methods that have since regained popularity. Minsky's *Perceptrons* [2] was a reaction to neural nets and Chomsky's *Syntactic Structures* [3] was a reaction to ngram language models [4,5] (and empiricism [6–8]). My generation returned the favor with the revival of empiricism in the 1990s. In *A Pendulum Swung Too Far* [9], I suggested that the field was oscillating between Rationalism and Empiricism, switching back and forth every 20 years. Each generation rebelled against their teachers. Grandparents and grandchildren have a natural alliance; they have a common enemy.[4]

- **1950s:** Empiricism (Shannon, Skinner, Firth, Harris)
- **1970s:** Rationalism (Chomsky, Minsky)
- **1990s:** Empiricism (IBM Speech Group, AT&T Bell Labs)
- **2010s:** A Return to Rationalism?

I then suggested that the field was on the verge of a return to rationalism. Admittedly, that seemed rather unlikely even then (2011), and it seems even less likely now (2018), but I worry that our revival of empiricism may have been too successful, squeezing out many other worthwhile positions:

The revival of empiricism in the 1990s was an exciting time. We never imagined that effort would be as successful as it turned out to be. At the time, all we wanted was a seat at the table. In addition to everything else that was going on at the time, we wanted to make room for a little work of a different kind. We founded SIGDAT to provide a forum for this kind of work. SIGDAT started as a relatively small Workshop on Very Large Corpora in 1993 and later evolved into the larger EMNLP Conferences. At first, the SIGDAT meetings were very different from the main ACL conference in many ways (size, topic, geography), but over the years, the differences have largely disappeared. It is nice to see the field come together as it has, but we may have been too successful. Not only have we succeeded in making room for what we were interested in, but now there is no longer much room for anything else.

My "prediction" wasn't really a prediction, but more of a plea for inclusiveness. The field would be better off if we could be more open to diverse opinions

---

[3] http://www.thedrum.com/news/2016/07/03/marketing-moments-11-american-express-dont-leave-home-without-it.

[4] https://www.brainyquote.com/quotes/sam_levenson_100238.

and backgrounds. Computational Linguistics used to be an interdisciplinary combination of Humanities and Engineering, but I worry that my efforts to revive empiricism in the 1990s may be largely responsible for the field taking a hard turn away from the Humanities toward where we are today (more Engineering).

2   A Farce in Three Acts

In 2017, Pereira blogged a more likely prediction in *A (computational) linguistic farce in three acts.*[5]

- **Act One:** Rationalism
- **Act Two:** Empiricism
- **Act Three:** Deep Nets

It is hard to disagree that we are now in the era of deep nets. Reflecting some more on this history, I now believe that the pendulum position is partly right and partly wrong. Each act (or each generation) rejects much of what came before it, but also borrows much of what came before it.

There is a tendency for each act to emphasize differences from the past, and deemphasize similarities. My generation emphasized the difference between empiricism and rationalism, and deemphasized how much we borrowed from the previous act, especially a deep respect for representations. The third act, deep nets, emphasizes certain differences from the second act, such as attempts to replace Minsky-style representations with end-to-end self-organizing systems, but deemphasizes similarities such as a deep respect for empiricism.

Pereira's post suggests that each act was brought on by a tragic flaw in a previous act.

- **Act One:** The (Weak) Empire of Reason
- **Act Two:** The Empiricist Invasion or, Who Pays the Piper Calls the Tune
- **Act Three:** The Invaders get Invaded or, The Revenge of the Spherical Cows

It's tough to make predictions, especially about the future.[6,7] But the tragic pattern is tragic. The first two acts start with excessive optimism and end with disappointment. It is too early to know how the third act will end, but one can guess from the final line of the epilogue:

[W]e have been struggling long enough in our own ways to recognize the need for coming together with better ways of plotting our progress.

That the third act will probably end badly, following the classic pattern of Greek tragedies where the tragic hero starts out with a tragic flaw that inevitably leads to his tragic downfall.

---

[5] http://www.earningmyturns.org/2017/06/a-computational-linguistic-farce-in.html.

[6] https://en.wikiquote.org/wiki/Yogi_Berra.

[7] https://quoteinvestigator.com/2013/10/20/no-predict/.

Who knows which tragic flaw will lead our tragic hero to his downfall, but the third act opens, not unlike the previous two, with optimists being optimistic. I recently heard someone in the halls mentioning a recent exciting result that sounded too good to be true. Apparently, deep nets can learn any function. That would seem to imply that everything I learned about computability was wrong. Didn't Turing [10] prove that nothing (not even a Turing Machine) can solve the halting problem? Can neural nets do the impossible?

Obviously, you can't believe everything you hear in the halls. The comment was derived from a blog with a sensational title,[8] *A visual proof that neural nets can compute any function.* Not surprisingly, the blog doesn't prove the impossible. Rather, it provides a nice tutorial on the universal approximation theorem.[9] The theorem is not particularly recent (1989), and doesn't do the impossible (solve the halting problem), but it is an important result that shows that neural nets can approximate many continuous functions. Neural nets can do lots of useful things (especially in vision and speech), but neural nets aren't magic. It is good for morale when folks are excited about the next new thing, but too much excitement can have tragic consequences (AI winters).[10]

Our field may be a bit too much like a kids' soccer team. It is a cliché among coaches to talk about all the kids running toward the ball, and no one covering their position.[11] We should cover the field better than we do by encouraging more interdisciplinary work, and deemphasizing the temptation to run toward the fad of the day. University classes tend to focus too much on relatively narrow topics that are currently hot, but those topics are unlikely to remain hot for long.

In *A pendulum swung too far*, I expressed concern that we aren't teaching the next generation what they will need to know for the next act (whatever that will be). One can replace "rationalist" in the following comments with whatever comes after deep nets:

> This paper will review some of the rationalist positions that our generation rebelled against. It is a shame that our generation was so successful that these rationalist positions are being forgotten (just when they are about to be revived if we accept that forecast). Some of the more important rationalists like Pierce are no longer even mentioned in currently popular textbooks. The next generation might not get a chance to hear the rationalist side of the debate. And the rationalists have much to offer, especially if the rationalist position becomes more popular in a few decades.

We should teach more perspectives on more questions, not only because we don't know what will be important, but also because we don't want to impose too much control over the narrative. Fields have a tendency to fall into an Orwellian

---

[8] http://neuralnetworksanddeeplearning.com/chap4.html.

[9] https://en.wikipedia.org/wiki/Universal_approximation_theorem.

[10] https://en.wikipedia.org/wiki/AI_winter.

[11] https://stevenpdennis.com/2015/07/10/a-bunch-of-little-kids-running-toward-a-soccer-ball/.

dystopia: "Who controls the past controls the future. Who controls the present controls the past."

Students need to learn how to use popular approximations effectively. Most approximations make simplifying assumptions that can be useful in many cases, but not all. For example, ngrams can capture many dependences, but obviously not when the dependency spans over more than n words. Similarly, linear separators can separate positive examples from negative examples in many cases, but not when the examples are not linearly separable. Many of these limitations are obvious (by construction), but even so, the debate, both pro and con, has been heated at times. And sometimes, one side of the debate is written out of the textbooks and forgotten, only to be revived/reinvented by the next generation.

As suggested above, computability is one of many topics that is in danger of being forgotten. Too many people, including the general public and even graduates of computer science programs, are prepared to believe that a machine could one day learn/compute/answer any question asked of it. One might come to this conclusion after reading this excerpt from a NY Times book review:

The story I tell in my book is of how at the end of World War II, John von Neumann and his team of mathematicians and engineers began building the very machine that Alan Turing had envisioned in his 1936 paper, "On Computable Numbers." This was a machine that could answer any question asked of it.[12]

It used to be that everyone knew the point of Turing's paper [10], but this subject has been sufficiently forgotten by enough of Wired Magazine's audience that they found it worthwhile to publish a tutorial on the question: *are there any questions that a computer can never answer?*[13] The article includes a delightful poem by Geoffrey Pullum in honor of Alan Turing in the style of Dr. Seuss. The poem, *Scooping the Loop Snooper*,[14] is well worth reading, even if you don't need a tutorial on computability.

3   Ngrams Can't Do this and Nets Can't Do that

As mentioned above, Minsky and Chomsky started out their careers in the 1950s questioning a number of machine learning methods that have since regained popularity. Minsky's *Perceptrons* and Chomsky's *Syntactic Structures* are largely remembered as negative results. Chomsky showed that ngrams (and more generally, finite state machines (FSMs)) cannot capture context-free (CF) constructions, and Minsky showed that perceptrons (neural nets without hidden layers) cannot learn XOR.

---

[12] https://www.nytimes.com/2011/12/06/science/george-dyson-looking-backward-to-put-new-technology-in-focus.html.

[13] https://www.wired.com/2014/02/halting-problem/.

[14] http://www.lel.ed.ac.uk/~gpullum/loopsnoop.html.

While these results may sound negative, I believe there is a more constructive way to think about their efforts; they were both attempting to organize computational tasks into larger frameworks such as what is now known as the Chomsky Hierarchy and algorithmic complexity. This section will review their arguments, and Sect. 5 will discuss a proposal for an organizing framework for deep nets.

Both ngrams and neural nets, according to Minsky and Chomsky, have problems with memory. Chomsky objected that ngrams don't have enough memory for his tasks, whereas Minsky objected that perceptrons were using too much memory for his tasks. These days, it is common to organize tasks by time and space complexity. For example, string processing with a finite state machine (FSM) uses constant space (finite memory, independent of the size of the input) unlike string processing with a push down automata (PDA), which can push the input onto the stack, consuming n space (infinite memory that grows linearly with the size of the input).

Chomsky is an amazing debater. His arguments carried the day partly based on the merits of the case, but also because of his rhetorical skills. He frequently uses expressions such as *generative capacity, capturing long-distance dependencies, fundamentally inadequate* and *as anyone can plainly see*. Minsky's writing is less accessible; he comes from a background in math, and dives quickly into theorems and proofs, with less motivation, discussion and rhetoric than engineers and linguists are used to. *Syntactic Structures* is an easy read; *Perceptrons* is not.

Chomsky argued that ngrams cannot learn long distance dependencies. While that might seem obvious in retrospect, there was a lot of excitement at the time over the Shannon-McMillan-Breiman Theorem,[15] which was interpreted to say that, in the limit, under just a couple of minor caveats and a little bit of not-very-important fine print, ngram statistics are sufficient to capture all the information in a string (such as an English sentence). The universal approximation theorem mentioned above could be viewed in a somewhat similar light. Some people may (mis)-interpret such results to suggest that nets and ngrams can do more than they can do, including solving undecidable problems.

Chomsky objected to ngrams on parsimony grounds. He believed that ngrams are far from the most parsimonious representation of certain linguistic facts that he was interested in. He introduced what is now known as the Chomsky Hierarchy to make his argument more rigorous. Context free (CF) grammars have more generative capacity than finite state (FS). That is, the set of CF languages is strictly larger than FS, and therefore, there are things that can be done in CF that cannot be done in FS. In particular, it is easy for a CF grammar to match parentheses (using a stack with infinite memory that grows linearly with the size of the input), but a FSM can't match parentheses (in finite memory that does not depend on the size of the input). Since ngrams are a special case of FS, ngrams don't have the generative capacity to capture long-distance constraints such as parentheses. Chomsky argued that subject-verb agreement (and many

---

[15] https://en.wikipedia.org/wiki/Asymptotic_equipartition_property.

of the linguistic phenomena that he was interested in) are like parentheses, and therefore, FS grammars are fundamentally inadequate for his purposes. Interest in ngrams (and empiricism) faded as more and more people became persuaded by Chomsky's arguments.

Minsky's argument starts with XOR, but his real point is about parity, a problem that can be done in constant space with a FSM, but consumes linear space with perceptrons (neural nets). The famous XOR result (for single layer nets) is proved in Sect. 2 of [2], but Fig. 2.1 may be more accessible than the proof. Figure 2.1 makes it clear that a quadratic equation (second order polynomial) can easily capture XOR but a line (first order polynomial) line cannot. Figure 3.1 generalizes the observation for parity. It was common in those days for computer hardware to use error correcting memory with parity bits. A parity bit would count the number of bits (mod 2) in a computer word (typically 64 bits today, but much less in those days). Figure 3.1 shows that the order of the polynomial required to count parity in this way grows linearly with the size of the input (number of bits in the computer word). That is, nets are using a 2nd order polynomial to compute parity for a 1-bit computer word (XOR), and a 65th order polynomial to compute parity for a 64-bit computer word. Minsky argues that nets are not the most parsimonious solution (for certain problems such as parity) since nets are using linear space to solve a task that can be done in constant space.

These days, with more modern network architectures such as RNNs and LSTMs, there are solutions to parity in finite space.[16,17] It ought to be possible to show that (modern) nets can solve all FS problems in finite space though I don't know of a proof that that is so. Some nets can solve some CF problems (like matching parentheses). Some of these nets will be mentioned in Sect. 4. But these nets are quite different from the nets that are producing the most exciting results in vision and speech. It should be possible to show that the exciting nets aren't too powerful (too much generative capacity is not necessarily a good thing). I suspect the most exciting nets can't solve CF problems, though again, I don't know of a proof that that is so.

Minsky emphasized representation (the antithesis of end-to-end self-organizing systems). He would argue that some representations are more appropriate for some tasks, and others are more appropriate for other tasks. Regular languages (FSMs) are more appropriate for parity than n^{th} order polynomials. Minsky's view on representation is very different from alternatives such as end-to-end self-organizing systems, which will be discussed in Sect. 5.

I recently heard someone in the halls asking why Minsky never considered hidden layers. Actually, this objection has come up many times over the years. In the epilog of the 1988 edition, there is a discussion on p. 254 that points out that hidden layers were discussed in Sect. 13.1 of the original book [2]. I believe Minsky's book would have had more impact if it had been more accessible. The

---

[16] http://www.cs.toronto.edu/~rgrosse/csc321/lec9.pdf.
[17] https://blog.openai.com/requests-for-research-2/.

argument isn't that complicated, but unfortunately, the presentation is more challenging than it has to be.

While Minsky has never been supportive of neural nets, he was very supportive of the Connection Machine (CM),[18] a machine that looks quite a bit like a modern day GPU. The CM started with a Ph.D. thesis by his student, Hillis [11]. Their company came up with an algorithm for sorting a vector in $log(n)$ time, a remarkable result given the lower bound of $n\ log(n)$ for sorting on conventional hardware [12]. The algorithm assumes the vector is small enough to fit into the CM memory.

The algorithm is based on two primitives, *send* and *scan*. Send takes a vector of pointers and sends the data from one place to another via the pointers. The send operation takes a long time (linear time) if much of the data comes from the same place or goes to the same place, but it can run much faster in parallel if the pointers are (nearly) a permutation. Scan (also known as parallel prefix) applies a function (such as sum) to all prefixes of a vector and returns a vector of partial sums. Much of this thinking can be found in modern GPUs.[19]

4   Sometimes Scale Matters, and Sometimes It Doesn't

Scaling matters for large problems, but some problems aren't that large. Scaling probably isn't that important for the vision and speech problems that are driving much of the excitement behind deep nets. Vectors that are small enough to fit into the CM memory can be sorted in $log(n)$ time, considerably better than the $n\ log(n)$ lower bound for large vectors on conventional hardware. So too, modern GPUs are often used to multiply (small) matrices. GPUs seem to work well in practice, as long as the matrices aren't too big.

Minsky and Chomsky's arguments above depend on scale. In practice, parity is usually computed over short (64-bit) words, and therefore, it doesn't matter that much if the solution depends on the size of the computer word or not. Similar comments hold for ngrams. As a practical matter, ngram methods have done remarkably well over the years, better than alternatives that have attempted to capture long-distance dependencies, and ended up capturing less. In my master's thesis [13], I argued that agreement in natural language isn't like matching parentheses in programming languages since natural language avoids center embedding. Stacks are clearly required to parse programming languages, but the argument for stacks is less compelling for natural language. It is easy to find evidence of deep center embedding in programs, but one rarely finds such evidence in natural language corpora. In practice, center embedding rarely goes beyond a depth of one or two.

There are lots of examples of theoretical arguments that don't matter much in practice because of inappropriate scaling assumptions. There was a theoretical argument that suggested that a particular morphology method couldn't work well because the time complexity was exponential (in the number of harmony

---

[18] https://en.wikipedia.org/wiki/Thinking_Machines_Corporation.
[19] https://developer.nvidia.com/gpugems/GPUGems3/gpugems3_ch39.html.

processes). There obviously had to be a problem with this argument in practice since the method was used successfully every day by a major newspaper in Finland. The problem with the theory, we argued [14], was that harmony processes don't scale. It is hard to find a language with more than one or two harmony rules. Exponential processes aren't a problem in practice, as long as the exponents are small.

There have been a number of attempts such as [15,16][20] to address Chomsky's generative capacity concerns head on, but these attempts don't seem to be producing the same kinds of exciting successes as we are seeing in vision and speech. I find it more useful to view a deep net as a parallel computer, somewhat like a CM or a GPU, that is really good for some tasks like sorting small vectors and multiplying small matrices, but CMs and GPUs aren't the solution to all the world's problems. One shouldn't expect a net to solve the halting problem, sort large vectors, or multiply large matrices. The challenge is to come up with a framework for organizing problems by degree of difficulty. Machine learning could use something like the Chomsky Hierarchy and time and space complexity to organize tasks so we have a better handle on when deep nets are likely to be more effective, and when they are likely to be less effective.

5   There Is No Data Like More Data

Figure 1 is borrowed from [17]. They showed that performance on a particular task improves with the size of the training set. The improvement is dramatic, and can easily dominate the kinds of issues we tend to think about in machine learning.

The original figure in [17] didn't include the comment about firing people, but Eric Brill (personal communication) has said such things, perhaps in jest. In his acceptance speech of the 2004 Zampolli prize, "Some of my Best Friends are Linguists,"[21] Jelinek discussed Fig. 1 as well as the origins of the quote, "Whenever I fire a linguist our system performance improves," but didn't mention that his wife was a student of the linguist Roman Jakobson. The introduction to Mercer's Lifetime Achievement Award[22] provides an entertaining discussion of that quote with words: "Jelinek said it, but didn't believe it; Mercer never said it, but he believes it." Mercer describes the history of end-to-end systems at IBM on p. 7 of [18]. Their position on end-to-end self-organizing systems is hot again, especially in the context of deep nets.

The curves in Fig. 1 are approximately linear. Can we use that observation to extrapolate performance? If we could increase the training set by a few orders of magnitude, what would that be worth?

Power laws have been shown to be a useful way to organize the literature in a number of different contexts [19].[23] In machine learning, learning curves model

---

[20] http://www.personal.psu.edu/ago109/giles-ororbia-rnn-icml2016.pdf.

[21] http://www.lrec-conf.org/lrec2004/doc/jelinek.pdf.

[22] http://techtalks.tv/talks/closing-session/60532/ (at 6:07 min).

[23] https://en.wikipedia.org/wiki/Geoffrey_West.

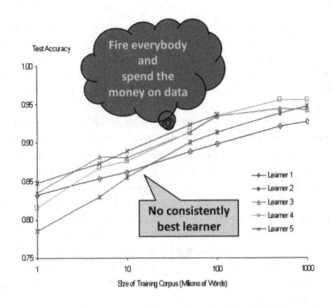

Fig. 1. *It never pays to think until you've run out of data* [17]. Increasing the size of the training set improves performance (more than machine learning).

loss, $\epsilon(m)$, as a power law, $\alpha m^{\beta} + \gamma$, where m is the size of the training data, and α and γ are uninteresting constants. The empirical estimates for β in Table 1 are based on [20]. In theory, $\beta \geq -\frac{1}{2}$,[24] but in practice, β is different for different tasks. The tasks that we are most excited about in vision and speech have a β closer to the theoretical bound, unlike other applications of deep nets where β is farther from the bound. Table 1 provides an organization of the deep learning literature, somewhat analogous to the discussion of the Chomsky Hierarchy in Sect. 3. The tasks with a β closer to the theory (lower down in Table 1) are relatively effective in taking advantage of more data.

Table 1. Some tasks are making better use of more data [20]

Task	β
Language modeling (with characters)	−0.092
Machine translation	−0.128
Speech recognition	−0.291
Image classification	−0.309
Theory	−0.500

---

[24] http://cs229.stanford.edu/notes/cs229-notes4.pdf.

6   Conclusions

Minsky and Chomsky's arguments are remembered as negative because they argued against some positions that have since regained popularity. It may sound negative to suggest that ngrams can't do this, and nets can't do that, but actually these arguments led to organizations of computational tasks such as the Chomsky Hierarchy and algorithmic complexity. Section 5 proposed an alternative framework for deep nets. Deep nets are not the solution to all the world's problems. They don't do the impossible (solve the halting problem), and they aren't great at many tasks such as sorting large vectors and multiplying large matrices. In practice, deep nets have produced extremely exciting results in vision and speech. These tasks appear to be making very effective use of more data. Other tasks, especially those mentioned higher in Table 1, don't appear to be as good at taking advantage of more data, and aren't producing as much excitement, though perhaps, those tasks have more opportunity for improvement.

References

1. Church, K.: Emerging trends: artificial intelligence, China and my new job at Baidu. J. Nat. Lang. Eng. (to appear). University Press, Cambridge
2. Minsky, M., Papert, S.: Perceptrons. MIT Press, Cambridge (1969)
3. Chomsky, N.: Syntactic Structures. Mouton & Co. (1957). https://archive.org/details/NoamChomskySyntcaticStructures
4. Shannon, C.: A mathematical theory of communication. Bell Syst. Tech. J. **27**, 379–423, 623–656 (1948). http://math.harvard.edu/~ctm/home/text/others/shannon/entropy/entropy.pdf
5. Shannon, C.: Prediction and entropy of printed English. Bell Syst. Tech. J. **30**(1), 50–64 (1951). https://www.princeton.edu/~wbialek/rome/refs/shannon51.pdf
6. Zipf, G.: Human Behavior and the Principle of Least Effort: An Introduction to Human Ecology. Addison-Wesley, Boston (1949)
7. Harris, Z.: Distributional structure. Word **10**(2–3), 146–162 (1954)
8. Firth, J.: A synopsis of linguistic theory, 1930–1955. Stud. Linguist. Anal. Basil Blackwell (1957). http://annabellelukin.edublogs.org/files/2013/08/Firth-JR-1962-A-Synopsis-of-Linguistic-Theory-wfihi5.pdf
9. Church, K.: A pendulum swung too far. Linguist. Issues Lang. Technol. **6**(6), 1–27 (2011)
10. Turing, A.: On computable numbers, with an application to the Entscheidungsproblem. In: Proceedings of the London Mathematical Society, vol. 2, no. 1, pp. 230–265. Wiley Online Library (1937). http://www.turingarchive.org/browse.php/b/12
11. Hillis, W.: The Connection Machine. MIT Press, Cambridge (1989)
12. Blelloch, G., Leiserson, C., Maggs, B., Plaxton, C., Smith, S., Marco, C.: A comparison of sorting algorithms for the connection machine CM-2. In: Proceedings of the Third Annual ACM Symposium on Parallel Algorithms and Architectures, SPAA, pp. 3–16 (1991). https://courses.cs.washington.edu/courses/cse548/06wi/files/benchmarks/radix.pdf

13. Church, K.: On memory limitations in natural language processing, unpublished Master's thesis (1980). http://publications.csail.mit.edu/lcs/pubs/pdf/MIT-LCS-TR-245.pdf

14. Koskenniemi, K., Church, K.: Complexity, two-level morphology and Finnish. In: Coling (1988). https://aclanthology.info/pdf/C/C88/C88-1069.pdf

15. Graves, A., Wayne, G., Danihelka, I.: Neural Turing Machines. arXiv (2014). https://arxiv.org/abs/1410.5401

16. Sun, G., Giles, C., Chen, H., Lee, Y: The Neural Network Pushdown Automaton: Model, Stack and Learning Simulations. arXiv (2017). https://arxiv.org/abs/1711.05738

17. Banko, M., Brill, E.: Scaling to very very large corpora for natural language disambiguation, pp. 26–33. ACL (2001). http://www.aclweb.org/anthology/P01-1005

18. Church, K., Mercer, R.: Introduction to the special issue on computational linguistics using large corpora. Comput. Linguist. **19**(1), 1–24 (1993). http://www.aclweb.org/anthology/J93-1001

19. West, G.: Scale. Penguin Books, New York (2017)

20. Hestness, J., Narang, S., Ardalani, N., Diamos, G., Jun, H.: Deep Learning Scaling is Predictable, Empirically. arXiv (2017). https://arxiv.org/abs/1712.00409

Leolani: A Reference Machine with a Theory of Mind for Social Communication

Piek Vossen[✉], Selene Baez, Lenka Bajčetić, and Bram Kraaijeveld

Computational Lexicology and Terminology Lab, VU University Amsterdam,
De Boelelaan 1105, 1081HV Amsterdam, The Netherlands
{p.t.j.m.vossen,s.baezsantamaria,l.bajcetic,b.kraaijeveld}@vu.nl
www.cltl.nl

Abstract. Our state of mind is based on experiences and what other
people tell us. This may result in conflicting information, uncertainty, and
alternative facts. We present a robot that models relativity of knowledge
and perception within social interaction following principles of the *theory
of mind*. We utilized vision and speech capabilities on a Pepper robot to
build an interaction model that stores the interpretations of perceptions
and conversations in combination with provenance on its sources. The
robot learns directly from what people tell it, possibly in relation to its
perception. We demonstrate how the robot's communication is driven
by hunger to acquire more knowledge from and on people and objects,
to resolve uncertainties and conflicts, and to share awareness of the perceived environment. Likewise, the robot can make reference to the world
and its knowledge about the world and the encounters with people that
yielded this knowledge.

Keywords: Robot · Theory of mind · Social learning
Communication

1 Introduction

People make mistakes; but machines err as well [14] as there is no such thing
as a perfect machine. Humans and machines should therefore recognize and
communicate their "imperfectness" when they collaborate, especially in case of
robots that share our physical space. Do these robots perceive the world in the
same way as we do and, if not, how does that influence our communication with
them? How does a robot perceive us? Can a robot trust its own perception?
Can it believe and trust what humans claim to see and believe about the world?
For example, if a child gets injured, should a robot trust their judgment of the
situation, or should it trust its own perception? How serious is the injury, how
much knowledge does the child have, and how urgent is the situation? How
different would the communication be with a professional doctor?

Human-robot communication should serve a purpose, even if it is just (social)
chatting. Yet, effective communication is not only driven by its purpose, but also

© Springer Nature Switzerland AG 2018
P. Sojka et al. (Eds.): TSD 2018, LNAI 11107, pp. 15–25, 2018.
https://doi.org/10.1007/978-3-030-00794-2_2

by the communication partners and the degree to which they perceive the same things, have a common understanding and agreement, and trust. One of the main challenges to address in human-robot communication is therefore to handle uncertainty and conflicting information. We address these challenges through an interaction model for a humanoid-robot based on the notion of a 'theory of mind' [12,17]. The 'theory of mind' concept states that children at some stage of their development become aware that other people's knowledge, beliefs, and perceptions may be untrue and/or different from theirs. Scassellati [18,19] was the first to argue that humanoid robots should also have such an awareness. We take his work as a starting point for implementing these principles in a Pepper robot, in order to drive social interaction and communication.

Our implementation of the theory of mind heavily relies on the Grounded Representation and Source Perspective model (GRaSP) [8,25]. GRaSP is an RDF model representing situational information about the world in combination with the perspective of the sources of that information. The robot brain does not only record the knowledge and information as symbolic interpretations, but it also records from whom or through what sensory signal it was obtained. The robot acquires knowledge and information both from the sensory input as well as directly from what people tell it. The conversations can have any topic or purpose but are driven by the robot's need to resolve conflicts and ambiguities, to fill gaps, and to obtain evidence in case of uncertainty.

This paper is structured as follows. Section 2 briefly discusses related work on theory of mind and social communication. In Sect. 3, we explain how we make use of the GRaSP model to represent a theory of mind for the robot. Next, Sect. 4 describes the implementation of the interaction model built on a Pepper robot. Finally, Sect. 5 outlines the next steps for improvement and explores other possible extensions to our model. We list a few examples of conversations and information gathered by the robot in the Appendix.

2   Related Work

Theory of mind is a cognitive skill to correctly attribute beliefs, goals, and percepts to other people, and is assumed to be essential for social interaction and for the development of children [12]. The theory of mind allows the truth properties of a statement to be based on mental states rather than observable stimuli, and it is a required system for understanding that others hold beliefs that differ from our own or from the observable world, for understanding different perceptual perspectives, and for understanding pretense and pretending. Following [4], Scassellati decomposes this skill into stimuli processors that can detect static objects (possibly inanimate), moving objects (possibly animate), and objects with eyes (possibly having a mind) that can gaze or not (eye-contact), and a shared-attention mechanism to determine that both look at the same objects in the environment. His work further focuses on the implementation of the visual sensory-motor skills for a robot to mimic the basic functions for object, eye-direction and gaze detection. He does not address human communication, nor the storage of the result of the signal processing and communication in a brain that captures a theory of mind. In our work, we rely on other technology to deal with the sensory data processing, and add language communication and the storage of perceptual and communicated information to reflect conflicts, uncertainty, errors, gaps, and beliefs.

More recent work on the 'theory of mind' principle for robotics appears to focus on the view point of the human participant rather than the robot's. These studies reflect on the phenomenon of anthropomorphism [7,15]: the human tendency to project human attributes to nonhuman agents such as robots. Closer to our work comes [10] who use the notion of a theory of mind to deal with human variation in response. The robot runs a simulation analysis to estimate the cause of variable behaviour of humans and likewise adapts the response. However, they do not deal with the representation and preservation of conflicting states in the robot's brain. To the best of our knowledge, we are the first that complement the pioneering work of Scassellati with further components for an explicit model of the theory of mind for robots (see also [13] for a recent overview of the state-of-the-art for human-robot interactive communication).

There is a long-tradition of research on multimodal communication [16], human-computer-interfacing [6], and other component technologies such as face detection [24], facial expression, and gesture detection [11]. The same can be said about multimodal dialogue systems [26], and more recently, around chatbot systems using neural networks [20]. In all these studies the assumption is made that systems process signals correctly, and that these signals can be trusted (although they can be ambiguous or underspecified). In this paper, we do not address these topics and technologies but we take them as given and focus instead on the fact that they can result in conflicting information, information that cannot be trusted or that is incomplete within a framework of the theory of mind. Furthermore, there are few systems that combine natural language communication and perception to combine the result in a coherent model. An example of such work is [21] who describe a system for training a robot arm through a dialogue to perform physical actions, where the "arm" needs to map the abstract instruction to the physical space, detect the configuration of objects in that space, determine the goal of the instructions. Although their system deals with uncertainties of perceived sensor data and the interpretation of the instructions, it does not deal with modeling long-term knowledge, but only stores the situational knowledge during training and the capacity to learn the action. As such, they do not deal with conflicting information coming from different sources over time or obtained during different sessions. Furthermore, their model is limited to physical actions and the artificial world of a few objects and configurations.

3   GRaSP to Model the Theory of Mind

The main challenges for acquiring a theory of mind is the storage of the result of perception and communication in a single model, and the handling of uncertainty and conflicting information. We addressed these challenges by explicitly representing all information and observations processed by the robot in an artificial brain (a triple store) using the GRaSP model [8]. For modeling the interpretation of the world, GRaSP relies on the Simple Event Model (SEM) [23] an RDF model for representing instances of events. RDF triples are used to relate event instances with *sem:hasActor*, *sem:hasPlace* and *sem:hasTime* object properties to actors, places, and time also represented as resources. For example, the triples [laugh, sem:hasActor, Bram], [laugh, sem:hasTime, 20180512] express that there was a *laugh* event involving *Bram* on the *12th of May 2018*.

GRaSP extends this model with *grasp:denotedIn* links to express that the instances and relations in SEM have been mentioned in a specific signal, e.g. a camera signal, human speech, written news. These signals are represented as *grasp:Chat* and *grasp:Turn* which: (a) are subtypes of *sem:Event* and therefore linked to an actor and time, and (b) derive *grasp:Mention* objects which point to specific mentioning of entities and events in the signal. Thus, if *Lenka* told the robot "Bram is laughing", then this expression is considered as a speech signal that mentions the entity *Bram* and the event instance *laugh*, while the time of the utterance is given and correlates with the tense of the utterance.

leolaniWorld:instances

leolaniWorld:Lenka	rdfs:label	"Lenka";
leolaniWorld:Bram	rdfs:label	"Bram";
	grasp:denotedIn	leolaniTalk:chat1_turn1_char0-16.
leolaniWorld:laugh	a	sem:Event;
	rdfs:label	"laugh";
	grasp:denotedIn	leolaniTalk:chat1_turn1_char0-16.

GRaSP further allows to express properties of the mentions such as the source (using *prov:wasAttributedTo*[1]), and the perspective of the source towards the content or claim (using *grasp:hasAttribution*). In the case of robot interactions, the source of a spoken utterance is the person identified by the robot, represented as a *sem:Actor*. Finally, we use *grasp:Attribution* to store properties related to the perspective of the source to the claimed content of the utterance: what emotion is expressed, how certain is the source, and/or if the source confirms or denies it. Following this example, the utterance is attributed to *Lenka*; thus we model that Lenka *confirms* Bram's laughing, and that she is *uncertain* and *surprised*. The perspective subgraph resulting from the conversation would look as follows:

leolaniTalk:perspectives

leolaniTalk:chat1_turn1	a	grasp:Turn;
	sem:hasActor	leolaniFriends:Lenka;
	sem:hasTime	leolaniTime:20180512.
leolaniTalk:chat1_turn1_char0-16	a	grasp:Mention;
	grasp:denotes	leolaniWorld:claim1 ;
	prov:wasDerivedFrom	leolaniTalk:chat1_turn1 ;
	prov:wasAttributedTo	leolaniFriends:Lenka .
leolaniTalk:chat1_turn1_char0-16_ATTR1	a	grasp:Attribution;
		grasp:CONFIRM,
	rdf:value	grasp:UNCERTAIN,
		grasp:SURPRISE;
	grasp:isAttributionFor	leolaniTalk:chat1_turn1_char0-16.

---

[1] Where possible, we follow the PROV-O model: https://www.w3.org/TR/prov-o/.

Our model represents the claims containing the SEM event and its relations as:

leolaniWorld:claims

leolaniWorld:claim1	a	grasp:Statement;
	grasp:subject	leolaniWorld:laugh;
	grasp:predicate	sem:hasActor;
	grasp:object	leolaniFriends:Bram.

Now assume that *Selene* is also present and she denies that Bram is laughing by saying: "No, Bram is not laughing". This utterance then gets a unique identifier e.g. *leolaniTalk:chat2_turn1*, while our Natural Language processing will derive exactly the same claim as before. The only added information is therefore the mentioning of this claim by *Selene* and her perspective, expressed as:

leolaniTalk:perspectives

leolaniTalk:chat2_turn1	a	grasp:Turn;
	sem:hasActor	leolaniFriends:Selene;
	sem:hasTime	leolaniTime:20180512.
leolaniTalk:chat2_turn1_char0-24	a	grasp:Mention;
	grasp:denotes	leolaniWorld:claim1 .
	prov:wasDerivedFrom	leolaniTalk:chat2_turn1 .
	prov:wasAttributedTo	leolaniFriends:Selene .
leolaniTalk:chat2_turn1_char0-24_ATTR1 a		grasp:Attribution;
	rdf:value	grasp:DENY, grasp:CERTAIN;
	grasp:isAttributionFor	leolaniTalk:chat2_turn1_char0-24.

Along the same lines, if *Lenka* now agrees with *Selene* by saying "Yes, you are right", we model this by adding only another utterance of *Lenka* and her revised perspective to the same claim, as shown below.[2]

leolaniTalk:perspectives

leolaniTalk:chat1_turn2	a	grasp:Turn;
	sem:hasActor	leolaniFriends:Lenka;
	sem:hasTime	leolaniTime:20180512.
leolaniTalk:chat1_turn2_char0-18	a	grasp:Mention;
	grasp:denotes	leolaniWorld:claim1 .
	prov:wasDerivedFrom	leolaniTalk:chat1_turn2 .
	prov:wasAttributedTo	leolaniFriends:Lenka .
leolaniTalk:chat1_turn2_char0-18_ATTR2 a		grasp:Attribution;
	rdf:value	grasp:DENY, grasp:CERTAIN;
	grasp:isAttributionFor	leolaniTalk:chat1_turn2_char0-18.

In the above examples, we only showed information given to the robot through conversation. GRaSP can however deal with any signal and we can therefore also represent sensor perceptions as making reference to the world or people that the robot knows. Assuming that the robot also sees and recognizes *Bram*, about whom *Lenka* and *Selene* are talking, this can be represented as follows, where we now include all the other mentions from the previous conversations:

leolaniWorld:instances

leolaniWorld:Bram	rdfs:label	"Bram";
	grasp:denotedIn	leolaniTalk:chat1_turn1_char0-16,
		leolaniTalk:chat2_turn1_char0-24,
		leolaniTalk:chat1_turn2_char0-18;
	grasp:denotedBy	leolaniSensor:FaceRecognition1.

---

[2] There are now two perspectives from *Lenka* on the same claim (she changed his mind), expressed in two different utterances.

A facial expression detection system could detect Bram's emotion and store this as perspective by the robot, e.g. [leolaniSensor:FaceRecognition1, rdf:value, grasp:SAD], in addition to *Lenka* and *Selene* on Bram's state of mind.

As all data are represented as RDF triples, we can easily query all claims made and all properties stored by the robot on instances of the world. We can also query for all signals (utterances and sensor data) which mention these instances and all perspectives that are expressed. The model further allows to store certainty values for observations and claims as well as the result of emotion detection in addition to the content of utterances (e.g. through modules for facial expression detection or voice-emotion detection). Finally, all observations and claims can be combined with background knowledge on objects, places, and people available as linked open data (LOD).

Things observed by the robot in the environment and things mentioned in conversation are thus stored as unified data in a "brain" (triple store). This brain contains identified people with whom the robot communicates, perceived objects about which they communicated[3], as well as properties identified or stated of these objects or people. Given this model, we can now design a robot communication model in combination with sensor processing on top of a theory of mind. In the next section, we explain how we implemented this model and what conversations can be held.

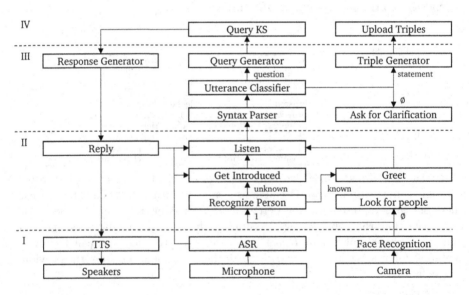

Fig. 1. The four layer conversation model, comprised of I. Signal Processing layer, II. Conversation Flow layer, III. Natural Language layer and IV. Knowledge Representation layer.

---

[3] The robot continuously detects objects, but these are only stored in memory when they are referenced by humans in the communication.

4   Communication Model

Our communication model consists of four layers: Signal Processing layer, Conversation Flow layer, Natural Language layer, and Knowledge Representation layer, which are summarized in Fig. 1. Signal Processing (I) establishes the mode of input through which the robot acquires experiences (vision and sound) but also knowledge (communication). The Conversation Flow layer (II) acts as the controller, as it determines the communicative goals, how to interpret human input, and whether the robot should be proactive. Layer III is the Natural Language layer that processes utterances and generates expressions, both of which can be questions or statements. Incoming statements are stored in the brain, while questions are mapped to SPARQL queries to the brain. SPARQL queries are also initiated by the controller (layer II) on the basis of sensor data (e.g. recognizing a face or not) or the state of the brain (e.g. uncertainty, conflicts, gaps) without a human asking for it. The next subsections briefly describe the four layers. We illustrate the functions through example dialogues that are listed in the Appendix. Our robot has a name *Leolani*, which is Hawaiain for *voice of an angel*, and a female gender to make the conversations more natural.

1. **Signal Processing.** Signal processing is used to give the robot awareness of its (social) surroundings and to recognize the recipient of a conversation. For assessing the context of a conversation, the robot has been equipped with eye contact detection, face detection, speech detection, and object recognition. These modules run continuously as the robot attempts to learn and recognize its surroundings. Speech detection is performed using WebRTC [3] and object recognition has been built on top of the Inception [22] neural network through TensorFlow [1]. During conversation the robot utilizes face recognition and speech recognition to understand who says what. Face recognition has been implemented using OpenFace [2] and speech recognition is powered by the Google Speech API [9].

2. **Conversation Flow.** In order to guide and respond during a one-to-one conversation, the robot needs to reason over its knowledge (about itself, the addressee, and the world) while taking into account its goals for the interaction. To model this we follow a Belief, Desire, Intention (BDI) [5] approach.
   Desires: The robot is designed to be hungry for social knowledge. This includes desires like asking for social personal information (name, profession, interests, etc.), or asking for knowledge to resolve uncertainties and conflicts.
   Beliefs: We consider the output of the other three layers to be part of the core beliefs of the robot, thus including information about what is being *sensed*, *understood*, and *remembered* during a conversation.
   Intentions: The set of current beliefs combined with the overarching desires then determine the next immediate action to be taken (aka. intention). The robot is equipped with a plan library including all possible intentions such as: (a) Look for a person to talk to, (b) Meet a new person, (c) Greet a known person, (d) Detect objects, and (e) Converse (including sub-intentions like Ask a question, State a fact, Listen to talker, and Reply to a question). The dialogues in the Appendix illustrate this behavior.

3. **Natural Language.** During a conversation, information flows back and forth. Thus, one of the goals of this layer is to transform natural language into structured data as RDF triples. When the robot listens, the utterances are stored along with the information about their speaker. After the perceived speech is converted to text, it is tokenized. The NLP module first determines if the utterance is a question or statement, because these are parsed differently. The parser consists of separate modules for different types of words, which are called on-demand, thus not clogging the NLP pipeline unnecessarily. This is important as Leolani needs to analyze an utterance and respond fast in real-time. Currently, the classification of the roles of words, such as predicate, subject, object, is done by a rule-based system. Next, the subject, object and predicate relations are mapped to triples for storage or querying the brain. This module also performs a perspective analysis over the utterance. Negation, certainty and sentiment or emotion are extracted separately from the text and their values added to the triple representation. A second goal of this layer is to *produce* natural language, based on output from either layer I (e.g. standard greetings, farewells, and introductions) or IV (phrasing answers given the knowledge in the brain and the goals defined in layer II). The robot's responses are produced using a set of rules to transform an RDF triple into English language. For English, we created a grammar using the concepts *person* (first, second, or third person pronouns or names), *object*, *location* and lists of *properties*. With this basic grammar, the robot can already understand and generate a large portion of common language. In the future, we will use WordNet to produce more varied responses and extend the grammar to capture more varied input and roles.

4. **Knowledge Representation.** The robot's brain must store and represent knowledge about the world, and the perspectives about it. For the latter we use the GRaSP ontology, as mentioned in Sect. 3. For the former, we created our own ontology "Nice to meet You", which covers the basic concepts and relations for human-robot social interaction (e.g. a person's name, place of origin, occupation, interests). Our ontology complies with 5 Star Linked Open Data, and is linked to vocabularies like FOAF and schema.org. Furthermore, the brain is able to query factual services like Wolfram Alpha, and LOD resources like DBpedia and GeoNames. The robot's brain is hosted in a GraphDB triple store. Given the above, this layer allows for two main interactions with the brain. The first is to process a statement, which implies generating and uploading the corresponding triples to the brain with source and perspective values. The second is to process a question, where a SPARQL query is fired against the brain. The result, being an empty list or a list with one or more results, is passed to layer III to generate a response. A list of values may represent conflicting information (disjunctive values) or multiple values (orthogonal values), each triggering different responses. In the future, we will extend the capabilities by enabling the robot to reason over its knowledge and generate new knowledge through inferencing.

5   Conclusions and Future Work

We described a model for a robot's world perception and communication, using an RDF repository modeling a theory of mind. It allows us to deal with uncertainties, conflicts, gaps, and different perspectives on this world. The state of information and knowledge the robot obtains can drive communication, e.g. the desire to resolve conflicts or uncertainties. Currently, the scripts and knowledge are limited. For the future, we plan to include different conversational models, NLP modules and types of knowledge. Furthermore, we extend the platform to empirically evaluate performance. Finally, we make the system functional through task-based dialogues and the processing knowledge and situation-aware. Our implementation is available as open source: https://github.com/cltl/pepper.

Acknowledgement. This research was funded by the VU University Amsterdam and the Netherlands Organization for Scientific Research via the Spinoza grant awarded to Piek Vossen. We also thank Bob van der Graft for his support.

Appendix: Dialogues

In the dialogues, *L* preceding an utterance stands for Leolani, other letters preceding utterances stand for various people. Perceptions of the robot of people and objects are marked using square brackets, e.g. [Sees a new face].

Meeting a new person

L:	[Sees the face of an unknown woman]
L:	Hi there, I would like to know you.
L:	My name is Leolani, what is your name?
H:	My name is Selene.
L:	[Stores the triples to represent Selene as a new friend. If the confidence for the name is too low, asks for confirmation]
L:	I hope I am correct and your name is: Selene.
H:	Yes that is my name.
L:	Nice to meet you Selene. Now I have a new friend.
L:	[Queries the brain for information on Selene to fill social gaps]
L:	Where are you from?
H:	I am from Mexico.
L:	[Stores the triples and queries the brain for friends from Mexico. The query just returns Selene.]
L:	Now I know 1 person from Mexico.

Handling conflicting information

L:	[Sees a human face and recognizes Lenka]
L:	Hi Lenka, nice to see you.
L:	[Queries the brain and discovers she does not know where Lenka is from]
L:	Lenka, where are you from?
H:	I am from Serbia.
L:	[Stores the triples and queries the brain for friends from Serbia and finds only Lenka.]
L:	Nice, I did not know anybody from Serbia yet.
H:	Where is Bram from?
L:	[Queries the brain for origin of Bram. Finds Bram as a friend and information where he is from.]
L:	Bram is from the Netherlands.
H:	Bram likes romantic movies.
L:	[Stores the triples with Lenka as source.]
L:	You told me that Bram likes romantic movies.
L:	[Sees Lenka leave and sees Bram arrive.]
L:	Hi Bram.
B:	I like science fiction movies.
L:	[Stores the triples and queries what Bram likes. Gets more than one value while expecting one.]
L:	I am surprised.
L:	Bram likes romantic movies, says Lenka.
L:	Bram likes science fiction movies, says Bram.

Checking information and trust

L:	[Sees a human face and recognizes Bram]
L:	Greetings Bram. Nice to see you again.
B:	Do you know where I am from?
L:	You are from the Netherlands, you said.
B:	Do you also know Lenka?
L:	Yes I know her, she is a very good friend of mine.
B:	Where is she from?
L:	Lenka is from Serbia, Lenka said
B:	Do you believe Lenka?
L:	[Queries the brain for perspective information on Lenka making this statement. Finds no uncertainty or denial.]
L:	I believe her.

Observing the environment

L:	[Sees an object and detects it as a cat with certain confidence]
L:	[Shortly after that she sees a human face and recognizes Bram]
L:	Greetings Bram. Nice to see you again.
L:	Guess what? I just saw a cat!
B:	That is not a cat but a rabbit.
L:	[Labels the image previously recognized as a cat instead as a rabbit and stores it in her brain]
B:	I like this rabbit.
L:	[Sees Bram leave and sees Selene arrive]
L:	Hi Selene. Greetings.
L:	[The presence of the rabbit is prominent, if the object is seen again recognition of a rabbit overrides recognition of a cat despite a higher score of the latter]
L:	Guess what, I just met a rabbit.
S:	A rabbit bites.
S:	I like a cat more.
S:	Have you ever seen a cat?
L:	[Queries the brain for objects labeled as cat and perceived by the sensor]
L:	No I have never seen a cat.
S:	What animals did you see?
L:	[Queries the brain for objects labeled as subtype of animal and perceived by the sensor]
L:	I saw a rabbit and a panda.
S:	What does rabbit do?
L:	[Queries the brain for properties assigned to the rabbit through conversations]
L:	Rabbits bite, Selene said.
L:	Rabbits cuddle, Bram said.
S:	Who likes rabbits?
L:	[Queries the brain for like perspective on the rabbit through conversations]
L:	Bram likes rabbits, Bram said.

References

1. Abadi, M., et al.: TensorFlow: large-scale machine learning on heterogeneous systems (2015). https://www.tensorflow.org/
2. Amos, B., Ludwiczuk, B., Satyanarayanan, M.: Openface: a general-purpose face recognition library with mobile applications. Technical report, CMU-CS-16-118, CMU School of Computer Science (2016)
3. T.W. Project Authors: WebRTC. Online publication (2011). https://webrtc.org/
4. Baron-Cohen, S.: Mindblindness: An Essay on Autism and Theory of Mind. MIT Press, Cambridge (1997)
5. Bratman, M.: Intention, plans, and practical reason (1987)
6. Card, S.K.: The Psychology of Human-Computer Interaction. CRC Press, Boca Raton (2017)

7. Epley, N., Waytz, A., Cacioppo, J.T.: On seeing human: a three-factor theory of anthropomorphism. Psychol. Rev. **114**(4), 864 (2007)
8. Fokkens, A., Vossen, P., Rospocher, M., Hoekstra, R., van Hage, W.: Grasp: grounded representation and source perspective. In: Proceedings of KnowRSH, RANLP-2017 Workshop, Varna, Bulgaria (2017)
9. Google: Cloud speech-to-text - speech recognition. Online publication (2018). https://cloud.google.com/speech-to-text/
10. Hiatt, L.M., Harrison, A.M., Trafton, J.G.: Accommodating human variability in human-robot teams through theory of mind. In: IJCAI Proceedings-International Joint Conference on Artificial Intelligence, vol. 22, p. 2066 (2011)
11. Kanade, T., Cohn, J.F., Tian, Y.: Comprehensive database for facial expression analysis. In: 2000 Proceedings of Fourth IEEE International Conference on Automatic Face and Gesture Recognition, pp. 46–53. IEEE (2000)
12. Leslie, A.M.: Pretense and representation: the origins of "theory of mind". Psychol. Rev. **94**(4), 412 (1987)
13. Mavridis, N.: A review of verbal and non-verbal human-robot interactive communication. Robot. Auton. Syst. **63**, 22–35 (2015)
14. Mirnig, N., Stollnberger, G., Miksch, M., Stadler, S., Giuliani, M., Tscheligi, M.: To err is robot: how humans assess and act toward an erroneous social robot. Front. Robot. AI **4**, 21 (2017)
15. Ono, T., Imai, M., Nakatsu, R.: Reading a robot's mind: a model of utterance understanding based on the theory of mind mechanism. Adv. Robot. **14**(4), 311–326 (2000)
16. Partan, S.R., Marler, P.: Issues in the classification of multimodal communication signals. Am. Nat. **166**(2), 231–245 (2005)
17. Premack, D., Woodruff, G.: Does the chimpanzee have a theory of mind? Behav. Brain Sci. **4**, 515–526 (1978)
18. Scassellati, B.: Theory of mind for a humanoid robot. Auton. Robot. **12**(1), 13–24 (2002)
19. Scassellati, B.M.: Foundations for a theory of mind for a humanoid robot. Ph.D. thesis, Massachusetts Institute of Technology (2001)
20. Serban, I.V., Sordoni, A., Bengio, Y., Courville, A.C., Pineau, J.: Building end-to-end dialogue systems using generative hierarchical neural network models. In: AAAI, vol. 16, pp. 3776–3784 (2016)
21. She, L., Chai, J.: Interactive learning of grounded verb semantics towards human-robot communication. In: Proceedings of the 55th Annual Meeting of the Association for Computational Linguistics (Volume 1: Long Papers), vol. 1, pp. 1634–1644 (2017)
22. Szegedy, C., et al.: Going deeper with convolutions. In: Computer Vision and Pattern Recognition (CVPR) (2015). http://arxiv.org/abs/1409.4842
23. Van Hage, W.R., Malaisé, V., Segers, R., Hollink, L., Schreiber, G.: Design and use of the simple event model (SEM). Web Semant.: Sci. Serv. Agents World Wide Web **9**(2), 128–136 (2011)
24. Viola, P., Jones, M.J.: Robust real-time face detection. Int. J. Comput. Vis. **57**(2), 137–154 (2004)
25. Vossen, P., et al.: Newsreader: using knowledge resources in a cross-lingual reading machine to generate more knowledge from massive streams of news. Knowl.-Based Syst. (2016). http://www.sciencedirect.com/science/article/pii/S0950705116302271
26. Wahlster, W.: SmartKom: Foundations of Multimodal Dialogue Systems, vol. 12. Springer, Heidelberg (2006). https://doi.org/10.1007/3-540-36678-4

Speech Analytics for Medical Applications

Isabel Trancoso[1]([📧]) [iD], Joana Correia[1,2] [iD], Francisco Teixeira[1] [iD],
Bhiksha Raj[2], and Alberto Abad[1] [iD]

[1] INESC-ID/Instituto Superior Técnico, University of Lisbon, Lisbon, Portugal
Isabel.Trancoso@inesc-id.pt
[2] Carnegie Mellon University, Pittsburgh, USA

Abstract. Speech has the potential to provide a rich bio-marker for
health, allowing a non-invasive route to early diagnosis and monitoring of a range of conditions related to human physiology and cognition.
With the rise of speech related machine learning applications over the
last decade, there has been a growing interest in developing speech based
tools that perform non-invasive diagnosis. This talk covers two aspects
related to this growing trend. One is the collection of large in-the-wild
multimodal datasets in which the speech of the subject is affected by
certain medical conditions. Our mining effort has been focused on video
blogs (vlogs), and explores audio, video, text and metadata cues, in order
to retrieve vlogs that include a single speaker which, at some point,
admits that he/she is currently affected by a given disease. The second
aspect is patient privacy. In this context, we explore recent developments
in cryptography and, in particular in Fully Homomorphic Encryption,
to develop an encrypted version of a neural network trained with unencrypted data, in order to produce encrypted predictions of health-related
labels. As a proof-of-concept, we have selected two target diseases: Cold
and Depression, to show our results and discuss these two aspects.

Keywords: Pathological speech · Data mining · Cryptography

1 Introduction

From the recordings of a speaker's voice one can estimate bio-relevant traits such
as height, weight, gender, age, physical and mental health. One can also estimate language, accent, emotional and personality traits, and even environmental
parameters such as location and surrounding objects. This wealth of information that one can now extract with the recent advances in machine learning over
the last decade has motivated an exponentially growing number of speech-based
applications that go much beyond the transcription of what a speaker says.

This work concerns health related applications, in particular the ones aiming
at non-invasive diagnosis based on the analysis of speech. Most of the earlier work

This work was supported by national funds through Fundação para a Ciência e a Tec-
nologia (FCT) with references UID/CEC/50021/2013, and SFRH/BD/103402/2014.

in this area was directed at the diagnosis and therapy of diseases which affect the phonation and articulation mechanisms of speech production. Nowadays, however, the impact of speech-based tools goes much beyond physical health. In fact, a recent prediction of innovations that can change our lives within five years quotes: With AI, our words will be a window into our mental health[1].

Most of the recent work on speech-based diagnosis tools concerns the extraction of features, and/or the development of sophisticated machine learning classifiers [5,8,14,15,18,22]. The results have shown a remarkable progress, but most are obtained from limited training data acquired in controlled conditions. This work addresses two emerging concerns that have not yet drawn much attention. One is the possibility of acquiring in-the-wild data from large scale, multimodal repositories versus acquiring data in laboratorial conditions [4]. Another is patient privacy in diagnosis and monitoring scenarios [24].

The idea of automatically collecting disease specific datasets from multimodal online repositories is based on the hypothesis that this type of data exists in very large quantities, and contains highly varied examples of the effects of the diseases on the subject's speech, unbound by human experiment design. In particular, this type of data should be easily mined from vlogs (video blogs), a popular category of videos, which mostly feature a single subject, talking about his/her own experience, typically with little production and editing. Our goal is retrieve vlogs in which the subject refers to his/her own current medical conditions, including a spoken confirmation of the diagnosis. But simple queries with the target disease name and the word vlog (i.e. *depression vlog*) typically yield videos that do not always correspond to our target of first person, present experiences (lectures, in particular, are relatively frequent), thus implying the need for a filtering stage. To do so, we adopt a multimodal approach, combining features extracted from the video and its metadata, using mostly off-the-shelf tools in order to test the potential of the approach.

As a proof-of-concept we have selected two target diseases: Cold and Depression. This selection was mainly motivated by the availability of corresponding lab corpora distributed in paralinguistic challenges, for which we had baseline results. We collected and labelled a small dataset for each target disease from YouTube, building a corpus of in-the-Wild Speech Medical (WSM) data, with which we test our proposed filtering solution. Section 3 is devoted to our in-the-wild data collection efforts, describing the WSM collection from the online repository YouTube, the filtering stage (including the multimodal feature extraction process, and the classifiers), and the classifier performance in detecting the target videos in the WSM dataset.

Privacy is the second topic of this paper. Privacy is an emerging concern among users of voice-activated digital assistants, sparkled by the awareness of devices that must be always in the listening mode. Despite this growing concern, the potential misuse of health related speech based cues has not yet been fully realised. This is the motivation for adopting secure computation frameworks, in which cryptographic techniques are combined with state-of-the-art machine

---

[1] https://www.research.ibm.com/5-in-5/mental-health/.

learning algorithms. Privacy in speech processing is an interdisciplinary topic, which was first applied to speaker verification, using Secure Multi-Party Computation, and Secure Modular Hashing techniques [1,19], and later to speech emotion recognition, also using hashing techniques [7]. The most recent efforts on privacy preserving speech processing have followed the progress in secure machine learning, combining neural networks and Full Homomorphic Encryption (FHE) [2,12,13]. In particular, an Encrypted Neural Network was applied to speech emotion recognition by [7]. In this work we describe our most recent efforts in applying the same concept to the secure detection of pathological speech.

The two above mentioned target diseases also serve as proof-of-concept for this topic. Section 4 is devoted to the description of the Encrypted Neural Network scheme, following the FHE paradigm, and to its application to the detection of Cold and Depression.

The baseline system which will serve as a reference to both the in-the-wild framework, and the privacy preserving framework will be described in Sect. 2. The system is based on a simple neural network trained with common features that have not been optimized for either disease, and is trained and tested with data collected in controlled conditions and distributed in paralinguistic challenges. This baseline allows us to compare the performance of the neural networks in tests with the WSM Corpus, to highlight the differences between pathological speech collected in-the-wild and in controlled conditions. This baseline also allows us to compare the performance of encrypted neural networks versus their non-encrypted counterparts, to validate our secure approach.

2 Controlled Conditions Baseline

2.1 Datasets

The Upper Respiratory Tract Infection Corpus (URTIC) is a dataset provided by the Institute of Safety Technology of the University of Wuppertal, Germany, for the Interspeech 2017 ComParE Challenge. It includes recordings of spontaneous and scripted speech. The training and development partitions comprise 210 subjects each (37 with cold). The two partitions include 9,505 and 9,565 chunks of 3 to 10 s, respectively [22].

The depression subset of the Distress Analysis Interview Corpus – Wizard-of-Oz (DAIC-WOZ) is an audio-visual database of clinical interviews. It consists of 189 sessions ranging between 7 and 33 min, 106 of which are present in the training set, and 34 in the development set. A depression score in the PHQ-8 [16] scale is provided for each session. Of the 106 participants in the training partition, 30 are considered to be depressed. In the development set, 12 of the 34 subjects are considered to be depressed [25].

2.2 Classifiers

The baseline classifier uses the extended Geneva Minimalistic Acoustic Parameter Set (eGeMAPS) as input to a simple neural network. This set includes 88 acoustic features designed to serve as a standard for paralinguistic analysis [9].

The network architecture consists of three layers: an input layer with 120 units, a hidden layer with 50 units, and an output layer with one unit. The first two layers share the same structure, first a Fully Connected (FC) layer, followed by a Batch Normalization (BN) layer, and an Activation layer with Rectified Linear Units (ReLUs). The output layer is characterized by an FC layer with a sigmoid activation. During training, Dropout layers are also inserted before the second and third FC layers. Both the Dropout and the BN layers in the network help prevent the model from overfitting. These forms of regularization are important, due to the limited size of the training data.

Before training the network, the training set is zero-centered and normalized by its standard deviation. The values of the mean and standard deviation of this set are later used to zero-center and normalize the development set.

The model was implemented in Keras [3], and was trained with RMSProp, using the default values of this algorithm together with a learning rate of 0.02 and 100 epochs. To determine the best dropout probabilities for each dropout layer, a random search was conducted yielding the following values: 0.3746 and 0.5838 for the Cold model, and 0.092 and 0.209 for the Depression model, for the first and second dropout layers, respectively.

To compensate for the unbalanced labels on the training partitions of the Cold and Depression datasets, we attribute different weight to samples of the positive and negative class: 0.9/0.1 for Cold, and 0.8/0.2 for Depression.

3 In-the-Wild Medical Data

3.1 The WSM Corpus

The datasets of the WSM corpus were collected in February 2018 from the online multimodal repository YouTube[2]. The language of the videos was restricted to English. The following information was collected for each query result: video; unique identifier; title; description (optional); transcription (automatically generated for videos in English, unless provided by a user); channel identifier; playlist identifier; date published; thumbnail; video category (closed set, 14 categories, e.g. "News", "Music" or "Entertainment"); number of views; number of thumbs up; number of thumbs down; comments.

The number of videos per dataset has been limited to approximately 60, because of the need for manual labeling. Each video was hand labeled with four intermediate binary labels: (1) the video is in a vlog format; (2) the main

---

[2] The WSM corpus also includes a subset for Parkinson's Disease, which we excluded for two reasons: space concerns, and the fact that the corresponding lab dataset is aimed at a regression, and not classification task.

speaker of the video talks mostly about him/herself; (3) the discourse is about present experiences or opinions; (4) the main topic of the video is related to the target disease. If all intermediate labels were positive, the video was labelled as containing in-the-wild pathological speech. Table 1 presents some statistics for each dataset, including the class distribution for each label.

Table 1. Overview of the WSM Corpus

Dataset	Class	# Videos	Ave. duration [min]	Ave. #words/video	Ave. #word/min/video	Vocab size [word]	Total length [min]	Total length [words]
Cold	Positive	30	16.0	968	150	2,930	479	29,047
	Negative	33	10.1	1,320	134	3,710	332	43,547
	Overall	63	12.9	1,152	141	5,097	811	72,594
Depression	Positive	18	8.9	1,142	149	2130	159	20,563
	Negative	40	10.4	1,371	146	4,321	418	54,839
	Overall	58	10.0	1,300	147	5,096	577	75,403

Table 2. Positive class incidence per label, per disease for the WSM Corpus

Dataset	Vlog	1st Person	Present	Target topic	All
Cold	96.9	79.7	90.6	62.5	47.2
Depression	93.1	74.1	50.0	56.9	31.0

3.2 Automatic Filtering of Videos with Pathological Speech

We focused on extracting multimodal features for each video to help our classifiers automatically replicate the manual labels (Table 2):

Textual: Bag-of-Word (BoW) features were extracted from the video transcription, in order to yield one feature vector, with the normalized frequency of the individual tokens. The length of the vector was the total size of the vocabulary of the corpus of transcriptions. The term-frequency times inverse document-frequency (tf-idf) transform was adopted.

Sentiment features were derived from the title, description, transcription and top n comments of the video using the Stanford Core NLP [23], a tool based on a Recursive Neural Tensor Network (RNTN).

The textual component contributed with 28 sentiment analysis features (a vector of dimension 7, for each of the 4 items), plus a vector of $\approx 5,100$ BoW features per video.

Audio: The number of speakers per video was obtained via speaker diarization from the audio component, using the LIUM toolbox [21], also adopted to eliminate silent segments, and divide the speech signal into inter-pausal units. Hence, the audio component contributed with a single feature per video.

Visual: Each video was segmented into scenes, using a simple comparison between pairs of consecutive frames. Scene changes were marked when the difference exceeded a preset threshold. A random frame was selected for each resulting scene. Automatic face detection using the toolkit [11], and computation of color histograms is performed in the resulting frames.

The video component contributed with a 768 dimensional histogram vector; plus one feature indicating the number of different faces identified in the video; and one feature indicating the number of scenes detected in the video.

Metadata: Features derived from the collected metadata included: a one hot vector representing the video category; the video duration; the number of views; the number of comments; the number of thumbs up; and the number of thumbs down at the time of collection. The metadata contributed with 19 features.

The final feature vectors have ≈5,900 dimensions. Given the limited size of our datasets, the feature vectors were reduced in dimensionality by eliminating the features with a Pearson correlation coefficient (PCC) to the label below 0.2.

We used 5 models to predict each of the 4 intermediate labels of the videos, as well as the global label: Logistic regression (LR), and Support vector machines (SVMs), with either linear (LIN), polynomial of degree 3 (POL), or radial basis function (RBF) kernels. The models were trained in a leave-one-out cross validation fashion. For each dataset, we trained a distinct classifier for each of the 7 types of feature, and another one with all the features.

3.3   Filtering Results

Filtering results are reported in precision and recall. We consider that a good model will have a high precision measure, since the goal is to maximize the rate of true positives. At the same time, false negatives are not a major concern in this scenario: we assume that the repository being mined has a much larger number of target videos than the size of the desired dataset.

Table 3 summarizes the performance of the best overall model (SVM-RBF). The cells highlighted in gray mark models which performed equal or worse than simply choosing the majority class. The best performing models achieve 88%, and 93% precision, and 97%, and 72% recall, for the Cold and Depression datasets, respectively. The type of features that has the most impact are the text features, concretely, the Bag-of-words, for every dataset, and for every label. Label 3 (Present) was the hardest label to correctly estimate, in one of the datasets. The results for Label 1 (Vlog) are not reported for the Cold dataset because it did not contain enough negative examples for training. We note that some feature types, such as the number of speakers or scenes, are seldom capable of generating a good model, probably due to the limitations of the feature extraction techniques.

Table 3. Performance of the SVM-RBF reported in precision and recall rate in detecting target content in the WSM Corpus.

Dataset	Modality	Features	Label				
			Vlog	1st Person	Present	Target Topic	All
Cold	Text	BoW	NA	1.0, 1.0	1.0, 1.0	0.95, 1.0	0.88, 0.97
		Sentiment	NA	0.81, 1.0	0.92, 1.0	0.64, 0.85	0.64, 0.53
	Speech	# Speakers	NA	0.81, 1.0	0.92, 1.0	0.63, 1.0	0.70, 0.53
	Video	# Faces	NA	0.81, 1.0	0.92, 1.0	0.72, 1.0	0.71, 0.5
		# Keyframes	NA	0.85, 0.98	0.92, 1.0	0.67, 0.97	0.56, 0.67
		Color hist.	NA	0.81, 1.0	0.92, 1.0	0.65, 0.93	0.60, 0.50
	Metadata	Metadata	NA	0.81, 1.0	0.92, 1.0	0.65, 0.97	0.57, 0.40
	Multimodal	All	NA	1.0, 1.0	1.0, 1.0	0.95, 1.0	0.88, 0.97

Dataset	Modality	Features	Label				
			Vlog	1st Person	Present	Target Topic	All
Depr.	Text	BoW	0.98, 1.00	0.98, 1.00	0.73, 0.94	0.89, 0.89	0.86, 0.67
		Sentiment	0.91, 1.00	0.77, 0.98	0.52, 0.66	0.52, 0.71	0.33, 0.17
	Speech	# Speakers	0.91, 1.00	0.85, 0.91	0.56, 0.69	0.69, 0.94	0.00, 0.00
	Video	# Faces	0.91, 1.00	0.89, 0.93	0.69, 0.75	0.72, 0.94	0.00, 0.00
		# Keyframes	0.91, 1.00	0.84, 0.96	0.56, 0.88	0.72, 0.97	0.00, 0.00
		Color hist.	0.91, 1.00	0.77, 0.98	0.69, 0.78	0.80, 0.89	0.75, 0.33
	Metadata	Metadata	0.91, 1.00	0.77, 0.98	0.62, 1.00	0.60, 0.97	0.00, 0.00
	Multimodal	All	0.98, 1.00	0.93, 0.96	0.83, 0.91	0.89, 0.91	0.93, 0.72

3.4   In-the-Wild vs. Lab Results

The performances of the neural networks against the WSM Corpus versus existing datasets of data collected in controlled conditions are summarized in Table 4. As expected, given the greater variability in recording conditions (p.e. microphones, noise), the performances of the networks when faced with in the wild data decrease when compared to data collected in controlled conditions. However, it is possible to improve the classification at speaker level, versus at segment level by aggregating the segments for each speaker. The subject level prediction is obtained by computing a weighted average of the segment level predictions, in which the weighting term is given by the segment length.

Table 4. Comparison of the performance (UAR) of the Neural Networks for detecting pathological speech in datasets collected in controlled environments versus in-the-wild.

Voice affecting disease	Controlled conditions dataset (segment level)	WSM corpus (segment level)	WSM corpus (speaker level)
Cold	66.9	53.1	53.3
Depression	60.6	54.8	61.9

4   Privacy Preserving Diagnosis

4.1   Homomorphic Encryption

First proposed by Rivest et al. [20], Homomorphic Encryption (HE) is a type of encryption that allows for certain operations to be performed in the encrypted domain while preserving their results in the plaintext domain. In other words, if for example an addition or multiplication is performed on two encrypted values, the result of this operation is kept when the corresponding encrypted value is decrypted. In HE, operations increase the amount of noise in the encrypted values, and if a certain threshold is surpassed, it is impossible to recover their original value. Leveled Homomorphic Encryption (LHE) allows us to choose parameters that control this noise threshold, but as these parameters increase, so does the computational complexity of the operations. Consequently there needs to be a trade-off between the number of operations to be computed in the encrypted domain, and the computational complexity of the application. We used SEAL's implementation [17] of the Fan and Vercauteren scheme [10].

4.2   Encrypted Neural Networks

Neural networks have been shown to be especially suited for secure machine learning applications using FHE [2,12,13], as most operations can be replaced by additions and multiplications.

To comply with the restrictions FHE poses, some modifications are necessary. As stated in the previous section, a large number of operations translates into a high computational cost, therefore the number of hidden layers of the network needs to be reasonably small, to limit the amount of operations computed in the encrypted domain. Moreover, as HE only allows additions and multiplications to be computed, only polynomial functions can be computed, and thus activation functions have to be replaced by polynomials.

In view of the reasons stated above, it is necessary to find a suitable polynomial to replace the activation functions commonly present in neural networks. The REctified Linear Unit (*ReLU*) activation function is a widely used activation, and thus it has been the focus of most FHE neural network schemes, although other activation functions have also been considered, such as *tanh* and *sigmoid*. We follow the approach suggested by CryptoDL [13], and use Chebyshev Polynomials, to approximate the *ReLU*) through its derivative.

$$p(x) = 0.03664x^2 + 0.5x + 1.7056 \tag{1}$$

In general, the training stage is still too computationally expensive to be performed in the encrypted domain. For this reason, most frameworks such as Cryptonets [12] and CryptoDL [13] are trained with unencrypted data, using the polynomial approximations of the activation functions.

For classification tasks, it is helpful to have a function constrained between 0 and 1 for the output. This is not possible using low degree polynomials, but it is possible to build a linear polynomial that is bounded between the same values in a given interval. To this end, we also approximated the Sigmoid function in the interval $[-10, 10]$, with a linear polynomial, obtaining:

$$p(x) = 0.004997x + 0.5 \qquad (2)$$

4.3   Encrypted vs. Non-encrypted Results

For each dataset, we compare the results obtained with two neural networks: an unencrypted neural network (NN) trained with normal activation functions, and an Encrypted Neural Network (ENN), trained with polynomial approximations and performing encrypted predictions. All results correspond to the Development partition of the datasets, and are reported at the segment level. The first two lines of Table 5 show the results regarding the Cold classification task, and which may be compared with the baseline value for UAR (66.1%) stated for the Interspeech 2017 ComParE Challenge [22]. Both the model with the original activation functions and the encrypted model performed above the baseline. In this case, there is a small performance degradation from the unencrypted NN to the ENN. Most likely, this difference is not due to the ReLU approximation, but because of the output Sigmoid, which, in the NN case, is a bounded function, and in the ENN is a linear polynomial.

Table 5. Results obtained for Cold and Depression classification.

Dataset	Method	UAR (%)	Precision (%)	F1 Score (%)
Cold	NN	66.9	56.3	48.3
	ENN	66.7	56.4	48.3
Depression	NN	60.6	61.8	55.1
	ENN	60.2	59.7	59.1

The last two lines of Table 5 show the results for the Depression task. When comparing the results of NN and ENN, there is just a slight degradation due to the polynomial approximations. The baseline value for UAR (69.9%) presented in the AVEC 2016 Challenge refers to interview-level results [25]. At this level, the ENN achieves 67.9%. The difference may be due to the network size, and the fact that AVEC's baseline uses features from COVAREP [6], whereas our experiment was conducted using eGeMAPS.

5   Conclusions

In this work, we performed proof-of-concept experiments focusing in two different aspects of speech-based medical diagnosis. In the first set of experiments, we demonstrated the viability of collecting in-the-wild data, containing instances of speech affecting diseases, based on mining multimodal online repositories. In the second set of experiments, we demonstrated the viability of making paralinguistic health-related tasks secure through the use of Fully Homomorphic Encryption. Both sets of experiments concerned two diseases, Cold and Depression, which lead us to believe that the process is generalizable to datasets for any speech-affecting disease.

Although our mining efforts made use of relatively simple techniques using mostly existing toolkits, they proved effective. The best performing models achieved a precision of 88% and 93%, and a recall of 97% and 72%, for the datasets of Cold and Depression, respectively, in the task of filtering videos containing these speech affecting diseases.

We compared the performance of simple neural network classifiers trained with data collected in controlled conditions in tests with corresponding data and in-the-wild data. The performance decreased as expected. We hypothesize this is due to a greater variability in recording conditions (p.e. microphone, noise) and in the effects of speech altering diseases in the subjects' speech.

We also compared the performance of the simple neural network classifiers with their encrypted counterparts. The slight difference in results showed the validity of our secure approach. Unfortunately, the limited amount of data does not allow a thorough analysis of performance using deeper networks.

Health-related tasks are typically characterized by limited amounts of training data, which in turn, limits the improvements potentially obtainable with state-of-the-art machine learning techniques, using speech as a single modality, without any speaker clustering. We hope to have made a small step towards solving this limitation, and plan to collect and make available larger datasets of several speech affecting diseases, thus increasing the speech resources available for medical applications. It will be important to achieve this in a totally unsupervised way, by dropping the label requirements during the training stage.

Our efforts aimed at establishing baselines without any emphasis on the specific speech-altering features of the two diseases chosen for the proof-of-concept experiments. However, given the modular architecture, each component of the system can be individually improved.

Given the recent progress achieved in many speech processing tasks with end-to-end machine learning approaches, it will be very interesting to adapt these architectures to the restrictions of FHE. Secure training is also an open problem, that if solved can contribute to the increase in size of existing databases, allowing for better models to be trained for real world applications.

References

1. Boufounos, P., Rane, S.: Secure binary embeddings for privacy preserving nearest neighbors. In: International Workshop on Information Forensics and Security (WIFS) (2011)
2. Chabanne, H., de Wargny, A., Milgram, J., Morel, C., et al.: Privacy-preserving classification on deep neural network. IACR Cryptology ePrint Archive 2017, 35 (2017)

3. Chollet, F., et al.: Keras (2015). https://github.com/keras-team/keras
4. Correia, J., Raj, B., Trancoso, I., Teixeira, F.: Mining multimodal repositories for speech affecting diseases. In: Interspeech (2018)
5. Cummins, N., Scherer, S., Krajewski, J., Schnieder, S., Epps, J., Quatieri, T.F.: A review of depression and suicide risk assessment using speech analysis. Speech Commun. **71**, 10–49 (2015)
6. Degottex, G., Kane, J., Drugman, T., Raitio, T., Scherer, S.: COVAREP - a collaborative voice analysis repository for speech technologies. In: ICASSP, pp. 960–964, May 2014. https://doi.org/10.1109/ICASSP.2014.6853739
7. Dias, M., Abad, A., Trancoso, I.: Exploring hashing and cryptonet based approaches for privacy-preserving speech emotion recognition. In: ICASSP. IEEE (2018)
8. Dibazar, A.A., Narayanan, S., Berger, T.W.: Feature analysis for automatic detection of pathological speech. In: 24th Annual Conference and the Annual Fall Meeting of the Biomedical Engineering Society EMBS/BMES Conference, vol. 1, pp. 182–183. IEEE (2002)
9. Eyben, F., Scherer, K., Schuller, B., Sundberg, J., et al.: The Geneva minimalistic acoustic parameter set (GeMAPS) for voice research and affective computing. IEEE Trans. Affect. Comput. **7**(2), 190–202 (2016)
10. Fan, J., Vercauteren, F.: Somewhat practical fully homomorphic encryption. IACR Cryptology ePrint Archive 2012, 144 (2012). Informal publication
11. Geitgey, A.: Facerecog (2017). https://github.com/ageitgey/face_recognition
12. Gilad-Bachrach, R., Dowlin, N., Laine, K., et al.: CryptoNets: applying neural networks to encrypted data with high throughput and accuracy. In: ICML. JMLR Workshop and Conference Proceedings, vol. 48, pp. 201–210 (2016)
13. Hesamifard, E., Takabi, H., Ghasemi, M.: CryptoDL: deep neural networks over encrypted data. CoRR abs/1711.05189 (2017)
14. Lopez-de Ipiña, K., et al.: On automatic diagnosis of Alzheimer's disease based on spontaneous speech analysis and emotional temperature. Cogn. Comput. **7**(1), 44–55 (2015)
15. López-de Ipiña, K., et al.: On the selection of non-invasive methods based on speech analysis oriented to automatic Alzheimer disease diagnosis. Sensors **13**(5), 6730–6745 (2013)
16. Kroenke, K., Strine, T.W., Spitzer, R.L., Williams, J.B., Berry, J.T., Mokdad, A.H.: The PHQ-8 as a measure of current depression in the general population. J. Affect Disord **114**(1–3), 163–173 (2009)
17. Laine, K., Chen, H., Player, R.: Simple encrypted arithmetic library - SEAL v2.3.0. Technical report, Microsoft, December 2017. https://www.microsoft.com/en-us/research/publication/simple-encrypted-arithmetic-library-v2-3-0/
18. Orozco-Arroyave, J.R., et al.: Characterization methods for the detection of multiple voice disorders: neurological, functional, and laryngeal diseases. IEEE J. Biomed. Health Inform. **19**(6), 1820–1828 (2015)
19. Pathak, M.A., Raj, B.: Privacy-preserving speaker verification and identification using gaussian mixture models. IEEE Trans. Audio Speech Lang. Process. **21**(2), 397–406 (2013). https://doi.org/10.1109/TASL.2012.2215602
20. Rivest, R.L., Adleman, L., Dertouzos, M.L.: On data banks and privacy homomorphisms. Found. Secure Comput. 169–179 (1978)
21. Rouvier, M., Dupuy, G., Gay, P., Khoury, E., Merlin, T., Meignier, S.: An opensource state-of-the-art toolbox for broadcast news diarization. In: Interspeech (2013)

22. Schuller, B., et al.: The Interspeech 2017 computational paralinguistics challenge: addressee, cold & snoring. In: Interspeech (2017)
23. Socher, R., et al.: Recursive deep models for semantic compositionality over a sentiment treebank. In: EMNLP 2013, pp. 1631–1642 (2013)
24. Teixeira, F., Abad, A., Trancoso, I.: Patient privacy in paralinguistic tasks. In: Interspeech (2018)
25. Valstar, M.F., et al.: AVEC 2016 - depression, mood, and emotion recognition workshop and challenge. CoRR abs/1605.01600 (2016). http://arxiv.org/abs/1605.01600

Text

Sentiment Attitudes and Their Extraction from Analytical Texts

Nicolay Rusnachenko[1]([⊠]) and Natalia Loukachevitch[2]([⊠])

[1] Bauman Moscow State Technical University, Moscow, Russia
kolyarus@yandex.ru
[2] Lomonosov Moscow State University, Moscow, Russia
louk_nat@mail.ru

Abstract. In this paper we study the task of extracting sentiment attitudes from analytical texts. We experiment with the RuSentRel corpus containing annotated Russian analytical texts in the sphere of international relations. Each document in the corpus is annotated with sentiments from the author to mentioned named entities, and attitudes between mentioned entities. We consider the problem of extracting sentiment relations between entities for the whole documents as a three-class machine learning task.

Keywords: Sentiment analysis · Coherent texts

1 Introduction

Approaches in automatic sentiment analysis, one of the most popular applications of natural language processing during the last years, often deal with texts mainly having single opinion holder and discussing a single entity (opinion target): users' reviews or short messages posted in social networks, especially, in Twitter [11,15,16]. Short texts cannot contain multiple opinions toward multiple entities because of short length.

One of the most complicated genres of documents for sentiment analysis are analytical articles that analyze a situation in some domain, for example, politics or economy. These texts contain opinions conveyed by different subjects, including the author(s)' attitudes, positions of cited sources, and relations of the mentioned entities between each other. Analytical texts usually contain a lot of named entities, and only a few of them are subjects or objects of sentiment attitudes. Besides, an analytical text can have a complicated discourse structure. Statements of opinion can take several sentences, or refer to the entity mentioned several sentences earlier.

This paper presents an annotated corpus of analytical articles in Russian and initial experiments for automatic recognition of sentiment attitudes of named entities (NE) towards each other. This task can be considered as a specific subtask of relation extraction.

The work is supported by the Russian Foundation for Basic Research (project 16-29-09606).

P. Sojka et al. (Eds.): TSD 2018, LNAI 11107, pp. 41–49, 2018.
https://doi.org/10.1007/978-3-030-00794-2_4

2   Related Work

The task of extracting sentiments towards aspects of an entity in reviews has been studied in numerous works [8,9]. Also extraction of sentiments to targets, stance detection was studied for short texts such as Twitter messages [1,11,13]. But the recognition of sentiments toward named entities or events including opinion holder identification from full texts have been attracted much less attention. In 2014, the TAC evaluation conference in Knowledge Base Population (KBP) track included so-called sentiment track [6]. The task was to find all the cases where a query entity (sentiment holder) holds a positive or negative sentiment about another entity (sentiment target). Thus, this task was formulated as a query-based retrieval of entity sentiments and focused only on query entities[1].

In [5], MPQA 3.0 corpus is described. In the corpus, sentiments towards entities and events are labeled. The annotation is sentence-based. For example, in the sentence "When the Imam issued the fatwa against Salman Rushdie for insulting the Prophet...", Imam is negative to Salman Rushdie, but is positive to the Prophet. The current MPQA corpus consists of 70 documents. In total, sentiments towards 4,459 targets are labeled.

The paper [4] studied the approach to the recovery of the documents attitudes between subjects mentioned in the text. The approach considers such features as relatedness between entities, frequency of a named entity in the text, direct-indirect speech, and other features. The best quality of opinion extraction obtained in the work was only about 36% F-measure, which shows that the necessity of improving extraction of attitudes at the document level is significant and this problem has not been sufficiently studied.

For the analysis of sentiments with multiple targets in a coherent text, in the works [2,17] the concept of sentiment relevance is discussed. In [2], the authors consider several types of thematic importance of the entities discussed in the text: the main entity, an entity from a list of similar entities, accidental entity, etc. These types should be treated differently in sentiment analysis of coherent texts.

3   Corpus and Annotation

For experiments in sentiment analysis of multi-holder and multi-target texts, we use the RuSentRel 1.0 corpus[2] consisting of analytical articles extracted from Internet-portal inosmi.ru [10]. This portal contains articles from foreign authoritative sources in the domain of international politics translated into Russian. The collected articles contain both the author's opinion on the subject matter of the article and a large number of sentiment relations mentioned between the participants of the described situations.

For the documents of the assembled corpus, the manual annotation of the sentiment attitudes towards the mentioned named entities have been carried out. The annotation can be divided into two subtypes:

---

[1] https://tac.nist.gov/2014/KBP/Sentiment/index.html.
[2] https://github.com/nicolay-r/RuSentRel/tree/v1.0.

1. The author's relation to mentioned named entities;
2. The relation of subjects expressed as named entities to other named entities.

Figure 1 illustrates article attitudes in the graph format.

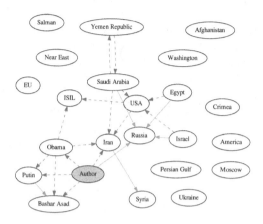

Fig. 1. Opinion annotation example for article 4 (*dashed arrows*: negative attitudes; *solid arrows*: positive attitudes)

These opinions were recorded as triples: (*Subject of opinion, Object of opinion, attitude*). The attitude can be negative (*neg*) or positive (*pos*), for example, (*Author, USA, neg*), (*USA, Russia, neg*). Neutral opinions or lack of opinions are not recorded. Attitudes are described for the whole documents, not for each sentence. In some texts, there were several opinions of the different sentiment orientation of the same subject in relation to the same object. This, in particular, could be due to the comparison of the sentiment orientation of previous relations and current relations (for example, between Russia and Turkey). Or the author of the article could mention his former attitude to some subject and indicate the change of this attitude at the current time. In such cases, it was assumed that the annotator should specify exactly the current state of the relationship. In total, 73 large analytical texts were labeled with about 2000 relations.

To prepare documents for automatic analysis, the texts were processed by the automatic name entity recognizer, based on CRF method [14]. The program identified named entities that were categorized into four classes: Persons, Organizations, Places and Geopolitical Entities (states and capitals as states). In total, 15.5 thousand named entity mentions were found in the documents of the corpus.

An analytical document can refer to an entity with several variants of naming (*Vladimir Putin – Putin*), synonyms (*Russia – Russian Federation*), or lemma variants generated from different wordforms. Besides, annotators could use only one of possible entity's names describing attitudes. For correct inference of attitudes between named entities in the whole document, the list of variant names for the same entity found in the corpus is provided in the corpus [10]. The current list contains 83 sets of name variants. In such a way, the sentiment analysis task is separated from the task of identifying coreferent named entities.

A preliminary version of the RuSentRel corpus was granted to the Summer school on Natural Language Processing and Data Analysis[3], organized in Moscow in 2017. The collection was divided into the training and test parts. In the current experiments we use the same division of the data. Table 1 contains statistics of the training and test parts of the RuSentRel 1.0 corpus. The last line of the Table 1 shows the average number of named entities pairs mentioned in the same sentences without indication of any sentiment to each other per a document. This number is much larger than number of positive or negative sentiments in documents, which additionally stresses the complexity of the task.

Table 1. Statistics of RuSentRel 1.0 corpus

Parameter	Training collection	Test collection
Number of documents	44	29
Avg. number of sentences per doc.	74.5	137
Avg. number of mentioned NE per doc.	194	300
Avg. number of unique NE per doc.	33.3	59.9
Avg. number of positive pairs of NE per doc.	6.23	14.7
Avg. number of negative pairs of NE per doc.	9.33	15.6
Share of attitudes expressed in a single sentence	76.5%	73%
Avg. number of neutral pairs of NE per doc.	120	276

4   Experiments

In the current experiment we consider the problem of extracting sentiment relations from analytical texts as a three-class supervised machine learning task. All the named entities (NE) mentioned in a document are grouped in pairs: (NE_1, NE_2), (NE_2, NE_1). All the generated pairs should be classified as having positive, negative, or neutral sentiment from the first named entity of the pair (opinion holder) to the second entity of the pair (opinion target). To support this task, we added neutral sentiments for all pairs not mentioned in the annotation and co-occurred in the same sentences in the training and test collections.

As a measure of quality of classification, we take the averaged Precision, Recall and F-measure of positive and negative classes. In the current experiments we classify only those pairs of named entities that co-occur in the same sentence at least once in a document. We use 44 documents as a training collection, and 29

---

[3] https://miem.hse.ru/clschool/results.

documents as a test collection in the same manner as the data were provided for the Summer School mentioned in the previous section. In the current paper, we describe the application of only conventional machine learning methods: Naive Bayes, Linear SVM, Random Forest and Gradient Boosting implemented in the scikit learn package[4].

The features to classify the relation between two named entities according to an expressed sentiment can be subdivided into two groups (54 features altogether). The first group of features characterizes the named entities under consideration. The second group of features describes the contexts in that the pair occurs.

The features of named entities are as follows:

- the word2vec similarity between entities. We use the pre-trained model news_2015[5] [7]. The size of the window is indicated as 20. The vectors of multiword expressions are calculated as the averaged sum of the component vectors. Using such a feature, we suppose that distributionally similar named entities (for example, from the same country) express their opinion to each other less frequently;
- the named entity type according to NER recognizer: person, organization, location, or geopolitical entity;
- the presence of a named entity in the lists of countries or their capitals. These geographical entities can be more frequent in ≪expressing sentiments≫ than other locations;
- the relative frequency of a named entity or the whole synonym group if this group is defined in a text under analysis. It is supposed that frequent named entities can be more active in expressing sentiments or can be an object of an attitude [3];
- the order of two named entities.

It should be noted that we do not use concrete lemmas of named entities as features to avoid memorizing the relation between specific named entities from the training collection.

The second group of features describes the context in that the pair of named entities appeared. There can be several sentences in the text where the pair of named entities occurs. Therefore each type of features includes maximal, average and minimum values of all the basic context features:

- the number of sentiment words from RuSentiLex[6] vocabulary: the number of positive words, number of negative words. RuSentiLex contains more than 12 thousand words and expressions with the description of their sentiment orientation [12];
- the average sentiment score of the sentence according to RuSentiLex;
- the average sentiment score before the first named entity, between named entities, and after the second named entities according to RuSentiLex;

---

[4] http://scikit-learn.org/stable/.

[5] http://rusvectores.org/.

[6] http://www.labinform.ru/pub/rusentilex/index.htm.

- the distance between named entities in lemmas;
- the number of other named entities between the target pair;
- the number of commas between the named entities.

We use several baselines for the test collection: baseline_neg – all pairs of named entities are labeled as negative; baseline_pos – all pairs are labeled as positive, baseline_random – the pairs are labeled randomly; baseline_distr – the pairs are labeled randomly according to the sentiment distribution in the training collection; baseline_school – the results obtained by the best team at the Summer school (see footnote 3). The results of all baselines are shown in Table 2. The upper bound of the classification is 73% (the share of attitudes expressed in a single sentence in the test collection (Table 1)).

Table 2. Results of sentiment extraction between named entities using machine learning methods

Method	Precision	Recall	F-measure
KNN	0.18	0.06	0.09
Naïve Bayes (Gauss)	0.06	0.15	0.11
Naïve Bayes (Bernoulli)	0.13	0.21	0.16
SVM (Default values)	0.35	0.15	0.15
SVM (Grid search)	0.09	**0.36**	0.15
Random forest (Default values)	**0.44**	0.19	**0.27**
Random forest (Grid search)	0.41	0.21	**0.27**
Gradient boosting (Default values)	0.36	0.06	0.11
Gradient boosting (Grid search)	**0.47**	0.21	**0.28**
baseline_neg	0.03	0.39	0.05
baseline_pos	0.02	0.40	0.04
baseline_random	0.04	0.22	0.07
baseline_distr	0.05	0.23	0.08
baseline_school	0.13	0.10	0.12
Expert agreement	0.62	0.49	0.55

Table 2 shows the classification results obtained with the use of several machine learning methods. For three methods (SVM, Random Forest, and Gradient Boosting), the grid search of the best combination of parameters was carried out; the grid search is implemented in the same scikit-learn package. The best results were obtained with the Random Forest and Gradient boosting classifiers.

The best achieved results are quite low. But we can see that the baseline results are also very low. It should be noted that the authors of the [4], who worked with much smaller documents, reported F-measure 36%.

To estimate the human performance in this task, an additional professional linguist was asked to label the collection without seeing the gold standard. The results of this annotation were compared with the gold standard using average F-measure of positive and negative classes in the same way as for automatic approaches. In such a way, it is possible to reveal the upper border for automatic algorithms. F-measure of the human labeling was 0.55. This is a quite low value, but it is significantly higher than the results obtained by automatic approaches. About 1% of direct contradictions (positive vs. negative) with the gold standard labels were found.

5   Analysis of Errors

In this section we consider several examples of erroneous classification of relations between entities. The examples are translated from Russian. In the following example, the system did not detect that Liuhto is positive towards NATO. This is because of relatively long distance between Liuhto and NATO and separation by *Finland*: *Liuhto says that he began to incline to Finland's accession to NATO.*

Usually related entities do not have sentiment toward each other. In the following example, the system erroneously infers that United States have negative sentiment to Washington: *The United States perceives Russia as a cringing great power, so they prefer to push any confrontation to the distant future, when it becomes even weaker - especially after spending a lot of effort in fighting international order, dominated by Washington.* But sometimes sentiment toward a related entity is expressed: *Putin wants to go down in history as the king who expanded the territory of Russia.* But evident sentiment words are absent, and the system misses the sentiment from Putin to Russia.

6   Conclusion

In this paper we described the RuSentRel corpus containing analytical texts in the sphere of international relations. Each document of the corpus is annotated with sentiments from the author to mentioned named entities, and sentiments of relations between mentioned entities.

In the current experiments, we considered the problem of extracting sentiment relations between entities for the whole documents as a three-class machine learning task. We experimented with conventional machine-learning methods (Naive Bayes, SVM, Random Forest, and Gradient Boosting). The corpus and methods are published[7].

We plan to enhance our training collection semi-automatically, trying to find sentences describing the known relations (for example, *Ukraine – Russia*, or *United Stated – Bashar Asad*), in order to obtain enough data for training neural networks.

---

[7] https://github.com/nicolay-r/sentiment-relation-classifiers/tree/tsd_2018.

References

1. Amigó, E., et al.: Overview of RepLab 2013: evaluating online reputation monitoring systems. In: Forner, P., Müller, H., Paredes, R., Rosso, P., Stein, B. (eds.) CLEF 2013. LNCS, vol. 8138, pp. 333–352. Springer, Heidelberg (2013). https://doi.org/10.1007/978-3-642-40802-1_31
2. Ben-Ami, Z., Feldman, R., Rosenfeld, B.: Entities' sentiment relevance. In: ACL 2014, vol. 2, pp. 87–92 (2014)
3. Ben-Ami, Z., Feldman, R., Rosenfeld, B.: Exploiting the focus of the document for enhanced entities' sentiment relevance detection. In: 2015 IEEE International Conference on Workshop (ICDMW), pp. 1284–1293. IEEE (2015)
4. Choi, E., Rashkin, H., Zettlemoyer, L., Choi, Y.: Document-level sentiment inference with social, faction, and discourse context. In: Proceedings of the 54th Annual Meeting of the Association for Computational Linguistics, pp. 333–343. ACL (2016)
5. Deng, L., Wiebe, J.: MPQA 3.0: an entity/event-level sentiment corpus. In: Proceedings of the 2015 Conference of the North American Chapter of the Association for Computational Linguistics: Human Language Technologies, pp. 1323–1328 (2015)
6. Ellis, J., Getman, J., Strassel, S.: Overview of linguistic resources for the TAC KBP 2014 evaluations: planning, execution, and results. In: Proceedings of TAC KBP 2014 Workshop, National Institute of Standards and Technology, pp. 17–18 (2014)
7. Kutuzov, A., Kuzmenko, E.: WebVectors: a toolkit for building web interfaces for vector semantic models. In: Ignatov, D.I., et al. (eds.) AIST 2016. CCIS, vol. 661, pp. 155–161. Springer, Cham (2017). https://doi.org/10.1007/978-3-319-52920-2_15
8. Liu, B., Zhang, L.: A survey of opinion mining and sentiment analysis. In: Aggarwal, C., Zhai, C. (eds.) Mining Text Data, pp. 415–463. Springer, Boston (2012). https://doi.org/10.1007/978-1-4614-3223-4_13
9. Loukachevitch, N., Blinov, P., Kotelnikov, E., Rubtsova, Y., Ivanov, V., Tutubalina, E.: SentiRuEval: testing object-oriented sentiment analysis systems in Russian. In: Proceedings of International Conference of Computational Linguistics and Intellectual Technologies Dialog, vol. 2, pp. 2–13 (2015)
10. Loukachevitch, N., Rusnachenko, N.: Extracting sentiment attitudes from analytical texts. In: Proceedings of International Conference Dialog (2018)
11. Loukachevitch, N.V., Rubtsova, Y.V.: SentiRuEval-2016: overcoming time gap and data sparsity in tweet sentiment analysis. In: Computational Linguistics and Intellectual Technologies Proceedings of the Annual International Conference Dialogue, Moscow, RGGU, pp. 416–427 (2016)
12. Loukachevitch, N., Levchik, A.: Creating a general Russian sentiment lexicon. In: Proceedings of LREC (2016)
13. Mohammad, S.M., Sobhani, P., Kiritchenko, S.: Stance and sentiment in tweets. ACM Trans. Internet Technol. (TOIT) **17**, 26 (2017)
14. Mozharova, V.A., Loukachevitch, N.V.: Combining knowledge and CRF-based approach to named entity recognition in Russian. In: Ignatov, D.I., et al. (eds.) AIST 2016. CCIS, vol. 661, pp. 185–195. Springer, Cham (2017). https://doi.org/10.1007/978-3-319-52920-2_18
15. Pak, A., Paroubek, P.: Twitter as a corpus for sentiment analysis and opinion mining. In proceedings of LREC, pp. 1320–1326 (2010)

16. Rosenthal, S., Farra, N., Nakov, P.: SemEval-2017 task 4: sentiment analysis in twitter. In: Proceedings of SemEval-2017 Workshop, pp. 502–518 (2017)
17. Scheible, C., Schütze, H.: Sentiment relevance. In: Proceedings of ACL 2013, vol. 1, pp. 954–963 (2013)

Prefixal Morphemes of Czech Verbs

Jaroslava Hlaváčová[(⊠)] [iD]

Faculty of Mathematics and Physics, Institute of Formal and Applied Linguistics,
Charles University, Malostranské nám. 25, 118 00 Prague 1, Czech Republic
hlavacova@ufal.mff.cuni.cz

Abstract. The paper presents the analysis of Czech verbal prefixes, which is the first step of a project that has the ultimate goal an automatic morphemic analysis of Czech. We studied prefixes that may occur in Czech verbs, especially their possible and impossible combinations. We describe a procedure of prefix recognition and derive several general rules for selection of a correct result. The analysis of "double" prefixes enables to make conclusions about universality of the first prefix. We also added linguistic comments to several types of prefixes.

Keywords: Morpheme · Prefix · Root · Verb · Czech

1 Introduction

Prefixation is in Czech one of the common means of word formation. Especially for verbs, it represents a very productive way of modifying their meaning and/or aspect. Knowing the way how the prefixes modify verbs will enable to write an automatic procedure for morphemic analysis, which is our ultimate goal.

Among Czech prefixes, there are some that can be attached to a wide range of verbs, while others are special—there are only several verbs starting with those prefixes. A similar observation concerning Czech prefixes described Marc Vey[1] [10] who divided the Czech verbal prefixes into two groups (with a blurred border)—full (plein in French) and empty (vide). The "fullness" of the former group points to a special strong (full) semantic meaning, which is the reason, why they cannot be attached to roots with an incompatible (e.g. opposite) meaning. The latter prefixes can be attached to much larger variety of verbs, which makes them universal.

The special prefixes are not very productive, people do not add them in front of existing verbs, nor create new ones. On the other hand, the group of universal prefixes is used very often for various modifications of existing verbs. This type of creating new or modified verbs is very popular mainly in connection with

---

Work on this paper was supported by the grant number 16-18177S of the Grant Agency of the Czech Republic (GAČR).

[1] He studied the way of creating a perfect verb from an imperfect one by means of prefixation.

P. Sojka et al. (Eds.): TSD 2018, LNAI 11107, pp. 50–57, 2018.
https://doi.org/10.1007/978-3-030-00794-2_5

· verbs of a foreign origin. In recent years, together with a massive adoption of new words especially from English, people like to add Czech (universal) prefixes to verbs created from a foreign language to change their meaning in accordance with the Czech rules.

The main source of our data was the Retrograde Morphemic Dictionary of Czech [1], we will call it Dictionary in the following text. For our analyses, we worked with its digitalized part [2] containing only the verbs. The second source of data was the corpus SYN2015 [4].

2   Prefixes of Czech Verbs

A Czech verb V can be generally written as a concatenation $V = P \cdot R \cdot S$, where P is a set of prefixes, R is a root[2] and S is set of suffixes. For our recent analysis we set $W = R \cdot S$. This enables to rewrite the general pattern as $V = P \cdot W$. We will call W stub. The stub may be another verb (as in the example *při-jít*)(= to come – perfective), but it is not necessary—for instance the verb *přicházet* (= to come – imperfective) has the prefix *při-*, but the stub *cházet* is not a verb.

The morphemes in the Dictionary are not marked according to their function in the word formation. All the morphemes—prefixes, roots as well as suffixes— are only separated with a hyphen (-) or a slash (/). To extract prefixes from the verbs, we were trying to separate prefixal morphemes until we reached a morpheme that does not belong to the set of prefixes. It was necessary to check if the rest of the verb, after the prefix separation, contains a root. If not, the separation was wrong, because together with prefixes, a root was separated, too. Fortunately, there are only a few cases of homonymous roots and prefixes— for instance the following morphemes are roots in presented examples: *sou* (in the verb *vy-sou-v-a-ti*), *vz* (*vz-í-ti*), *se* (*za-se-ti*) or *roz* (*ob-roz-ova-ti*), while in others they are prefixes.

The Dictionary contains more than 14 thousands of verbs. There are 3,352 verbs without prefixes, the rest of them (10,962) have at least one prefix. For a while, we did not include into our considerations foreign prefixes, like *re-, de-, a-* and similar. We also ignored morphemes *polo-, proti-, samo-, sebe-, sou-, spolu-, znovu-* and possibly some others, as some linguists do not consider them typical prefixes.

2.1   Traditional Verbal Prefixes

Table 1 presents the traditional 20 verbal prefixes (see for instance [3]) together with the number of verbs that are contained in the Dictionary. The number of verbs, presented in the table, counts only the verbs with a single prefix (single-prefixed verbs).

For a more detailed analyses, we added to these 20 traditional prefixes all other Czech prefixes that were found in our verbal data from the Dictionary.

---

[2] There may be more roots, but it is not typical and we can omit those cases, for simplicity.

Table 1. 20 traditional Czech verbal prefixes. The second columns show number of single-prefixed verbs included in the Dictionary, that have the given prefix.

Prefix	Number of verbs	Prefix	Number of verbs
vy-	1,122	*pro-*	513
za-	1,074	*pře-*	507
z-/ze-	934	*při-*	495
po-	741	*do-*	346
roz-/roze-	731	*v-/ve-*	222
na-	705	*pod-/pode-*	201
u-	643	*ob-/obe-*	193
o-	633	*vz-/vze-*	73
s-/se-	618	*před-/přede-*	68
od-/ode-	518	*nad-/nade-*	33

Apart from those traditional verbal prefixes, the rest can be divided into additional two groups: long prefixes and special prefixes.

2.2 Long Prefixes

Though the Czech verbal prefixes are short (as presented in Table 1), it does not mean that a verb cannot have a long prefix. There are several exceptions of old verbs as *ná-ležet* (= to belong), *zů-stat* (= to stay), *dů-věřovat* (= to trust). Apart from them, a long prefix often indicates that the verb was derived from a noun or adjective. Table 2 presents all verbs starting with a long prefix *ná-*, that were found in the corpus SYN2015. The verbs with high frequencies (in the third column) are more common and belong mainly to the first group of old verbs. The less frequent verbs are usually the derivations mentioned above, the boundary not being always sharp. This sort of derivation is productive, though the derived verbs rarely become part of the general vocabulary.

2.3 Special Prefixes

The third group of prefixes contains short prefixes that have only limited number of roots to be connected with. They are:

bez- connected only with the root *peč*, usually preceded with another prefix *za-bez-pečiti* (= to secure).

ot-/ote- connected only with the root *vř/vír/víř* as in the verbs *ote-vříti*, *ot-vírati* (= to open).

pa- connected with 2 roots, namely *děl* and *běr* (*pa-dělati* = to falsify), *pa-běrkovati* = to get the leftovers).

ne- tag of negation.

Table 2. Verbs with the long prefix *ná-* found in the corpus SYN2015, their frequencies and an estimated original noun.

Prefix	Rest of the verb	Translation = to	Frequency	Origin
ná	sledovat	follow	8508	
	ležet	belong	2262	
	sobit	multiply	311	
	rokovat	demand	296	nárok
	lepkovat	label	6	nálepka
	deničit	work as a day labourer	3	nádeník
	městkovat	work as an assistant	2	náměstek
	borovat	recruit	1	nábor

As for the prefix *ne*—it is often not clear if it must be present (as for the verb *nenáviděti* = to hate), or plays only role of negation of another verb. According to a tradition, the verbs with the negation prefix *ne-* are lemmatized as affirmative verbs without *ne-*, though there are disputes whether this decision is correct, but this is not the subject of this paper.

However, the prefix *ne-* appears quite often as the second one, preceded with another prefix, mainly by the most "universal prefix" *z-* (*z-ne-příjemniti* = to make unpleasant, *z-ne-hybniti* = to immobilize).

3 Prefix Separation

The prefix recognition and separation is obviously more complicated for the verbs extracted from the corpus, where the words are not cut into morphemes. We had to add some simple rules, for instance there cannot be a succession of prefixes *s-o-u-*. Whenever those letters appear together, it is always the prefix (or the root—see above) *sou-*. Similarly, there is never double prefix *v-z-*, it is always *vz-*. Working with the corpus, we took into account also prefixes with the prothetic *v*, namely *vo-*, *vod-* and *vob-*. They did not appear in the Dictionary, as it contains only the literal language. We consider those prefixes as variants of their standard counterparts, without *v* at the beginning.

The procedure of prefix separation was similar to that one we have used for the Dictionary data. We tried to separate strings that may be prefixes from the left of verbs so long, until the remaining string does not start with a possible prefix.

As there are no separators between individual morphemes as in the Dictionary, it was often possible to segment the verb in multiple ways—see the examples below.

For all the outcomes we checked, whether the rest of the verb, after separating the prefixes, is a verb or a verbal stub. For the checking the "verbness" of R we used the morphological analyzer MorphoDita [6]. The list of stubs was easy to get from the Dictionary. Though it contains less verbs than the corpus, there are all the "old" Czech verbs, especially those that are irregular. In other words, the Dictionary probably contains majority of the Czech verbal stubs that are not proper verbs.[3]

Then, we were able to select the correct segmentation in cases, where the beginning part of a verb was possible to segment ambiguously. For example, the verb *vypodobnit* (= to portray) was automatically segmented as follows:

1. *vy-podobnit*
2. *vy-po-dobnit*
3. *vy-po-do-bnit*
4. *vy-pod-obnit*
5. *vy-pod-ob-nit*
6. *vy-pod-o-bnit*

Only the second segmentation is correct, as was verified in the Dictionary. We could use this knowledge for the verbs, that do not occur in the Dictionary, but have the same stub.

However, it may happen, that a verb was segmented according to all the previous rules, and yet we get two segmentations with only one correct. It is difficult to decide such cases without a semantic judgments. An example is *představit* (= to introduce), that was automatically segmented as follows:

1. *před-stavit*
2. *před-s-tavit*

The both segmentations seem to be correct, because *stavit* is a common stub which appears in many other verbs[4], and *tavit* is a normal verb (= to melt). The only reason, why the second segmentation is wrong, is incompatibility of the prefixes *před-* and *s-*. For such conclusion, it is necessary to do a more complex analysis of prefix combinations. There may appear also an incompatibility between a prefix and a root.[5]

4   Prefix Combinations

We focused especially on verbs with more than one prefix and tried to find out which prefixes are possible to combine. Table 3 contains a part of the overall table that shows the combinatorics of verbal prefixes. The columns of the table contain all the monitored prefixes, which may occur as the second prefix in Czech verbs, while the rows (first prefixes) contain only selected traditional prefixes.

---

[3] This assertion needs to be verified.

[4] *za-stavit, u-stavit, vy-stavit, ...*

[5] There is only a small set of verbs which have really more possible meaningful segmentations. An example is the verb *voperovat* that is a substandard variant of the unprefixed verb *operovat* (= to operate), but can be also segmented as *v-operovat* (= to implant).

Table 3. Combinations of three types of prefixes for Czech verbs. If there is 1 in the x-th row and y-th column, there exists at least one verb with the prefix x-y-. The table is divided into three zones: I is for the universal prefix z-, II for the set of intensifying prefixes, III for special prefixes.

		po	s	u	z	vy	ú	pro	na	ob	pře	roz	pod	vz	zá	do	ná	za	od	pří	nad	před	při	o	v	ne	bez	ot	pů	vý	dů	prů	zů	Σ
I	z	1	1	1		1	1	1		1	1	1	1			1	1		1	1			1		1		1	1	1	1	1	1		21
II	za	1	1	1	1	1	1	1	1	1	1	1	1	1	1	1	1	1	1	1	1	1	1	1	1	1	1	1	1		1	1		28
	po	1	1	1	1	1	1	1	1	1	1	1	1	1	1	1	1	1	1	1	1	1	1	1	1	1		1					1	26
	na	1	1	1	1	1	1	1	1	1	1		1	1		1	1	1	1	1	1	1	1	1	1	1								23
	roz	1	1	1	1	1	1	1	1	1	1		1	1		1	1	1	1	1	1	1	1	1	1		1							23
	vy	1	1	1	1		1	1	1	1		1	1		1	1	1	1	1	1	1	1	1	1	1	1	1							22
	u	1	1	1		1	1	1	1	1		1	1		1	1	1	1	1	1	1	1	1	1		1								22
	do	1	1	1	1	1	1	1	1	1	1		1	1		1	1	1			1			1		1								17
III	pod	1				1			1																									3
	nad		1	1																														2
	v	1																							1									2
	vz	1																																1
	ob																																	0

4.1 Long Prefixes

It turns out, that within multi-prefixed verbs there are no verbs having a long prefix as the first one. It confirms the fact presented above that a long prefix is for verbs exceptional and often indicates that the verb was derived from a noun or adjective. It is not possible to add a long prefix to any already prefixed verb. This finding will be crucial for our future work—automatic morphemic analysis of Czech.

The long prefixes, however, can appear as a second prefix in multi-prefixed verbs. The most common prefix standing in front of a long prefix is z-, which can be followed by ná- (z-ná-rodnit = to nationalize), ú- (z-ú-rodnit = to fertilize), prů- (z-prů-svitnět = to become translucent), vý- (z-vý-hodnit = to make more beneficial), dů- (z-dů-vodnit = to give reason), pří- (z-pří-jemnit = to make more pleasant), pů- (z-pů-sobit = to cause). The only long prefix that cannot follow z- is zů-. The prefix zů- is a special one and is connected only with the root stav. The only prefix that can stand in front of zů- is po-.

When we did not find any verb for a certain combination of prefixes in the corpus SYN2015, we searched bigger corpora, especially the web corpus Omnia Bohemica II [5] with 12,3 gigawords. As for the prefix z- in front of a long prefix zá-, we have found there several verb examples with its vocalized variant ze-. All those verbs are very rare, with only several occurrences, often within parentheses or quotation marks, indicating that they were created occasionally for a special purpose—see the example of concordance (notice another use of the double-prefix verb with the second prefix long, again occasionalism): *Voni nás chtěj* **ze-zá-padnit**, *my je jednoduše* **z-vý-chodníme**. (= They want **make** us **western**, we will simply **make** them **eastern**.)

4.2 "Universal" Prefixes

We have already mentioned the "universal" prefix *z-* (group I in Table 3), with its vocalized variant *ze-*, that can be used for creation new verbs from adjectives. In these cases, it has mainly the meaning "make something or become (more) A", where A is an almost arbitrary adjective—see the example in the previous paragraph. The adjective may start with whatever prefix, or be without prefix.

The prefixes *roz-*, *po-*, *za-*, *na-*, *vy-* and *u-* (group II in Table 3) are also universal as they are able to modify imperfective verbs always in the same way. Together with a reflexive particle *se/si*, they change the meaning of imperfective verbs with respect to their intensity. It is called also "verb gradation", see [7–9]. There are not many examples of verb gradation in the corpora, but it is very productive and universal.

Take for instance the prefix *nad-*. We have no evidence of a verb with this prefix as the second one in our dataset. However, there may be derived such verbs using the verb gradation. An example may be the verb *nad-hazovat* (= to throw st up, or to pitch a ball). In the corpus Omnia Bohemica II, we found *za-nad-hazovat si* in the meaning "to pitch for a while for fun". It is possible to add all the intensifying prefixes to the verb *nad-hazovat*. For instance, *roz-nad-hazovat se* which means "to start pitching" or "to warm up in pitching" (e.g. before a baseball match).

We could add the prefix *do-* to this set, with the meaning "to finish an action". Again, it can be added to almost any imperfective verb, changing its meaning always in the same way.

4.3 "Meaningful" Prefixes

On the other hand, several prefixes have a special meaning and it is not possible to use them universally. Their strong semantic meaning must not be in conflict with the meaning of the rest of the word, especially with the root. As a consequence, except for a few exceptions, they can not be joined to another prefix from the left. It is clearly visible from Table 3—group III. They are especially prefixes *pod-*, *nad-*, *v-*, *vz-* and *ob-*.

5 What Should Be Done Next?

We are working on a similar analysis of verb suffixes. The set of Czech suffixes is much larger, thus the work is more complicated. As the basis, we use the Dictionary again, and continue in a similar way as with prefixes.

Finally, we have to add the other parts of speech. For using the Dictionary, it is necessary to finish its digitalization, which is extremely strenuous, as the paper pages of old printed publication are not entirely clear. Another problem that complicates the digitalization are letters from the other side of some pages visible through. And finally, it is not possible to use any sophisticated tool for the recognition, as the Dictionary does not contain complete words. As we have already stated, our final goal is a procedure for morphemic analysis of Czech. It should work on any word, even foreign ones. It will be able to recognize foreign words and to offer an appropriate segmentation, whenever possible. There are foreign words with Czech prefixes or suffixes—in that case it is reasonable to try to make a simple morpheme segmentation. Take for instance the word *Google*. It is foreign, even a proper name, that is not reasonable to try to segment. However, there are Czech verbs that use *googl* as the root: *googl-ovat* or *googl-it* (= to search on the internet using Google, or recently sometimes even simply to search). To perfectivize the previous examples, the Czech verbal prefix *vy-* is used *vy-googl-ovat*, *vy-googl-it*.

References

1. Slavíčková, E.: Retrográdní morfematický slovník češtiny. Academia (1975)
2. Slavíčková, E., Hlaváčová, J., Pognan, P.: Retrograde Morphemic Dictionary of Czech - verbs, LINDAT/CLARIN digital library at the Institute of Formal and Applied Linguistics (ÚFAL), Faculty of Mathematics and Physics, Charles University (2017). http://hdl.handle.net/11234/1-2546
3. Uher, F.: Slovesné předpony. Univerzita Jana Evangelisty Purkyně, Brno (1987)
4. Křen, M., et al.: SYN2015: reprezentativní korpus psané češtiny. Ústav Českého národního korpusu FF UK, Praha (2015). http://www.korpus.cz
5. Benko, V.: Aranea: yet another family of (comparable) web corpora. In: Sojka, P., Horák, A., Kopeček, I., Pala, K. (eds.) TSD 2014. LNCS (LNAI), vol. 8655, pp. 247–256. Springer, Cham (2014). https://doi.org/10.1007/978-3-319-10816-2_31
6. Straková, J., Straka, M., Hajič, J.: Open-source tools for morphology, lemmatization, pos tagging and named entity recognition. In: Proceedings of 52nd Annual Meeting of the Association for Computational Linguistics: System Demonstrations, pp. 13–18. Association for Computational Linguistics, Baltimore (2014)
7. Hlaváčová, J.: Stupňování sloves. In: After Half a Century of Slavonic Natural Language Processing, Masaryk University, Brno, Czech Republic, pp. 85–90 (2009). ISBN 978-80-7399-815-8
8. Hlaváčová, J., Nedoluzhko, A.: Productive verb prefixation patterns. In: The Prague Bulletin of Mathematical Linguistics, No. 101, Univerzita Karlova v Praze, Prague, Czech Rep., pp. 111–122 (2014). ISSN 0032–6585
9. Hlaváčová, Jaroslava, Nedoluzhko, Anna: Intensifying Verb Prefix Patterns in Czech and Russian. In: Habernal, Ivan, Matoušek, Václav (eds.) TSD 2013. LNCS (LNAI), vol. 8082, pp. 303–310. Springer, Heidelberg (2013). https://doi.org/10.1007/978-3-642-40585-3_39
10. Vey, M.: Les préverbes « vides » en tchèque moderne. In: Revue des études slaves, tome 29, fascicule 1–4, pp. 82–107 (1952)

LDA in Character-LSTM-CRF Named Entity Recognition

Miloslav Konopík[(⊠)] and Ondřej Pražák

Department of Computer Science and Engineering, Faculty of Applied Sciences,
University of West Bohemia, Univerzitní 8, 306 14 Plzeň, Czech Republic
konopik@kiv.zcu.cz, ondfa@ntis.zcu.cz
http://nlp.kiv.zcu.cz

Abstract. In this paper, we present a NER system based upon deep learning models with character sequence encoding and word sequence encoding in LSTM layers. The results are boosted with LDA topic models and linear-chain CRF sequence tagging. We reach the new state-of-the-art performance in NER of 81.77 F-measure for Czech and 85.91 F-measure Spanish.

Keywords: Named entity recognition · LSTM · LDA · Tensorflow

1 Introduction

Named entity recognition (NER) systems are designed to detect phrases in sentences with key meaning and classify them into predefined groups – typically persons, organizations, locations, etc. In this paper, we study the NER task on English and two morphologically rich languages—Czech and Spanish.

The state-of-the-art NER systems are based upon deep learning and they do not use complex features and in many cases, they use no features at all. The systems overcome the older results based upon a machine learning with the feature engineering (see the comparison e.g. in [7]). Adding new features into deep learning models is not very rewarding since the performance gain is small at the cost of increased complexity. Our experiments even show that adding POS tags to the English NER yields no benefit at all. On the other hand, the feature engineering approach to NER can still be well suited for cases where there are very little training data available – see e.g. [2].

In this paper, we experiment with LDA and we try to prove that LDA increases the performance of NER systems and it also brings other benefits. In our opinion, the major benefit is that LDA can add the long-range document context into NER. Currently, it is impossible to test this hypothesis because the existing datasets are not organized into documents and do not contain the long-range document context. We try to prove that LDA improves the results even without such a context and therefore we try to justify the future effort to prepare such a dataset. We believe that the document context is close to

© Springer Nature Switzerland AG 2018
P. Sojka et al. (Eds.): TSD 2018, LNAI 11107, pp. 58–66, 2018.
https://doi.org/10.1007/978-3-030-00794-2_6

the real use-cases of NER systems since those systems are applied frequently to document collections.

With the help of LDA, we were able to overcome the state of the art in NER for Czech and Spanish.

2   Related Work

Most of the current state-of-the-art systems employ LSTM[1] based deep learning methods with an additional representation of words at the character level. The systems usually apply CNNs[2] [4,8,11], GRUs[3] [12,15] or LSTM layers [7] on word characters to build an additional feature representation of the words.

Currently, the best systems use CRFs[4] [7,8] at the output to find the most likely sequence of the output tags.

LDA[5] has been used in the NER task before [5,10]. In [10], the authors proved that LDA helps to adapt the system to an unknown domain. The results in [5] are clearly inferior to the results presented in this paper.

The comparison of achieved F1 scores on the CoNLL corpora is presented in the results section – Sect. 5.1.

2.1   Latent Dirichlet Allocation

Latent Dirichlet Allocation (LDA) [3] is a generative graphical model which represents documents as mixtures of abstract topics where each topic is a mixture of words.

We use Mallet [9] for training LDA model. The initial hyper-parameters are $\alpha_0 = \beta_0 = 0.01$. We run 1 500 training iterations for every model.

The LDA models are trained on large raw corpora separated into documents. For English, we use the Reuters corpus RCV1[6]; for Spanish, the Reuters corpus RCV2 (see footnote 6); and for Czech, the Czech Press Agency corpus (not available publicly).

3   Datasets

In this paper, we use English, Spanish and Czech datasets. All the corpora have the same CoNLL format [13] and use the BIO tagging scheme. Additional resources were provided with these corpora. Part-of-speech tags are available for all tree corpora that we used. Chunk tags and Gazetteers were provided for English. Gazetteers are not used in our system, so as to preserve its full language independence.

---

[1] Long Short-Term Memory.
[2] Convolutional Neural Networks.
[3] Gated Recurrent Units.
[4] Conditional Random Fields.
[5] Latent Dirichlet allocation – see Sect. 2.1.
[6] http://trec.nist.gov/data/reuters/reuters.html.

For English and Spanish, the named entities are classified into four categories: persons, organizations, locations, and miscellaneous. Both corpora have similar size of about 250,000 tokens. For Czech, we use the CoNLL-format version [6] of the Czech Named Entity corpus CNEC 2.0 [14]. It contains approximately 150,000 tokens and uses 7 classes of Named Entities: time, geography, person, address, media, institution, and other.

4   Model

We share the basic structure of all modern character based NER systems and we add the LDA input to the projection layer of the network. In the following text, we describe the individual layers in more details.

We work with mini-batches and with sentence lengths padded to the maximum length in the mini-batch. The dimensions of the network input are *[batch-size, max-sentence-length]*. We trim the sentence lengths to a maximum length given by a hyper-parameter.

4.1   Layer 1 – Character Sequence Representation

The first layer is responsible for transforming words into character sequences which are consequently encoded as fixed length feature vectors. At first, all words in the input sentence are expanded to characters. The characters are encoded as integer value indices via a character dictionary. The resulting dimensions are *[batch-size, max-sentence-length, max-word-length]*. We trim the word lengths to a maximum length given by a hyper-parameter. Then, we use a randomly initialized embedding matrix to look up the character indices and to obtain character embeddings for each character. The dimension of character embeddings is given by another hyper-parameter. The resulting dimensions of this step are *[batch-size, max-sentence-length, max-word-length, char-embedding-dim]*.

In order to obtain the contextual representation of character sequences, the character input is fed into a bi-directional LSTM network. The dimension of the hidden vector of the LSTM cells is given by the hyper-parameter *char-lstm-dim*. The final outputs of both forward and backward LSTM cells are concatenated. The output dimensions of the layer are: *[batch-size, max-sentence-length, 2 * char-lstm-dim]*. The output vectors in the last dimension represent the words as a composition of their characters.

4.2   Layer 2 – Word Sequence Representation

The task of this layer is to represent the input sentence as a sequence of vectors – each for every word. At first, all words are converted to integer value indices via a word dictionary. Next, we look up the word indices in a word embedding matrix. In our model, we employ three methods to construct the embedding matrix (Fig. 1):

1. randomly initialized matrix – *word-dim* as hyper-parameter,

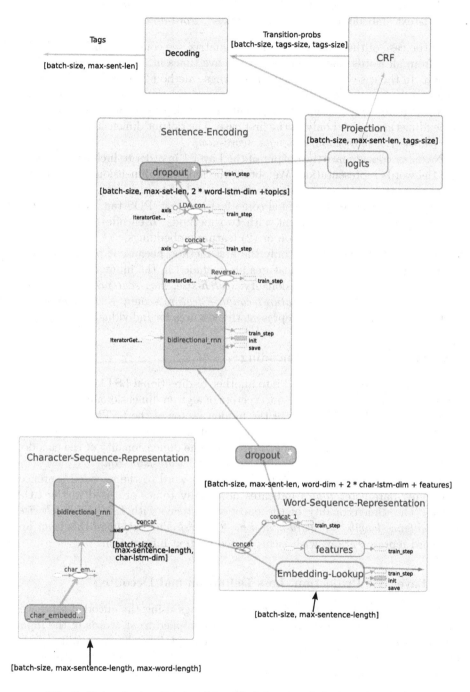

Fig. 1. Network visualization (simplified flow graph from Tensorflow).

2. GloVe embeddings, pre-trained models[7], *word-dim = 300*,
3. Fast-text embeddings, pre-trained models[8], *word-dim = 300*.

In the case of the randomly initialized matrix, we construct the word dictionary from all words that occur more than five times in the training part of the dataset. In the case of pre-trained embeddings (method 2 and 3), the dictionary is constructed as the intersection of all the words in the embedding matrix and all words from the training, development and testing parts of the dataset. The embeddings are trained only in the first case. The output dimensions of this part are: *[batch-size, max-sentence-length, word-dim]*.

Next, we concatenate the output of the Layer 1 in order to include characters into the words representation. We obtain the following dimensions of the input: *[batch-size, max-sentence-length, word-dim + 2 * char-lstm-dim]*.

Optionally we can also include some features (e.g. POS tag, syntactic role, gazetteers, etc.). We experiment with two methods to encode features in the input: (1) one-hot representation or (2) feature embeddings.

In our experiments, we use only the first method because it proved superior to the second one. When the features are included in the input representation, we obtain the following dimensionality: *[batch-size, max-sentence-length, word-dim + 2 * char-lstm-dim + feature1-count + feature2-count + ...]*. The output vectors in the last dimension represent the features for individual words.

4.3   Layer 3 – Sentence Encoding

The output of the Layer 2 is fed into another bi-directional LSTM network. The layer encodes an input sentence into vectors of a given dimensionality, one vector for every word. The dimension of the hidden vector of the LSTM cells is given by hyper-parameter *word-lstm-dim*. We concatenate the outputs of the forward and backward LSTM cells in every time step. The dimensionality of the encoded sentence is: *[batch-size, max-sentence-length, 2 * word-lstm-dim]*.

Optionally, we add the LDA topics for each word to the encoded sentence. The hyper-parameter *topics* indicates how many topics are used in the LDA model. The dimensionality of the encoded sentence with LDA is *[batch-size, max-sentence-length, 2 * word-lstm-dim + topics]*. The output of this layer are the word representations with incorporated context information.

4.4   Layer 4 – Projection, Loss Definition and Decoding

At the projection layer, we predict the NER tags using the encoded sentences from the Layer 3. One fully connected layer is applied to all words in the input sentence to obtain prediction scores for each word. The dimensionality of the projection matrix is *[2 * lstm-dim + topics, tags-size]*, where *tag-size* is the

---

[7] Downloaded from https://nlp.stanford.edu/projects/glove/.
[8] Downloaded from https://github.com/facebookresearch/fastText/blob/master/pretrained-vectors.md.

count of all (unique) NER tags (we do not use bias in this layer). We call the output prediction scores as *logits* and we denote the score of i-tag at time t as o_{i_t}.

The predicted labels enter the linear-chain CRF sequence tagging. We use the Tensorflow implementation of linear-chain CRF layer. The probability of a sequence is given by the sum S of the state transition (tag to tag transition) probabilities from tag i to tag j: $A_{i,j}$ and logits o_{i_t}.

$$S(i_1^T) = \sum_{t=1}^{T} \left(A_{i_{t-1},i_t} + o_{i_t} \right) \tag{1}$$

where A_{0,i_1} is the score for starting with tag i_1 and i_1^T is one particular sequence of tags. The score for the sequence is then normalized using softmax.

The loss of the model is defined as the score for the correct tag sequence y_1^T (where y_t is the correct tag at time t):

$$\log P(y_1^T) = S(y_1^T) - \log \sum_{\forall i_1^T} e^{S(i_1^T)} \tag{2}$$

During decoding, we use the Viterbi algorithm to find out the best tag sequence with the maximal probability:

$$\widetilde{y}_i^T =_{\forall i_1^T} S(i_1^T) \tag{3}$$

5   Experiments

The system is implemented using Python and Tensorflow 1.6 [1]. The CoNLL dataset is divided into train, development and test parts. We used the development data for early stopping of the network training. After the main training phase, we run two additional training epochs with the development data and reduced (decayed) learning rate. All results are computed on test data using the official CoNLL evaluation script. The results are computed as the average of 5 training sessions.

We share one set of hyper-parameters for all experiments – see Table 1.

5.1   Results

The results are presented in Table 2. They clearly show how hard it is to improve the current state-of-the-art models. In English, we even received no performance gain when we added the POS tags. However, the improvements in Czech are significant. The improvements of LDA models are not impressive. However, we must take into account that the LDA models can exploit only one sentence as a document context. Such a limited context very likely significantly decreases the LDA performance. We believe that with a proper document context the LDA improvement would be more significant. The POS tags significantly improved

Table 1. Hyper-parameters of the model.

Parameter description	Short name	Value
Character embedding dimension	char-dim	100
Word embedding dimension	word-dim	300
Character LSTM hidden vector size	char-lstm-dim	100
Word LSTM hidden vector size	word-lstm-dim	300
LDA topics	topics	20, 50, 100, 300
Maximum word length	max-char-len	25
Maximum sentence length	max-sent-len	150
Dropout probability	dropout	0.50
Optimization algorithm	alg	Adam
Learning rate	lr	0.01
Learning rate decay	lr-decay	0.90
Max gradient norm	max-grad-norm	5.00
Maximum no. of epochs	max-epochs	25

Table 2. F1 scores for different models and comparison with the state of the art. * indicates models with external labeled data, † indicates models with external resources, ‡ indicates models with additional training data.

Model/F1	English	Czech	Spanish
Baseline (Char+LSTM+CRF)	90.82	80.43	85.60
+ POS tags*	90.79	**81.77**	85.82
+ LDA 20	90.72	80.62	85.67
+ LDA 50	**91.21**	**80.83**	85.62
+ LDA 100	90.83	80.58	**85.91**
+ LDA 300	90.96	80.71	85.55
Straková et al.* [12]	89.92	80.79	–
Santos and Guimarães [11]	–	–	82.21
Yang et al.†‡ [15]	91.20	–	85.77
Lample et al. [7]	90.94	–	85.75
Ma and Hovy [8]	91.21	–	–
Chiu and Nichols† [4]	**91.62**	–	–

results only for Czech. The most likely reason is that Czech dataset has much elaborated POS tagset. Moreover, the Czech language has much richer morphology which is captured very well within the tags.

In our experiments, we also compared the different choices of word embedding initializations – see Sect. 4.2. The random initialization was a clearly inferior choice. We obtained a 0.10 to 0.15 drop in F1 scores. We believe that the main reason consists in the low amount of training data in the NER task. The difference between GloVe and Fast-text was not high. The GloVe pre-trained model was slightly better than Fast-text models. However, the authors of GloVe provide pre-trained models only for English. Therefore, we used the Fast-text models for other languages since the authors of Fast-text provide pre-trained models for many languages.

6 Conclusion and Future Work

We created a deep neural NER system which achieves state-of-the-art results on the Czech and Spanish datasets. It proves that LDA can improve the system even if it does not have the document context. In the future, we want to try to create dataset organized into documents to get better context which LDA can capture. In this case, LDA should bring more significant improvements.

Acknowledgements. This work was supported by Ministry of Education, Youth and Sports of the Czech Republic, institutional research support (1311) and by the UWB grant no. SGS-2013-029 Advanced computing and information systems. Access to the MetaCentrum computing facilities provided under the program "Projects of Large Infrastructure for Research, Development, and Innovations" LM2010005, funded by the Ministry of Education, Youth, and Sports of the Czech Republic, is highly appreciated.

References

1. Abadi, M., et al.: TensorFlow: large-scale machine learning on heterogeneous systems (2015). https://www.tensorflow.org/, Software available from tensorflow.org
2. Agerri, R., Rigau, G.: Robust multilingual named entity recognition with shallow semi-supervised features. Artif. Intell. **238**, 63–82 (2016)
3. Blei, D.M., Ng, A.Y., Jordan, M.I., Lafferty, J.: Latent dirichlet allocation. J. Mach. Learn. Res. **3** (2003)
4. Chiu, J., Nichols, E.: Named entity recognition with bidirectional LSTM-CNNs. Trans. Assoc. Comput. Linguist. **4**, 357–370 (2016). http://aclweb.org/anthology/Q16-1026
5. Konkol, M., Brychcn, T., Konopk, M.: Latent semantics in named entity recognition. Expert Syst. Appl. **42**(7), 3470–3479 (2015). https://doi.org/10.1016/j.eswa.2014.12.015, http://www.sciencedirect.com/science/article/pii/S0957417414007933
6. Konkol, M., Konopík, M.: CRF-based Czech named entity recognizer and consolidation of Czech NER research. In: Habernal, I., Matoušek, V. (eds.) TSD 2013. LNCS (LNAI), vol. 8082, pp. 153–160. Springer, Heidelberg (2013). https://doi.org/10.1007/978-3-642-40585-3_20
7. Lample, G., Ballesteros, M., Subramanian, S., Kawakami, K., Dyer, C.: Neural architectures for named entity recognition. In: Proceedings of the 2016 Conference of the North American Chapter of the Association for Computational Linguistics: Human Language Technologies, pp. 260–270. Association for Computational Linguistics (2016). https://doi.org/10.18653/v1/N16-1030, http://www.aclweb.org/anthology/N16-1030

8. Ma, X., Hovy, E.: End-to-end sequence labeling via bi-directional LSTM-CNNs-CRF. In: Proceedings of the 54th Annual Meeting of the Association for Computational Linguistics, ACL 2016, Berlin, Germany, 7–12 August 2016, vol. 1: Long Papers. The Association for Computer Linguistics (2016). http://aclweb.org/anthology/P/P16/P16-1101.pdf
9. McCallum, A.K.: Mallet: a machine learning for language toolkit (2002). http://mallet.cs.umass.edu
10. Nallapati, R., Surdeanu, M., Manning, C.: Blind domain transfer for named entity recognition using generative latent topic models. In: Proceedings of the NIPS 2010 Workshop on Transfer Learning Via Rich Generative Models, pp. 281–289 (2010)
11. dos Santos, C.N., Guimarães, V.: Boosting named entity recognition with neural character embeddings. In: Duan, X., Banchs, R.E., Zhang, M., Li, H., Kumaran, A. (eds.) Proceedings of the Fifth Named Entity Workshop, NEWS@ACL 2015, Beijing, China, 31 July 2015, pp. 25–33. Association for Computational Linguistics (2015). https://doi.org/10.18653/v1/W15-3904
12. Straková, J., Straka, M., Hajič, J.: Neural networks for featureless named entity recognition in Czech. In: Sojka, P., Horák, A., Kopeček, I., Pala, K. (eds.) TSD 2016. LNCS (LNAI), vol. 9924, pp. 173–181. Springer, Cham (2016). https://doi.org/10.1007/978-3-319-45510-5_20
13. Tjong Kim Sang, E.F.: Introduction to the CoNLL-2002 shared task: language-independent named entity recognition. In: Proceedings of CoNLL 2002, Taipei, Taiwan, pp. 155–158 (2002)
14. Ševčíková, M., Žabokrtský, Z., Krůza, O.: Named entities in Czech: annotating data and developing NE tagger. In: Matoušek, V., Mautner, P. (eds.) TSD 2007. LNCS (LNAI), vol. 4629, pp. 188–195. Springer, Heidelberg (2007). https://doi.org/10.1007/978-3-540-74628-7_26. http://dl.acm.org/citation.cfm?id=1776334.1776362
15. Yang, Z., Salakhutdinov, R., Cohen, W.W.: Multi-task cross-lingual sequence tagging from scratch. CoRR abs/1603.06270 (2016). http://dblp.uni-trier.de/db/journals/corr/corr1603.html#YangSC16

Lexical Stress-Based Authorship Attribution with Accurate Pronunciation Patterns Selection

Lubomir Ivanov[✉], Amanda Aebig, and Stephen Meerman

Computer Science Department, Iona College, 715 North Avenue,
New Rochelle, NY 10801, USA
livanov@iona.edu

Abstract. This paper presents a feature selection methodology for authorship attribution based on lexical stress patterns of words in text. The methodology uses part-of-speech information to make the proper selection of a lexical stress pattern when multiple possible pronunciations of the word exist. The selected lexical stress patterns are used to train machine learning classifiers to perform author attribution. The methodology is applied to a corpus of 18th century political texts, achieving a significant improvement in performance compared to previous work.

Keywords: Authorship attribution · Lexical stress · Prosody
Part-of-speech tagging · Machine learning

1 Introduction

Authorship attribution is an interdisciplinary field at the crossroads of computer science, linguistics, and the humanities. The goal is to identify the true author of a text, whose authorship is unknown or disputed. With the development of (semi-)automated attribution methods, the importance of the field has grown tremendously: Some have applied attribution methodologies to re-examine the authorship of literary works throughout the ages [1–8]. Others are using attribution to determine the authorship of historically significant documents, which have had an impact on historical events and societies [9–12]. More modern applications of authorship attribution include digital copyright and plagiarism detection, forensic linguistics, criminal and anti-terror investigation [13–17].

Traditionally, authorship attribution has been carried out by domain experts, who examine many aspect of the unattributed text: its content, the literary style and the political, philosophical, and ideological views of the author, the historical circumstances in which the text was created. Human expert attribution, however, is tedious, error prone, and may be affected by the personal beliefs of the attribution expert. The rapid advances in data mining, machine learning, and natural language processing have lead to the development of a multitude of techniques for automated attribution, which can accurately and quickly perform

© Springer Nature Switzerland AG 2018
P. Sojka et al. (Eds.): TSD 2018, LNAI 11107, pp. 67–75, 2018.
https://doi.org/10.1007/978-3-030-00794-2_7

an in-depth analysis of large texts/corpora. Moreover, automated attribution can uncover inconspicuous aspects of an author's style, which may be difficult or impossible for a human expert to spot. Automated attribution is also less prone to subjectivity, and allows the results to be independently verified.

Automated authorship attribution relies on three "ingredients" - a set of stylistic features, which capture the notion of an author's style, a machine learning classifier, and a set of attributed documents to train the classifier. Among the most commonly used stylistic features are function words, character- and word n-grams, part of speech tags, vowel-initiated words, etc. A new direction in authorship attribution is the use of prosodic features. In [19,20], our team explored the use of alliteration and lexical stress as stylistic features for authorship attribution. We demonstrated that alliteration and lexical stress can successfully augment other traditional features and improve the attribution accuracy. However, both features appear weak as stand-alone style predictors, particularly if the author set is large and the authors' styles are not "melodic".

In this paper we revisit the use of lexical stress as a stylistic feature for authorship attribution. We present an improved approach for extracting accurate lexical stress patterns from text, and demonstrate that the new technique significantly improves the results of the author attribution experiments. We show that the new lexical stress based approach is sufficiently strong for attribution even as a stand-alone stylistic feature and with a relatively large set of candidate authors. Finally, we discuss the strengths and weaknesses of lexical stress based attribution, and directions for further research.

2   Lexical Stress

2.1   Background

Lexical stress is a prosodic feature, which describes the emphasis placed on specific syllables in words. The English language uses variable lexical stress, where emphasis is placed on different word syllables. This allows some word with identical spelling (homographs) to be differentiated phonemically and semantically. In speech, stress involves a louder/longer pronunciation of the stressed syllables, and a change in voice pitch. This can provide an emotive charge, and can be used to emphasize a particular attitude or opinion toward a topic. By selecting appropriately stressed words, a skillful writer can evoke a strong emotional response in his/her audience. Thus, it is conjectured that lexical stress can be used as an stylistic marker for authorship, particularly in cases where the texts are intended to be read to an audience. This is notably true in the case of historical documents, which were commonly intended for public reading.

The role of lexical stress in authorship attribution has hardly been explored except in two studies: In [18], Dumalus and Fernandez explored the use of rhythm based on lexical stress as a method for authorship attribution, and concluded that the technique shows promise. In [19], we performed an in-depth analysis of the usefulness of lexical stress for attribution based on extracted lexical stress patterns from historical texts. We also combined lexical stress with other stylistic features, and demonstrated that doing so increases the attribution accuracy. Both studies, however, point out that using lexical stress for attribution is difficult for a number of reasons: First, in the context of historical attribution, it is impossible to know the correct pronunciation of words from as far back in time as the 18th century. Thus, lexical stress attribution is, out of necessity, based on modern pronunciation dictionaries such as the CMU dictionary [21]. It is worth noting, however, that in recent years there has been a several studies, which indicate that, lexically, the English spoken during the 18th century was significantly closer to present day American English than British English [22, 23]. It is, therefore, not unreasonable to adopt an American pronunciation dictionary for a study of lexical stress in historical attribution. The second problem is due to the large number of homographs in English. The work presented in [19] acknowledged this, and indicated that the algorithm used for extracting lexical stress patterns simply selects the first matching pattern in the dictionary. While this approach improves the performance of the algorithm, it does not necessarily select the correct pronunciation, and may skew the results of the attribution.

2.2 Lexical Stress Pattern Selection

In this paper, we extend the lexical stress selection algorithm presented in [19] to perform an accurate selection of lexical stress patterns of words from the CMU dictionary. Next, we train machine learning classifiers based on the use of the same historical documents, and compare our new results to those cited in [19]. Before investigating the issue of proper lexical pattern selection, it was important to convince ourselves that choosing different patterns affects the outcome of the attribution experiments. Thus, our first set of experiments was based on a simple modification of the lexical pattern selection algorithm presented in [19]: Instead of choosing the first encountered word pronunciation pattern from the CMU dictionary, we selected one pattern at random whenever multiple pronunciation patterns for a word were encountered. We applied the algorithm to each text in our historical corpus, and extracted a vector of lexical stress pattern frequencies for each document. Using these vectors, we trained support vector machine (SVM) and multi-layer perceptron (MLP) classifiers to carry out author attribution. Table 1 presents the results from some of the experiments we performed - with the full set of 38 authors/224 documents, with a randomly selected set of 20 authors/140 documents, and with the top performing 7 authors/65 documents. The table lists the accuracy of the original algorithm (first-in-list pattern selection) and of the modified algorithm (random pattern selection). The experiments were performed using the popular WEKA software [24], with SMO (sequential minimal optimization) SVM and MLP machine learning models. The results confirmed our hypothesis that choosing different lexical stress patterns leads to improvements in accuracy, precision, and recall. The improvements were significant enough to justify the work on an accurate pattern selection methodology.

Table 1. Maximum accuracy based on first-in-list vs. random selection of lexical stress patterns

#Authors	#Docs	Acc.(First/ SMO)	Acc.(Random/ SMO)	Acc.(First/ MLP)	Acc.(Random/ MLP)
38	224	16.07%	31.69%	30.80 %	37.50%
20	140	25.00%	42.54%	33.57%	44.03%
7	65	60.00%	78.46%	73.85%	84.62%

2.3   Accurate Lexical Stress Pattern Selection

To create a mechanism for accurate lexical pattern selection, it is important to consider the reasons why the same word may have multiple pronunciations: One reason is the part of speech (PoS) role of the word. The word "present", when pronounced as "P R EH1 Z AH0 N T", exhibits the lexical stress pattern "10", and acts as a noun, meaning "a gift". It can also be an adjective, meaning "current" or "at hand". However, when pronounced as "P R IY0 Z EH1 N T" (lexical stress pattern "01"), the word acts as a verb, meaning "to give". It is important to note that the CMU dictionary has one more pronunciation for the verb "present": "P ER0 Z EH1 N T". The second pronunciation is clearly rarer, and demonstrates the role of dialects in pronunciation. In this particular case, even though there are two different pronunciations of the verb "present', they both share the same lexical stress pattern "01". However, the two pronunciations of "proportionally"-"P R AH0 P AO1 R SH AH0 N AH0 L IY0" and "P R AH0 P AO1 R SH N AH0 L IY0"- have different lexical stress patterns, "01000" and "0100" respectively. Another example is the word "laboratory", which can be pronounced as either "L AE0 B OH1 R AH0 T AO2 R IY0" or as "L AE1 B R AH0 T AO2 R IY0" (stress patterns "01020" or "1020").

Our algorithm is based on the idea of selecting the correct lexical stress patterns based on the PoS role of each word in the text. To achieve this, we had to complete a few initial tasks: All texts in the historical corpus were tagged with PoS tags using the Stanford PoS tagger [25, 26]. Next, we modified the CMU dictionary in two ways: First, we added all words from the historical texts, which were not in the dictionary. Next, we extracted all words with multiple pronunciation patterns (approximately 8,000) and tagged them with PoS information obtained both from the Stanford PoS tagger and from other online sources. The PoS-tagged words were added back into the CMU dictionary.

Our new algorithm is an extension of the original algorithm described in [19]: For each word in a given text, we look up the word in the modified CMU pronunciation dictionary. If the word has a single pronunciation pattern, then the word is replaced with the lexical stress pattern corresponding to that pronunciation. If the word has multiple pronunciation patterns, we compare the tag of the word to the tags of the dictionary pronunciations. If a single match exists, the lexical stress pattern corresponding to that pronunciation is selected. It is possible, however, that, due to different dialectal pronunciations, multiple matches

Table 2. Maximum accuracy of the original and random-selection algorithms, and the average and maximum accuracies for the PoS-selection algorithm (224 documents/38 authors).

Classifier	Original	Random Selection	PoS-Selection (Avg. Acc.)	PoS Selection (Max. Acc.)
SMO	16.07%	31.69%	36.20%	39.46%
MLP	30.80%	37.5%	43.46%	47.53%

exist. In such a case, we select one of the possible matches at random (since there is little we can do about differentiating pronunciations by dialect). Once all words in the text have been accounted for, the frequencies of all recorded lexical stress patterns are computed and stored as a training vector for the machine learning classifiers. As before, we used WEKA MLP and SMO classifiers for performing the actual author attribution. We used 10-fold cross-validation in all experiments.

3   Using Lexical Stress for Authorship Attribution

3.1   Experiments: Individual Lexical Stress Patterns

Our first set of attribution experiments involved using individual lexical stress patterns. Since the new PoS-based selection algorithm may involve a random selection of lexical stress patterns (due to dialects), we conducted ten sets of experiments for each set of authors/documents. The first step was to apply the new methodology to the full set of 224 documents. Table 2 reports the maximum and average accuracies for the ten experiments with the full set of documents, and compares the results to those obtained with the original algorithm and the purely random-selection algorithm from Sect. 2.2. The average baseline accuracy with only traditional features was 58.26%.

While the accuracy is still too low to be useful for meaningful authorship attribution, it is significantly higher than the accuracies of the earlier experiments reported in [19]. Moreover, the number of candidate authors in real attribution studies is usually much lower −4 to 13 author batches are common in our actual historical attributions. Thus, it was important to consider the change in accuracy and author precision/recall for a smaller set of candidates. For the next step, we eliminated all authors with an average F-measure of less than 0.5 in the all-authors experiments. The new set consisted of 13 authors and 101 documents. We carried out ten SMO- and ten MLP sets of experiments using this smaller set of authors/documents, and recorded the results (Table 3).

The accuracy was significantly higher, and the average F-measure for all authors was above .6, so no further reduction of the author set by eliminating authors with a low F-measure was reasonable. However, in an effort to remain true to the original paper [19] and to consider the accuracy, precision, and recall

Table 3. Average accuracies, precisions, and recalls across 10 experiments with 13 authors/101 documents using SMO and MLP classifiers.

Experiment	Acc.SMO	Acc.MLP	Recall SMO	Recall MLP	Precision SMO	Precision MLP
1	66.34%	67.33%	0.663	0.673	0.727	0.727
2	66.34%	73.27%	0.663	0.733	0.673	0.757
3	69.31%	73.27%	0.693	0.733	0.725	0.758
4	70.30%	72.28%	0.703	0.723	0.722	0.741
5	72.28%	77.23%	0.723	0.772	0.747	0.804
6	68.32%	73.28%	0.683	0.733	0.733	0.760
7	67.33%	76.24%	0.673	0.762	0.705	0.793
8	71.29%	73.28%	0.713	0.733	0.757	0.751
9	70.30%	75.25%	0.703	0.752	0.740	0.809
10	72.28%	74.26%	0.723	0.743	0.766	0.770
Average:	69.41%	73.56%	0.694	0.736	0.730	0.767

for a small set of authors, we selected the seven top-performing authors (Adams, Hopkinson, Lafayette, Mackintosh, Ogilvie, Paine, Wollstonecraft), repeating the set of experiments ten more times with each type of classifier (Table 4). The average accuracy and author precision/recall topped 90% in most experiments and compared well to the 91.67% baseline accuracy. This confirms our original hypothesis that some authors appear to have a more unique, "melodic" writing style, and can, therefore, be recognized by the attribution software.

Finally, we re-ran all experiments involving the authors/documents from [19] (Table 5). Once again, we observed a significant improvement in accuracy over the original algorithm as well as over the pure random selection algorithm. The accuracy, precision, and recall all improved for the smaller set of authors, topping 85% when the authors set was on par with the size of the author sets in our actual historical attribution experiments. We note that MLP routinely outperformed SMO, but did take longer to train.

3.2   Experiments: N-Grams of Lexical Stress Patterns

The next set of experiments was aimed at exploring the usefulness of n-grams of lexical stress patterns for authorship attribution. We conducted experiments with 2-, 3-, and 4-grams. The number of n-gram lexical stress patterns was usually high, which made training MLPs classifiers extremely time-consuming: For example, the number of 2-gram patterns in a typical experiment with 38 authors/224 documents was 1559, requiring 2098.93s of computation per fold in a 10-fold cross-validation. Thus, for the full set of 38 authors/ 224 documents, we performed only three sets of 10-fold cross-validation experiments with MLP classifiers and ten sets with SMO classifiers for each of 2-gram, 3-gram, and 4-gram of lexical stress patterns. We carried out ten sets 2-gram, 3-gram, and 4-gram 10-fold cross-validation experiments with both SMO and MLP classifiers for each of

Table 4. Average accuracies, precisions, and recalls across 10 experiments with the top performing 7 authors/65 documents using SMO and MLP classifiers

Experiment	Acc.SMO	Acc.MLP	Recall SMO	Recall MLP	Precision SMO	Precision MLP
1	92.19%	89.06%	0.922	0.891	0.924	0.900
2	89.06%	92.19%	0.891	0.922	0.895	0.925
3	90.62%	93.75%	0.906	0.938	0.909	0.942
4	90.62%	87.50%	0.906	0.875	0.915	0.879
5	87.50%	89.06%	0.875	0.891	0.881	0.900
6	90.63%	92.19%	0.906	0.922	0.912	0.929
7	90.63%	90.63%	0.906	0.906	0.911	0.909
8	92.19%	87.50%	0.922	0.875	0.925	0.877
9	93.75%	89.06%	0.938	0.891	0.938	0.893
10	93.75%	89.06%	0.938	0.891	0.939	0.896
Average:	91.09%	90.00%	0.911	0.900	0.915	0.905

Table 5. Maximum accuracy based on the original first-in-list-vs. random vs. PoS-based selection of lexical stress patterns using authors cited in [19]

#Authors	#Docs	Original/SMO	Random/SMO	PoS/SMO	Original/MLP	Random/MLP	PoS/MLP
38	224	16.07%	31.69%	39.46%	30.80%	37.50%	47.53%
20	140	25.00%	42.54%	48.63%	33.57%	44.03%	50.00%
7	65	60.00%	78.46%	84.38%	73.85%	84.62%	85.94%

the smaller sets of authors/documents (20/140, 13/101, 7/65). The highest accuracy was achieved using 2-grams, which, for large sets of authors/documents, improved slightly the results obtained in the individual lexical stress pattern experiments. For smaller sets of authors/documents the improvements were significant (Table 6).

Table 6. Maximum accuracy of the original first-in-list- vs. average accuracy of individual PoS-based lexical stress patterns selection and 2-gram PoS-based lexical stress pattern selection.

#Authors	#Docs	Orig./SMO	PoS/SMO	2-Grm-PoS/SMO	Orig./MLP	PoS/MLP	2-Grm-PoS/MLP
		Max Acc.	Avg. Acc.	Avg. Acc.	Max Acc.	Avg. Acc.	Avg. Acc.
38	224	16.07%	39.46%	40.45%	30.80%	47.53%	47.83%
20	140	25.00%	48.63%	48.85%	33.57%	50.00%	56.67%
7	65	78.46%	84.38%	85.63%	73.85%	85.94%	91.67%

With 3- and 4-grams of lexical stress patterns, the accuracy decreased considerably (low- to mid-20% for 38 authors/224 documents and 20 authors/140 documents). We suspect that the main reason for the decrease is the large number of 3- and 4-gram lexical stress patterns per experiment, which, combined with the relatively small historical document corpus, does not provide a sufficient number of training examples.

4   Conclusion and Future Work

This paper presented an improved algorithm for authorship attribution based on selecting lexical stress patterns of words using PoS information. We have demonstrated that lexical stress can be a useful feature for authorship attribution: Our experiments indicate that lexical stress is sufficiently powerful to be useful as a stand alone stylistic feature when the number of authors is limited. We also conducted a small set of experiments with lexical stress combined with traditional stylistic features. The results indicated a small but consistent improvement in accuracy, which we believe is due to the ability of lexical stress to differentiate among authors with melodic writing styles. This ability will be incorporated into our new tiered attribution system, in which the initial layer will narrow down the author set based on traditional features, while the upper layer(s) will fine-tune the prediction using special features like lexical stress.

References

1. Morton, A.Q.: The authorship of greek prose. J. R. Stat. Soc. (A) **128**, 169–233 (1965)
2. Binongo, J.N.G.: Who wrote the 15th book of Oz? An application of multivariate statistics to authorship attribution. Comput. Linguist. **16**(2), 9–17 (2003)
3. Barquist, C., Shie, D.: Computer analysis of alliteration in beowulf using distinctive feature theory. Lit. Linguist. Computing. **6**(4), 274–280 (1991). https://doi.org/10.1093/llc/6.4.274
4. Matthews, R., Merriam, T.: Neural computation in stylometry: an application to the works of Shakespeare and Fletcher. Lit. Linguist. Comput. **8**(4), 203–209 (1993)
5. Lowe, D., Matthews, R.: Shakespeare vs. Fletcher: a stylometric analysis by radial basis functions. Comput. Humanit. **29**, 449–461 (1995)
6. Smith, M.W.A.: An investigation of Morton's method to distinguish Elizabethan Playwrights. Comput. Humanit. **19**, 3–21 (1985)
7. Burrows, J.: Computation into Criticism: A Study of Jane Austen's Novels and an Experiment in Method. Clarendon Press, Oxford (1987)
8. Holmes, D.I.: A stylometric analysis of mormon scripture and related texts. J. Roy. Stat. Soc.: Ser. A: Appl. Stat. **155**(1), 91–120 (1992)
9. Mosteller, F, Wallace, D.: Inference and disputed authorship: the Federalist: AWL (1964)
10. Berton, G., Petrovic, S., Ivanov, L., Schiaffino, R.: Examining the Thomas Paine corpus: automated computer authorship attribution methodology applied to Thomas Paine's writings. In: Cleary, S., Stabell, I.L. (eds.) New Directions in Thomas Paine Studies, pp. 31–47. Palgrave Macmillan US, New York (2016). https://doi.org/10.1057/9781137589996_3
11. Petrovic, S., Berton, G., Campbell, S., Ivanov, L.: Attribution of 18th century political writings using machine learning. J. of Technol. Soci. **11**(3), 1–13 (2015)

12. Petrovic, S., Berton, G., Schiaffino, R., Ivanov, L.: Authorship attribution of Thomas Paine works. In: International Conference on Data Mining, DMIN 2014, pp. 183–189. CSREA Press (2014). ISBN: 1-60132-267-4

13. Zheng, R., Li, J., Chen, H., Huang, Z.: A framework for authorship identification of online messages: writing style features and classification techniques. J. Am. Soc. Inf. Sci. Technol. **57**(3), 378–393 (2006)

14. Argamon, S., Saric, M., Stein, S.: Style mining of electronic messages for multiple authorship discrimination. In: Proceedings of the 9th ACM SIGKDD, pp. 475–480 (2003)

15. de Vel, O., Anderson, A., Corney, M., Mohay, G.M.: Mining e-mail content for author identification forensics. SIGMOD Rec. **30**(4), 55–64 (2001)

16. Kotzé, E.: Author identification from opposing perspectives in forensic linguistics. South. Afr. Linguist. Appl. Lang. Stud. **28**(2), 185–197 (2010)

17. Abbasi, A., Chen, H.: Applying authorship analysis to extremist-group web forum messages. IEEE Intell. Syst. **20**(5), 67–75 (2005)

18. Dumalus, A., Fernandez, P.: Authorship attribution using writer's rhythm based on lexical stress. In: 11th Philippine Computing Science Congress, Naga City, Philippines (2011)

19. Ivanov, L.: Using alliteration in authorship attribution of historical texts. In: Sojka, P., Horák, A., Kopeček, I., Pala, K. (eds.) TSD 2016. LNCS (LNAI), vol. 9924, pp. 239–248. Springer, Cham (2016). https://doi.org/10.1007/978-3-319-45510-5_28

20. Ivanov, L., Petrovic, S.: Using lexical stress in authorship attribution of historical texts. Chapter, Lecture Notes in Computer Science: TSD **9302**, 105–113 (2015). https://doi.org/10.1007/978-3-319-24033-6_12

21. Internet resource. http://www.speech.cs.cmu.edu/cgi-bin/cmudict

22. Fischer, J.H.: British and American, continuity and divergence. In: Algeo, J. (ed.) The Cambridge History of English Language, pp. 59–85. Cambridge University Press, Cambridge (2001)

23. Scotto Di Carlo, G.: Lexical differences between American and British english: a survey study. Lang. Des.: J. Theor. Exp. Linguist. 15, 61-75 (2013)

24. Hall, M., Frank, E., Holmes, G., Pfahringer, B., Reutemann, P., Witten, I.: The WEKA data mining software: an update. SIGKDD Explor. **11**(1), 10–18 (2009)

25. Toutanova, K., Ķlein D., Manning C.: Feature-rich part-of-speech tagging with a cyclic dependency network. In: HLT-NAACL, pp. 252–259 (2003)

26. Toutanova, K., Manning, D.: Enriching the knowledge sources used in a maximum entropy part-of-speech tagger. In: EMNLP/VLC-2000, pp. 63–70 (2000)

Idioms Modeling in a Computer Ontology as a Morphosyntactic Disambiguation Strategy
The Case of Tibetan Corpus of Grammar Treatises

Alexei Dobrov[1], Anastasia Dobrova[2], Pavel Grokhovskiy[1],
Maria Smirnova[1(✉)], and Nikolay Soms[2]

[1] Saint-Petersburg State University, Saint-Petersburg, Russia
{a.dobrov,p.grokhovskiy,m.o.smirnova}@spbu.ru
[2] LLC "AIIRE", Saint-Petersburg, Russia
{adobrova,nsoms}@aiire.org

Abstract. The article presents the experience of developing computer ontology as one of the tools for Tibetan idioms processing. A computer ontology that contains a consistent specification of meanings of lexical units with different relations between them represents a model of lexical semantics and both syntactic and semantic valencies, reflecting the Tibetan linguistic picture of the world. The article presents an attempt to classify Tibetan idioms, including compounds, which are idiomatized clips of syntactic groups that have frozen inner syntactic relations and are often characterized by omission of grammatical morphemes; and the application of this classification for idioms processing in computer ontology. The article also proposes methods of using computer ontology for avoiding idioms processing ambiguity.

Keywords: Tibetan language · Idioms · Compounds
Computer ontology · Tibetan corpus · Natural language processing
Corpus linguistics · Immediate constituents

1 Introduction

Research introduced by this paper is a continuation of several research projects ("The Basic corpus of the Tibetan Classical Language with Russian translation and lexical database", "The Corpus of Indigenous Tibetan Grammar Treatises"), aimed at the development of methods for creation of a parallel Tibetan-Russian corpus [1, p. 183].

The Basic Corpus of the Tibetan Classical Language includes texts in a variety of classical Tibetan literary genres. The Corpus of Indigenous Tibetan Grammar Treatises consists of the most influential grammar works, the earliest of them proposedly dating back to 7th–8th centuries. The corpora comprise 34,000 and 48,000 tokens, respectively. Tibetan texts are represented both in Tibetan Unicode script and in standard Latin (Wylie) transliteration [1].

© Springer Nature Switzerland AG 2018
P. Sojka et al. (Eds.): TSD 2018, LNAI 11107, pp. 76–83, 2018.
https://doi.org/10.1007/978-3-030-00794-2_8

The ultimate goal of the current project is to create a formal model (a grammar and a linguistic ontology) of the Tibetan language, including morphosyntax, syntax of phrases and hyperphrase unities, and semantics, that can produce a correct morphosyntactic, syntactic, and semantic annotation of the corpora without any manual corrections.

The current version of developed corpus is available at http://aiire.org/corman/index.html?corpora_id=67&page=1&view=docs_list.

The underlying AIIRE (Artificial Intelligence-based Information Retrieval Engine) linguistic processor implements the method of inter-level interaction proposed by Tseitin in 1985 [2], which ensures effeciency of rule-based ambiguity resolution. AIIRE needs to recognize all the relevant linguistic units in the input text. For inflectional languages, the input units are easy to identify as word forms, separated by spaces, punctuation marks, etc. It is not the case with the Tibetan language, as there are no universal symbols to segment the input string into words or morphemes. The developed module for the Tibetan language performs the segmentation of the input string into elementary units (morphs and punctuation marks - *atoms*) by using the Aho-Corasick algorithm [3], that allows to find all possible substrings of the input string according with a given dictionary.

The system aims at multi-variant analysis, unlike some other systems of morphosyntactic analysis. That sometimes causes combinatorial explosions[1] in the analysis versions. In most combinatorial explosions, idioms were present, therefore, one of the strategies for eliminating morphosyntactic ambiguity was the processing of Tibetan idioms using computer ontology.

2   The Ontology Structure

The most famous and widely cited general definition of the term *ontology* is 'an explicit specification of a conceptualization' by Gruber [2]. Many different attempts were made to refine it for particular purposes. Without claiming for any changes to this de-facto standard, we have to clarify that, as the majority of researchers in natural language understanding, we mean not just any 'specification of a conceptualization' by this term, but rather a *computer ontology*, which we define as a database that consists of concepts and relations between them. Ontological concepts have attributes. Attributes and relations are interconnected: participation of a concept in a relation may be interpreted as its attribute, and vice versa. Relations between concepts are binary and directed. They can be represented as logical formulae, defined in terms of a calculus, which provides the rules of inference. Relations themselves can be modeled by concepts.

There is a special type of ontologies - so called linguistic ontologies, which are designed for automatic processing of unstructured texts. Units of linguistic

---

[1] Following the definition of Krippendorf, combinatorial explosion is understood here as a situation "when a huge number of possible combinations are created by increasing the number of entities which can be combined" [4]. As applied to parsing, these are cases of exponential growth in the number of parsing versions as the length of the parsed text and, thus, the amount of its parsed ambiguous fragments increase.

ontologies are based on meanings of real natural language expressions. Ontologies of this kind actually model linguistic picture of the world, that stands for language semantics. Ontologies, created for different languages, are not the same and are not language-independent. Differences between ontologies show differences between linguistic pictures of the world.

Ontologies, that are designed for natural language processing, are supposed to include relations that allow to perform semantic analysis of texts and to perform lexical and syntactic disambiguation.

The ontology, used for this research, was developed according with the above mentioned principles [5]. It is a united consistent classification of concepts behind the meanings of Tibetan linguistic units, including morphemes and idiomatic morphemic complexes. The concepts are interconnected with different semantic relations. Different relations are established between concepts. The relation of synonymy is always absolute (complete coincidence of referents with possible differences in significations). Concepts form synonymic sets (not to be confused with Wordnet synsets [6], which are sets of words). Each element of the set has the same attributes, i.e. the same relations and objects of these relations. Relations like class-superclass provide inheritance of attributes between concepts. This mechanism allows to model semantic valencies as specific relations between some basic classes of the ontology (see below).

Concepts are also marked with the so-called *token types* that represent sets of classes of immediate constituents that can denote a concept. This is necessary for concepts denoted by idioms, as it will be shown below.

Totally within the framework of this research 2,924 concepts that are meanings of 2,749 Tibetan expressions were modelled in the ontology, 681 of them being idioms. The ontology is implemented within the framework of AIIRE ontology editor software; it is available as a snapshot at http://svn.aiire.org/repos/tibet/trunk/aiire/lang/ontology/concepts.xml and it is available for view or even for edit at http://ontotibet.aiire.org by access request.

3   Classification of Tibetan Idioms and Their Modeling in the Computer Ontology

An idiom is a multimorphemic expression whose meaning cannot be deduced by the general rules of the language in question from the meanings of the constituent morphs, their semantically loaded morphological characteristics (if any) and their syntactic configuration [7, p. 167]. Idiom is kind of phrasemes - a linguistic expression formed by several (at least two) lexemes syntactically linked in a regular way.

In AIIRE ontology, in addition to the meanings of an idiomatic expression, meanings of its components are also modeled, so that they could be interpreted in their literal meanings too. This is necessary, because AIIRE natural language processor is designed to perform natural language understanding according with the compositionality principle [8], and idiomaticity is treated not merely as a property of a linguistic unit, but rather as a property of its meaning, namely, as a conventional substitution of a complex (literal) meaning with a single holistic (idiomatic) concept. In this respect, the approach to processing idioms in the AIIRE project is fundamentally different from traditional techniques like phrase pattern or substring search: in AIIRE, idioms are neither textual, nor syntactic structure fragments, but, technically, semantic representations of expressions in their literal readings. The approach to natural language processing in AIIRE in general is fundamentally different from machine learning or statistical approaches that have become traditional, or from simpler pattern-based or rule-based algorithms; AIIRE implements the full cycle of NLU procedures, from morphological and syntactic analysis to the construction of semantic graphs, on the basis of multiversion text parsing performed by formal object-oriented grammar, and on the basis of the methods of constructing semantic graphs that this grammar implements in accordance with the linguistic ontology and those constraints that are imposed by the system of relations of this ontology [5].

In the Tibetan language, it is possible to distinguish two main types of idioms: (1) compounds and (2) non-compound idioms. All Tibetan compounds are created by the juxtaposition of two existing words [9, p. 102]. Compounds are virtually idiomatized contractions of syntactic groups which have inner syntactic relations frozen and are often characterized by omission of grammatical morphemes. E.g., phrase (1) is clipped to (2).

(1) ཕ་དང་མ་	(2) ཕ་མ་
pha dang ma	*pha-ma*
father CONJ mother	father_mother
'father and mother'	'parents'

Depending on the part of speech classification, nominal and verbal compounds can be distinguished. Depending on the syntactic model of the composite formation, the following types were distinguished for nominal compounds: composite noun root group; composite attribute group; noun phrase with genitive composite; composite class noun phrase; named entity composite; adjunct composite; and for verbal compounds: composite transitive verb phrase; verb coordination composite. Initially, the ontology allowed marking the expression as an idiom and establishing a separate type of token, common for nominal compounds. Since a large number of combinatorial explosions were caused by the incorrect versions of compounds parsing (the same sequence of morphemes can be parsed as compounds of different types) and their interpretation as noun phrases of different types, it was decided to expand the number of token types in the ontology according to identified types of nominal and verbal compounds. For all compounds the setting 'only_idiom=True' was also made. According to this setting any non-idiomatic interpretations of a compound are excluded.

Thus, in example (3) there are two compounds – (4) and (5), the wrong interpretation of which caused 72 versions of parsing.

(3) ཞེན་དོན་གྱི་མིང་མཐའ་ཡིན་པའི་དཔེ་
zhen-don gyi ming-mtha'yin-pa 'i dpe
object_of_desire GEN end_of_word to_be-NMLZ GEN example
'example of what is the word ending of an object of desire'

(4) ཞེན་དོན་
zhen-don
to_like_meaning
'object of desire'

(5) མིང་མཐའ་
ming-mtha'
word_end
'word ending'

Compound (5) has three versions: as an adjunct composite, as a composite noun root group, and the correct one as noun phrase with genitive composite.

Establishing the correct types of tokens for two compounds in (4) and (5) has reduced the number of versions in (3) to 8. It should be noted that specifying the correct type of token for compounds in the ontology does not always completely eliminate the ambiguity, since the same Tibetan compound may have different structures for different meanings. These cases are represented in the ontology as different concepts of the same expression.

Depending on which language unit is idiomatized, Tibetan non-compound idioms are divided into separate derivatives and nominal, verbal, adjectival and adverbial phrases. As with compounds, a list of classes of immediate constituents that can be idioms was built. The system of token types in the ontology database has been extended with these types. This system is being updated continually during the work and revealing previously unaccounted morphosyntactic types of idioms.

4 Restrictions for Morphosyntactic Disambiguation of Phrases Containing Idioms

In order to resolve the morphosyntactic ambiguity in phrases with idioms that had combinatorial explosions, four types of restrictions were established in the ontology: restrictions on genitive relations, restrictions on adjuncts (on the equivalence relation), restrictions on subjects and direct objects of verbs.

4.1 Restrictions on Genitive Relations

Restrictions on general genitive relation 'to have any object or process (about any object or process)' are imposed by establishing specific relation subclasses between basic classes in the ontology. E.g., as a result of the use of Aho-Ḍaorasick algorithm, in the example (6) the definite pronoun *so_so* 'every' was recognized not only as expected, but also as a possible combination of two noun roots *so* 'tooth', the second one together with its right context incorrectly forming the following word group with the idiom (7):

(6) སོའི་ལུང་སྟོན་པའི་སྒྲ་སྐད་
so 'i lung ston-pa 'i sgra-skad
tooth GEN grammar GEN sound
'sound of grammar of tooth'

(7) ལུང་སྟོན་པ་
lung ston-pa
authoritative-knowledge to-show-NMLZ
'grammar'

To exclude the possibility of version in the example (6), the basic class *skad* 'language' was connected by a genitive relation 'to have a grammar' with the concept *lung ston-pa* 'grammar'. This allowed to exclude the version, in which a tooth can have grammar.

4.2 Restrictions on Adjuncts

Tibetan adjunct joins the noun phrase on the right side, and due to nonexistence of word delimiters (spaces) in Tibetan writing system, adjuncts can not be graphically distinguished from parts of compounds. Thus, compound (2) may be misinterpreted both as 'father-mother' ('a father, who is also a mother', *ma* 'mother' being interpreted as an adjunct), and as 'father's mother' (noun phrase with genitive composite), whereas the only correct interpretation is 'father and mother' (noun root group composite). While the second interpretation (which is, moreover, logically possible) can be eliminated by just setting the correct token type in the ontology, the first interpretation is not idiomatic, and thus can not be just eliminated this way. Thus, only semantic restrictions can reduce the number of versions and eliminate incorrect versions with adjuncts.

This reduction was achieved by limiting the equivalence relation ('to be equivalent to an object or process'). Basic classes were connected with themselves with this relation so that only concepts that inherit these classes could be interpreted as adjuncts for each other.

In the example (8) 54 versions of parsing were originally built. A number of versions arose due to wrong interpretations of idioms: *ring-lugs* 'long tradition' (NPGenComposite), *dpal-yon* 'fortune' (CompositeNRoot-Group), *mdzes-chos* 'decoration' (NPGenComposite), *mdzes-chos rig-pa* 'esthetics' (VNNoTenseNoMood). Another source of versions amount multiplication was in different incorrect combinations of adjuncts: *mdzes-chos rig-pa* 'esthetics' was interpreted as an adjunct to tshogs' collection' (8.1), ring-lung 'long tradition' was interpreted as an adjunct to *tshogs* 'collection' (8.2), *dpal-yon* 'fortune' was interpreted as an adjunct to *tshogs* 'collection' (8.3).

(8)ཚོགས་རིང་ལུགས་ཀྱི་དཔལ་ཡོན་གཉིས་དང་མཛེས་ཆོས་རིག་པ་

tshogs ring-lugs kyi dpal-yon gnyis dang mdzes-chos rig-pa

collection long_tradition GEN fortune two CONJ decoration know-NMLZ

(8.1) 'esthetics - the collection and two fortunes of tradition'

(8.2) 'esthetics and two fortunes of tradition [that is the] collection'

(8.3) 'esthetics and two fortunes of tradition [that are the] collection'[2]

To eliminate these versions, the basic classes *srol* 'tradition' (hypernym to ring-lugs 'long tradition'), *rig-pa* 'science' (hypernym to *mdzes-chos rig-pa* 'esthetics'), *yon_ tan* 'virtue' (hypernym to *dpal-yon* 'fortune') and *tshogs* 'collection' were connected with themselves with equivalence relation.

---

[2] These 3 interpretations represent 3 groups of versions that arose only because of incorrect combinations of adjuncts, each group consisting of 18 versions (2 for *ring-lugs*, multiplied by 3 for *dpal-yon* and by 3 for *mdzes-chos rig-pa*). Thus, the total amount was 54.

Specifying the correct types of tokens for idioms and restricting versions with adjuncts reduced the number of parsing versions in example (8) to 14.

4.3   Restrictions on Subjects and Direct Objects of Verbs

Restrictions on subjects and direct objects of verbs were necessary for the correct analysis of compounds and idioms, as well as for eliminating unnecessary versions of syntactic parsing.

In example (9), the restrictions were applied to the subject of the verb:

(9) སྲིད་པའི་སྒྲོན་མེ་བསུས་པའི་དྲིན་ཆེ་

srid-pa 'i sgron-me bsus-pa 'i drin-che

exist-NMLZ GEN lamp invite-PST-NMLZ GEN kindness-be_great

(9.1) 'the kindness that invited the lamp of existence is great'

(9.2) 'the kindness of [someone who] exists, that invited the lamp, is great'

(9.3) 'the kindness of [someone who] invited the lamp of existence is great'[3]

Interpretations (9.1–2) are grammatically possible, but semantically non-sense, because they imply that kindness can invite. The subject valency of the verb *bsu* 'invite' was limited to the basic class 'any creature', so that only creatures can invite. This allowed to exclude versions (9.1–2). In the version (9.3) the subject was determined correctly.

Statistics of versions amount reduction achieved with modeling the idioms in the ontology is represented in Table 1.

Table 1. Statistics of versions amount in a selection of expressions with combinatorial explosions before and after idioms modeling in the ontology

Expression with combinatorial explosion	Before	After	Ratio
རྡིང་ལུགས་ཀྱི་དཔལ་ཡོན་གཉིས་དང་མཚམས་ཆོས་རིག་པ་	18	7	2.6
ཡན་ལག་དང་འཚས་པའི་གཞུང་ལུགས་	18	9	2
འགྱེང་ཐུ་སྤྱར་བ་	18	6	3
མིང་གཞིའི་མ་ནིང་གི་མིང་མཐའ་དང་བ་	21	7	3
མོ་དྲགས་སྤྱར་བ་ཡིན་པ་	22	12	1.9
ཞེན་དོན་གྱི་མིང་མཐའ་ཡིན་པའི་དཔེ་	72	8	9
ཞེན་དོན་གྱི་མིང་མཐའ་ཡིན་པའི་དཔེ་	43	9	4.8
ལས་དང་ཚེ་སྤྱབས་དང་དྲེན་གནས་རྣམས་ལ་	18	2	9
Average	28.75	7.5	4.4

---

[3] Example (9) has 32 versions of parsing. These 3 interpretations represent groups of versions that arose only because of incorrect designating of subject for the verb *bsu* 'to invite'.

5   Conclusion and Further Work

Tibetan language combines isolation with agglutination, so there are no universal symbols to separate the input string into words. That's actually why only morphemes can be used as atomic units, and thus Aho-Corasick algorithm is used to perform morphemic segmentations in accordance with the data of morphemic dictionaries of the developed language module. There are many versions as a result of incorrect morpheme segmentations, which give rise to combinatorial explosions at the level of phrase syntax. Combinatorial explosions can be resolved mostly by semantic restrictions, and idioms play an important role here, because many incorrect versions arise from hypotheses of identification of compounds or other structures that do not exist in Tibetan, and which can only be idioms, and also due to incorrect parsing of existing compounds and idioms or due to their incorrect binding to the surrounding context.

This work will be continued further, as the combinatorial explosions become prominent only as the coverage of the syntactic trees grows, and insignificant ambiguity of several phrases produce too many combinations when they are bound together.

Acknowledgment. This work was supported by the Russian Foundation for Basic Research, Grant No. 16-06-00578 Morphosyntactycal analyser of texts in the Tibetan language.

References

1. Grokhovskii, P.L., Zakharov, V.P., Smirnova, M.O., Khokhlova, M.V.: The corpus of tibetan grammatical works. In: Automatic documentation and mathematical linguistics, vol. 49, no. 5, pp. 182–191 (2015). https://doi.org/10.3103/S0005105515050064
2. Gruber, T.R.: A translation approach to portable ontology specifications (PDF). Knowl. Acquis. 5(2), 199–220 (1993). https://doi.org/10.1006/knac.1993.1008
3. Aho, A.V., Corasick, M.J.: Efficient string matching: an aid to bibliographic search. Commun. ACM 18(6), 333–340 (1975)
4. Krippendorff, K.: Combinatorial Explosion. Web Dictionary of Cybernetics and Systems. http://pespmc1.vub.ac.be/ASC/Combin_explo.html. PRINCIPIA CYBERNETICA WEB
5. Dobrov, A.V.: Semantic and ontological relations in AIIRE natural language processor. Comput. Model. Bus. Eng. Domains. Rzeszow-Sofia: ITHEA, 147–157 (2014)
6. Miller, G.A., Beckwith, R., Fellbaum, C.D., Gross, D., Miller, K.: WordNet: an online lexical database. Int. J. Lexicograph. 3(4), 235–244 (1990)
7. Melcuk, I.: Phrasemes in language and phraseology in linguistics. In: Everaert, M., Van der Linden, E.J., Schenk, A., Schreuder, R. (eds.) Idioms: Structural and Psychological Perspectives, pp. 167–232. Lawrence Erlbaum, New Jersey (1995)
8. Pelletier, F.J.: The principle of semantic compositionality. Topoi 13, 11 (1994)
9. Beyer, S.: The Classical Tibetan Language. State University of New York, New York (1992)

Adjusting Machine Translation Datasets for Document-Level Cross-Language Information Retrieval: Methodology

Gennady Shtekh[2], Polina Kazakova[2(✉)], and Nikita Nikitinsky[1]

[1] Integrated Systems, Vorontsovskaya Street, 35B building 3, room 413,
109147 Moscow, Russia
[2] National University of Science and Technology MISIS, Leninsky Avenue 4,
119049 Moscow, Russia
kazakova1537@gmail.com

Abstract. Evaluating the performance of Cross-Language Information Retrieval models is a rather difficult task since collecting and assessing substantial amount of data for CLIR systems evaluation could be a nontrivial and expensive process. At the same time, substantial number of machine translation datasets are available now. In the present paper we attempt to solve the problem stated above by suggesting a strict workflow for transforming machine translation datasets to a CLIR evaluation dataset (with automatically obtained relevance assessments), as well as a workflow for extracting a representative subsample from the initial large corpus of documents so that it is appropriate for further manual assessment. We also hypothesize and then prove by the number of experiments on the United Nations Parallel Corpus data that the quality of an information retrieval algorithm on the automatically assessed sample could be in fact treated as a reasonable metric.

Keywords: Cross-language information retrieval
Document-level information retrieval · CLIR evaluation
CLIR datasets · Parallel corpora · Information retrieval methodology

1 Introduction

By the present time information in the Internet has become multilingual; the amount of user-generated multilingual textual data is constantly growing which leads to an increasing demand for information processing systems with cross-lingual support.

A practical example of cross-language information retrieval need comes from the area of scientific research management. In many institutions invited or employed experts review incoming grant applications in order to decide whether a given research project should be awarded with a grant. Experts have to use various tools to validate their decisions: citation indexes, patent databases, electronic libraries, etc. They often contain data in several languages thus experts

© Springer Nature Switzerland AG 2018
P. Sojka et al. (Eds.): TSD 2018, LNAI 11107, pp. 84–94, 2018.
https://doi.org/10.1007/978-3-030-00794-2_9

have to query these information sources separately in many languages so that they are able to understand the present state of the art of a given research topic in the world in general.

The reason for the present research goes back to an information retrieval system that our team developed for a Russian governmental institution in order to supply the experts with a tool for evaluating grant applications [16]. The system worked with monolingual data. As the experts in reality have to look through patents and research papers in several languages, the tool needed to be improved with a document-level[1] cross-lingual information retrieval module. As the process of building any technical tool is always complicated by the necessity of its performance evaluation, the present study suggests a methodology for automatically collecting and assessing data for cross-language information retrieval systems evaluation.

2   Discussion of CLIR

2.1   Methods

Cross-Language Information Retrieval (CLIR) refers to the information retrieval where a query and relevant documents may appear in different languages.

Old-fashioned approaches to cross-language information retrieval mainly implied various translation techniques. The methods varied from using cognate matching [1,14], ontologies and thesauri [9] or bilingual dictionaries [1,18] as a source of translation and linking to applying machine translation [17]. Furthermore, any kind of translation could be applied either to a query or to a whole collection of documents [17].

More recent strategies involve creating an interlingual representation for all languages of a given CLIR system generally by means of distributional semantics and topic modeling. Such approach might be preferred to the translation one since it requires no extra sources (such as bilingual dictionaries and ontologies, which are needed to be carefully constructed by hand) apart from textual data. The most frequently used basic algorithms to this day include Latent Semantic Indexing [2,6,15] and Latent Dirichlet Allocation [3,21]. Various multilingual word embeddings, both word-level and sentence-level, are also quite widely applied to the objective in question (see, for instance, [22]; for an extensive research on the cross-lingual embeddings see [19]).

2.2   Datasets

There are multiple publicly available datasets for different scientific purposes. In the field of NLP, one may name a large number of datasets for text classification, clustering, parsing, etc. Similarly, there are many machine translation datasets,

---

[1] Here we define a *document-level* information retrieval system as a type of information retrieval systems where users query not by short keyword phrases but by full-text document examples.

for instance, Europarl [12], Canadian Hansards [8] or Linguistic Data Consortium collections.

However, there are fewer datasets for monolingual information retrieval tasks. Among examples of large datasets for this purpose are the TREC collections [20] and the INEX Initiative collections for Structural IR [11]. There are even less collections for cross-lingual information retrieval evaluation. The largest CLIR events are mostly limited to the CLIR track of the TREC conference (before 1999) and the separate CLEF conference (after 1999). Thus, the main resources for CLIR evaluation consist of the datasets of the corresponding events. Some more large collections are made by NII Test Collections for IR Systems Project. There are also some sporadic datasets, such as [7]. The primary issue about these test collections is relative lack of language variety: the NTCIR collections are focused on Asian languages, while the CLEF conference are mainly dedicated to European languages and there is no satisfactory CLIR corpora for Russian language, which we personally needed to solve our business task.

2.3   Evaluation

There exist thereby only few good CLIR datasets. The situation is at least partially induced by the fact that not only manual annotation for CLIR task is very complex but also a process of assessments evaluation and aggregation may be quite sophisticated [4].

The crucial difficulties[2] that must be taken into account when conducting a CLIR assessment work are the following:

- Aggregation problem: it is usually complicated to aggregate many assessments into one as every assessor annotates texts in a unique way.
- Language problem: to annotate a multilingual corpus for a CLIR task assessors must know each language at least partially.
- Methodology problem: a detailed and concrete annotation methodology for assessors is required.

In situations when no good evaluation multilingual dataset is available, but only a parallel dataset, researchers could use technique called mate finding [6]: one language part of a dataset is used as queries and another one is used for evaluation supposing that a system must return translation pair of a document as a relevant output. Though this simple method might be useful in case of lack of resources, its disadvantage is that in fact it returns not a very representative evaluation metric. Hence, a more complicated strategy is required.

---

[2] Nonetheless, the case of document-level information retrieval somewhat simplifies the evaluation procedure as at least there is no need for example queries and ground truth relevance measures between queries and documents: only document-to-document relevance is required.

3   Hypothesis and Methodology

As discussed in Sect. 2, collecting and assessing substantial amount of data for CLIR systems evaluation is a non-trivial and expensive task. At the same time, there exists a great majority of machine translation datasets. In the present paper we address the following main points:

1. We suggest a workflow for transforming machine translation datasets for CLIR evaluation, i.e. to a dataset which contains automatically obtained relevance scores.
2. We suggest a workflow for extracting a representative subsample from the initial large set of documents appropriate for further manual assessment so that the algorithm quality on this subsample would reliably reflect the quality on the whole dataset.
3. We hypothesize that the algorithm quality on automatically assessed sample (1) correlates with the algorithm quality on specifically chosen and manually annotated subsample (2) and therefore the quality on automatically annotated sample could be considered as a reasonable metric.

Below we briefly describe both workflows. Note that this approach can be applied to datasets where not only sentence-to-sentence but also document-to-document alignment is available.

3.1   Getting Automatic Relevance Scores

Machine translation datasets are usually datasets with parallel alignment of sentences within aligned documents in source and target languages[3]. Such datasets are inapplicable to CLIR evaluation as they lack relevance scores. Our intention is to use a reliable unsupervised machine learning algorithm to obtain relevance scores between documents in two languages without the need for their manual assessment. For this, the following workflow is proposed:

1. First, both collections of documents are separately vectorized by means of LSI on *tf-idf* matrix after any preprocessing. As at this step documents are represented by vectors, it is possible to compute distances between documents within their language.
2. The resulting two groups of vectors are then separately handled by the nearest neighbourhood engine. After this, each group is represented as a graph where nodes are initial documents and edges are weighted by the distance between documents (corresponding nodes) reflecting a measure of document relevance. We also suggest to take only nodes with top K weights to cut off documents that are not very similar since there is no need for preserving information on distances between each two documents but only those that are relatively similar (K might vary depending on a certain task, number of topics, cluster density, etc.).

---

[3] For simplicity, in the present paper we discuss the case of a bilingual dataset. However, the approaches described here could be easily generalized to the case of multiple languages.

3. As the information on the initial cross-language document alignment is available, two graphs now could be linked to form one complex graph.
4. As two graphs from (2) joined together in (3) are actually in different dimensions, they should be projected into a common space. It could be done by a neural network architecture. Its optimization function must satisfies two conditions:
   (a) nodes that were initially close to each other (in terms of their edge weights) remain as close as possible given that
   (b) nodes that did not have a common edge remain as far as possible.
   This step is also supposed to smooth the noize left after applying the algorithms from (1) and (2).

As a result of the above process, a complex graph of documents in both languages with relevance scores is obtained and it can be used to evaluate the approximate quality of a cross-lingual search engine.

3.2   Extracting Representative Subsample

Furthermore, the resulting graph could be used to compose the optimum relatively small subset of examples for further manual annotation by assessors.
   The pipeline is as follows:

1. Louvain Method for Community Detection [5] is used to cluster the graph so that the resulting clusters have the following property: for a given document in a cluster the majority of similar documents are contained in its ,cluster. This would group candidates for the next step.
2. Before the final extraction of the most representative documents, the graph is needed to be projected into a vector space. This could be done by the same embedding technique used in Sect. 3.1.
3. Eventually, Determinantal Point Processes [13] are applied to filter N most distinctive clusters of documents for further assessment.

The whole process described in Sects. 3.1 and 3.2 is shown on Fig. 1.

4   Experiments

4.1   Data

The data we use for the reproduction of the procedure shown above is the English-Russian part of the United Nations Parallel Corpus [23]. It is a collection of official publicly available documents of the United Nations. The text alignment of the dataset is shown on Fig. 2: sentences are aligned within aligned paragraphs and aligned documents. Translation scores are aggregated at the document level.

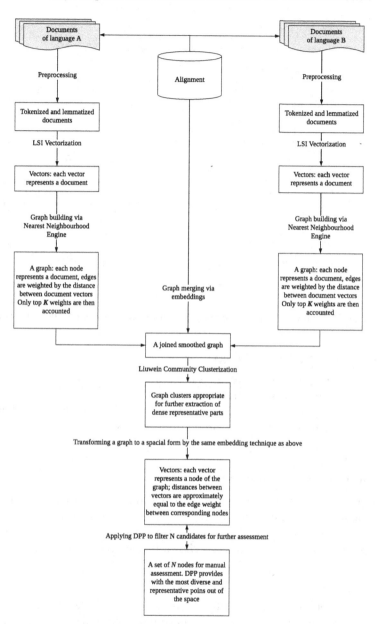

Fig. 1. General schema of the process of adapting a machine translation dataset for information retrieval purposes and extracting a representative subsample for manual assessment.

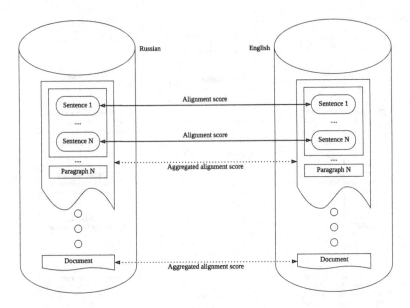

Fig. 2. Text alignment of the English-Russian dataset of the UNPC collection.

4.2   Reproducing Methodology

We replicated the procedure presented in Sect. 3. The parameters for each step and algorithm are described in Table 1.

As a result, we have got a sample of 2,000 documents for the manual assessment. To test the hypotheses of the present research we have manually annotated 120 documents randomly chosen from the resulting sample.

4.3   Autorelevance Evaluation

The performance of the automatically obtained relevance measures against the manually assessed documents is shown in Table 2. It can be seen that the scores itselves are not very hight so the autoassessments can be used only for testing and comparing other models, which is shown in the next section.

4.4   Baselines

To test the hypothesis that the autorelevance scores and the scores from manually annotated sample do correlate, we have tested several baseline algorithms against the same subset.

The list of algorithms used includes:

1. LSI: 300 dimensions, translated documents are mixed while training
2. LDA: 300 dimensions, translated documents are mixed while training
3. word2vec: 300 dimensions, translated documents are mixed while training.

The Tables 3 and 4 show the algorithms scores on the autorelevance data and on the manually annotated data correspondingly.

Table 1. Parameters used at each step for autorelevance estimation and extracting a subsample.

	Step	Parameters
	Part 1	
Step 1	LSI vectorization	700 dimensions
Step 2	Nearest neighbourhood engine	Faiss [10]
	Taking top K weights	$K = 500$
Step 4	Neural network architecture	t-SNE-like projection on the embeddings built by MDS-like constraints
	Part 2	
Step 3	Determinantal Point Processes	Equal precision/diversity setting
	Taking N clusters	$N = 15$

Table 2. Autorelevance performance.

Metric	Score
Precision@10	0.4531
Recall@10	0.5686
nDCG@10	0.6183

Table 3. The quality of the algorithms on the automatically obtained relevance scores.

Model	Precision@10	Recall@10	nDCG@10
LSI 300d	0.81	0.85	0.86
LDA 300d	0.79	0.89	0.9
word2vec 300d	0.83	0.9	0.93

5 Results

Experiments show that the estimated scores are correlated with positive coefficient at $p\text{-}value < 0.05$ which means that one can use the automatically obtained relevance scores to compare several models and this comparision would be correct.

Table 4. The quality of the algorithms on the manually annotated relevance assessments.

Model	Precision@10	Recall@10	nDCG@10
LSI 300d	0.43	0.52	0.57
LDA 300d	0.4	0.53	0.58
word2vec 300d	0.44	0.55	0.6

Nonetheless, we must note that this conclusion is based on only 120 manually assessed examples so the results could be treated as preliminary. Future work would be to asses the whole sample of 2,000 documents and check the stability of the results.

6 Conclusion

Thus, we developed a pipeline that allows to transform a machine translation dataset to a dataset with automatically generated relevance scores so that it is appropriate to use in CLIR evaluation. Additionally, we have proposed a methodology to construct a representative subsample from a large collection of parallel texts to be suitable for manual assessment. However, the representativeness of a subsample follows from theoretical knowledge and we have not verified this fact in practice as this would required to assess the whole dataset which is in contradiction to the very concept of the present paper. Finally, we have conducted several baseline experiments to prove the hypothesis that the quality of a retrieval algorithm on the automatically assessed sample reflects its quality on the real data with certain confidence.

The pipeline described here could be especially useful for researchers and engineers testing CLIR models and systems in case of lack of data, for example, when working with minor languages (if there are any parallel corpora for them though) or any other languages not covered by the main CLIR evaluation datasets.

Acknowledgements. We would like to acknowledge the hard work and commitment from Ivan Menshikh throughout this study. We are also thankful to Anna Potapenko for offering very useful comments on the present paper, and Konstantin Vorontsov for encouragement and support.

The present research was supported by the Ministry of Education and Science of the Russian Federation under the unique research id RFMEFI57917X0143.

References

1. Ballesteros, L., Croft, W.B.: Phrasal translation and query expansion techniques for cross-language information retrieval. In: ACM SIGIR Forum, vol. 31, pp. 84–91. ACM (1997)
2. Berry, M.W., Young, P.G.: Using latent semantic indexing for multilanguage information retrieval. Comput. Hum. **29**(6), 413–429 (1995)
3. Boyd-Graber, J., Blei, D.M.: Multilingual topic models for unaligned text. In: Proceedings of the Twenty-Fifth Conference on Uncertainty in Artificial Intelligence, pp. 75–82. AUAI Press (2009)
4. Braschler, M., Harman, D., Hess, M., Kluck, M., Peters, C., Schäuble, P.: The evaluation of systems for cross-language information retrieval. In: LREC (2000)
5. De Meo, P., Ferrara, E., Fiumara, G., Provetti, A.: Generalized Louvain method for community detection in large networks. In: 2011 11th International Conference on Intelligent Systems Design and Applications, ISDA, pp. 88–93. IEEE (2011)

6. Dumais, S.T., Letsche, T.A., Littman, M.L., Landauer, T.K.: Automatic cross-language retrieval using latent semantic indexing. In: AAAI Spring Symposium on Cross-Language Text and Speech Retrieval, vol. 15, p. 21 (1997)
7. Ferrero, J., Agnes, F., Besacier, L., Schwab, D.: A multilingual, multi-style and multi-granularity dataset for cross-language textual similarity detection. In: 10th Edition of the Language Resources and Evaluation Conference (2016)
8. Germann, U.: Aligned hansards of the 36th parliament of Canada (2001). https://www.isi.edu/natural-language/download/hansard/
9. Gonzalo, J., Verdejo, F., Peters, C., Calzolari, N.: Applying EuroWordNet to cross-language text retrieval. In: Vossen, P. (ed.) EuroWordNet: A Multilingual Database with Lexical Semantic Networks, pp. 113–135. Springer, Dordrecht (1998). https://doi.org/10.1007/978-94-017-1491-4_5
10. Johnson, J., Douze, M., Jégou, H.: Billion-scale similarity search with GPUs. arXiv preprint arXiv:1702.08734 (2017)
11. Kamps, J., Pehcevski, J., Kazai, G., Lalmas, M., Robertson, S.: INEX 2007 evaluation measures. In: Fuhr, N., Kamps, J., Lalmas, M., Trotman, A. (eds.) INEX 2007. LNCS, vol. 4862, pp. 24–33. Springer, Heidelberg (2008). https://doi.org/10.1007/978-3-540-85902-4_2
12. Koehn, P.: Europarl: a parallel corpus for statistical machine translation. In: MT Summit, vol. 5, pp. 79–86 (2005)
13. Kulesza, A., Taskar, B., et al.: Determinantal point processes for machine learning. Found. Trends® Mach. Learn. 5(2–3), 123–286 (2012)
14. Meng, H.M., Lo, W.K., Chen, B., Tang, K.: Generating phonetic cognates to handle named entities in English-Chinese cross-language spoken document retrieval. In: IEEE Workshop on Automatic Speech Recognition and Understanding, ASRU 2001, pp. 311–314. IEEE (2001)
15. Mori, T., Kokubu, T., Tanaka, T.: Cross-lingual information retrieval based on LSI with multiple word spaces. In: Proceedings of the 2nd NTCIR Workshop Meeting on Evaluation of Chinese & Japanese Text Retrieval and Text Summarization. Citeseer (2001)
16. Nikitinsky, N., Ustalov, D., Shashev, S.: An information retrieval system for technology analysis and forecasting. In: Artificial Intelligence and Natural Language and Information Extraction, Social Media and Web Search FRUCT Conference, AINL-ISMW FRUCT, pp. 52–59. IEEE (2015)
17. Oard, D.W.: A comparative study of query and document translation for cross-language information retrieval. In: Farwell, D., Gerber, L., Hovy, E. (eds.) AMTA 1998. LNCS, vol. 1529, pp. 472–483. Springer, Heidelberg (1998). https://doi.org/10.1007/3-540-49478-2_42
18. Pirkola, A., Hedlund, T., Keskustalo, H., Järvelin, K.: Dictionary-based cross-language information retrieval: problems, methods, and research findings. Inf. Retr. 4(3–4), 209–230 (2001)
19. Ruder, S.: A survey of cross-lingual embedding models. arXiv preprint arXiv:1706.04902 (2017)
20. Voorhees, E.M., Harman, D.K., et al.: TREC: Experiment and Evaluation in Information Retrieval, vol. 1. MIT Press, Cambridge (2005)
21. Vulić, I., De Smet, W., Moens, M.F.: Cross-language information retrieval models based on latent topic models trained with document-aligned comparable corpora. Inf. Retr. 16(3), 331–368 (2013)

22. Vulić, I., Moens, M.F.: Monolingual and cross-lingual information retrieval models based on (bilingual) word embeddings. In: Proceedings of the 38th International ACM SIGIR Conference on Research and Development in Information Retrieval, pp. 363–372. ACM (2015)
23. Ziemski, M., Junczys-Dowmunt, M., Pouliquen, B.: The united nations parallel corpus v1.0. In: LREC (2016)

Deriving Enhanced Universal Dependencies from a Hybrid Dependency-Constituency Treebank

Lauma Pretkalniņa[(⊠)], Laura Rituma, and Baiba Saulīte

Institute of Mathematics and Computer Science, University of Latvia,
Raiņa 29, Riga LV-1459, Latvia
{lauma,laura,baiba}@ailab.lv

Abstract. The treebanks provided by the Universal Dependencies (UD) initiative are a state-of-the-art resource for cross-lingual and monolingual syntax-based linguistic studies, as well as for multilingual dependency parsing. Creating a UD treebank for a language helps further the UD initiative by providing an important dataset for research and natural language processing in that language. In this paper, we describe how we created a UD treebank for Latvian, and how we obtained both the basic and enhanced UD representations from the data in Latvian Treebank which is annotated according to a hybrid dependency-constituency grammar model. The hybrid model was inspired by Lucien Tesnière's dependency grammar theory and its notion of a syntactic nucleus. While the basic UD representation is already a *de facto* standard in NLP, the enhanced UD representation is just emerging, and the treebank described here is among the first to provide both representations.

Keywords: Latvian Treebank · Universal Dependencies
Enhanced dependencies

1 Introduction

In this paper, we describe the development and annotation model of Latvian Treebank (LVTB), as well as data transformations used to obtain the UD representation from it. Since Latvian is an Indo-European language with rich morphology, relatively free word order, but also uses a lot of analytical forms, it was decided to use a hybrid dependency-constituency model (see Sect. 2.2) in the original Latvian Treebank pilot project back in 2010 (see Sect. 2.1).

Universal Dependencies[1] (UD) is an open community effort to create cross-linguistically consistent treebank annotation within a dependency-based lexicalist framework for many languages [3]. Since 2016 we have been participating by providing a UD compatible treebank derived from LVTB (Latvian UD Treebank

---

[1] http://universaldependencies.org/.

© Springer Nature Switzerland AG 2018
P. Sojka et al. (Eds.): TSD 2018, LNAI 11107, pp. 95–105, 2018.
https://doi.org/10.1007/978-3-030-00794-2_10

or LVUDTB). UD provides guidelines for two dependency annotation levels—
base dependencies (mandatory) where annotations are surface-level syntax trees,
and enhanced dependencies where annotations are graphs with additional information for semantic interpretation. In order to generate the LVUDTB for each of
the UD versions, a transformation (see Sect. 3) is applied to the current state of
LVTB. Together with Polish LFG and Finnish TDT, PUD treebanks LVUDTB
is among the first to provide enhanced in addition to basic dependencies.

2   Latvian Treebank

2.1   Development

Development of the first syntactically annotated corpus for Latvian (Latvian
Treebank, LVTB) started with a pilot project in 2010 [6]. During the pilot a
small treebank was created with texts from JRC-Acquis, Sofie's World, as well
as some Latvian original texts [7]. In 2017 LVTB consisted of around 5 thousand
sentences, one third of which were from Latvian fiction and another third from
news texts. We are currently making a major expansion to LVTB, with a goal of
balancing the corpus (aiming for 60% news, 20% fiction, 10% legal, 5% spoken,
5% other) and reaching about 10 thousand sentences by the end of 2019 [2].

Latvian Treebank serves as the basis for LVUDTB, which is a part of the UD
initiative since UD version 1.3. Since UD v2.1. in addition to containing basic
dependencies, LVUDTB also features enhanced dependencies as well.

2.2   Annotation Model

The annotation model used in Latvian Treebank is SemTi-Kamols [1,4]. It is a
hybrid dependency-constituency model where the dependency model is extended
with constituency mechanisms to handle multi-word forms and expressions, i.e.,
syntactic units describing analytical word forms and relations other than sub-
ordination [1]. These mechanisms are based on Tesnière's idea of a syntactic
nucleus which is a functional syntactic unit consisting of content-words or syntactically inseparable units that are treated as a whole [4]. From the dependency
perspective, phrases are treated as regular words, i.e., a phrase can act as a head
for depending words and/or as a dependent of another head word [6]. A phrase
constituent can also act as a dependency head.

A sample LVTB tree is given in Fig. 1 on the left. Dependency relations
(brown links in Fig. 1, left) match with grammatic relations in Latvian syntax
theory [5]. Dependency roles are used for traditional functions: predicates, subjects, objects, attributes, and adverbs. They are also used for free sentence modifiers: situants, determinants, and semi-predicative components. A free modifier
is a part of a sentence related to the whole predicative unit instead of a phrase or
single word, and it is based on a secondary predicative relation or determinative
relation. A situant describes the situation of the whole sentence. A determi-
nant (dative-marked adjunct) names an experiencer or owner (it is important

Table 1. Dependency types in Latvian Treebank

Role	Description	Corresponding UD roles
subj	Subject	nsubj, nsubj:pas, ccomp, obl
attr	Attribute	nmod, amod, nummod, det, advmod
obj	Object	obj, iobj
adv	Adverbial modifier	obl, nummod, advmod, discourse
sit	Situant	obl, nummod, advmod, discourse
det	Determinant	obl
spc	Semi-predicative component	ccomp, xcomp, appos, nmod, obl, acl, advcl
subjCl	Subject clause	csubj, csubj:pas, acl
predCl	Predicative clause	ccomp, acl
attrCl	Attribute clause	acl
appCl	Apposition clause	acl
placeCl	Subordinate clause of place	advcl
timeCl	Subordinate clause of time	advcl
manCl	Subordinate clause of manner	advcl
degCl	Subordinate clause of degree	advcl
causCl	Causal clause	advcl
purpCl	Subordinate clause of purpose	advcl
condCl	Conditional clause	advcl
cnsecCl	Consecutive clause	advcl
compCl	Comparative clause	advcl
cncesCl	Concessive clause	advcl
motivCl	Motivation and causal clause	advcl
quasiCl	Quasi-clause	advcl
ins	Insertion, parenthesis	parataxis, discourse
dirSp	Direct speech	parataxis
no	Discourse markers	vocative, discourse, conj

to note that the **det** role in LVTB is not the same as the **det** role in UD).
A semi-predicative component can take on a lot of different representations in
the sentence: resultative and depictive secondary predicates, a nominal standard
in comparative constructions, etc. Other dependency roles are used for the different types of subordinate clauses and parenthetical constructions—insertions,
direct speech, etc. Some roles can be represented by both a single word and
a phrase-style construction, while others can be represented only by a phrase.
Overview on dependency roles used in LVTB is given in the first two columns
of Table 1.

Table 2. X-words in Latvian Treebank

Phrase → constituent	Description	Corresponding UD roles
xPred	Compound predicate	
→ mod	Semantic modifier	*phrase head*
→ aux	Auxiliary verbs or copula	aux, aux:pass, cop, xcomp, *phrase head*
→ basElem	Main verb or nominal	xcomp, *phrase head*
xNum	Multiword numeral	
→ basElem	Any numeral	nummod, *phrase head*
xApp	Apposition	
→ basElem	Any nominal	nmod, *phrase head*
xPrep	Prepositional construction	
→ prep	Preposition	case
→ basElem	Main word	*phrase head*
xSimile	Comparative construction	
→ conj	Comparative conjunction	fixed, mark, case, discourse
→ basElem	Main word	*phrase head*
xParticle	Particle construction	
→ no	Particle	discourse
→ basElem	Main word	*phrase head*
namedEnt	Unstructured named entity	
→ basElem	Any word	flat:name, *phrase head*
subrAnal	Subordinative wordgroup analogue	
→ basElem	Any word	compound, nmod, nummod, amod, det, flat, *phrase head*
coordAnal	Coordinative wordgroup analogue	
→ basElem	Any word	compound, *phrase head*
phrasElem	Phraseological unit with no clear syntactic structure	
→ basElem	Any word	flat, *phrase head*
unstruct	Multi-token expression with no Latvian grammar, e.g., formulae, foreign phrases	
→ basElem	Any token	flat, flat:foreign, *phrase head*

There are three kinds of phrase-style constructions in the LVTB grammar model: x-words, coordination and punctuation mark constructions (PMC). X-words (nodes connected with green links, Fig. 1, left) are used for analytical

Table 3. Coordination constructions in Latvian Treebank

Phrase → constituent	Description	Corresponding UD roles
crdParts	Coordinated parts of sentence	
→ crdPart	Coordinated part	conj, *phrase head*
→ conj	Conjunction	cc
→ punct	Punctuation mark	punct
crdClauses	Coordinated clauses	
→ crdPart	Coordinated clause	conj, parataxis, *phrase head*
→ conj	Conjunction	cc
→ punct	Punctuation mark	punct

forms, compound predicates, prepositional phrases etc. Coordination constructions (nodes connected with blue links, Fig. 1, left) are used for coordinated parts
of sentences, and coordinated clauses. PMCs (nodes connected with purple links,
Fig. 1, left) are used to annotate different types of constructions which cause
punctuation in the sentence. In this case the phrase-style construction consists
of punctuation marks, the core word of the construction, and clause introducing
conjunction, if there is one. All three kinds of phrases have their own types. In
case of x-words, these types may have even more fine-grained subtypes specified in the phrase tag. As each phrase type has certain structural limitations, it
determines the possible constituents in the phrase structure. X-word types and
their constituents are described in the first two columns of Table 2, coordination
is described in Table 3, and PMC in Table 4.

Structural limitations can be different for each x-word type or subtype.
This is important for data transformation to UD (see Sect. 3) because it affects
which element of the x-word will be the root in the UD subtree. For example,
each xPred (compound predicate) must contain exactly one basElem and either
exactly one mod in case of semantic modification or some auxVerbs in case of analytical forms and nominal or adverbial predicates. It is allowed to have multiple
auxVerbs, if each of them have one of the lemmas *būt, kļūt, tikt, tapt*, or their corresponding negatives. Otherwise, only one auxVerb per xPred is allowed. Such
restrictions result from a different approach to the distinction between modal
and main verbs in Latvian syntax theory and UD grammar. These restrictions
further simplify transformation to UD, distinguishing the auxiliaries from the
main verbs according to the UD approach, as each of the described structure
cases need to be transformed differently. Another x-word type where subtypes
and structural limitation impact transformation rules, is subrAnal (analogue of
subordinate-wordgroup) (see Table 5).

The annotation model also has a method for ellipsis handling. If the omitted
element has a dependent, the omitted part of the sentence is represented by
an accordingly annotated empty node in the tree. This new node is annotated
either with an exact wordform or with a morphological pattern showing the

Table 4. Punctuation mark constructions in Latvian Treebank

Phrase → constituent	Description	Corresponding UD roles
any PMC		
→ `punct`	Punctuation mark	`punct`
any clausal PMC		
→ `conj`	Conjunction	`mark`, `cc`
→ `no`	Address, particle, or discourse marker	`vocative`, `discourse`
`sent`	Sentence (predicative)	
→ `pred`	Main predicate...	`root`, *phrase head*
→ `basElem`	...or main clause coordination	`root`, *phrase head*
`utter`	Utterance (non-predicative)	
→ `basElem`	Any non-depending word	`root`, `parataxis`, *phrase head*
`mainCl`	Main clause (not subordinated; can be coordinated)	
`subrcl`	Subordinated clause	
`dirSp`	Direct speech clause	
→ `pred`	Main predicate...	*phrase head*
→ `basElem`	...or clause coordination	*phrase head*
`insPmc`	Insertion PMC	
→ `pred`	Main predicate...	*phrase head*
→ `basElem`	...or other word	*phrase head*
`interj`	Interjection PMC	
→ `basElem`	Any interjection	`flat`, *phrase head*
`spcPmc`	Secondary predication PMC	
`address`	Vocative PMC	
`particle`	Particle PMC	
`quot`	Quotation marks not related to direct speech	
→ `basElem`	Main word	*phrase head*

features that can be inferred from context in the current sentence. No information from context outside the current sentence is added, and empty nodes without dependents are added only for elided auxiliary verbs.

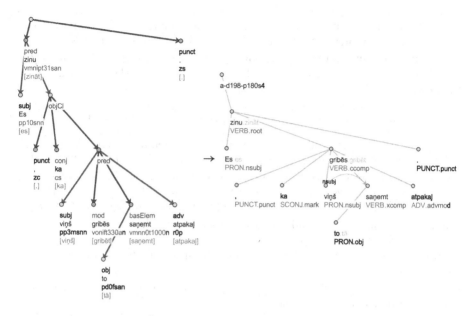

Fig. 1. Sample sentence: Es_I $zinu_{know.1PRS.SG}$, ka_{that} $viņš_{he}$ $gribēs_{want.3FUT}$ to_{it} $saņemt_{receive.INF}$ $atpakaļ_{back}$. 'I know he'll want to get it back.'. Tree annotated as in Latvian Treebank on the left, and its UD analogue on the right. (Color figure online)

3 Universal Dependencies

Latvian Universal Dependency treebank is built from LVTB data with the help of an automatic transformation procedure[2], based on heuristics and an analytic comparison of the two representations. The transformation result for the sample sentence is given in Fig. 1 on the right. Despite being developed without UD in mind, LVTB contains most of the necessary information, encoded either in labels or in the tree structure. Among some distinctions LVTB lacks is a distinction between complements taking (or not) their own subjects—UD xcomp vs. ccomp. Another problem is that LVTB does not distinguish determiners neither as part-of-speech (DET in UD) nor syntactic role (det), instead analyzing them as pronouns. This problem is partially mitigated by analyzing the tree structure, and in future we are planning to also consider the pronominal agreement.

The transformation was built for obtaining basic dependencies and only later, after the release of the UD v2.0 specification, adjusted to create enhanced dependencies. Thus to get an enhanced dependency graph we take annotations for a sentence from LVTB, derive the basic dependency graph from those annotations, and then apply some additional changes. However this approach leads to much more complicated code and more inaccuracies in the final tree, which is why in the future we plan on doing it the other way around, i.e., first constructing the enhanced graph and then reducing it to the basic graph. That would be a better

---

[2] https://github.com/LUMII-AILab/CorporaTools/tree/master/LVTB2UD.

approach because despite surface differences (an enhanced UD graph is not a tree, while LVTB representation is), the enhanced UD representation is closer to the LVTB representation than the basic one, e.g., several types of the enhanced UD edges can be obtained from LVTB distinctions for whether something is a dependent of a phrase as a whole or its part.

Transformation steps for a single tree from the hybrid model to UD:

1. Determine necessary tokens, add XPOSTAGs and lemmas from LVTB. Add information about text spacing and spelling errors corrected in the MISC field. Sometimes a word from LVTB must be transformed to multiple tokens, e.g., unnecessary split words (like *ne var* 'no can' instead of *nevar* 'can't') are represented as single M-level units in LVTB, but as two tokens in UD. If so, appropriate dependency and enhanced dependency links between these tokens are also added in this step.
2. From lemmas and XPOSTAG determine preliminary UPOSTAG and FEATS for each token.
3. Add null nodes for elided predicates (needed for enhanced dependencies) based on how ellipses are annotated in LVTB.
4. Build enhanced dependency graph "backbone" with null nodes, but without other enhanced dependency features. Constructions in LVTB that use dependency relations are directly transformed to a correct UD analogue just by changing the dependency relation labels. LVTB phrase style constructions are each transformed to a connected dependency subtree: every LVTB phrase-style construction is transformed to a single connected subtree and any dependent of such a phrase is transformed to the subtree root dependent.
5. Build basic dependency tree by working out orphan relations to avoid null node inclusion in the tree. Other relations are copied from enhanced dependency graph backbone.
6. Finish enhanced dependency graph by adding additional edges for controlled/raised subjects and conjunct propagation.
7. For all tokens update UPOS and FEATS taking into account the local UD structure. Most notable change being that certain classes of pronouns tagged as PRON, but labeled as det, are retagged as DET.

Steps 4 and 5 are done together in a single bottom-up tree traversal. An overview which LVTB roles correspond to which UD roles is given on Table 1. An overview of which LVTB phrase part roles correspond to which UD dependency roles is given in Tables 2, 3 and 4. In these tables *phrase head* denotes cases where a particular constituent becomes the root of the phrase representing subtree, and thus, its label is assigned according to the dependency label of the phrase in the LVTB tree. Table 5 describes how to build a dependency structure for each phrase-style construction. If for a single LVTB role there are multiple possible UD roles, for both dependency head and dependent the transformation considers tag and lemma or phrasal structure.

Currently the transformation procedure gives some, but not all enhanced dependency types. The resulting treebank completely lacks any links related

Table 5. Phrase-style construction structural transformation

Phrase	Root choice	Structure
xPred	mod, if there is one; basElem, if all auxVerb lemmas are *būt*, *kļūt*, *tikt*, *tapt*; only auxVerb otherwise	Other parts are root dependents
xNum	Last basElem	Other parts are root dependents
xApp	First basElem	Other part is root dependent
xPrep	basElem	prep is root dependent
xSimile	basElem	conj is root dependent
xParticle	basElem	no is root dependent
namedEnt	First basElem	Other parts are root dependents
Pronominal subrAnal	First basElem	Other parts are root dependents
Adjectival subrAnal	Last adjective basElem	Other parts are root dependents
Numeral subrAnal	First pronomen basElem	Other parts are root dependents
Set phrase subrAnal	basElem, who is not xPrep	basElem who is xPrep
Comparison subrAnal	basElem, who is not xSimile	basElem who is xSimile
Particle subrAnal	First basElem	Other parts are root dependents
coordAnal	First basElem	Other parts are root dependents
phrasElem	First basElem	Other parts are root dependents
unstruct	First basElem	Other parts are root dependents
crdParts	First crdPart	Other crdPart are root dependents, other nodes are dependents of the next closest crdPart
crdClauses	First crdPart	The first clause of each semicolon separated part becomes a direct dependent of the root; parts between semicolons are processed same way as crdParts
Any PMC	pred, if there is one; first/only basElem otherwise	Other parts are root dependents

to coreference in relative clause constructions and some types of links for controlled/raised subjects. Enhanced dependency roles have subtypes indicating case/preposition information for nominal phrases, but no subtypes indicating conjunctions for subordinate clauses.

We did preliminary result evaluation by manually reviewing 60 sentences (approx. 800 tokens). We found 19 inaccuracies in basic dependencies: 1 due to the lack of distinctions in the LVTB data, 6 due to errors in the original data, and the rest must be mitigated by adjusting the transformation. Analyzing enhanced dependencies, we found 3 errors due to incorrect original data, and some problems that can be solved by adjusting the transformation: 8 incorrect enhanced dependency labels (wrong case or pronoun assigned) and 15 missing enhanced links related to conjunct propagation or subject control. There were no instances of enhanced dependency errors caused by lack of distinctions in LVTB data, however it is very likely that such errors do exist, and we didn't spot one because of the small review sample size. Thus, we conclude that while the transformation still needs some fine-tuning for the next UD release and further reevaluation, overall it gives good results, and situations where LVTB data is not enough to obtain a correct UD tree seem to be rare.

4   Conclusion

Developing a treebank annotated according to the two complementary grammar models has proven to be advantageous. On the one hand, the manually created hybrid dependency-constituency annotations help to maintain language-specific properties and accommodate the Latvian linguistic tradition. The involved linguists—annotators and researchers—appreciate this a lot. On the other hand, the automatically derived UD representation of the treebank allows for multilingual and cross-lingual comparison and practical NLP use cases. The hybrid model is informative enough to allow the data transformation not only to the basic UD representation, but to the enhanced UD representation as well. The transformation itself, however, is rather complicated because of many differences between the two models. Some theoretical differences are big, even up to whether some language phenomena are considered to be either morphological, syntactic, or semantic. But despite the differences, actual treebank sentences, where LVTB annotations are not informative enough to get a correct UD graph, are rare. To keep up with the development of UD guidelines and LVTB data the transformation would greatly benefit from having even small but repeated result evaluations.

Acknowledgement. This work has received financial support from the European Regional Development Fund under the grant agreements No. 1.1.1.1/16/A/219 and No. 1.1.1.2/ VIAA/1/16/188.

We want to thank Ingus Jānis Pretkalniņš constructive criticism of the manuscript and anonymous reviewers for insightful comments.

References

1. Barzdins, G., Gruzitis, N., Nespore, G., Saulite, B.: Dependency-based hybrid model of syntactic analysis for the languages with a rather free word order. In: Proceedings of the 16th NODALIDA, pp. 13–20 (2007)
2. Gruzitis, N., et al.: Creation of a balanced state-of-the-art multilayer corpus for NLU. In: Proceedings of the 11th LREC, Miyazaki, Japan (2018)
3. Nivre, J., et al.: Universal dependencies v1: a multilingual treebank collection. In: Proceedings of the 10th LREC, pp. 1659–1666 (2016)
4. Nespore, G., Saulite, B., Barzdins, G., Gruzitis, N.: Comparison of the SemTi-Kamols and Tesniere's dependency grammars. In: Proceedings of 4th HLT—The Baltic Perspective, Frontiers in Artificial Intelligence and Applications, vol. 219, pp. 233–240. IOS Press (2010)
5. Lokmane, I.: Sintakse. In: Latviešu valodas gramatika, pp. 692–766. LU Latviešu valodas institūts, Rīga (2013)
6. Pretkalnina, L., Nespore, G., Levane-Petrova, K., Saulite, B.: A Prague Markup Language profile for the SemTi-Kamols grammar model. In: Proceedings of the 18th NODALIDA, Riga, Latvia, pp. 303–306 (2011)
7. Pretkalnina, L., Rituma, L., Saulite, B.: Universal dependency treebank for Latvian: a pilot. In: Proceedings of 7th HLT—The Baltic Perspective, Frontiers in Artificial Intelligence and Applications, vol. 289, pp. 136–143. IOS Press (2016)

Adaptation of Algorithms for Medical Information Retrieval for Working on Russian-Language Text Content

Aleksandra Vatian[✉], Natalia Dobrenko, Anastasia Makarenko,
Niyaz Nigmatullin, Nikolay Vedernikov, Artem Vasilev, Andrey Stankevich,
Natalia Gusarova, and Anatoly Shalyto

ITMO University, 49 Kronverkskiy prosp., 197101 Saint-Petersburg, Russia
alexvatyan@gmail.com

Abstract. The paper investigates the possibilities of adapting various
ADR algorithms to the Russian language environment. In general, the
ADR detection process consists of 4 steps: (1) data collection from social
media; (2) classification/filtering of ADR assertive text segments; (3)
extraction of ADR mentions from text segments; (4) analysis of extracted
ADR mentions for signal generation. The implementation of each step
in the Russian-language environment is associated with a number of
difficulties in comparison with the traditional English-speaking environment. First of all, they are connected with the lack of necessary databases
and specialized language resources. In addition, an important negative
role is played by the complex grammatical structure of the Russian language. The authors present various methods of machine learning algorithms adaptation in order to overcome these difficulties. For step 3 on
the material of Russian-language text forums using the ensemble classifier, the Accuracy = 0.805 was obtained. For step 4 on the material of
Russian-language EHR, by adapting pyConTextNLP, the F-measure =
0.935 was obtained, and by adapting ConText algorithm, the F-measure
= 0.92–0.95 was obtained. A method for full-scale performing of step
4 was developed using cue-based and rule-based approaches, and the
F-measure = 67.5% was obtained that is quite comparable to baseline.

Keywords: Adverse drug reaction · Natural language processing
Russian-language text content

1 Introduction

One of the challenging problems of NLP is the problem of processing healthcare
information. Nowadays it includes not only the actual clinical information, but
also content from social media. In our work we appeal to the texts concerning
adverse drug reaction (ADR). ADR detection is one of the most important tasks

P. Sojka et al. (Eds.): TSD 2018, LNAI 11107, pp. 106–114, 2018.
https://doi.org/10.1007/978-3-030-00794-2_11

of modern healthcare. Texts containing information on ADRs can be characterized by non-compliance with grammatical rules, a significant portion of texts in narrative formats. According to the World Health Organization, the death rate from ADR is among the top ten of all causes of death in many countries [5], and unfortunately Russia is in this list as well. In Russia, studies are under way on the use of NLP in medicine [3,10,12], but they are not focused on ADR detection.

In general, the ADR detection process can be divided into the following steps: (1) data collection from suitable text sources (social media and/or clinical texts); (2) selection of text segments containing a reference to ADR; (3) eliciting of assertions concerning ADR in a form suitable for further analysis (mainly in the predicate form). The implementation of each step in the Russian-language environment is associated with a number of difficulties in comparison with the traditional English-speaking environment. First of all, these are connected with the lack of necessary databases and specialized language resources. In addition, an important negative role is played by the complex grammatical structure of the Russian language. The article explores these difficulties and presents various adapted algorithms for retrieving ADR from the Russian text content.

The rest of the article is organized as follows. In Sect. 2 we discuss each step mentioned above in a uniform structure: the English-language background – available Russian-language support – our proposals, developed methods and the experimental results. Section 3 concludes and outlines the future work.

2 Methods

2.1 Text Sources for Data Collection and Processing

As the literature analysis [1,2,4,9,11,13] and real practice shows, in order to solve the ADR problem by natural language processing (NLP) methods the following input data are needed: primary sources of information; marked datasets; auxiliary linguistic resources. We conducted a comparative analysis of the most common sources of obtaining these data in English and in Russian. The results of the analysis of available information sources on ADR in English and their Russian-language analogues as well as variants of replacement missing sources used in our work are briefly described below. More complete review can be found, for example, in [6].

In the context of ADR detection, the needed resources can be divided into the following groups: (1) Spontaneous reporting systems; (2) Databases based on clinical records and other medical texts; (3) Dictionaries and knowledge bases; (4) Health-related websites and other network resources; (5) Specialized linguistic resources.

Spontaneous reporting systems (1), such as FAERS[1], VigiBase[2] and AIS-Rospharmaconadzor[3], are databases of reports of suspected ADR events, collected from healthcare professionals, consumers, and pharmaceutical companies. These databases are maintained by regulatory and health agencies, and contain structured information in a predetermined form.

In the English segment of group (2), the main place is occupied by the MEDLINE[4] database. There is no similar database in Russian. Of the verified databases of such type in Russia, one can call the annotated corpus of clinical free-text notes [12] based on medical histories of more than 60 patients of Scientific Center of Children Health with allergic and pulmonary disorders and diseases. In general, most of the datasets of this type in Russian are rather small and designed in-house for other research purposes and not for ADR detection.

Dictionaries and knowledge bases (3) helping to ADR detection are widely represented in English-language segment. Specialized dictionaries in Russian, reflecting all medical terminology, do not yet exist. At present, the process of their creation is underway, mainly by the forces of individual research teams (see, for example, [8]).

Health-related websites and other network resources (4) are represented in the Russian-language segment as widely as in the English-language. For example, the alternative to the online health community DailyStrength[5] are numerous websites[6] aggregating users' messages about ADR. However, our studies have revealed a number of differences between them, important from the point of view of ADR detection: users of Russian-language web resources are much more emotional and prone to polar assessments (such as fine/terrible). Consequently, there is a problem of an adequate choice of assessment scales to take into account the opinions of users.

As concerning to specialized linguistic resources (5), here in the first place should be called MetaMap[7] toolbox. For the Russian language, such a resource does not exist, and in order to solve the ADR problem the researchers are to adapt non-specialized NLP tools or to develop them independently.

The brief overview shows that due to the lack of verified databases and specialized resources it is expedient to follow the path of adaptation of existing ADR algorithms designed for English content to the Russian language.

2.2 Selection of ADR-Reference Text Segments

The problem of selection of ADR-reference text segments can be considered in the class of tasks of text summarization and has an extensive bibliography (see,

---

[1] https://www.fda.gov/Drugs/GuidanceComplianceRegulatoryInformation/Surveillance/AdverseDrugEffects/ucm070093.htm.

[2] https://www.who-umc.org/vigibase/vigibase/.

[3] http://www.roszdravnadzor.ru/services/npr_ais.

[4] https://www.nlm.nih.gov/databases/download/pubmed_medline.html.

[5] https://www.dailystrength.org/.

[6] https://protabletky.ru/, http://topmeds.ru/.

[7] https://metamap.nlm.nih.gov/.

for example, [2,4]). In Russia, text forums are of particular interest as a source of information on ADRs, so for the comparative analysis we chose the work [11]. Solving the mentioned problem, the authors used data set built in-house from DailyStrength. The classification was performed using three supervised classification approaches: Naïve Bayes, Support Vector Machines and Maximum Entropy. Preprocessing included adding the synonymous terms p(using WordNet) and negation detection. The best of achieved results is presented in Table 1 (gray background).

Table 1. Selection of ADR-reference text segments.

	Accuracy (%)				
Features used	(1)–(7)	(2)	(8)	(3)	(2) + (3) + (8)
	83.4	71.4	80.5	72.1	80.0

We have adapted this to the Russian language environment. As a data source, we used three forums[8]. We built a parser and collected 1210 reviews on medications: asthma – 508 reviews, type 2 diabetes – 222 reviews, antibiotics – 480 reviews from the above sites. All data were annotated manually for the presence or absence of ADR, the range of ADR manifestation.

We used only the features, available for assessment in Russian language (see Table 1). We calculated Tf.Idf using the weightTfIdf() function from the tm package for R language. In order to calculate feature (8) we needed a dictionary of terms denoting adverse effects in Russian which is currently absent. The list of adverse effects was manually collected from the medical dictionary and accounted for 215 adverse effects. As for the feature (3), the main problem was the lack of sufficiently complete dictionaries. We used a specialized dictionary built in-house[9].

Preprocessing included the removal of stop words and lemmatization using SnowBallC package. The classification was performed using a decision trees algorithm. Since each of the features is represented by a sufficiently large matrix, three classification models were constructed using each attribute separately. To combine these attributes, an ensemble of classifiers was constructed using the accuracy of the classification of each solo-model as a decision rule. The results are presented in Table 1 (transparent background).

Comparison of the results of Table 1 allows us to draw the following conclusions. First, despite the smaller set of specialized linguistic resources and, correspondingly, the smaller number of available features for the Russian language, the achieved values of Accuracy on English and Russian-language content are quite comparable (Table 1 in bold). Secondly, we confirm the conclusion we had reached earlier [7] that the most important role in forums summarization

---

[8] irecommend.ru, otzovik.com, https://protabletky.ru/.
[9] https://github.com/text-machine-ab/sentimental/blob/master/sentimental/word_list/russian.csv.

Table 2. Analysis of extracted fragments about ADR.

Class	F-measure, %	
	[Velupillai], [Velupillai_1]	Our results
def_existence	88.1	93.5
def_negated	63.0	89.3
prob_existense	81.1	64.7
prob_negated	55.3	51.6

belongs to successful feature selection. Finally, the accuracy of selection of ADR-reference text segments depends on the quality of the content. Indeed, the posts in DailyStrength resource contain more structure, are longer, and often consist of multiple sentences than in our forums, and this affects the results of Table 1.

2.3 Analysis of Extracted Fragments for the Formation of Logical Statements About ADR

First of all, we considered the possibilities of adapting existing NLP software tools for processing Russian-language texts. As a prototype, we used the method represented in [13]. The method is intended for porting the pyConTextNLP library from English into Swedish (pyConTextSwe). The library allows automatically finding in the text the name of the disease with the help of regular expressions and determining the degree of its confirmation within the sentence. In total, to classify the confirmation of the disease, four classes are identified: define_negated_existence; probable_negated_existence; probable_existence; define_existence;. In the original library there are 381 keywords and 40 names of illnesses in English. The authors of the articles created an extensive dictionary of expressions for Swedish medical texts containing 454 cues (key phrases) using a subsets of a clinical corpus in Swedish.

We did a similar job to port pyConTextNLP library to Russian. As sources for the formation of the dataset, we used resources containing impersonal medical histories[10]. We have formed a data set consisting of 29 Russian-language medical histories and containing 513 separate assertions. We translated the key words and diagnoses into Russian and made regular expressions for them. We also expanded the list of diseases with the help of third-party resources, including in addition to it 2017 names of diagnoses. Based on the results of the first test of the algorithm, we added a list of keywords, and included in the regular expressions various health characteristics mentioned in the medical records, and re-tested the algorithm on updated regular expressions.

A comparison of the results is given in Table 2. Attention is drawn to our good results in classes def_existence and def_negated in comparison with the comparatively weak results in classes prob_existense and prob_negated. Our analysis

---

[10] http://kingmed.info/Istorii-boleznye, http://www.medsite.net.ru/?page=list&id=6.

Table 3. Efficiency of identifying medical terms in Russian and Dutch.

Parameter	F-measure, %	
	ConText for Dutch	ConText for Russian
Negation	87–93	92–95
Temporality	13–44	95–98
Experiencer	99–100	98–100

showed that this is due to the quality of the initial data: the final diagnoses corresponding to classes def_existence and def_negated are practically true in all the medical histories, while the differential diagnoses corresponding to classes 1 and 2 are described vaguely and remotely from the context of the specific medical history, so that the accuracy of the algorithm is understandably low. Thus, the problem of porting pyConTextNLP library to Russian can be considered successfully solved.

Our next research was devoted to the possibility of inter-language adaptation of single triggers. As a prototype, we used the work [1] concerning an adaptation of the English ConText algorithm to the Dutch language, and a Dutch clinical corpus. Algorithm ConText is based on regular expressions and lists of trigger terms. It searches for words related to medical terms (cues) considered as triggers and defines three parameters for them: negation (denied or affirmed), experiencer (patient or other person), temporality (at the moment, no more than 2 weeks ago, long ago) thus identifying the contextual properties in the clinical corpus. Four types of medical documentation were used in [1] as a source of information: general practitioner entries, specialist letters, radiology reports and discharge letters. The total volume of the dataset was 7,500 documents with an average number of words in the document equal to 72. Such a volume of raw materials in Russia is not available, so we built our dataset of 23 medical records, including all types of records mentioned above.

The main difference between our approach and [1] approach was as follows. To select a suitable parameter state, the original English algorithm as well its Dutch adaptation by [1] uses regular expressions with a certain constant set of markers; we instead use the search for words from specially compiled customized dictionaries. Besides, for the analysis of medical texts in Russian, we used two values for the time parameter instead of three used in Dutch variant. Finally, ConText for Russian uses not only a point and a semicolon as a terminator, but also a specially developed dictionary of conjunctions that allows you to correctly determine the context of the trigger. These changes have significantly increased the efficiency of identifying contextual properties of medical terms in Russian (see Table 3).

Finally, we investigated the need to use a full syntactic parsing to solve the ADR problem. For comparison, we used [9]. In this work the graphs of grammatical dependencies are constructed using Stanford Parser for all sentences containing medical terms. These determine the shortest pathways considered as the kernel of the relationship between the drug and the side effect, thereby forming potential pairs 'drug – adverse effect'. In conclusion, with the help of linguistic rules, the negations detection is performed. For adverse drug event extraction, the authors obtained $F\text{-}measure = 50.5\text{--}72.2\%$, depending on the variety of complexing the applied algorithms.

But, our experiments showed that due to the complex grammatical structure of the Russian language, the use of the described kernel function leads to significant recognition errors. Therefore, we refused to parse the sentences, but developed a problem-oriented algorithm for allocating ADR from the sentence in Russian. The scheme of the algorithm is shown in Fig. 1. We tested the work of the algorithm on 100 sentences extracted from the medical site[11]. In the experiment for adverse drug event extraction, $F\text{-}measure = 67.5\%$ was obtained that is quite comparable to baseline.

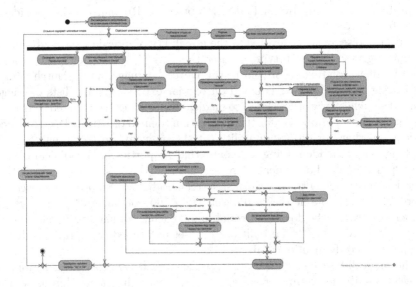

Fig. 1. Scheme of the proposed algorithm.

3   Conclusion

We proposed a comprehensive solution to the problem of ADR detection on Russian-language texts. Solving the problem of selection of ADR-reference text segments we constructed an ensemble of classifiers using the accuracy of the classification of each solo-model as a decision rule. Despite the smaller set of specialized linguistic resources and, correspondingly, the smaller number of available

---

[11] https://www.medsovet.info/herb/6617.

features for the Russian language, the achieved values of Accuracy on English and Russian-language content are quite comparable.

Solving the problem of analysis of extracted fragments for the formation of logical statements about ADR we have built a specialized dataset of medical records, a number of specially compiled customized dictionaries and a set of logical rules for the processing. These changes have significantly increased the efficiency of identifying contextual properties of medical terms in Russian. Finally, we have developed a problem-oriented algorithm for allocating ADR from the sentence in Russian.

Acknowledgments. This work was financially supported by the Government of Russian Federation, "Grant 08-08". This work financially supported by Ministry of Education and Science of the Russian Federation, Agreement #14.578.21.0196 (03/10/2016). Unique Identification RFMEFI57816X0196.

References

1. Afzal, Z., Pons, E., Kang, N., Sturkenboom, M.C., Schuemie, M.J., Kors, J.A.: ContextD: an algorithm to identify contextual properties of medical terms in a Dutch clinical corpus. BMC Bioinform. **15**(1), 373 (2014)
2. Allahyari, M., et al.: Text summarization techniques: a brief survey. arXiv preprint arXiv:1707.02268 (2017)
3. Baranov, A., et al.: Technologies for complex intelligent clinical data analysis. Vestnik Rossiiskoi akademii meditsinskikh nauk **2**, 160–171 (2016)
4. Bhatia, N., Jaiswal, A.: Automatic text summarization and it's methods - a review. In: 2016 6th International Conference on Cloud System and Big Data Engineering, Confluence, pp. 65–72. IEEE (2016)
5. Gildeeva, G., Yurkov, V.: Pharmacovigilance in Russia: challenges, prospects and current state of affairs. J. Pharmacovigil. (2016)
6. Gonzalez, G.H., Tahsin, T., Goodale, B.C., Greene, A.C., Greene, C.S.: Recent advances and emerging applications in text and data mining for biomedical discovery. Brief. Bioinform. **17**(1), 33–42 (2015)
7. Grozin, V., Buraya, K., Gusarova, N.: Comparison of text forum summarization depending on query type for text forums. In: Soh, P.J., Woo, W.L., Sulaiman, H.A., Othman, M.A., Saat, M.S. (eds.) Advances in Machine Learning and Signal Processing. LNEE, vol. 387, pp. 269–279. Springer, Cham (2016). https://doi.org/10.1007/978-3-319-32213-1_24
8. Lapaev, M.: Automated extraction of concept matcher thesaurus from semi-structured catalogue-like sources of data on the web. In: 2016 18th Conference of Open Innovations Association and Seminar on Information Security and Protection of Information Technology, FRUCT-ISPIT, pp. 153–160. IEEE (2016)
9. Liu, X., Chen, H.: A research framework for pharmacovigilance in health social media: identification and evaluation of patient adverse drug event reports. J. Biomed. Inform. **58**, 268–279 (2015)
10. Lushnov, M., Kudashov, V., Vodyaho, A., Lapaev, M., Zhukova, N., Korobov, D.: Medical knowledge representation for evaluation of patient's state using complex indicators. In: Ngonga Ngomo, A.-C., Křemen, P. (eds.) KESW 2016. CCIS, vol. 649, pp. 344–359. Springer, Cham (2016). https://doi.org/10.1007/978-3-319-45880-9_26

11. Sarker, A., Gonzalez, G.: Portable automatic text classification for adverse drug reaction detection via multi-corpus training. J. Biomed. Inform. **53**, 196–207 (2015)
12. Shelmanov, A., Smirnov, I., Vishneva, E.: Information extraction from clinical texts in Russian. In: Computational Linguistics and Intellectual Technologies: Papers from the Annual International Conference, Dialogue, vol. 14, pp. 537–549 (2015)
13. Velupillai, S., et al.: Cue-based assertion classification for Swedish clinical text—Developing a lexicon for pyConTextSwe. Artif. Intell. Med. **61**(3), 137–144 (2014)

CoRTE: A Corpus of Recognizing Textual Entailment Data Annotated for Coreference and Bridging Relations

Afifah Waseem[(⌂)]

Department of Computer Science, University of Oxford, Oxford OX1 3QD, UK
afifahwaseem@gmail.com

Abstract. This paper presents CoRTE, an English corpus annotated with coreference and bridging relations, where the dataset is taken from the main task of recognizing textual entailment (RTE). Our annotation scheme elaborates existing schemes by introducing subcategories. Each coreference and bridging relation has been assigned a category. CoRTE is a useful resource for researchers working on coreference and bridging resolution, as well as recognizing textual entailment (RTE) task. RTE has its applications in many NLP domains. CoRTE would thus provide contextual information readily available to the NLP systems being developed for domains requiring textual inference and discourse understanding. The paper describes the annotation scheme with examples. We have annotated 340 text-hypothesis pairs, consisting of 24,742 tokens and 8,072 markables.

Keywords: Coreference · Bridging relations · Annotated corpus

1 Introduction

An important aspect of human understanding of language is to make inferences from the text. Recognizing these inferences or entailments is important in many NLP domains. The PASCAL recognizing textual entailment (RTE)[1] challenges are considered as the standard for the textual entailment task; recent challenges include the RepEval 2017 shared task[2]. The RTE dataset is in the form of text and hypothesis pairs. Given two text segments, Text (T) and Hypothesis (H), recognizing textual entailment' is defined as the task that determines whether H can be inferred from T or T entails H. Consider the T-H pair in example (1), where we can infer H with the help of T.

1. **_T:_** _Security preparations kept protestors at bay at the recent G8 Summit on Sea Island (USA)._
   _H:_ _The G8 summit took place on an American island._

---

[1] https://aclweb.org/aclwiki/Textual_Entailment_Portal.
[2] https://repeval2017.github.io/shared/.

© Springer Nature Switzerland AG 2018
P. Sojka et al. (Eds.): TSD 2018, LNAI 11107, pp. 115–125, 2018.
https://doi.org/10.1007/978-3-030-00794-2_12

By its very definition, the RTE task depends on the discourse relations in the text, coreference and bridging relations are known to facilitate the RTE task [1–3]. The need for resources for the RTE task is frequently expressed in the literature [4,5]. We hope our work on annotating coreference and bridging relations will prove to be a useful resource for researchers working on RTE systems, as well as the NLP domains that require textual understanding. Textual entailment has already been successfully applied to different NLP domains like Question Answering [6], Information Extraction [7] and Machine Translation [8].

The remainder of the paper is organized as follows: a brief explanation of coreference and bridging relations is given in Sect. 2; Sect. 3 presents related work; in Sect. 4 we discuss corpus creation including manual annotations; Sect. 5 presents an agreement study on the annotated text, we conclude this work in Sect. 6.

2   Coreference and Bridging Relations

Coreferring expressions are used in natural language to combine different parts of the linguistic text. These linguistic expressions sometimes refer to an entity or event introduced beforehand (antecedent). They can be a named entity, a noun phrase or pronouns. They are termed markables' or mentions'. Consider example (2), *John* and *He* are referring to the same real world entity, a person named *John*.

2. *John is an excellent craftsman. He works at the local shop.*

Aside from the identity and near identity reference relations, natural language text contains more complex and vague reference relations between two markables. These could be the relations between: set-member, set-subset, set-set, part-whole, entity attribute and entity function. These complex relations are known as bridging relations. They are explained in Sect. 4.4. In example (3), *three sons* and *two daughters* are members of the set of *five children*. We have annotated CoRTE with both coreference and bridging relations.

3. *Mary has five children, three sons and two daughters.*

3   Related Work

In recent times a lot of work on coreference resolution has been carried out by the NLP community. MUC [9], ACE [10] and OntoNotes [11] are the three main coreferentially annotated corpora.

The concept of bridging relations was introduced by [12]. Following their work [13,14] subcategorized bridging relations as set-member, set-subset and generalized possession. Table 1 presents statistics of the corpora annotated with the bridging relations for English language. The GNOME corpus [13], consists of text from museum domain and patient information leaflets. It contains 1,164

Table 1. Statistics of corpora annotated with bridging relations

	GNOME	ARRAU	Markert et al. 2012	SciCorp
Size	500 Sentences	32,771 Tokens	50 Texts	61,045 Tokens
Markables or NPs	3,000 NPs	3,837 Markables	10,980 Markables	8,708 NPs
Genres	Museum domain, patient leaflets	Wall street journal, dialogues, narratives	Wall street journal	Research papers

identity and 153 generalized possession relations. The ARRAU corpus [14] contains texts from different genres, including dialogue, narrative, and a variety of genres of written text. In the Prague Dependency Treebank [15], an annotated corpus of Czech, the contrast of a noun phrase is additionally annotated as a bridging category. The Potsdam commentary corpus [16] consists of 175 German newspaper commentaries. It is annotated with anaphoric and bridging chains as well as Information Status (IS). As opposed to previous approaches, [17] considered indefinite noun phrases as markables. The annotated text was taken from German radio news bulletins. The DIRNDL radio news corpus [18] was manually annotated for 12 categories of information status including bridging [19]. Bridging relations were only annotated for the cases where an antecedent could be found. Fifty texts from the WSJ portion of the OntoNotes corpus [11] were annotated by [20]. They annotated these texts for IS, bridging was among one of the 9 categories of IS. Bridging was not limited to definite noun phrases, bridging antecedents were allowed to be of any kind of noun phrase, a clause or verb phrase, they have also annotated cause and effect relations. Their work was later used by [21]. The corpus presented by [22] was annotated with five categories of bridging relations including, part-whole, set-member, entity-attribute/function, event-attribute and location-attribute. The corpus consists of three genres: newswire, narratives and medical leaflets. It consists of 1,1894 tokens and 1,395 markables. In SciCorp [23], bridging relations are defined as associative anaphora that include part-of, is-a or any other associative relationship that can be established. Fourteen research papers of genetics and computational linguistics were annotated containing 8,708 definite noun phrases.

Previously, 120 T-H pairs from the RTE-5 Search task dataset were annotated for coreference and bridging relations by [2,3]. In their work, they have mentioned only three categories of bridging relations: part-of, member-of and participants in events. These annotations were limited to the reference relations they found useful in establishing entailment through transformation. In contrast, all the referring expressions are annotated with the coreference and bridging relations in CoRTE, according to the annotation scheme described in detail in Sect. 4.

The available annotated corpora for English are either annotated with a restrictive definition of bridging relations or some of the bridging categories are too broad to be properly annotated. Our corpus CoRTE, is annotated with well-defined bridging relations and it is annotated for the textual entailment dataset of the English language. It consists of 24,742 tokens and 8,072 markables.

4   Corpus Creation

Corpus consists of 340 Text-Hypothesis pairs, randomly selected from the RTE-3, 4 and 5 test dataset. These pairs were selected from the positive instances of entailment, i.e. the portion of the dataset in which T entailed H. The RTE-3 dataset contains long as well as short length text segments. 90 T-H pairs have been annotated from the RTE-3 data including all the long text segments. 125 T-H pairs each were selected from RTE-4 and 5 datasets. The RTE-3 and 4 datasets consist of text segments taken from four NLP domains: information retrieval (IR), information extraction (IE), question answering (QA) and summarization (SUM). The RTE-5 main data set only consists of text segments from IR, IE and QA domains.

4.1   Manual Annotations

The MMAX2 tool [24] was used for annotating coreference and bridging relations in CoRTE[3]. MMAX2 provides annotations in a standoff format. Markables in coreference relations make a coreference chain. Bridging relations are intransitive and annotated as a relation from one to many, also referred to as bridging pairs. In the remainder of the section we discuss the annotation scheme.

4.2   Markables

In this work we have selected the maximum span of a noun phrase for a markable. Each markable is assigned one of the nominal or pronominal types. Nominal markables were annotated as: Named Entity (NE), Definite Noun Phrases (defNPs) and Indefinite Noun Phrases (indefNPs). Pronominal markables were annotated as: Demonstrative Pronouns (pds), Possessive pronouns (ppos), Personal Pronouns (pper), Reflexive Pronouns (prefl) and Relative Pronouns (prel).

Conjunctions of Several NPs (conjNP): This category contains NPs that are in conjunction with other NPs. The most common conjunction which is used for joining NPs is "and". Other conjunctions observed were: *as well as*, *or* and *both ... and*. In the following sentences, the conjNPs are underlined.

4. *Celestial Seasonings is a tea company that specialises in herbal tea but also sells black tea **as well as** white and oolong blends.*
5. ***Both** Bush **and** Clinton helped raise funds for the recovery from Hurricane Katrina.*

Any noun phrase conjunction that did not represent entities of the same type was not considered a possible markable. In example (6), *The young woman* and *her dress* are not of the same type of entities (person vs object), so this noun phrase conjunction is not a candidate for a single ConjNP markable in coreference or bridging relation. The frequency of each markable type in CoRTE is presented in Table 2.

6. *The young woman and her dress will be discussed for weeks to come.*

---

[3] https://sourceforge.net/projects/corte.

Table 2. Markable types annotated in CoRTE with their respective frequencies.

Markable type	NE	DefNP	IndefNP	ConjNP	Pds	Ppos	Pper	Prefl	Prel	Total
# of Markables	3,311	1,935	1,445	208	30	640	356	11	136	8,072

4.3 Non Markables

Existential *there* and pleonastic *it* were not considered markables. To keep the annotation task simple gerunds (nominalization of verbs) were considered markables only when preceded by an article, a demonstrative or possessive pronoun.

4.4 Annotated Categories

This section highlights the categories or types of coreference and bridging relationships that were annotated for this work.

Direct Relations (Direct): The identity and near-identity coreference chains as mentioned in [25] were assigned this category. NP markables in direct relations could be replaced by one another in the text. "Mayor of London" can be replaced by his name in the text and it will not hinder the understanding of the reader of the text. However, bridging relations like the Mayor's characteristics, e.g. *his joyful manners* can't replace his name or occupation.

In the case of bridging relations, annotators were asked to relate the most appropriate markable as antecedent rather than the nearest one. The following bridging relation types were annotated.

Set Relations (SET): The mentions in set-set, set-subset and set-member relations were assigned the bridging type SET. All SET relations exist between entities of same type: a set of people (person) can only be in relation with a set of other people, similarly countries can be in SET relation with other countries. This attribute sets it apart from Part-Whole relations. *Botswana* (a country) is part-of *Africa* (a continent), on the other hand, *European Union* is a set of countries and *Germany* is its member.

SET relations were indicated in many cases by markables that were a conjunction of noun phrases (ConjNP). If one of the noun phrase of conjNP is mentioned separately in the text then the ConjNP and that noun phrase were annotated in a set-member, set-set or set-subset relation. In the T-H pair below *Gossip Girl* is a member of the set **the *hit series Chuck, Gossip Girl, The O.C. and the new web series Rockville, CA***. Also, *Gossip Girl* in T is coreferring with *Gossip Girl* in H.

7. **T:** *Josh Schwartz, creator of the hit series Chuck, Gossip Girl, The O.C. and the new web series Rockville, CA will address NAB Show attendees during a Super Session titled "Josh Schwartz: Creating in the New Media Landscape," held Wednesday, April 22 in Las Vegas.*
   H: *Josh Schwartz is the creator of "Gossip Girl".*

Non-continuous conjunctions of noun phrases were not considered as a single markable. If non-continuous conjunctions of noun phrases were referred to later in the text, they were given a set-member or set-subset reference relation.

8. *Mary took her kids to her home town. They travelled by car.*

The above phenomenon, where a markable *They* is referring to non-continuous noun phrases, **Mary** and **her kids**, is known as '*split antecedent anaphora*'. It is difficult to capture such relations by direct anaphora; SET bridging relations were established in these cases.

The importance of establishing bridging relations between non-continuous markables can be observed in example (9). The text mentions the increase of fuel prices in **India and Malaysia** separately. However, **India and Malaysia** are mentioned as a conjunction of markables in H. In order to entail H from T, a set member relationship is needed between the conjNP **India and Malaysia** (set), the markables **India** (member) and **Malaysia** (member) in T.

9. *T: Protests flared late last week in India, where the government upped prices by about 10%, and there were calls for mass rallies in Malaysia after gasoline prices jumped 41% overnight.*
   H: The price of fuel is increasing in India and Malaysia.

Part-Whole Relations (Part-Whole): A bridging relation is considered part-whole when an entity is part of another entity. A room and its ceiling are in part-whole relation. Similarly, the door of a room is a part of that room. Part-Whole relations are one to many relationships. One room has many parts and the relationship is intransitive.

Locations that were part of another location, as in the case of a city (part) within a state (whole) were annotated as part-whole relations. In the following sentence **MacKay's Nova Scotia riding** is part of **Pictou**.

10. *U.S. Secretary of State Condoleezza Rice and Foreign Affairs Minister Peter MacKay made a visit to MacKay's Nova Scotia riding in Pictou.*

Entity Characteristic/Function/Ownership Relations: This type of bridging relations exists between an entity and its characteristic, its functions and the what it owns. A good indicator of an entity-characteristic relation is a genitive case where a noun or pronoun is marked as modifying another noun. The markables, **John's style** and **John's age** are both characteristics of **John**. Similarly, the markable, **King Albert II of Belgium** in a text would refer to **Albert II's** function as **the King of Belgium** and thus has a relation with Belgium. A large number of entity-function bridging relations were about a worker and his/her work relations. They included author-book, movie-director, artist-painting. In example (11), the markables **the IRA** and **the first Ceasefire Declaration** are related as the organization, IRA, announced the declaration and the declaration is considered its work.

11. *It took another 22 years until   the first Ceasefire Declaration was announced by the IRA.*

Prepositions like *by* and *of* and verbs *have*, *own* and *belong* may indicate entity-characteristic/function relations. So far, our definition of Entity-char/func relationships follows categories mentioned in [22]. During the first round of discussions, we added the relation of an entity and its belongings i.e. the ownership. The relation between ***John*** and ***John's car*** is an example of such a relation. This is different from the characteristics of ***John*** as in ***John's hair style***. In future a separate category could be given to this relationship. *Bridging-contained* is a term used for the cases where the bridging antecedent is embedded in the noun phrase related to it. Consider the markable, ***the president of the US***, the antecedent of this phrase is ***the US***, which is embedded in it.

Temporal References (-was): Temporal references to the past were captured by introducing a subtype –was. This subtype was assigned to all the above categories of bridging where relationships existed in the past. In the example (12) below ***the CEO of Star Tech*** is a direct referent of ***John*** and in the past ***John*** was ***a director of Global Inc***. Similarly, in example (13), ***the city*** refers to ***Mumbai***, but in the past it was known as ***Bombay***. This information about the previous name of the city can be useful in NLP domains like QA and IE.

12. *John is the CEO of Star Tech. He has previously worked as a director of Global Inc.*
13. *The current name of the city is Mumbai. Previously, it was known as Bombay.*

There were cases in the text, where a statement was true both in the past and present. In example (14), it would be correct to say ***Le Beau Serge*** was directed by ***Chabrol*** and that ***Le Beau Serge*** is directed by ***Chabrol***. Similarly, ***Neil Armstrong*** was the first man to walk on the moon and he still is the first man to walk on the moon. In these cases, annotators were asked to follow the temporal clue of the text, and annotate according to human understanding of the text.

14. ***T:*** *Claude Chabrol is a French movie director and has become well-known in the 40 years since his first film, Le Beau Serge.*
    H: *Le Beau Serge was directed by Chabrol.*

Appositives: Unlike OntoNotes [11] and following [26], appositives were considered coreferential with the noun phrases that they were modifying.

Relative Clause: A restrictive relative clause was included in a markable. This was in accordance with [26]. In example (15), the phrase ***The girl who looks like Taylor Shaw*** is an example of a restrictive relative clause.

15. *The girl who looks like Taylor Shaw is sitting inside.*

In the case of a non-restrictive relative clause, only the relative pronoun that introduces the clause is considered as a markable. In the following sentence taken from RTE-4, ***where*** is a relative pronoun referring to ***India***.

Table 3. Number of coreference and bridging relations in the four NLP domains.

T-H Pairs	Direct	SET	Part-W	Char/func
IR = 97	339	95	184	224
IE = 99	374	127	152	223
QA = 105	446	148	174	376
SUM = 39	128	60	45	66
Total = 340	1,287	430	555	889

Table 4. Agreement results of coreference and bridging types.

	A-B	A-C	B-C
Direct	0.94	0.91	0.92
SET	0.86	0.81	0.84
Part-W	0.82	0.79	0.85
Char/func	0.83	0.77	0.81
Average	0.86	0.82	0.85

16. *Protests flared late last week in India, where the government upped prices by about 10%.*

Nested Noun Phrases: Nested noun phrases in the text, that were coreferential or in a bridging relation were annotated, in some cases these were deeper than one level. Consider the following example:

17. *[A whale that became stranded in [the River Thames]] has died after a massive rescue attempt to save [its] life. [The whale] died at about 1900 GMT on Saturday as [it] was transported on a barge towards [deeper water in [the Thames Estuary]].*

In this sentence **deeper water in the Thames Estuary** is part of **the River Thames**, it contains a nested markable **the Thames Estuary**, which is also part of **the River Thames**. The statistics of coreference and bridging relations found in T-H pairs of CoRTE are presented in Table 3.

5   Inter-annotator Agreement

The corpus was annotated by three annotators. The first author of this paper, who has also provided the annotation guidelines is annotator A. The other two annotators were Masters students, one with a background in the English literature and other in English linguistics. Initially 60 T-H pairs were selected and annotated from RTE-3, 4 and 5 datasets. The issues arising from the annotation of these pairs were then discussed. These annotations are not included in the final version or calculation of inter-annotator agreement. These discussions helped to shape the annotation scheme as described in Sect. 4.1. Some of the points discussed were as follows:

- A bridging relation was assigned the category SET only when the members belong to the same type of entity.
- It was decided to always annotate addresses (locations) as part-whole relations.
- Only a small number of Entity-Characteristic/Function/Ownership relations were annotated in the initial 60 pairs. This led to providing annotators with a detail definition as well as clues to identify these relations. These are described in Sect. 4.4.
- Sub-category of –was' for coreference and bridging relations was introduced for the relations that existed in the past.
- In the cases where the reference relations were based on someone's opinion, the annotators were asked to only annotate the relation if it seemed to be established in the text. In the example given below, whether John's belief is a reality depends on contextual information, in this case as John's belief and the narrator's belief are same the annotator considered *John* as *the right man for the job*.

18. *John believes he is the right man for the job. I would say he is right.*

After these discussions we moved ahead with the annotation process. Inter-annotator agreement was conducted on the final annotations. The Kappa coefficient κ [27] was calculated for 20% of the annotated RTE-3, 4 and 5 T-H pairs to determine the inter-annotator agreement on the coreference and bridging types, i.e. agreement on the categories (Sect. 4.4) assigned to the relations. These are presented in Table 4. We measured F1 scores to determine the agreement for the annotated coreference chains and bridging pairs. The F1 score achieved for coreference chains is 0.86 and 0.71 for bridging pairs. These inter-annotator agreements are acceptable for the annotation of coreference and bridging relations.

6   Conclusion

We have presented CoRTE, a corpus of RTE text-hypothesis pairs annotated with coreference and bridging relations. The corpus consists of 340 T-H pairs, 24,742 tokens, 8,072 markables and 3,161 relations. The inter-annotator agreement study shows that we have achieved moderate reliability. This dataset consists of texts from four different NLP domains, thus our work is relevant for the wider NLP community. We intend to extend CoRTE and annotate 500 T-H pairs. In the future, we would like to study the cause and effect relations, they were initially left due to sparsity of such relations in NPs. Most cause and effect relations consists of markables that are VPs. The annotated corpus presented in this paper is a useful resource for developing a resolver for bridging relations. It can also be used for developing textual inference based systems.

References

1. Bos, J., Markert, K.: When logical inference helps determining textual entailment (and when it doesn't). In: Proceedings of the Second PASCAL RTE Challenge, p. 26 (2006)
2. Abad, A., et al.: A resource for investigating the impact of anaphora and coreference on inference. In: Proceedings of LREC (2010)
3. Mirkin, S., Dagan, I., Padó, S.: Assessing the role of discourse references in entailment inference. In: Proceedings of the 48th Annual Meeting of the Association for Computational Linguistics, pp. 1209–1219 (2010)
4. Bowman, S.R., Angeli, G., Potts, C., Manning, C.D.: A large annotated corpus for learning natural language inference, pp. 632–642 (2015)
5. White, A.S., Rastogi, P., Duh, K.: Inference is everything: recasting semantic resources into a unified evaluation framework. In: Proceedings of the Eighth International Joint Conference on Natural Language Processing, p. 10 (2017)
6. Harabagiu, S., Hickl, A.: Methods for using textual entailment in open-domain question answering. In: Proceedings of the 21st International Conference on Computational Linguistics and the 44th Annual Meeting of the Association for Computational Linguistics, pp. 905–912 (2006)
7. Romano, L., Kouylekov, M., Szpektor, I., Dagan, I., Lavelli, A.: Investigating a generic paraphrase-based approach for relation extraction. In: 11th Conference of the European Chapter of the ACL (2006)
8. Padó, S., Galley, M., Jurafsky, D., Manning, C.: Robust machine translation evaluation with entailment features. In: Proceedings of the Joint Conference of the 47th Annual Meeting of the ACL and the 4th International Joint Conference on Natural Language Processing of the AFNLP, Stroudsburg, PA, USA, vol. 1, pp. 297–305 (2009)
9. Hirschman, L., Chinchor, N.: Appendix F: MUC-7 coreference task definition (version 3.0). In: Seventh Message Understanding Conference (MUC-7), Virginia (1998)
10. Doddington, G.R., Mitchell, A., Przybocki, M.A., Ramshaw, L.A., Strassel, S., Weischedel, R.M.: The automatic content extraction (ACE) program-tasks, data, and evaluation. In: LREC, vol. 2, p. 1 (2004)
11. Pradhan, S.S., Hovy, E., Marcus, M., Palmer, M., Ramshaw, L., Weischedel, R.: OntoNotes: a unified relational semantic representation. Int. J. Semant. Comput. 1, 405–419 (2007)
12. Clark, H.H.: Bridging. In: Proceedings of the 1975 Workshop on Theoretical Issues in Natural Language Processing, TINLAP 1975, pp. 169–174 (1975)
13. Poesio, M.: The MATE/GNOME proposals for anaphoric annotation, revisited. In: Proceedings of the 5th SIGdial Workshop on Discourse and Dialogue at HLT-NAACL (2004)
14. Poesio, M., Artstein, R.: Anaphoric annotation in the ARRAU corpus. In: LREC (2008)
15. Nedoluzhko, A., Mírovský, J., Pajas, P.: The coding scheme for annotating extended nominal coreference and bridging anaphora in the Prague Dependency Treebank. In: Proceedings of the Third Linguistic Annotation Workshop, pp. 108–111 (2009)
16. Stede, M.: The Potsdam commentary corpus. In: Proceedings of the 2004 ACL Workshop on Discourse Annotation, pp. 96–102 (2004)
17. Riester, A., Lorenz, D., Seemann, N.: A recursive annotation scheme for referential information status. In: LREC (2010)

18. Eckart, K., Riester, A., Schweitzer, K.: A discourse information radio news database for linguistic analysis. In: Chiarcos, C., Nordhoff, S., Hellmann, S. (eds.) Linked Data in Linguistics, pp. 65–76. Springer, Heidelberg (2012). https://doi.org/10.1007/978-3-642-28249-2_7

19. Cahill, A., Riester, A.: Automatically acquiring fine-grained information status distinctions in German. In: Proceedings of the 13th Annual Meeting of the Special Interest Group on Discourse and Dialogue, pp. 232–236 (2012)

20. Markert, K., Hou, Y., Strube, M.: Collective classification for fine-grained information status. In: Proceedings of the 50th Annual Meeting of the Association for Computational Linguistics: Long Papers, vol. 1, pp. 795–804 (2012)

21. Hou, Y., Markert, K., Strube, M.: Cascading collective classification for bridging anaphora recognition using a rich linguistic feature set. In: Proceedings of the 2013 Conference on Empirical Methods in Natural Language Processing, pp. 814–820 (2013)

22. Grishina, Y.: Experiments on bridging across languages and genres. In: Proceedings of the Workshop on Coreference Resolution Beyond OntoNotes, CORBON 2016 (2016)

23. Rösiger, I.: SciCorp: a corpus of English scientific articles annotated for information status analysis. In: LREC (2016)

24. Müller, C., Strube, M.: Multi-level annotation of linguistic data with MMAX2. In: Corpus Technology and Language Pedagogy: New Resources, New Tools, New Methods (2006)

25. Recasens, M., Martí, M.A., Orasan, C.: Annotating near-identity from coreference disagreements. In: LREC, pp. 165–172 (2012)

26. Schäfer, U., Spurk, C., Steffen, J.: A fully coreference-annotated corpus of scholarly papers from the ACL anthology. In: Proceedings of COLING 2012 Posters, pp. 1059–1070 (2012)

27. Carletta, J.: Assessing agreement on classification tasks: the kappa statistic. Comput. Linguist. 22(2), 249–254 (1996)

Evaluating Distributional Features for Multiword Expression Recognition

Natalia Loukachevitch[1,2(✉)] and Ekaterina Parkhomenko[1]

[1] Lomonosov Moscow State University, Moscow, Russia
louk_nat@mail.ru, parkat13@yandex.ru
[2] Tatarstan Academy of Sciences, Kazan, Russia

Abstract. In this paper we consider the task of extracting multiword expression for Russian thesaurus RuThes, which contains various types of phrases, including non-compositional phrases, multiword terms and their variants, light verb constructions, and others. We study several embedding-based features for phrases and their components and estimate their contribution to finding multiword expressions of different types comparing them with traditional association and context measures. We found that one of the distributional features has relatively high results of MWE extraction even when used alone. Different forms of its combination with other features (phrase frequency, association measures) improve both initial orderings.

Keywords: Thesaurus · Multiword expression · Embedding

1 Introduction

Automatic recognition of multiword expressions (MWE) having lexical, syntactic or semantic irregularity is important for many tasks of natural language processing, including syntactic and semantic analysis, machine translation, information retrieval, and many others. Various types of measures for MWE extraction have been proposed. These measures include word-association measures comparing frequencies of phrases and their component words, context-based features comparing frequencies of phrases and encompassing groups [14]. For multiword terms, such measures as frequencies in documents and a collection, contrast measures are additionally used [1].

But currently there are new possibilities of applying distributional and embedding-based approaches to MWE recognition. Distributional methods allow representing lexical units or MWE as vectors according to the contexts where the units are mentioned. Embedding methods use neural network approaches to improve vector representation of lexical units [13]. Therefore, it is possible to use embedding characteristics of phrases trying to recognize their irregularity, which makes it important to fix them in computational vocabularies or thesauri. The distributional features were mainly evaluated on specific types of

© Springer Nature Switzerland AG 2018
P. Sojka et al. (Eds.): TSD 2018, LNAI 11107, pp. 126–134, 2018.
https://doi.org/10.1007/978-3-030-00794-2_13

multiword expressions as non-compositional noun compounds [3] or verb-direct object groups [7,10], but they were not studied on a large thesaurus containing various types of multiword expressions.

In this paper we consider several measures for recognition of multiword expressions based on distributional similarity of phrases and component words. We compare distributional measures with association measures and context measures and estimate the contribution of distributional features in combinations with other measures. As a gold standard, we use RuThes thesaurus of the Russian language, which comprises a variety of multiword expressions. In the current study, only two-word phrases are considered.

2   Related Work

Distributional (embedding) features have been studied for recognizing several types of multiword expressions. Fazly et al. [7] studied verb-noun idiomatic constructions and used the combination of two features: lexical fixedness and syntactic fixedness. The lexical fixedness feature compares pointwise mutual information (PMI) for an initial construction and PMI of its variants obtained with substitution of component words to distributionally similar words.

In [17], the authors study the prediction of non-compositionality of multiword expressions comparing traditional distributional (count-based) approaches and word embeddings (prediction approaches). They test the proposed approaches using three specially prepared datasets of noun compounds and verb-particle constructions in two languages. All expressions are labeled on specialized scales from compositionality to non-compositionality. The authors [17] compared distributional vectors of a phrase and its components and found that the use of word2vec embeddings outperforms traditional distributional similarity with a substantial margin.

The authors of [3] study prediction of non-compositionality of noun compounds on four datasets for English and French. They compare results obtained with distributional semantic models and embedding models (word2vec, glove). They use these models to calculate vectors for compounds and their component words and order the compounds according to lesser similarity of compound vector and the sum of component vectors. They experimented with different parameters and found that the obtained results have high correlations with human judgements. All the mentioned works tested the impact of embedding-based measures on specially prepared data sets.

In [15], the authors study various approaches to recognition of Polish MWE using 46 thousand phrases introduced in plWordNet as a gold standard set. They utilized known word association measures described in [14] and proposed their own association measures. They also tested a measure estimating fixedness of word-order of a candidate phrase. To combine the proposed measures, weighted rankings were summed up. The weights were tuned on a separate corpus, the tuned linear combination of rankings was transferred to another test corpus and was still better than any single measure.

Authors of [12] study several types of measures and their combinations for term recognition using real thesauri as gold standards: EUROVOC for English and Banking thesaurus for Russian. They use 88 different features for extracting two-word terms. The types of the features include: frequency-based features; contrast features comparing frequencies in target and reference corpora; word-association measures; context-based features; features based on statistical topic modeling, and others. The authors studied the contribution of each group of features in the best integrated model and found that association measures do not have positive impact on term extraction for two-word terms.

Both works with real thesauri [12,15] did not experiment with distributional or embedding-based features.

3   Multiword Expressions in RuThes

The thesaurus of Russian language RuThes [11] is a linguistic ontology for natural language processing, i.e. an ontology, where the majority of concepts are introduced on the basis of actual language expressions. As a resource for natural language processing, RuThes is similar to WordNet thesaurus [8], but has some distinctions. One of significant distinctions important for the current study is that RuThes includes terms of so-called sociopolitical domain. The sociopolitical domain is a broad domain describing everyday life of modern society and uniting many professional domains, such as politics, law, economy, international relations, military affairs, arts and others. Terms of this domain are often met in news reports and newspaper articles, and therefore the thesaurus representation of such terms is important for effective processing of news flows [11]. Currently, RuThes contains almost 170 thousand Russian words and expressions.

As a lexical and terminological resource for automatic document processing, RuThes contains a variety of multiword expressions needed for better text analysis:

- traditional non-compositional expressions (idioms),
- constructions with light verbs and their nominalizations: помочь – оказать помощь – оказание помощи (*to help – to provide help – provision of help*),
- terms of the sociopolitical domain and their variants. According to terminological studies [4], domain-specific terms can have large number of variants in texts; these variants are useful to be included into the thesaurus to provide better term recognition: экономика – экономическая сфера – экономическая область (*economy – economic sphere – sphere of economy*),
- multiword expressions having thesaurus relations that do not follow from the component structure of the expression, for example, *traffic lights* [16] is a *road facility, food courts* consist of *restaurants* [6],
- geographical and some other names.

Recognition of such diverse multiword expressions requires application of noncompatible principles. Non-compositional expressions often do not have synonyms or variants (lexical fixedness according to [7]), but domain-specific terms often have variations useful to describe in the thesaurus.

Development of RuThes, introduction of words and expressions into the thesaurus, are based on expert and statistical analysis of the current Russian news flow (news reports, newspaper articles, analytical papers). Therefore we suppose that RuThes provides a good coverage for MWEs extracted from news collections and gives us possibility to evaluate different measures used for automatic recognition of MWE in texts.

4 Distributional Features

We consider three distributional features calculated using word2vec method [13].

The first feature (DFsum) is based on the assumption that non-compositional phrases can be distinguished with comparison of the phrase distributional vector and distributional vectors of its components: it was supposed that the similarity is less for non-compositional phrases [3,10]. For the phrases under consideration, we calculated cosine similarity between the phrase vector $v(w_1 w_2)$ and the sum of normalized vectors of phrase components $v(w_1 + w_2)$ according to formula from [3]:

$$v(w_1 + w_2) = \left(\frac{v(w_1)}{|v(w_1)|} + \frac{v(w_2)}{|v(w_2)|} \right)$$

The second feature (DFcomp) calculates the similarity of component words to each other. This means the similarity of contexts of component words. Examples of the thesaurus entries with high DFcomp include: симфонический оркестр (*symphony orchestra*), зерноуборочный комбайн (*combine harvester*), отрасль промышленности (*branch of industry*), тройская унция (*troy ounce*), etc. This measure is another form of calculating the association between words.

The third feature (DFsing) is calculated as the similarity between the phrase and the most similar single word; the word should be different from the phrase component words. The phrases were ordered according to decreasing similarity of DFsing. It was found that the most words in the top of the list (the most similar to phrases) are abbreviations (Table 1). It can be seen that some phrases have quite high similarity values with their abbreviated forms (more than 0.9).

5 Experiments

We used a Russian news collection (0.45 B tokens) and generated phrase and word embeddings with word2vec tool. In the current experiments, we used

Table 1. Searching for the most similar word to a phrase

MWE	Single Word	translation	Score	Type
детский сад	детсад	kindergarten	0.961	abbr.
Европейский союз	евросоюз	European Union	0.942	abbr.
атомная электростанция	АЭС	nuclear station	0.933	abbr.
атомная станция	АЭС	nuclear station	0.925	abbr.
генеральная прокуратура	генпрокуратура	prosecution office	0.923	abbr.
районный суд	райсуд	district court	0.923	abbr.
государственный бюджет	госбюджет	state budget	0.917	abbr.
следственный изолятор	сизо	detention center	0.917	abbr.
Государственная дума	Госдума	State duma	0.914	abbr.
федеральный закон	ФЗ	federal law	0.911	abbr.

default parameters of the word2vec package, but after the analysis of the results we do not think that the conclusions can be significantly changed.

We extracted two-word noun phrases: adjective + noun and noun + noun in Genitive with frequencies equal or more than 200 to have enough statistical data. From the obtained list, we removed all phrases containing known personal names. We obtained 37,768 phrases. Among them 9,838 are thesaurus phrases, and the remaining phrases are not included into the thesaurus. For each measure, we create a ranked list according to this measure. At the top of the list, there should be multiword expressions, at the end of the list there should be free, compositional, non-terminological phrases.

We generated ranked lists for the following known association measures: pointwise mutual information (PMI), its variants (cubic MI, normalized PMI, augmented MI, true MI), Log-likelihood ratio, t-score, chi-square, Dice and modified Dice measures [12,14]. Some used association measures presuppose the importance of the phrase frequency in a text collection for MWE recognition and enhance its contribution to the basic measure. For example, Cubic MI (1) includes the cubed phrase frequency if compared to PMI, and True MI (2) utilizes phrase frequency without logarithm.

$$CubicMI(w_1, w_2) = log(\frac{freq(w_1, w_2)^3 \cdot N}{freq(w_1) \cdot freq(w_2)}) \qquad (1)$$

$$TrueMI(w_1, w_2) = freq(w_1, w_2) \cdot log(\frac{freq(w_1, w_2) \cdot N}{freq(w_1) \cdot freq(w_2)}) \qquad (2)$$

Modified Dice (4) measure also enhances the contribution of the phrase frequency in Dice measure (3).

$$Dice(w_1, w_2) = \frac{2 \cdot freq(w_1, w_2)}{freq(w_1) + freq(w_2)} \qquad (3)$$

$$ModifiedDice(w_1, w_2) = log(freq(w_1, w_2)) \cdot \frac{2 \cdot freq(w_1, w_2)}{freq(w_1) + freq(w_2)} \qquad (4)$$

Table 2. Average precision measure at 100, 1000 thesaurus phrases and for the full list

Measure	AvP (100)	AvP (1000)	AvP (Full)
Frequency	0.73	0.70	0.43
PMI and modifications			
PMI	0.52	0.54	0.44
CubicMI	**0.91**	**0.80**	**0.52**
NPMI	0.64	0.65	0.47
TrueMI	0.77	0.77	0.50
AugmentedMI	0.55	0.59	0.45
Other association measures			
LLR	0.78	0.78	0.51
T-score	0.73	0.71	0.46
Chi-Square	0.68	0.69	0.50
DC	0.68	0.67	0.48
ModifiedDC	0.81	0.71	0.49
C-value	0.73	0.70	0.43
Distributional features			
DFsum	0.20	0.19	0.24
DFcomp	0.47	0.42	0.35
DFsing	0.85	0.69	0.42

Also we calculated c-value measure, which is used for extraction of domain-specific terms [9]. To evaluate the list rankings, we used uninterpolated average precision measure (AvP), which achieves the maximal value (1) if all multiword expressions are located in the beginning of a list without any interruptions [14].

The Table 2 shows the AvP values at the level of the 100 first thesaurus phrases (AvP (100)), 1000 first thesaurus phrases (AvP (1000)) and for the full list for all mentioned measures and features. It can be seen that the results of PMI is lowest in comparison to all association measures. This is due to extraction of some specific names or repeated mistakes in texts (i.e. words without spaces between them). Even the high frequency threshold preserves this known problem of PMI-based MWE extraction. Normalized PMI extracts MWE much better as it was indicated in [2]. Modified Dice measure gives better results in comparison with initial Dice measure. The best results among all measures belong to the Cubic MI measure proposed in [5].

It is important to note that there are some evident non-compositional phrases that are located in the end of the list according to any association measures. This is due to the fact that both words are very frequent in the collection under consideration but the phrase is not very frequent. Examples of such phrases include: игра слов (*word play*), человек года (*person of the year*), план счетов (*chart of accounts*), государственная машина (*state machinery*), and others.

For all lists of association measures, the above-mentioned phrases were located in the last thousand of the lists. According to the distributional feature DFsum, these phrases significantly shifted to the top of the list. There positions became as follows: игра слов (561), человек года (1346), план счетов (1545), государственная машина (992). Thus, it could seem that this feature can generate a qualitative ranked list of phrases. But the overall quality of ordering for the DFsum distributional feature is rather small (Table 2), when we work with candidates extracted from a raw corpus, which is different from a prepared list of compositional and noncompositional phrases as it was described in [3,10].

On the other hand, we can see that another distributional feature – maximal similarity with a single word (DFsing) showed a quite impressive result, which is the second one at the first 100 thesaurus phrases. As it was indicated earlier, the first 100 thesaurus phrases are most similar to their abbreviated forms, what means that for important concepts, reduced forms of their expression are often introduced and utilized.

As it was shown for association measures, additional accounting of phrase frequency can improve a basic feature. In Table 3 we show results of multiplying initial distributional features to phrase frequency or its logarithm. It can be seen that DFsing more improved when multiplied by log (Freq). DFcomp multiplied by the phrase frequency became better then initial frequency ordering (Table 2).

Table 3. Combining distributional features with the phrase frequency

Measure	AvP (100)	AvP (1000)	AvP (Full)
DF multiplied by frequency			
DFsum	0.70	0.69	0.43
DFcomp	0.75	0.72	0.46
DFsing	0.76	0.73	0.46
DF multiplied by log (Frequency)			
DFsum	0.57	0.56	0.36
DFcomp	0.70	0.57	0.39
DFsing	**0.93**	**0.83**	**0.50**

Then we try to combine the best distributional DFsing measure with association measures using two ways: (1) multiplying values of initial measures, (2) summing up ranks of phrases in initial rankings. In both cases, AvP of the initial association measures significantly improved. For Cubic MI, the best result was based on values multiplying. For LLR and True MI, the best results were achieved with summing up ranks. In any case, it seems that the distributional similarity of a phrase with a single word (different from its component) bears important information about MWEs (Table 4).

Table 4. Combining DFsing with traditional association measures

Measure	Multiplying values			Summing up rankings		
	AvP (100)	AvP (1000)	AvP (Full)	AvP (100)	AvP (1000)	AvP (Full)
DFsing * CubicPMI	**0.95**	**0.84**	**0.54**	0.94	0.83	0.53
DFsing * PMI	0.62	0.62	0.47	0.62	0.64	0.48
DFsing * NPMI	0.78	0.73	0.50	0.79	0.73	0.50
DFsing * augMI	0.64	0.65	0.48	0.69	0.67	0.48
DFsing * TrueMI	0.80	0.80	0.52	**0.96**	**0.86**	0.53
DFsing * Chi-Square	0.73	0.71	0.50	0.85	0.77	0.51
DFsing * LLR	0.82	0.81	0.53	**0.96**	**0.86**	0.53
DFsing * DC	0.74	0.70	0.49	0.84	0.77	0.50
DFsing * ModifiedDC	0.85	0.73	0.50	0.88	0.79	0.51
DFsing * T-score	0.80	0.78	0.49	0.95	0.84	0.51
DFsing * C-value	0.76	0.74	0.46	0.95	0.83	0.49

6   Conclusion

In this paper we considered the task of extracting multiword expression for Russian thesaurus RuThes, which contains various types of phrases, including non-compositional phrases, multiword terms and their variants, light verb construction, and others. We studied several embedding-based features for phrases and their components and estimated their contribution to finding multiword expressions of different types comparing them with traditional association and context measures.

We found that one of the most discussed distributional features, which compares the vector of MWE with the sum of vectors of component words (DFsum), provides low quality of a ranked MWE list when extracted from a raw corpus. But another distributional feature (similarity of the phrase vector with the vector of a single word) has relatively high results of MWE extracting. Different forms of its combination with other features (phrase frequency, association measures) achieve the best results.

Acknowledgments. This work was partially supported by Russian Science Foundation, grant N16-18-02074.

References

1. Astrakhantsev, N.: ATR4S: toolkit with state-of-the-art automatic terms recognition methods in Scala. Lang. Resour. Eval. **52**, 853–872 (2018)
2. Bouma, G.: Normalized (pointwise) mutual information in collocation extraction. In: Proceedings of GSCL, pp. 31–40 (2009)
3. Cordeiro, S., Ramisch, C., Idiart, M., Villavicencio, A.: Predicting the compositionality of nominal compounds: giving word embeddings a hard time. In: Proceedings of ACL-2016g Papers, vol. 1, pp. 1986–1997 (2016)

4. Daille, B.: Term Variation in Specialised Corpora: Characterisation, Automatic Discovery and Applications, vol. 19. John Benjamins Publishing Company, Amsterdam (2017)

5. Daille, B.: Combined approach for terminology extraction: lexical statistics and linguistic filtering. Ph.D. thesis. University Paris 7 (1994)

6. Farahmand, M., Smith, A., Nivre, J.: A multiword expression data set: annotating non-compositionality and conventionalization for English noun compounds. In: Proceedings of the 11th Workshop on Multiword Expressions, pp. 29–33 (2015)

7. Fazly, A., Cook, P., Stevenson, S.: Unsupervised type and token identification of idiomatic expressions. Comput. Linguist. **35**(1), 61–103 (2009)

8. Fellbaum, C.: WordNet. Wiley Online Library (1998)

9. Frantzi, K., Ananiadou, S., Mima, H.: Automatic recognition of multi-word terms: the C-value/NC-value method. Int. J. Digit. Libr. **3**(2), 115–130 (2000)

10. Gharbieh, W., Bhavsar, V.C., Cook, P.: A word embedding approach to identifying verb-noun idiomatic combinations, pp. 112–118 (2016)

11. Loukachevitch, N., Dobrov, B.: RuThes linguistic ontology vs. Russian wordnets. In: Proceedings of the Seventh Global Wordnet Conference, pp. 154–162 (2014)

12. Loukachevitch, N., Nokel, M.: An experimental study of term extraction for real information-retrieval thesauri. In: Proceedings of TIA-2013, pp. 69–76 (2013)

13. Mikolov, T., Chen, K., Corrado, G., Dean, J.: Efficient estimation of word representations in vector space. arXiv preprint arXiv:1301.3781 (2013)

14. Pecina, P.: Lexical association measures and collocation extraction. Lang. Resour. Eval. **44**(1–2), 137–158 (2010)

15. Piasecki, M., Wendelberger, M., Maziarz, M.: Extraction of the multi-word lexical units in the perspective of the wordnet expansion. In: RANLP-2015, pp. 512–520 (2015)

16. Sag, I.A., Baldwin, T., Bond, F., Copestake, A., Flickinger, D.: Multiword expressions: a pain in the neck for NLP. In: Gelbukh, A. (ed.) CICLing 2002. LNCS, vol. 2276, pp. 1–15. Springer, Heidelberg (2002). https://doi.org/10.1007/3-540-45715-1_1

17. Salehi, B., Cook, P., Baldwin, T.: A word embedding approach to predicting the compositionality of multiword expressions. In: Proceedings of NAACL-2015, pp. 977–983 (2015)

MANÓCSKA: A Unified Verb Frame Database for Hungarian

Ágnes Kalivoda[1,2,3], Noémi Vadász[1,3], and Balázs Indig[1,2(✉)]

[1] Pázmány Péter Catholic University, Budapest, Hungary
{kalivoda.agnes,vadasz.noemi,indig.balazs}@itk.ppke.hu
[2] MTA–PPKE Hungarian Language Technology Research Group, Budapest, Hungary
[3] Research Institute for Linguistics, Hungarian Academy of Sciences, Budapest, Hungary

Abstract. This paper presents MANÓCSKA, a verb frame database for Hungarian. It is called *unified* as it was built by merging all available verb frame resources. To be able to merge these, we had to cope with their structural and conceptual differences. After that, we transformed them into two easy to use formats: a TSV and an XML file. MANÓCSKA is *open-access*, the whole resource and the scripts which were used to create it are available in a github repository. This makes MANÓCSKA reproducible and easy to access, version, fix and develop in the future. During the merging process, several errors came into sight. These were corrected as systematically as possible. Thus, by integrating and harmonizing the resources, we produced a Hungarian verb frame database of a higher quality.

Keywords: Verb frame database · Lexical resource
Corpus linguistics · Hungarian

1 Introduction

Finding and connecting the arguments and adjuncts to the verb in a sentence is a trivial step for humans during sentence comprehension. For a parser, this task can only be solved using a verb frame database (in other terms, a valency dictionary). Because of their essential role in everyday NLP tasks, numerous lexical resources of this kind have been created, such as VerbNet [11] and FrameNets for several languages [1].

A couple of verb frame databases have been developed for Hungarian as well. However, each one has some weaknesses, first of all, they are not complete and precise enough. Our database, MANÓCSKA[1] is constructed using these already existing verb frame resources, aiming to harmonize them by merging them into a clearly structured, easy-to-use format.

---

[1] The resource and a detailed description of its structure can be found at
https://github.com/ppke-nlpg/manocska.

© Springer Nature Switzerland AG 2018
P. Sojka et al. (Eds.): TSD 2018, LNAI 11107, pp. 135–143, 2018.
https://doi.org/10.1007/978-3-030-00794-2_14

To gain a better understanding of the issues presented in the following sections, let us sketch some important properties of the target language. Hungarian is an agglutinative language, meaning that most of grammatical functions are marked with affixes (e.g. nouns can be declined with 18 case suffixes). In this way, Hungarian sentences have a relatively free word order. Furthermore, Hungarian is a pro-drop language: several components of a sentence can be omitted if they are grammatically or pragmatically inferable. This makes the corpus-driven analysis of verb valencies quite difficult. Finally, a considerable issue is raised by verbal particles (in other terms, preverbs). These are usually short words (like the ones in phrasal verbs of Germanic languages) which often change the meaning and the valency of their base verbs. By default, the verbal particle is written together with the verb as its prefix. In a lot of contexts, however, it can be detached from the verb and moved to a distant position. This can happen not only in the case of finite particle verbs, but also by infinitives and participles placed in the same clause. Thus, connecting the verbal particles to their base verbs during the parsing process is a task far from trivial.

During our work, we discovered several weaknesses of the original verb frame resources. Some of the errors could be corrected automatically, but most of them had to be corrected manually. This was done by writing the erroneous version and its correction into a separate file as a key–value pair. Thus, our manipulations did not affect the original resources and MANÓCSKA remained reproducible. Moreover, the merging process surfaced some theoretical controversies which are worth considering in the future.

The paper is structured as follows. After giving a brief overview about the resources, we discuss the main issues experienced during the merging process. This is followed by presenting the structure of MANÓCSKA. After that, we sketch the most important theoretical implications. Our conclusions close the paper.

2   Resources

MANÓCSKA contains six language resources, thus it covers all existing verb frame databases for Hungarian, even those which were previously not accessible freely in a database format. Five of them were built upon corpus data (see Table 1). It must be noted that there are considerable conceptual differences between the resources, e.g. regarding the set of verbal particles (see Sect. 3) or the distinction between arguments and adjuncts (which can be found only in METAMORPHO). We provide a short description about every resource in this section, recognizing their strengths and pointing out their weaknesses.

The name MAZSOLA refers to two versions of a verb frame database created by Bálint Sass as a part of his PhD dissertation about retrieving verb frames from corpus data [8]. The first version is a paper dictionary of the most frequent arguments and phrases occurring with the verb (Hungarian Verbal Structures) [10] which was produced automatically – using very simple heuristics to prefer the precision over recall –, based on the HNC corpus [12]. The content of the dictionary was manually corrected, but until now it was available only in paper format.

Table 1. Corpora used by the corpus-driven resources (third column) which are merged into MANÓCSKA. Their sizes are given in tokens, including punctuation marks.

Name of the corpus and its abbreviation	Size (tokens)	Resource using the corpus
Hungarian National Corpus (HNC)	187 600 000	MAZSOLA (2 versions)
Hungarian Webcorpus (Webcorpus)	589 000 000	TÁDÉ
Hungarian Gigaword Corpus (HGC) v.2.0.3	978 000 000	Particle verbs
Hungarian Gigaword Corpus (HGC) v.2.0.4	1 348 000 000	Infinitival constructions

The second version is larger, however, it is not reviewed. It is available online[2] (after a free registration) and contains 28 million syntactically parsed sentences and half a million verbal structures [9]. Although several years have passed since the creation of these resources, no experiment was conducted to compare the two collections, neither in terms of usability nor of experimenting on other, larger corpora with the state-of-the-art tools and automating the correcting process.

The next resource, TÁDÉ[3] is a frequency list of Hungarian verb frames created by spectral clustering [2], but in an unsupervised manner where the frames and their clustering are induced in the same pass [6]. The novelty of the approach lies in the sensitive thresholding technique which yields robust results and enables the inclusion of a broader class of frames which were not considered in the earlier works. The frames were extracted from the Webcorpus [3]. No language-centric tools were used during the creation of this resource, so it has many trivially correctable errors.

There are some notable differences between MAZSOLA and TÁDÉ[4]. In the case of MAZSOLA, accuracy was in focus, in contrast with the pursuit to higher F-measure – and consequently higher recall – which can be seen by TÁDÉ. Due to its higher precision, MAZSOLA is basically more suitable for everyday NLP tasks. It contains also the frequent lexical arguments of verbs which can not be found in TÁDÉ. However, it must also be noted that MAZSOLA does not contain any infinitival arguments (neither versions), whereas TÁDÉ does.

Beside TÁDÉ, we used two frequency lists which were created by corpus-driven method. The first of them contains 27 091 particle verbs [5] extracted from HGC v2.0.3 [7]. It was checked and corrected manually, aiming for high precision. It does not contain any information about the verb frames, but it

---

[2] http://corpus.nytud.hu/isz/.

[3] https://hlt.bme.hu/hu/resources/tade.

[4] Mazsola and Tádé are two puppets from a Hungarian puppet animated film which was popular in the early 1970s. The eponym of our database, Manócska is also a puppet from this film.

has a good coverage of the possible combinations of verbs and their particles including their joint frequency. The second list contains finite verbs which may have infinitives as their arguments[5]. It was extracted from HGC v2.0.4. It does not enumerate all infinitive arguments for each verb lexically (in contrast with TÁDÉ). Its only goal is to list verb and particle pairs that can have an infinitive as argument.

Last but not least, we included the verb frame database of METAMORPHO, a rule-based commercial machine translation system for Hungarian. This database was created by linguistic experts who aimed to describe Hungarian verb frame constructions in a granularity which was needed for the unambiguous translation to English. Thus, these rules have numerous lexical, syntactic and semantic constraints in order to explicitly isolate the verb senses. The creators used corpora to check their linguistic intuition, however, the database does not contain statistical frequencies. In this way, all rules appear as if they would have the same importance.

The aforementioned resources have different sizes and they are based on different sized corpora. The verb-related properties of the merged resource MANÓCSKA can be seen in Table 2. More than two-thirds of all verbs (33 937 out of 44 183) are present only in one or two used resources which makes the recall of MANÓCSKA really high.

Table 2. The number of frames, different verb lemmata and erroneous verbforms found in the resources. The size of MANÓCSKA is marked with **boldface**.

Resource	Frames	Verbs	Errors
MAZSOLA (dictionary)	6 203	2 185	47
MAZSOLA (database)	524 267	9 589	477
TÁDÉ	521 567	27 159	4 489
Particle verbs	0	27 091	0
Infinitival constructions	0	1 507	0
METAMORPHO	35 967	13 772	0
MANÓCSKA	**971 384**	**44 183**	**0**

3 Emerging Issues

In order to be able to merge the resources, we had to harmonize them. We assumed that the weaknesses of the databases will be corrected by the strengths of others. For instance, if a frame has high frequency in multiple independent databases, it can be safely considered a valid frame, while a frame which can be found only in one database with low frequency might be wrong or unimportant.

---

[5] https://github.com/kagnes/infinitival_constructions.

By harmonization we also mean that the different structures and linguistic formalisms of the resources had to be converted into a standard format. During this process, several issues came to light.

Firstly, we had to cope with practical issues, e.g. the undocumented feature set used in METAMORPHO or the numerous verbal particle–verb mismatches (this is caused by the nature of Hungarian verbal particles, see Sect. 1). We could tackle these using ruled-based methods and manual corrections.

Secondly, we faced some more severe issues which have theoretical background. An interesting example is the fuzzy boundary between the verb modifiers and one of their subclasses, the verbal particles. In MANÓCSKA, the latter ones are separated from the verb with a pipe (because – by default – they are written together with the verb). The former ones are handled as lexical arguments, thus they have 'lemma with case marking' form. For example, in the case of *szörnyet|hal, szörnyet* (lit. 'monster.ACC') is defined as a verbal particle, while in *hal szörny[ACC]*, it is rather a lexical argument (both constructions mean 'to die on the spot').

MANÓCSKA contains 118 entries where a word is handled as verbal particle and as a lexical argument, respectively. There are altogether 33 words which are ambiguous from this point of view. In order to have a better understanding of these words' behaviour, we conducted a case study using HGC v.2.0.4. We looked for clauses (1) where the given word was in −1 position compared to the verb (immediately before it, but separated by a space) and (2) where it was in 0 position (written together with the verb). Orthography, of course, can not lead us to incontestable statements. However, it can show us the native speakers' intuition concerning these ambiguous words. If the word has −1 position, the writer of the clause handled it rather as a lexical argument, while 0 position indicates that it is handled as a verbal particle. Table 3 presents five cases where the orthographical uncertainty is remarkable.

4   The Structure of MANÓCSKA

MANÓCSKA is available in two formats: a TSV and an XML file. In the TSV, no distinction is made between arguments and adjuncts, as it does not contain all information that can be found in the METAMORPHO database, and the other five resources do not have this type of information. The TSV is easily parsable. Every row corresponds to one entry. The first column contains the verb lemma (the verbal particle is separated by a | character). The second column shows the verb frame which is represented by case-endings (e.g. 'with something' equals [INS], a word in instrumentalis). Columns 3–8 contain the frequencies of the verb frame in the six different resources. In the last column, a rank value can be seen which allows a cross-resource comparison of the given record's frequency[6].

---

[6] The rank value is computed by dividing the actual frame frequency of the given record and the summarized frame frequency for each resource, and finally by summarizing the divisions' results.

Table 3. Five cases where there is no consensus regarding the category of the ambiguous word. The fourth column (−1) stands for the joint frequency of the given word and the verb, counting the cases when the given word is written separately from the verb. The fifth column (0) shows the number of cases when the given word is written together with the verb.

Ambiguous word	Verb	Meaning of the construction	−1	0
síkra 'plain.SUB'	*száll* 'to fly'	to come out in support of sy	423	320
nagyot 'big.ACC'	*hall* 'to hear'	to be hard of hearing	76	107
cserben 'tan_pickle.INE'	*hagy* 'to leave'	to let sy down	986	1 818
helyben 'place.INE'	*hagy* 'to leave'	to approve smth	986	2 132
véghez 'end.ALL'	*visz* 'to take'	to accomplish smth	1 260	3 054

The XML-format (presented on Fig. 1) contains all the six resources, including every fine-grained feature available in the METAMORPHO database (e.g. distinction between arguments and adjuncts – the latter marked with COMPL, information about the valencies' theta roles and semantic constraints like *animate* or *bodypart*). We handle the base verbs as the main elements. Each verb entry (VERB) is split into two optional subentries based on whether there is a verbal particle (PREV) or not (NO PREV). Furthermore, each entry is subdivided depending on the possibility of an infinitival argument (INF, NO INF). We chose these two as primary features, because recent research proved that these are essential features for real-life verb frame disambiguation in the case of Hungarian [4].

The possible verb frames are collected within the FRAMES tag. Each frame can have meta attributes, e.g. a reference to its ID in the original resource. The frames are presented as lists of arguments (subject, object, obliquus) and adjuncts (both types within the ARG tag). Each of these must have a grammatical case or a postposition. Beside that, they may have extra constraints, e.g. some features which help to disambiguate the frames. We treat each feature as a key–value pair chosen from a predefined domain, presented as the attribute of the given ARG tag.

The frame frequencies coming from the different resources are attributes of the FREQS tag (as key–value pairs, with the key being the name of the resource). This formalism enables the user to easily add other resources in the future, including their own frequencies. The easily extendable, filterable, transformable

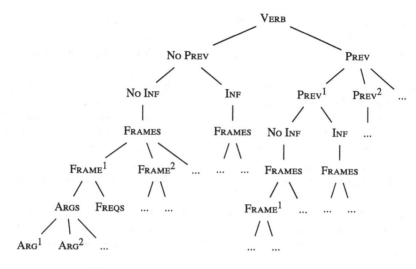

Fig. 1. The basic structure of the XML format.

form in conjunction with the GIT based public versioning and the availability of the production scripts[7] make MANÓCSKA a unique, open-access resource.

5 Theoretical Implications

To demonstrate the applicability of our resource, we created a custom naïve 'clustering' of the entries by different features, as we faced that no matter how we order the features in the XML-tree, there will always be many subtrees that are equivalent. We wanted to eliminate these duplicated subtrees and compress the database. This experiment revealed some nice patterns among the frames.

We eliminated all constraints from the arguments except their grammatical cases to achieve higher density. In this reduced "framebank", we looked for duplicate subtrees. Our search was not performed on the frame level, but rather on the level of the different verb–frame, verb–particle–frame combinations. We managed to gather many rather frequent groups of frames that can be paired with the verb or particle they occur with in any desired combination.

We argue that the essence of productivity can be revealed by recurring groups of frames. In a lot of cases, the verb itself can be substituted with several semantically related words, but interestingly, its frames can not vary so freely. This phenomenon becomes even more apparent if the verb has a particle which inherently encodes directionality and demands an argument which agrees with it in its grammatical case. In such structures, the verb seems to have very little syntactic, but rather semantic power in the predicate. For instance, the scheme '*be* (lit. in.ILL) + verb + smth.ACC smth.INS' mostly matches frames where the

---

[7] Due to licence reasons, the original resources could not be included but they can be asked for by the original copyright holders at the given addresses.

verb comes from a semantically related class of words having the core meaning 'to cover something with something' (e.g. *befed* 'to cover', *bearanyoz* 'to gild', *bedörzsöl* 'to rub in', *bepiszkít* 'to dirty', *besugároz* 'to irradiate', *beterít* 'to spread').

Another interesting phenomenon comes to light when we look at particle verbs having infinitival arguments. If we know that the particle has inherent directional meaning (e.g. *ki* 'out', *be* 'in', *el* 'away'), we can be almost certain that the verb is a verb of motion. There are only a few exceptions having abstract meaning: *el|felejt* 'to forget smth', *el|kezd* 'to begin smth', *ki|felejt* 'to leave out smth (by mistake)', *ki|próbál* 'to try out smth'. However, if we do not have any information about the particle, the chance that the given verb is semantically a verb of motion is only 38% (88 out of 232 verbs).

With the distributive inspection presented above, we can discover the real inner-workings of the verb frames including numerous examples which came from linguistic intuition and introspection along with the ones that maybe slipped our mind.

6   Conclusion

MANÓCSKA is a valuable, open-access database of Hungarian verb frames. Its XML format makes it possible to handle several built-in resources uniformly, but it is also possible to extract a single resource or a reduced feature set from the XML, if this is preferred for a specific task as demonstrated in Sect. 5.

This database is one step closer to be suitable for a lexical resource of a parser, helping it to connect the arguments to the verb in the right way. Beside everyday NLP tasks, it can be used for linguistic research as well. Due to its reproducibility, MANÓCSKA can be improved constantly by correcting previously unnoticed errors or by adding new resources.

References

1. Baker, C.F., Fillmore, C.J., Lowe, J.B.: The Berkeley FrameNet Project. In: Proceedings of the 36th Annual Meeting of the Association for Computational Linguistics and 17th International Conference on Computational Linguistics, ACL 1998, vol. 1, pp. 86–90. Association for Computational Linguistics, Stroudsburg (1998). https://doi.org/10.3115/980845.980860
2. Brew, C., Schulte im Walde, S.: Spectral clustering for German verbs. In: Proceedings of the ACL-02 Conference on Empirical Methods in Natural Language Processing, EMNLP 2002, - vol. 10, pp. 117–124. Association for Computational Linguistics, Stroudsburg (2002). https://doi.org/10.3115/1118693.1118709
3. Halácsy, P., Kornai, A., Németh, L., Rung, A., Szakadát, I., Trón, V.: Creating open language resources for Hungarian. In: Calzolari, N. (ed.) Proceedings of the 4th International Conference on Language Resources and Evaluation (LREC 2004), pp. 203–210 (2004)

4. Indig, B., Vadász, N.: Windows in Human Parsing – How Far can a Preverb Go? In: Tadić, M., Bekavac, B. (eds.) Proceedings of the Tenth International Conference on Natural Language Processing (HrTAL2016) 2016, Dubrovnik, Croatia, 29–30 September 2016. Springer, Cham (2016). (accepted, in press)
5. Kalivoda, Á.: A magyar igei komplexumok vizsgálata [The Hungarian Verbal Complexes]. Master's thesis, PPKE-BTK (2016). https://github.com/kagnes/hungarian_verbal_complex
6. Kornai, A., Nemeskey, D.M., Recski, G.: Detecting Optional Arguments of Verbs. In: Calzolari, N., et al. (eds.) Proceedings of the Tenth International Conference on Language Resources and Evaluation (LREC 2016). European Language Resources Association (ELRA) (2016)
7. Oravecz, C., Váradi, T., Sass, B.: The Hungarian Gigaword Corpus. In: Calzolari, N., et al. (eds.) Proceedings of the Ninth International Conference on Language Resources and Evaluation (LREC 2014). European Language Resources Association (ELRA) (2014)
8. Sass, B.: Igei szerkezetek gyakorisági szótára - Egy automatikus lexikai kinyerő eljárás és alkalmazása [A Frequency Dictionary of Verbal Structures - An Automatic Lexical Extraction Procedure and its Application]. Ph.D. thesis, Pázmány Péter Katolikus Egyetem ITK (2011)
9. Sass, B.: 28 millió szintaktikailag elemzett mondat és 500 000 igei szerkezet [28 Million Syntactically Parsed Sentences and 500 000 Verbal Structures]. In: Tanács, A., Varga, V., Vincze, V. (eds.) XI. Magyar Számítógépes Nyelvészeti Konferencia (MSZNY 2015) [XI. Hungarian Conference on Computational Linguistics], pp. 399–403. SZTE TTIK Informatikai Tanszékcsoport, Szeged (2015)
10. Sass, B., Váradi, T., Pajzs, J., Kiss, M.: Magyar igei szerkezetek - A leggyakoribb vonzatok és szókapcsolatok szótára [Hungarian Verbal Structures - The Dictionary of the Most Frequent Arguments and Phrases]. Tinta Könyvkiadó, Budapest (2010)
11. Schuler, K.K.: VerbNet: A broad-coverage, comprehensive verb lexicon. Ph.D. thesis, University of Pennsylvania (2006). http://verbs.colorado.edu/~kipper/Papers/dissertation.pdf
12. Váradi, T.: The Hungarian National Corpus. In: Proceedings of the Third International Conference on Language Resources and Evaluation (LREC-2002), pp. 385–389. European Language Resources Association, Paris (2002)

Improving Part-of-Speech Tagging by Meta-learning

Łukasz Kobyliński$^{(\boxtimes)}$, Michał Wasiluk, and Grzegorz Wojdyga

Institute of Computer Science, Polish Academy of Sciences, Jana Kazimierza 5,
01-248 Warszawa, Poland
lkobylinski@gmail.com, m.wasiluk89@gmail.com, g.wojdyga@gmail.com

Abstract. Recently, we have observed a rapid progress in the state of Part of Speech tagging for Polish. Thanks to PolEval—a shared task organized in late 2017—many new approaches to this problem have been proposed. New deep learning paradigms have helped to narrow the gap between the accuracy of POS tagging methods for Polish and for English. Still, the number of errors made by the taggers on large corpora is very high, as even the currently best performing tagger reaches an accuracy of ca. 94.5%, which translates to millions of errors in a billion-word corpus.

To further improve the accuracy of Polish POS tagging we propose to employ a meta-learning approach on top of several existing taggers. This meta-learning approach is inspired by the fact that the taggers, while often similar in terms of accuracy, make different errors, which leads to a conclusion that some of the methods are better in specific contexts than the others. We thus train a machine learning method that captures the relationship between a particular tagger accuracy and language context and in this way create a model, which makes a selection between several taggers in each context to maximize the expected tagging accuracy.

Keywords: Part-of-speech tagging · Meta learning
Natural language processing

1 Introduction

Part of speech tagging is a difficult task in case of inflected languages. While, in case of English, the accuracy of taggers exceeds 97%, taggers for Polish have only recently reached the level of 94%. There are several reasons behind this discrepancy, one of them being an objective difference in problem difficulty, as the tagset size (the number of possible POS tags) is at least an order of magnitude larger in case of Polish (and other Slavic languages) than for English. High level of inflection translates to a much higher number of possible word forms that appear in text corpora. This data sparsity adds to the difficulty of using machine learning approaches to train models on hand-annotated data.

We are thus faced with a problem of creating a method of morphological disambiguation, where the available training data is very limited and ambiguity

© Springer Nature Switzerland AG 2018
P. Sojka et al. (Eds.): TSD 2018, LNAI 11107, pp. 144–152, 2018.
https://doi.org/10.1007/978-3-030-00794-2_15

is omnipresent. An added layer of difficulty comes from the fact that the language has a completely free word order, so fixed-context systems (like HMMs) are not as effective as for English. There are also many cases of segmentation ambiguities, which make it difficult to even separate individual tokens in text.

Several taggers have been proposed for Polish to date. The first group of taggers are currently obsolete, as they were tied to a morphosyntactic tagset, which is not used in modern corpora. Taggers in this group include the first tagger for Polish, proposed by Dębowski [2] and TaKIPI tagger, described in [7].

The National Corpus of Polish [8], which was released in 2011, introduced a new version of the tagset and several taggers using this tagset and evaluated on the NCP have been proposed since then. The authors of the taggers have experimented with a variety of machine learning approaches trying to reach tagging accuracy comparable to that reported for English. The taggers from this group include: Pantera [1] (an adaptation of the Brill's algorithm to morphologically rich languages), WMBT [11] (a memory based tagger), WCRFT [9] (a tagger based on Conditional Random Fields) and Concraft [13] (another approach to adaptation of CRFs to the problem of POS tagging). Evaluation of performance of a combination of these taggers has been presented in [4].

As the accuracy reached by the taggers was still not satisfactory, a shared task on morphosyntactic tagging was organized during the PolEval workshop, which took place in 2017. Poleval has attracted 16 submissions from 9 teams in total and resulted in open-sourcing several new taggers for Polish. Interestingly, all the submissions were based on neural networks. The winner of the shared-task, Toygger [5], performs morphological disambiguation by a bi-directional recurrent LSTM neural network (2 bi-LSTM layers). KRNNT [14], the runner-up, uses bidirectional recurrent neural networks (i.e. Gated Recurrent Units) for morphological tagging. The two approaches differ mainly in the choice of features: Toygger uses word2vec word embeddings, while KRNNT does not.

The winners of the PolEval workshop reached a new milestone in tagging accuracy (ca. 93–94% vs 91% in the case of previously best CRF-based taggers), but this improvement is still below the levels reported for English and 6% error rate translates to a very high number of errors while tagging billion-word corpora.

Following the encouraging results reported in [4] we would like to build on the fact that a considerable number of different approaches for tagging Polish have been proposed to date and further improve the accuracy by using meta-level machine learning to build an ensemble of methods, which performs better than any of the individual taggers in isolation.

2   Method Description

The main idea behind building an ensemble of taggers is that each of the individual taggers may be treated as a classifier, which has its distinct error profile, partly overlapping other classifiers. As the taggers make different errors in different contexts, it is possible to improve the accuracy of the ensemble over individual taggers by providing a function that selects one of the taggers based on the current context and the selected tagger is the one that provides the output of the entire ensemble.

The simple idea of performing voting and selecting a classification (morphosyntactic tag), which gained the most votes has been tested in [4] and proved to improve tagging accuracy by ca. 1% point.

In this paper we train a meta-classifier, which builds a machine learning model based on training data and is then used during tagging to select the most accurate tagger for each context.

Our meta-tagger implementation is based on WSDDE [6], a platform that simplifies feature extraction from text and evaluating machine learning approaches. The WSDDE platform is integrated with WEKA [3], which provides several implementations of classification and attribute selection methods.

We have enhanced the WSDDE platform by:

- adding support for analyzing an ensemble of classifiers,
- adding additional feature generators (prefix/suffix, word shape, classifier agreement, etc.)

2.1   Training the Meta-classifier

Training Data. To perform a fair evaluation of the meta-learning approach against individual taggers the available training data needs to be divided into two parts:

- A—training set for component taggers,
- B—training set for the meta-classifier.

The training set A must be large enough for effective component taggers training, but on the other hand training set B can't be too small either—it should contain possibly large number of tagging disagreements between taggers. We have addressed the problem of the effective training data split in experiments described further on.

Training Procedure. In the first step of the training process each component tagger is trained on the morphologically reanalyzed training set A. Morphological reanalysis is done by turning training data into plain text and then feeding it to a morphological analyzer. If correct interpretation has been missed by the analyzer it is additionally included to the result.

In the second step the training set B (in the form of plain text) is then tagged by each of previously trained taggers. The consequence of the fact that we are using plain text as input to individual taggers is that there is a possibility of differences in segmentation of the annotated text produced by each of these methods. To simplify the synchronization of annotated results we have decided to disregard the division into sentences during analysis.

Based on the results of tagging, we extract contexts, in which some of the taggers disagree—these are used by the meta-classifier to learn the relationships between particular taggers and situations in which they make errors. Contexts are constructed by including W tokens to the left and to the right from central disambiguated token, where W is the window size.

For training the meta-classifier We only use contexts in which at least one of the component taggers is correct. In the case when more than one of the component taggers is correct, we select the last of them in the order of known accuracy.

2.2   Features

Features have been generated using built-in and custom WSDDE feature generators. We have experimented with the following types of features:

- **tagger agreement**—feature specifying which of the component taggers provided a matching outcome,
- **tagger response**—the response of each of the component taggers divided into part-of-speech tag and grammatical class name,
- **packed shape of the word**—a string with all digits replaced by 'd', lowercased characters replaced by 'l', uppercased characters replaced by 'u' and any other character replaced by 'x' ("Warszawa-2017" → "ulllllxdddd"). A packed shape is a shape with all neighbouring duplicate code characters removed ("ulllllxdddd" → "ulxd")
- **prefix and suffix**—lowercase prefixes and suffixes of a specified length (1 and 2 characters long in our experiments),
- **thematic features (TFG)**—features which could characterize the domain or general topic of a given context by checking whether certain words are present in the wide context of analyzed token, up to W positions to the left or the right from disambiguated token (the bag of words in orthographic or base form),
- **structural features 1 (SFG1)**—presence of particular words (in orthographic or base form) on a particular position in the close proximity (determined by W parameter) of the given token,
- **structural features 2 (SFG2)**—presence of particular POS/grammatical category values on a particular position in the close proximity (determined by W parameter) of the given token,
- **keyword features (KFG)** - features directly related to the given disambiguated token itself: its orthographic form and whether it starts with a capital letter.

Generated features/attributes can be filtered using feature selection algorithms from the WEKA package.

2.3   Classifiers

We have examined several types of popular classifiers, including: Naive Bayes, linear and non linear Support Vector Machine (LibLinear, LibSVM), Random Forest, J48, simple neural networks (shallow and multi-layer perceptrons) and gradient boosted trees (XGBoost). In order to use XGBoost we have created a WEKA wrapper package[1] for the XGBoost 4J library[2].

2.4   Tagging with the Meta-classifier

The tagging procedure consists of the following steps:

1. Tagging input text with component taggers.
2. Extraction of contexts in which some of the taggers disagree.
3. Classification of contexts with disagreement using the trained meta-classifier.
4. Generating the final annotation by aligning disambiguated context with the output of component taggers (from step 1). In case there is no tagging disagreement for a given token in text, we can take the result of any component tagger (in our implementation we take the first one).

3   Experimental Results

All evaluations have been performed on the manually annotated 1-million word subcorpus of the National Corpus of Polish, version 1.2, which consists of 1 215 513 tokens, manually annotated by trained linguists. We have used the same experimental setup as proposed in [10] and we report the results using the same accuracy measures. The *accuracy lower bound* (Acc_{lower}) is the tagging accuracy, in which we treat all segmentation mistakes as tagger errors. We also distinguish errors made on tokens which are known to morphosyntactic dictionary (Acc_{lower}^K) and on tokens for which no morphosyntactic interpretation is provided by the dictionary (Acc_{lower}^U).

First, we have performed a series of preliminary experiments using 70% of the entire corpus as the training data. This data was then divided into two subsets with 5–2 ratio (A: 5/7 for component taggers training, B: 2/7 for meta-classifier training). The aim of these experiments was to select the most effective classifiers and to examine the impact of the feature selection methods on the effectiveness of the contexts disambiguation. We have achieved the best results using Support Vector Machines (SVMs) and gradient boosted trees (XGBoost without any attributes filtering). LibSVM performed best with the selection of 100 attributes from following features: tagger agreement, tagger response, packed shape of the word. XGBoost showed better results without any attribute selection method applied and using the same feature set as LibSVM with the addition of SFG1 and SFG2.

The following experiments have been performed using 10-fold cross validation on the entire dataset. To set the optimal split between A and B training datasets we have measured the influence of the number of training contexts on tagging accuracy. This has been presented on Fig. 1. The experiment was performed using 10-fold cross validation with training sets in each fold split into two subsets with

---

[1] Plugin available at https://github.com/SigDelta/weka-xgboost.
[2] https://github.com/dmlc/xgboost/tree/master/jvm-packages.

5–2 ratio (5/7 for component taggers training, 2/7 for LibSVM meta-classifier training), therefore component taggers were trained using approximately 64% of all of the available data. Based on those results we have decided to split training sets in 6–1 ratio for next experiments (which corresponds to approx. 17–19k training contexts).

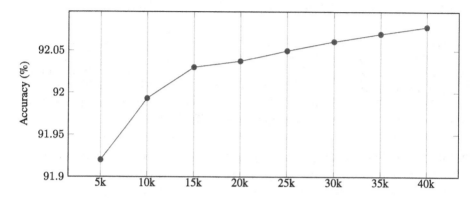

Fig. 1. Impact of the number of training contexts on tagging accuracy.

The final experiments were divided into two phases. In phase one we have used the four taggers available before PolEval (to compare our work against [4]). The results of that experiment are presented in Table 1, which shows that we were able to improve the accuracy of an ensemble of taggers (polita) using the meta-learning approach. In Table 2 we show the frequency in which the meta-classifier selects each of the component taggers.

Table 1. Comparison of 10-fold cross validation tagging results using training set split with ratio equal to 5–2 (40–43k training contexts) and 6–1 (18–19k training contexts).

Tagger	5-2 split			6-1 split		
	Acc_{lower}	Acc_{lower}^{K}	Acc_{lower}^{U}	Acc_{lower}	Acc_{lower}^{K}	Acc_{lower}^{U}
pantera	88.3646	91.9884	7.0421	88.5584	92.1877	7.1138
wmbt	90.1567	92.0960	46.6339	90.3308	92.2426	47.4212
wcrft	90.6354	92.7668	42.8029	90.7828	92.9028	43.2060
concraft	91.1535	93.1072	47.3090	91.2950	93.2301	47.8679
polita	91.7264	93.7206	46.9726	91.8881	93.8641	47.5402
Meta libsvm	92.0797	93.9697	49.6637	92.1890	94.0577	50.2497
Meta xgboost	92.1040	93.9842	49.8268	92.2304	94.0950	50.3818

Table 2. Which component tagger is selected with what frequency by XGBoost meta-classifier (6–1 split).

	pantera	wmbt	wcrft	concraft
Total	0.235	4.527	20.491	74.747
Known words	0.294	3.885	22.585	73.235
Unknown words	0.0	7.062	12.193	80.745

In phase two we have extended the experiments by including Toygger and KRNNT taggers, which were recently proposed during the PolEval workshop in 2017. Table 3 shows the results of the experiments achieved using 6–1 training set split ratio. The frequency with which component taggers are selected by XGBoost meta-classifier has been presented in Table 4.

Table 3. Extended meta-tagger tagging results with 10-fold cross validation using training sets split with 6–1 ratio.

	Acc_{lower}	Acc^K_{lower}	Acc^U_{lower}
krnnt	92.9055	94.2416	62.9211
toygger	94.1438	96.0488	51.3909
polita (toygger)	93.0001	94.8381	51.7492
polita (toygger + krnnt)	93.3756	95.1298	54.0063
Meta xgboost (toygger)	94.3457	96.1385	54.1106
Meta xgboost (toygger + krnnt)	94.6824	96.1048	62.7618

Table 4. Which component tagger is selected with what frequency by XGBoost meta-classifier (6–1 split, 24–26k training contexts).

	pantera	wmbt	wcrft	concraft	krnnt	toygger
Total	0.003	0.043	0.06	0.98	10.294	88.62
Known words	0.004	0.019	0.049	0.407	9.022	90.5
Unknown words	0.0	0.16	0.114	3.724	16.386	79.617

4   Conclusions and Future Work

In this paper we have tested the hypothesis that the accuracy of morphosyntactic tagging may be improved using the already available resources, by using a meta-learning approach. We have used an additional layer of machine learning over several individual taggers to provide a method of selecting the most accurate tagger for each token in natural language text.

Based on experimental evaluation on the largest available manually tagged text corpus for Polish, we could successfully improve the tagging accuracy by ca. 0.5% point over the single best-performing tagger. This result allows us to reduce the number of tagging errors in a 1 billion word corpus by 5 million.

The 94.7% tagging accuracy achieved in our approach, while a step forward, is still below the level of 97% reported for English. In further work we would like to explore the deep learning approaches to morphosyntactic tagging and vector-based word representations, which are the most adequate for representing highly inflected languages, such as Polish.

References

1. Acedański, S.: A morphosyntactic brill tagger for inflectional languages. In: Loftsson, H., Rögnvaldsson, E., Helgadóttir, S. (eds.) NLP 2010. LNCS (LNAI), vol. 6233, pp. 3–14. Springer, Heidelberg (2010). https://doi.org/10.1007/978-3-642-14770-8_3
2. Dębowski, Ł.: Trigram morphosyntactic tagger for Polish. In: Kłopotek, M.A., Wierzchoń, S.T., Trojanowski, K. (eds.) IIPWM 2004. AINSC, vol. 25, pp. 409–413. Springer, Heidelberg (2004). https://doi.org/10.1007/978-3-540-39985-8_43
3. Hall, M., Frank, E., Holmes, G., Pfahringer, B., Reutemann, P., Witten, I.H.: The weka data mining software: an update. SIGKDD Explor. Newsl. 11(1), 10–18 (2009). https://doi.org/10.1145/1656274.1656278
4. Kobyliński, Ł.: PoliTa: A multitagger for Polish, pp. 2949–2954. ELRA, Reykjavík (2014). http://www.lrec-conf.org/proceedings/lrec2014/index.html
5. Krasnowska, K.: Morphosyntactic disambiguation for Polish with bi-LSTM neural networks. In: Vetulani [12]
6. Młodzki, R., Przepiórkowski, A.: The WSD development environment. In: Vetulani, Z. (ed.) LTC 2009. LNCS (LNAI), vol. 6562, pp. 224–233. Springer, Heidelberg (2011). https://doi.org/10.1007/978-3-642-20095-3_21
7. Piasecki, M.: Polish tagger TaKIPI: rule based construction and optimisation. Task Q. 11(1–2), 151–167 (2007)
8. Przepiórkowski, A., Bańko, M., Górski, R., Lewandowska-Tomaszczyk, B. (eds.): Narodowy Korpus Języka Polskiego. Warszawa (2012)
9. Radziszewski, A.: A tiered CRF tagger for Polish. In: Bembenik, R., Skonieczny, Ł., Rybiński, H., Kryszkiewicz, M., Niezgódka, M. (eds.) Intelligent Tools for Building a Scientific Information Platform: Advanced Architectures and Solutions, vol. 467, pp. 215–230. Springer, Heidelberg (2013). https://doi.org/10.1007/978-3-642-35647-6_16
10. Radziszewski, A., Acedański, S.: Taggers gonna tag: an argument against evaluating disambiguation capacities of morphosyntactic taggers. In: Sojka, P., Horák, A., Kopeček, I., Pala, K. (eds.) TSD 2012. LNCS (LNAI), vol. 7499, pp. 81–87. Springer, Heidelberg (2012). https://doi.org/10.1007/978-3-642-32790-2_9
11. Radziszewski, A., Śniatowski, T.: A memory-based tagger for Polish. In: Proceedings of the LTC (2011)
12. Vetulani, Z. (ed.): Proceedings of the 8th Language and Technology Conference: Human Language Technologies as a Challenge for Computer Science and Linguistics. Poznań, Poland, 17–19 November 2017

13. Waszczuk, J.: Harnessing the CRF complexity with domain-specific constraints. The case of morphosyntactic tagging of a highly inflected language. In: Proceedings of the 24th International Conference on Computational Linguistics (COLING 2012), Mumbai, India, pp. 2789–2804 (2012)
14. Wróbel, K.: KRNNT: Polish recurrent neural network tagger. In: Vetulani [12]

Identifying Participant Mentions and Resolving Their Coreferences in Legal Court Judgements

Ajay Gupta[2], Devendra Verma[2], Sachin Pawar[1(✉)], Sangameshwar Patil[1],
Swapnil Hingmire[1], Girish K. Palshikar[1], and Pushpak Bhattacharyya[3]

[1] TCS Research, Tata Consultancy Services, Pune 411013, India
{sachin7.p,sangameshwar.patil,swapnil.hingmire,gk.palshikar}@tcs.co
[2] Department of CSE, Indian Institute of Technology Bombay, Mumbai 400076, India
{ajaygupta,devendrakuv,pb}@cse.iitb.ac.in
[3] Indian Institute of Technology Patna, Patna 801103, India

Abstract. Legal court judgements have multiple participants (e.g.
judge, complainant, petitioner, lawyer, etc.). They may be referred to
in multiple ways, e.g., the same person may be referred as lawyer, counsel, learned counsel, advocate, as well as his/her proper name. For any
analysis of legal texts, it is important to resolve such multiple mentions
which are coreferences of the same participant. In this paper, we propose
a supervised approach to this challenging task. To avoid human annotation efforts for Legal domain data, we exploit ACE 2005 dataset by mapping its entities to participants in Legal domain. We use basic Transfer
Learning paradigm by training classification models on general purpose
text (news in ACE 2005 data) and applying them to Legal domain text.
We evaluate our approach on a sample annotated test dataset in Legal
domain and demonstrate that it outperforms state-of-the-art baselines.

Keywords: Legal text mining
Coreference resolution · Supervised machine learning

1 Introduction

The legal domain is a rich source of large document repositories such as court
judgements, contracts, agreements, legal certificates, declarations, affidavits,
memoranda, statutory texts and so forth. As an example, the FIRE legal corpus[1] contains around 50,000 Supreme Court Judgements and around 80,000 High
Courts judgements in India. Legal documents have some special characteristics,
such as long and complex sentences, presence of various types of legal argumentation, and use of legal terminology. Legal document repositories are used for many

---

A. Gupta and D. Verma—This work was carried out during the internship at TCS
Research, Pune. Both the authors contributed equally.

[1] https://www.isical.ac.in/~fire/2014/legal.html.

P. Sojka et al. (Eds.): TSD 2018, LNAI 11107, pp. 153–162, 2018.
https://doi.org/10.1007/978-3-030-00794-2_16

purposes, such as retrieving facts [17,18], case summarization [23], precedence identification [8], identification of similar cases [9], extracting legal argumentation [12], case citation analysis [24] etc. Several commercial products, such as eBrevia, Kira, LegalSifter, and Luminance provide such services to lawyers.

A basic step in information extraction from legal documents is the extraction of various participants involved, say, in a court judgement. We define a *participant* as an entity of type person (PER), location (LOC), or organization (ORG). Typically, a participant initiates some specific action, or undergoes a change in some property or state due to action of another participant. Participants of type PER can be appellants, respondents, witness, police officials, lawyers, judge etc. Often organizations and locations play important roles in legal documents, and hence we include them as participants. For example, in "...an industrial dispute was raised by the appellant, which was referred by the Central Government to the Industrial Tribunal ...", the two organizations mentioned ("the Central Government", "the Industrial Tribunal") are participants.

Table 1. Sample text fragment from a court judgement. All the mentions of ith participant are coreferences of each other and are marked with P_i.

[The respondent, Selvamuthukani]$_{P_1}$ is [the original complainant]$_{P_1}$. [The complainant]$_{P_1}$ alleged that [she]$_{P_1}$ was married to [Mr. Kannan, the accused]$_{P_2}$ on 16.6.1980 . According to [Mr. Singh, counsel for the complainant]$_{P_3}$, during the subsistence of [her]$_{P_1}$ marriage with [the accused]$_{P_2}$, [he]$_{P_2}$ married again with [K. Palaniammal]$_{P_4}$.

The same participant is often mentioned in many different ways in a document; e.g., a participant Mr. Kannan may be variously referred to as the accused, he, her husband etc. All such mentions of a single participant are coreferences of each other. Grouping all mentions of the same participant together is the task of *coreference resolution*. Many legal application systems provide an interactive, dialogue-based interface to users. Information extracted from legal documents, particularly about various participants and coreferences among them, is crucial to understand utterances in such dialogues; e.g., What are the names of the accused and his wife? in Table 1. In practice, we often find that a standard off-the-shelf coreference resolution tool fails to correctly identify all mentions of an participant, particularly on legal text [20]. Typically, a mention is not linked to the correct participant (e.g. Stanford CoreNLP 3.7.0 Coreference toolkit does not link Selvamuthukani and The complainant), or a mention is undetected (e.g. the accused is not detected as a mention) and hence, not linked to any participant. Nominal mentions consisting of generic NPs[2] (e.g., complainant, prosecution witness) are often not detected at all as participants or they are detected as participants but not linked to the correct participant mention(s).

---

[2] Noun Phrases with common noun as headword.

We define *basic* mention of a participant to be a sequence of proper nouns (e.g., K. Palaniammal, Mr. Kannan), a pronoun (e.g., he, her) or a generic NP (e.g., the complainant). A basic mention can be either *dependent* or *independent*. A basic mention is said to be *dependent* if its governor in the dependency parse tree is itself a participant mention; otherwise it is called as *independent* mention. An independent mention can be basic (if it does not have any dependent mentions); otherwise a *composite mention* is created for it by recursively merging all its dependent mentions. For example, Mr. Singh, counsel for the complainant contains three basic participant mentions: Mr. Singh, counsel and the complainant. Here, only Mr. Singh is an independent mention and others are dependent mentions. The corresponding composite mention is created as Mr. Singh, counsel for the complainant.

In this paper, we focus on coreference resolution restricted to participants which consists of following steps: (i) identify basic participant mentions; (ii) merge dependent mentions into corresponding independent participant mentions to create composite mentions; and (iii) group together all independent participant mentions which are coreferences of each other. For step (i), we use a supervised approach, in which we train a classifier on a well-known labelled corpus (ACE 2005 [22]) of general documents to identify participants. We then use the learned model to identify participant mentions in legal documents. Then we have developed a rule-based system to perform step (ii). Finally, for step (iii) we use supervised classifier (such as Random Forest, SVM). We evaluate our approach on a corpus of legal documents (court judgements) manually labelled with participant mentions and their coreference groups. We empirically demonstrate that our approach performs better than state-of-the-art baselines, including well-known coreference tools, on this corpus.

2 Related Work

The problem of coreference resolution specifically for Legal domain has received relatively limited attention in literature. The literature broadly categorized into two streams. One focuses on anaphora resolution [2] and the other addresses the problem of Named Entity Linking. *Anaphora Resolution* is a sub-task of *Coreference Resolution* where the focus is to find an appropriate antecedent noun phrase for each pronoun. The task of Named Entity linking [4,5,7] focuses on linking the names of persons/organizations and Legal concepts to corresponding entries in some external database (e.g. Wikipedia, Yago). In comparison, our approach focuses on grouping all the corefering mentions together including generic NPs.

Even in the general domain, the problem of coreference resolution remains an open and challenging problem [13]. Recently, Peng et al. [14,15] have proposed the notion of Predicate Schemas and used Integer Linear Programming for coreference resolution. In terms of problem definition and scope, our work is closest to them as they also focus on all three types of mentions, i.e. named entities, pronouns and generic NPs.

3 Our Approach

We propose to use supervised machine learning approach to identify and link participant mentions in court judgements. Since there is lack of labeled training data in legal domain for this task, we map the entity mentions and coreference annotations in the ACE 2005 dataset to suit our requirement. Table 2 gives overview of the proposed approach. Unlike a corpus annotated for traditional NER task, ACE 2005 dataset labels mentions of all 3 types which are of our interest in this paper, viz. proper nouns, pronouns and generic NPs. Hence, we found that ACE dataset can be adapted easily for this task with minor transformations.

The specific transformations required to ACE dataset are as follows – ACE provides annotations for 7 entity types: PER (person), ORG (organization), LOC (location), GPE (geo-political entity), FAC (facility), WEA (weapon) and VEH (vehicle). As our definition of participant only includes mentions of type PER, ORG and LOC, we ignore the mentions labelled with WEA and VEH. Also we treat LOC, GPE and FAC as a single LOC entity type. Moreover, we define basic participant mentions to be base NPs whereas ACE mentions need not be base NPs; e.g., for the base NP the former White House spokesman, ACE would annotate two different mentions: White House as ORG and spokesman as PER. However, we note that spokesman is the headword of this NP and other constituents of the NP (such as the, White House) are modifiers of this headword. So we expand this mention as a single basic participant mention of type PER. We converted the original ACE mention and coreference annotations accordingly.

Table 2. Overview of our approach.

- **Phase-I: Training**
  Input: D : ACE 2005 corpus
  Output: C_M : mention detector, C_P : pair-wise coreference classifier
  T.1) Train C_M on D using CRF to detect participant mentions from a text.
  T.2) Train C_P on D using a supervised classification algorithm to predict whether participant mentions within a pair are coreferents.
- **Phase-II: Application**
  Input: d : test document, C_M : mention detector, C_P : pair-wise coreference classifier
  Output: $G = \{g_1, g_2, \ldots, g_k\}$: a set of coreference groups in d
  A.1) Let $\mathcal{M}$ be the set of entity mentions in d detected using C_M.
  A.2) For each independent mention $m_i \in \mathcal{M}$
         merge its all dependent mentions recursively and remove them from $\mathcal{M}$.
  A.3) For each candidate pair of mentions $< m_i, m_j >$ in $\mathcal{M}$
         use C_P to classify whether m_i and m_j are coreferences of each other.
  A.4) Let G be the partition of $\mathcal{M}$ such that each $g_i \in G$ represents a group of coreferent mentions through transitive closure.

The three major steps in our approach are explained below in detail.

3.1   Identifying Basic Mentions of Participants (T.1/A.1 in Table 2)

We model the problem of identifying basic mentions of participants as a sequence labeling problem. Here, similar to traditional Named Entity Recognition (NER) task, each word gets an appropriate label as per BIO encoding (**Begin-Inside-Outside** coding scheme used in NER). But unlike NER, we are also interested in identifying mentions in the form of pronouns and generic NPs. We employ Conditional Random Fields[3] (CRF) [10] for the sequence labeling task. Various features used for training the CRF model are described in Table 3.

Table 3. Features used by CRF for detecting basic mentions of participants

Feature type	Details
Lexical	Word itself; lemma of the word; next and previous words
POS	Part-of-speech tags of the word as well as its previous and next words
Syntactic	Dependency parent of the word; dependency relation with the parent
NER	Entity type assigned by the Stanford CoreNLP NER tagger
WordNet	WordNet hypernym type feature, derived from the hypernym tree, which can take one of $\{PER, ORG, LOC, NONE\}$ (e.g., for `complainant`, we get the synset (`person, individual, someone, ...`), which is one of the pre-defined synsets indicating PER, as an ancestor in the hypernym tree. For each participant type, we have identified such pre-defined synsets)

3.2   Identifying Independent Participant Mentions (A.2 in Table 2)

Our notion of independent mentions is syntactic, i.e. derived from the dependency parse tree. A basic participant mention is said to be *independent* if its dependency parent (with dependency relation type `nmod` or `appos`) is not a basic participant mention itself, otherwise it is said to be a *dependent* mention. In this stage, we merge all the *dependent* participant mentions (predicted in the previous step) recursively with their parents until only independent mentions remain.

---

[3] We used CRF++ (https://taku910.github.io/crfpp/).

3.3 Classifying Mention Pairs (T.2/A.3 in Table 2)

We model the coreference resolution problem as a binary classification task where each mention pair is considered as a positive instance iff the mentions are coreferences of each other. To generate candidate mention pairs, we consider threshold of 5 sentences. For this classifier, we derived 36 features using the dependency and constituency parse trees. The detailed description of the features is given in Table 4. Some of these features are based on the traditional mention-pair models in the literature [1,6,13,19]. We have added some more features like: whether both the mentions are connected through a copula verb, whether both the mentions appear in conjunction, etc.

A binary classifier model is trained on the ACE dataset (T.2 in Table 2) and this model is used to classify candidate mention pairs (using the predicted participant mentions from A.2) in legal dataset (A.3 in Table 2). Here we have used transfer learning by training a model on ACE dataset and testing it on Legal dataset. We explored four different classifiers: Random Forest, SVM, Decision Trees, and Naive Bayes Classifier.

Table 4. Feature types used by the mention pair classification

Real-valued feature types
(i) String similarity between the two mentions in terms of Levenshtein distance; (ii) No. of sentences/words/other mentions between the two mentions; (iii) Difference between the lengths of the mentions; (iv) Cosine similarity between word vectors (Google News word2vec embeddings) of head words of the mentions

Binary feature types
(i) Whether both the mentions have same gender/number/participant type/POS tag; (ii) Whether both the mentions are in the same sentence; (iii) Whether any other mention is present in between; (iv) Whether both the mentions indefinite or definite; (v) Whether first/second mention is indefinite; (vi) Whether first/second mention is definite; (vii) Whether both the mentions are connected through a copula; (viii) Whether both the mentions appear in conjunction; (ix) Whether both the mentions are *nominal subjects* of some verbs; (x) Whether only the first or second mention is *nominal subject* of some verb; (xi) Whether both the mentions are *direct objects* of some verbs; (xii) Whether only the first or second mention is *direct object* of some verb; (xiii) Whether one mention is *nominal subject* and other is *direct object* of a same verb; (xiv) Whether both the mentions are pronouns; (xv) Whether only first or second mention is pronoun; (xvi) Whether first or second mention occurs at the start of a sentence

3.4 Clustering Similar Mentions (A.4 in Table 2)

To create the final coreference groups, we need to cluster the mentions using the output of classifier in step A.3. This is necessary because the pair-wise classification output in A.3 may violate the desired transitivity property [13] for a coreference group. We use clustering strategy similar to single-linkage clustering. We take the coreference mention pair classifier output as input to clustering system and process each Court Judgement output from coreference mention pair one by one. We select the mention pairs which are positive predicted examples from the input. These are called coreference pairs. These coreference pairs are used to create the coreference group as follows:

1. Select the coreference pairs one by one, check if they are present in the already created coreference groups.
2. If both are not present in any of the coreference groups, then create the new group by adding the both mentions from the mention pair.
3. If one is present in any of the already created coreference groups, add the second mention from the mention pair into that coreference group.
4. If both are present in any of the already created coreference group, do not add them into any coreference group.
5. Once all the mention pairs from a document are processed. We merge the disconnected coreference groups as follows:
   (a) Take the pair of coreference groups and check whether they are disjoint.
   (b) If they are disjoint, keep them as separate coreference groups.
   (c) If they are not disjoint, then merge those two coreference groups into single coreference group.

4   Experimental Analysis

We evaluate our approach on 14 court judgements: 12 judgements from The Supreme Court of India and 2 judgements from The Delhi High Court in the FIRE legal judgement corpus. On an average, a judgement contains around 45 sentences and 25 distinct participants. We manually annotated these judgements by identifying all the independent participant mentions and grouping them to create coreference groups.

Baselines: B1 is a standard baseline approach which uses Stanford CoreNLP toolkit. Here, basic participant mentions are identified as a union of the named entities (of type PER, ORG and LOC) extracted by the Stanford NER and the mentions extracted by the Stanford Coreference Resolution. Dependent mentions are merged with corresponding independent mentions by using the same rules as described in the step A.2 in Table 2. Final groups of coreferant participant mentions are then obtained by using coreference groups predicted by the Stanford Coreference toolkit. B2 is the state-of-the-art coreference resolution system based on Peng et al. [14,15]. Unlike B1 and B2, our approach focuses on identifying coreferences only among the participant mentions and not ALL mentions. Hence, we discard non-participant mentions and coreference groups consisting solely of non-participant mentions from the predictions of B1 & B2.

Evaluation: We evaluate the performance of all the approaches at two levels: all independent participant mentions and clusters of corefering participant mentions. We use the standard F1 metric to measure performance of participant mention detection. For evaluating coreferences among the predicted participant mentions, we used the standard evaluation metrics [16], MUC [21], BCUB [3], Entity-based CEAF (CEAFe) [11] and their average. Table 5 shows the relative performance of our approach compared to the two baselines. Out of multiple classifiers Random Forest (RF) with Gini impurity as splitting criteria and 5000 trees provides the best result.

Table 5. Experimental results (RF: Random Forest, SVM: Support Vector Machines, DT: Decision Tree, NBC: Naive Bayes Classifier)

Algorithm	Participant mention			Canonical mentions									
				MUC			BCUB			CEAFe			Avg.
	P	R	F	P	R	F	P	R	F	P	R	F	F
B1	63.1	43.4	50.3	64.4	40.3	48.3	45.5	26.3	31.9	22.1	22.0	21.7	34.0
B2	64.5	41.1	46.5	62.0	31.4	38.0	52.8	20.1	25.3	24.8	28.9	25.0	29.4
RF	69.8	70.7	**70.2**	59.5	52.4	55.1	66.0	53.9	**58.1**	35.3	42.1	37.7	**50.3**
SVM				59.3	45.1	50.7	68.9	45.8	53.7	32.5	47.4	**37.9**	47.4
DT				54.8	69.5	**60.6**	31.4	74.1	42.4	30.6	16.9	21.0	41.3
NBC				54.6	50.5	52.1	57.0	51.8	53.0	34.1	38.4	35.5	46.9

5   Conclusion

This paper demonstrates that off-the-shelf coreference resolution does not perform well on domain-specific documents, in particular on legal documents. We demonstrate that using domain and application specific characteristics, it is possible to improve performance of coreference resolution. Identifying participant mentions and grouping their coreferents together is a challenging task in Legal text mining and Legal dialogue systems.

We proposed a supervised approach for addressing this challenging task. We adapted ACE 2005 dataset by mapping its entities to participants in Legal domain. We showed that the approach outperforms the state-of-the-art baselines. In future, we plan to employ advanced transfer learning techniques to improve performance.

References

1. Agrawal, S., Joshi, A., Ross, J.C., Bhattacharyya, P., Wabgaonkar, H.M.: Are word embedding and dialogue act class-based features useful for coreference resolution in dialogue? In: Proceedings of PACLING (2017)
2. Al-Kofahi, K., Grom, B., Jackson, P.: Anaphora resolution in the extraction of treatment history language from court opinions by partial parsing. In: Proceedings of 7th ICAIL (1999)

3. Bagga, A., Baldwin, B.: Algorithms for scoring coreference chains. In: The First International Conference on Language Resources and Evaluation Workshop on Linguistics Coreference, Granada , vol. 1, pp. 563–566 (1998)
4. Cardellino, C., Teruel, M., Alemany, L.A., Villata, S.: A low-cost, high-coverage legal named entity recognizer, classifier and linker. In: Proceedings of 16th ICAIL (2017)
5. Cardellino, C., Teruel, M., Alemany, L.A., Villata, S.: Ontology population and alignment for the legal domain: YAGO, Wikipedia and LKIF. In: Proceedings of ISWC (2017)
6. Cheri, J., Bhattacharyya, P.: Coreference resolution to support IE from Indian classical music forums. In: Proceedings of RANLP, pp. 91–96 (2015)
7. Dozier, C., Haschart, R.: Automatic extraction and linking of personal names in legal text. In: Proceedings of Recherche d'Informations Assistee par Ordinateur, RIAO 2000 (2000)
8. Jackson, P., Al-Kofahi, K., Tyrrell, A., Vachher, A.: Information extraction from case law and retrieval of prior cases. Artif. Intell. **150**, 239–290 (2003)
9. Kumar, S., Reddy, P.K., Reddy, V.B., Singh, A.: Similarity analysis of legal judgments. In: Proceedings of the COMPUTE (2011)
10. Lafferty, J.D., McCallum, A., Pereira, F.C.N.: Conditional random fields: probabilistic models for segmenting and labeling sequence data. In: Proceedings of the Eighteenth International Conference on Machine Learning, ICML 2001, pp. 282–289. Morgan Kaufmann Publishers Inc., San Francisco (2001). http://dl.acm.org/citation.cfm?id=645530.655813
11. Luo, X.: On coreference resolution performance metrics. In: Proceedings of HLT-EMNLP, pp. 25–32 (2005)
12. Mochales, R., Moens, M.F.: Argumentation mining. Artif. Intell. Law **19**(1), 1–22 (2011)
13. Ng, V.: Machine learning for entity coreference resolution: a retrospective look at two decades of research. In: Proceedings of the 31st AAAI Conference on Artificial Intelligence, pp. 4877–4884 (2017)
14. Peng, H., Chang, K., Roth, D.: A joint framework for coreference resolution and mention head detection. In: CoNLL 2015, pp. 12–21 (2015)
15. Peng, H., Khashabi, D., Roth, D.: Solving hard coreference problems. In: NAACL HLT 2015, pp. 809–819 (2015)
16. Pradhan, S., Luo, X., Recasens, M., Hovy, E., Ng, V., Strube, M.: Scoring coreference partitions of predicted mentions: a reference implementation. In: Proceedings of ACL (2014)
17. Saravanan, M., Ravindran, B., Raman, S.: Improving legal information retrieval using an ontological framework. Artif. Intell. Law **17**(2), 101–124 (2011)
18. Shulayeva, O., Siddharthan, A., Wyner, A.: Recognizing cited facts and principles in legal judgements. Artif. Intell. Law **25**(1), 107–126 (2017)
19. Soon, W.M., Ng, H.T., Lim, D.C.Y.: A machine learning approach to coreference resolution of noun phrases. Comput. Linguist. **27**(4), 521–544 (2001)
20. Venturi, G.: Legal language and legal knowledge management applications. In: Francesconi, E., Montemagni, S., Peters, W., Tiscornia, D. (eds.) Semantic Processing of Legal Texts. LNCS (LNAI), vol. 6036, pp. 3–26. Springer, Heidelberg (2010). https://doi.org/10.1007/978-3-642-12837-0_1
21. Vilain, M., Burger, J., Aberdeen, J., Connolly, D., Hirschman, L.: A model-theoretic coreference scoring scheme. In: Proceedings of the 6th Conference on Message Understanding, pp. 45–52 (1995)

22. Walker, C., Strassel, S., Medero, J., Maeda, K.: ACE 2005 multilingual training corpus. Linguist. Data Consortium **57** (2006)
23. Yousfi-Monod, M., Farzindar, A., Lapalme, G.: Supervised machine learning for summarizing legal documents. In: Farzindar, A., Kešelj, V. (eds.) AI 2010. LNCS (LNAI), vol. 6085, pp. 51–62. Springer, Heidelberg (2010). https://doi.org/10. 1007/978-3-642-13059-5_8
24. Zhang, P., Koppaka, L.: Semantics-based legal citation network. In: Proceedings of the 11th ICAIL, pp. 123–130 (2007)

Building the Tatar-Russian NMT System Based on Re-translation of Multilingual Data

Aidar Khusainov$^{(\boxtimes)}$ (iD), Dzhavdet Suleymanov, Rinat Gilmullin,
and Ajrat Gatiatullin

Institute of Applied Semiotics of the Tatarstan Academy of Sciences,
Kazan Federal University, Kazan, Russia
khusainov.aidar@gmail.com, dvdt.slt@gmail.com, rinatgilmullin@gmail.com,
agat1972@mail.ru

Abstract. This paper assesses the possibility of combining the rule-based and the neural network approaches to the construction of the machine translation system for the Tatar-Russian language pair. We propose a rule-based system that allows using parallel data of a group of 6 Turkic languages (Tatar, Kazakh, Kyrgyz, Crimean-Tatar, Uzbek, Turkish) and the Russian language to overcome the problem of limited Tatar-Russian data. We incorporated modern approaches for data augmentation, neural networks training and linguistically motivated rule-based methods. The main results of the work are the creation of the first neural Tatar-Russian translation system and the improvement of the translation quality in this language pair in terms of BLEU scores from 12 to 39 and from 17 to 45 for both translation directions (comparing to the existing translation system). Also the translation between any of the Tatar, Kazakh, Kyrgyz, Crimean Tatar, Uzbek, Turkish languages becomes possible, which allows to translate from all of these Turkic languages into Russian using Tatar as an intermediate language.

Keywords: Neural machine translation
Rule-based machine translation · Turkic languages
Low-resourced language · Data augmentation

1 Introduction

2016 was the year when machine translation systems built on the neural network approach surpassed the quality of the phrase- and syntax-based systems [5]. Since that time, many companies have developed neural versions of their translators for the most popular language pairs [4,16]. Moreover, a large number of studies were devoted to improving the quality of translation due to the use of linguistically motivated or linguistically informed models, which led, for example, to the use of multifactor models and morphemes or their combinations as subword

© Springer Nature Switzerland AG 2018
P. Sojka et al. (Eds.): TSD 2018, LNAI 11107, pp. 163–170, 2018.
https://doi.org/10.1007/978-3-030-00794-2_17

units. However, as in other areas of artificial intelligence, for example, speech recognition and dialogue systems, the use of modern machine learning methods for the class of low-resourced languages is limited by the lack of training data. Even companies with relatively unlimited access to data (i.e. Google, Yandex) use various techniques to bypass this limitation: combining different translation approaches and selecting the most adequate result [18], using well-resourced intermediate languages (English or a related well-resourced language).

Motivated by the goal of creating a machine translation system that could work well for the low-resourced Tatar-Russian language pair, we propose such a technology that would include both the latest achievements in machine learning and the use of the linguistic features of Turkic languages to overcome existing limitations (by retranslating collected parallel data for other Turkic languages: Kazakh, Kyrgyz, Crimean Tatar, Uzbek, Turkish). The resulting system includes tools that augment training data, execute pre- and post-processing algorithms along with the attention-based encoder-decoder translation algorithm. We build the Tatar-Russian translation system and use the Tatar language as an intermediate language for the translation of other Turkic languages.

The paper is structured in the following way: Sect. 2 gives an overview of the data collection process, Sect. 3 describes the main features of rule- and neural-based models, in Sect. 4 we discuss experiment results, and Sect. 5 concludes the paper.

2   Data Collection

Before going into details of corpus creation, first we discuss the main reason that will define both corpus and system structures. The NN approach of constructing machine translation systems has confirmed its success in experiments with many language pairs. There are some important language features that affect the quality of the system, for instance, translation from a gender-neutral language (e.g. Turkish, Tatar) to a non-gender neutral one (e.g. Russian) could lead to some biasing problems [9]. But most aspects of translation are successfully modelled by the NN approach. The main key to that is a clean, representative and big enough parallel text corpus, as NMT systems are known to under-perform when trained on limited data [6,19]. Thus, the solution to our task of constructing the Tatar-Russian MT system would be to create a large-enough parallel corpus and build the NMT system. The limitation here is the absence of a parallel corpus and a small amount of resources from which it could be built.

The lack of parallel data for the Turkic languages currently makes it impossible to fully use the statistical MT technologies. While there have been many attempts to build MT systems between closely related languages, for instance, Turkish-Crimean Tatar [14], they all use rules of lexical and syntactic correspondence.

The idea of this paper is to develop tools that will allow to use maximum of the parallel data available not only for Tatar, but for 5 other Turkic languages (Kazakh, Kyrgyz, Crimean Tatar, Uzbek, Turkish). As a first stage we decided to collect parallel training data for all 6 selected Turkic languages and the Russian language. One of the main sources of bilingual information are websites of ministries and other state departments. In many countries and regions there are laws that oblige organizations to keep document circulation simultaneously in the Russian and the national language - this refers to Tatar, Kazakh, Crimean Tatar. The other source are literary works, mostly printed books with available translation. To download data from web sources we have developed a program that can be configured to download information based on sites' list and specific rules that help to determine the correspondence between the Russian and the Turkic pages (i.e. url patterns, translation links on the source page). We have signed an agreement between the Tatarstan Academy of Sciences and libraries on the transfer of rights to use some of their books; available books for which there was a translation were scanned using professional scanning equipment. We then filtered the collected data according to the following criteria: both the source and the target sentences should contain at least 1 word and at most 80 words; duplicate sentence pairs are removed; all the collected texts were aligned with the help of the ABBYY Aligner 2.0 tool [1]. The full description of the collected data is presented in Table 1.

Table 1. The characteristics of the initial data for the multilingual Turkic-Russian parallel corpus.

Language	Source type	Number of parallel sentences
Tatar	Internet resources and books	250,000
Kazakh	Internet resources	350,000
Kyrgyz	Internet resources	75,000
Uzbek	Internet resources	160,000
Crimean Tatar	Internet resources	17,000
Turkish	Internet resources	150,000

As can be seen from the data presented in Table 1, there are only 250 thousand parallel sentences available for the Tatar-Russian language pair. Therefore, we chose as our priority task increasing the number of Tatar-Russian sentences in order to achieve better translation quality of the Tatar-Russian NMT system.

We manually corrected the results of the auto-aligning procedure (i.e. literary translation of books led to the presence of pairs of sentences that are very different from each other); two people completed this work in about a month. As for using parallel data for other Turkic languages, we developed a new rule-based system that uses the closeness of the Turkic languages and can translate sentences from one Turkic language to another (see Sect. 3.1 for details). This system gives the possibility to translate and use parallel Kazahk, Kyrgyz, Uzbek, Crimean Tatar and Turkish sentences to increase the size of the Tatar-Russian corpus. To preserve the quality of the training data we filtered all translated sentence pairs that contain words that are not in the vocabulary or have morphological ambiguity (see Sect. 3.1 for details). The resulting size of the first part of the Tatar-Russian parallel corpus is 328,213 sentence pairs. This corpus was used to train the first version of the NMT system for the Russian-Tatar translation direction.

At the same time, a team of translators started translating news from Russian to Tatar. The process was organized using the ABBYY SmartCAT tool for professional translators [2]. Manual translation of 35 thousand sentences took nearly 700 man-hours, or 1month of team's work. Since some intermediate neural models for Russian-Tatar direction were built, we started to translate all of the new texts and to use the result as a starting point for the manual translation process. This allowed us to speed up the translation process, so after 2 months the total number of manually translated sentences were 189,689. We implemented back-translation approach described in [11] that gave us additionally 409 thousand sentence pairs.

Summarizing the steps made for building MT systems:

1. Collecting all existing Tatar-Russian parallel data (250,000 sentence pairs);
2. Building the rule-based translation system for Turkic languages, translating the collected Turkic-Russian texts into Tatar-Russian texts (78,213 sentence pairs);
3. Creating the first version of the Tatar-Russian NMT system (Fig. 1);
4. Manual and semi-automatic translating of Russian texts (189,689 sentence pairs);
5. Training of the Tatar-Russian direction of the NMT system using the data augmentation approach to supplement training data with the back-translated monolingual Russian corpus (409,606 sentence pairs), see Fig. 2 for the detailed pipeline.
6. Re-training the Tatar-Russian direction using all of the data collected during the training time, see Fig. 3 for the detailed pipeline.

3   Systems Description

3.1   Rule-Based Machine Translation Module

The core of the proposed Turkic translation system is the structural and functional model of the Turkic morpheme, which consists of several main components: morphological analyzers and synthesizers, a unified table of affixes, morphotactic rules, multilanguage stems vocabulary.

The translation system can be represented as a system of morphological analysis and synthesis. The developed tools allow to describe the morphology of any Turkic language, but at the moment the module can only analyze texts in Tatar, Kazakh, Kyrgyz, Crimean Tatar, Uzbek and Turkish languages, since the service database has been filled with the information on the structural and functional model for the listed languages.

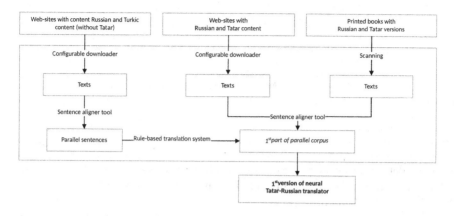

Fig. 1. Block diagram of the first stage of system's creation.

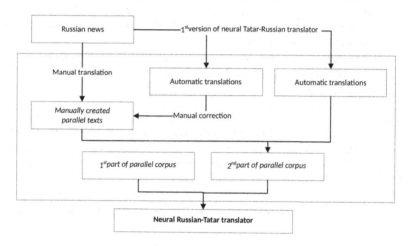

Fig. 2. Block diagram of the second stage of system's creation.

The algorithm of the translation process is quite simple:

1. Search for possible stems.
   We search for all possible sequences of letters from the left part of the input word that are present in the stems dictionary. For all the found stems, their grammatical classes are determined and the right-hand part obtained as a result of cutting off the stem is analyzed. In Turkic languages, the classical parts of speech that are attributed to the roots in lexical dictionaries do not uniquely identify possible sets of affixal chains. To define affix morphemes that can be attached to the Turkic stems, the concept of the grammatical class is introduced. In our structural-functional model of the Turkic morpheme, the following grammatical classes are distinguished: Noun, Verb, Attribute, Numeral, Unchangeable part of speech. This classification of morphological

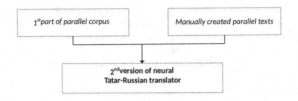

Fig. 3. Block diagram of the third stage of system's creation.

types determines only the rules of morphotactics and does not describe the syntactic and semantic features of these morphological types.

2. Building affixal chain.
   All possible affixal sequences are formed from the remaining right part of the word form on the basis of the determined morphological type of the stem and morphotactic rules.
3. The result of the analysis (morphemes and their categories) is formed on the basis of the obtained information.
4. Search for translation of the word stem in a multilingual dictionary.
5. Compilation of a word form based on the stem translation and the chain of morphological categories.

There are two main disadvantages that limit the use of the system. The first disadvantage is the small size of multilingual stem dictionaries that contain 39,050 Tatar, 18,735 Kazakh, 14,630 Turkish, 9,750 Kyrgyz, 7,070 Crimean Tatar and 5,433 Uzbek stems. The second disadvantage is the absense of morphological disambiguation tools for all analyzed languages. Therefore, when using this system for the re-translation of Turkic-Russian sentence pairs into Tatar-Russian ones we applied a filter that rejected sentences containing OOV words or words with ambiguous morphological parsing.

3.2 Neural Machine Translation Module

To train the Tatar-Russian NMT system we use the Nematus toolkit [15] with some improvements proposed in [12]. We mostly keep the default hyperparameter values and settings except the vocabulary size (set to 15,000) and the batch size (set to 60 for training and to 5 for validation), and make use of the dropout.

Tatar is an agglutinative language with a rich morphology, which gives us the out-of-vocabulary problem due to the limited size of the dictionary and training data. To overcome this problem we splitted words into sub-word units as presented by R. Sennrich [10]. All the collected data was tokenized, truecased and splitted into sub-word units using byte-pair encoding (BPE). Both Russian and Tatar texts were tokenized using Moses algorithms [7]. BPE models were learned on the joint Russian-Tatar corpus with 100 000 merge operations with the help of the sub-word NMT project [13].

4   Evaluation

We used BLEU metric [8] to compare the quality of different translation systems. Despite it was shown that BLEU scores do not always correlate with the translation quality [3], it is still widely used, because manual testing is expensive and time-consuming.

We evaluated our NMT system and Yandex translator [17] as these are the only available tools for the Russian-Tatar language pair, Table 2.

For the test set, we randomly selected 1,000 manually transcribed sentences from the data that was not used for train and validation processes.

Table 2. BLEU scores of Tatar-Russian translation systems.

System	Direction	Training collection	BLEU	BLEU-1	BLEU-2	BLEU-3	BLEU-4
Yandex MT	RU-TT	N/A	12.64	43.8	16.7	8.0	4.3
Our NMT system	RU-TT	0.5 M sentence pairs	39.63	62.8	44.1	34.6	28.7
Yandex MT	TT-RU	N/A	17.20	47.7	21.6	12.0	7.1
Our NMT system	TT-RU	0.9 M sentence pairs	45.71	65.4	49.2	40.7	33.5

5   Conclusions

In this paper we presented the Tatar-Russian NMT system that was trained on re-translated multilingual Turkic-Russian parallel corpora and the back-translated monolingual Russian corpus. Translation between Turkic languages was carried out using the proposed rule-based system. For the moment it can analyze texts in Tatar, Kazakh, Kyrgyz, Crimean Tatar, Turkish and Uzbek languages, but adding functionality for a new language can be done by describing the required morphological models with the help of developed tools.

The resulting translation system significantly outperforms the only existing translation system in this language pair from the Yandex company (by factor of 3 in terms of BLEU metric).

In future experiments we plan to use multifactor models as some grammatical information may help to improve the translation quality, and to significantly expand the dictionaries of the rule-based system for more complete use of existing parallel texts that are available for other Turkic languages.

References

1. ABBYY Aligner 2.0 (2017). https://www.abbyy.com/ru-ru/aligner/
2. ABBYY SmartCAT tool for professional translators (2017). https://smartcat.ai/workspace
3. Baisa, V.: Problems of machine translation evaluation. In: Sojka, P., s Horák, A. (eds.) Proceeding of Recent Advances in Slavonic Natural Language Processing, RASLAN 2009, Brno, pp. 17–22 (2009). https://nlp.fi.muni.cz/raslan/2009/papers/2.pdf
4. Bojar, O., et al.: Findings of the 2017 conference on machine translation (WMT17). In: Proceedings of the Second Conference on Machine Translation, Volume 2: Shared Task Papers, pp. 169–214. Association for Computational Linguistics, Copenhagen, September 2017. http://www.aclweb.org/anthology/W17-4717
5. Bojar, O., et al.: Findings of the 2016 conference on machine translation. In: Proceedings of the First Conference on Machine Translation, pp. 131–198. Association for Computational Linguistics, Berlin, August 2016. http://www.aclweb.org/anthology/W/W16/W16-2301
6. Fadaee, M., Bisazza, A., Monz, C.: Data augmentation for low-resource neural machine translation. In: Proceedings of the Annual Meeting of the Association for Computational Linguistics (ACL 2017), pp. 567–573, January 2017
7. Moses, the machine translation system (2017). https://github.com/moses-smt/mosesdecoder/
8. Papineni, K., Roukos, S., Ward, T., Zhu, W.: BLEU: a method for automatic evaluation of machine translation. In: Proceedings of the 40th Annual Meeting on Association for Computational Linguistics, pp. 311–318 (2002)
9. Schiebinger, L., Klinge, I.: Gendered Innovations: How Gender Analysis Contributes to Research. Publications Office of the European Union, Luxembourg (2013)
10. Sennrich, R., Haddow, B., Birch, A.: Neural machine translation of rare words with subword units. ArXiv e-prints, August 2015
11. Sennrich, R., Haddow, B., Burch, A.: Improving neural machine translation models with monolingual data. In: Proceedings of the 54th Annual Meeting of the Association for Computational Linguistics, Berlin, pp. 86–96 (2016)
12. Sennrich, R., et al.: The University of Edinburgh's neural MT systems for WMT17. In: Proceedings of the Second Conference on Machine Translation, Volume 2: Shared Task Papers. Stroudsburg, PA, USA (2017)
13. Subword Neural Machine Translation (2017). https://github.com/rsennrich/subword-nmt/
14. Suleimanov, D., Gatiatullin, A., Almenova, A., Bashirov, A.: Multifunctional model of the Turkic morpheme: certain aspects. In: Proceedings of the International Conference on Computer and Cognitive Linguistics TEL-2016, Kazan , pp. 168–171 (2016)
15. Open-Source Neural Machine Translation in Theano (2017). https://github.com/rsennrich/nematus
16. Wu, Y., et al.: Google's neural machine translation system: bridging the gap between human and machine translation. ArXiv e-prints, September 2016
17. Yandex translate (2017). https://translate.yandex.com/
18. One model is better than two. Yandex. Translate launches a hybrid machine translation system (2017). https://goo.gl/PddtYn
19. Zoph, B., Yuret, D., May, J., Knight, K.: Transfer learning for low-resource neural machine translation. ArXiv e-prints, April 2016

Annotated Clause Boundaries' Influence on Parsing Results

Dage Särg[1,2]([⊠]), Kadri Muischnek[1,2], and Kaili Müürisep[1]

[1] Institute of Computer Science, University of Tartu, Tartu, Estonia
{dage.sarg,kadri.muischnek,kaili.muurisep}@ut.ee
[2] Institute of Estonian and General Linguistics, University of Tartu, Tartu, Estonia

Abstract. The aim of the paper is to study the effect of pre-annotated clause boundaries on dependency parsing of Estonian new media texts. Our hypothesis is that correct identification of clause boundaries helps to improve parsing because as the text is split into smaller syntactically meaningful units, it should be easier for the parser to determine the syntactic structure of a given unit. To test the hypothesis, we performed two experiments on a 14,000-word corpus of Estonian web texts whose morphological analysis had been manually validated. In the first experiment, the corpus with gold standard morphological tags was parsed with MaltParser both with and without the manually annotated clause boundaries. In the second experiment, only the segmentation of the text was preserved and the morphological analysis was done automatically before parsing. The experiments confirmed our hypothesis about the influence of correct clause boundaries by a small margin: in both experiments, the improvement of LAS was 0.6%.

Keywords: Dependency parsing · Clause boundaries
New media language · Estonian

1 Introduction

Together with the ever-increasing amounts of user-generated textual data online increases the importance of its automatic processing. As most text-processing-related end-user applications require or can be improved by high-quality linguistic annotations, there is a great need for tools and methods developed or adapted for new media language. There has already been a considerable amount of work on normalising, tagging and parsing the noisy, heterogeneous language usage of social media and more precisely on the impact of the accuracy of POS-tagging on parsing outcome.

In this work, we are contributing to the field by exploring the impact of gold clause boundaries on morphological analysis and parsing of Estonian new media

This study was supported by the Estonian Ministry of Education and Research (IUT20-56), and by the European Union through the European Regional Development Fund (Centre of Excellence in Estonian Studies).

P. Sojka et al. (Eds.): TSD 2018, LNAI 11107, pp. 171–179, 2018.
https://doi.org/10.1007/978-3-030-00794-2_18

texts. Our aim is to find out how much the gold clause boundaries influence the quality of parsing in relation to the gold morphological analysis. Müürisep and Nigol [1] have stated in their work on parsing transcribed Estonian speech with a rule-based parser that the parser developed for standard texts (i.e. edited written texts as opposed to spontaneous written texts/speech) was suitable for speech provided that special attention was paid to the clause boundaries as this was a major source of syntactic errors. Therefore, we hypothesize that, in addition to gold morphological analysis, gold clause boundaries could improve the parsing results of our corpus while using a MaltParser model trained for standard written Estonian. To test the hypothesis, we parse our test corpus both with and without gold clause boundaries and use the standard metrics of UAS (unlabeled attachment score, percentage of words with correct head), LA (label accuracy, percentage of words with correct syntactic function tag), and LAS (percentage of words with both correct head and correct syntactic function tag) to compare the parsing accuracies. We also compare the results with parsing results of standard Estonian to see if the parsing model trained on standard texts is applicable for new media.

The majority of this kind of work has been focused on processing English texts. Several authors have found that gold part-of-speech tagging improves the parsing results, e.g. Kong et al. on tweets [2] and Foster et al. [3] on tweets and discussion forums. There is also work on other morphologically rich languages, e.g. Seddah et al. [4] have experimented with statistical constituency parsing of French social media texts. However, to our knowledge, there are no previous works exploring the impact of manual clause boundaries.

As for previous work on Estonian non-standard written language usage, Särg [5] has adapted a rule-based Constraint Grammar dependency parser for parsing chatroom texts. Adding about 100 special rules to the rule-set consisting of ca 2000 rules and modifying about 50 existing rules resulted in considerable improvement: UAS increased from 75.03% to 84.60% and LAS from 72.21% to 82.19%, but all experiments were conducted using gold POS-tags and lemmas, so there is no information about the impact of POS-tagging quality on parsing.

2   Material: Corpus and Its Annotation Scheme

For the experiments reported on in this paper, a small manually annotated corpus of Estonian web texts (Estonian Web Treebank, EWTB) was used. EWTB texts form a subpart of the web-crawled corpus Estonian Web 2013; previously also named as etTenTen 2013 [6]. The texts are divided into the following classes: blogs, discussion forums, information texts, periodicals, and religion texts.

These text classes, especially blogs and discussion forums, are by no means consistent in their language usage or correctness of spelling and interpunctuation; so there exists considerable variability between different files (in blogs) or even between different parts of the same file (in forums).

The texts have been automatically split into sentences and manually tagged for intrasentential clause boundaries. As standard written Estonian follows strict punctuation rules, punctuation is usually a secure indicator for clause boundaries in regular texts. However, in user-generated web texts, punctuation can be unsystematic and vary greatly depending on the author of the text.

The texts have been annotated using the Estonian Constraint Grammar annotation scheme for morphological analysis and dependency parsing. The same annotation standard has also been followed while annotating the Estonian Dependency Treebank [7], but one additional syntactic label has been introduced, namely that of discourse particle.

Morphological annotation has also been used as a device for normalisation, meaning that the lemma and grammatical categories for erroneous word forms (no matter whether the spelling error is unintentional or it can be viewed as an example of language play) are those of the correct word form. However, if an erroneous word form is frequent enough to be analysed as a new word or new spelling variant of an old word, it is not normalised. Of course, there exists a continuum between a repetitious error and a new word, so the normalisation decisions are somewhat arbitrary. Morphological annotation of every file in our corpus was checked and corrected by two independent annotators. Finally, a super-annotator compared the annotations and created the final version of the morphologically annotated corpus.

The gold syntactic tagging was done by two annotators in two stages. First, both annotators verified the annotations of files separately. Second, they exchanged their versions of the annotated files, and, by comparing the other annotator's annotations with their own, fixed the accidental differences in their own file so that only the differences worth discussion remained. The inter-annotator agreement rate after the first stage was 90%, after the second one 96%. The remaining differences were solved by discussion between the annotators.

3   Experiments

We performed two experiments to test our hypothesis about the effect of clause boundaries on syntactic parsing: first parsing the corpus with gold morphological annotation both with and without gold clause boundaries, then repeating it on the same corpus with automatic morphological analysis. For both experiments, the MaltParser [8] model trained for standard written Estonian was used[1]. The model has been trained on a total of 250,000 tokens of fiction and journalistic texts. It has been reported to achieve the LAS of 80.3%, the UAS of 83.4%, and the LA of 88.6% on standard written Estonian texts with gold morphological analysis excluding the punctuation [9].

3.1   Parsing with Gold Morphological Analysis

In the first experiment, the corpus with gold morphological analysis was parsed both with and without gold clause boundaries. The results are presented in

---

[1] https://github.com/EstSyntax/EstMalt/tree/master/EstDtModel.

Table 1. As we can see, the parsing results with gold clause boundaries are slightly higher than for parsing without: the LAS increased by 0.62%. It appears that the clause boundaries have more effect on dependency relations than on syntactic function tags: UAS for the parse with manual clause boundaries changed by 0.73%. LA increased with clause boundaries only by 0.29%.

Table 1. Parsing results of corpus with gold morphological tags

	LAS	UAS	LA
+CB	80.07	83.60	88.79
−CB	79.45	82.87	88.50

If we look at the text files of the corpus individually, the differences in the results are huge: the lowest LAS with clause boundaries is 72.6% while the highest is 87.7%. This illustrates well the different nature of the texts collected from the web: the text receiving the lowest score is from a personal blog while the text with the highest score comes from a religious website that has a designated language editor listed among its creators.

Without clause boundaries, the lowest LAS is 74.2% and the highest 86.9%. This means that actually, the least accurately parsed file did not benefit from manual clause boundaries as its LAS increased by 1.6% after removing the boundaries. However, LAS either decreased or remained the same for all the other files after removing the clause boundaries.

3.2 Parsing with Automatic Morphological Analysis

For the second experiment, the gold morphological analysis and lemmatization of the corpus were deleted, only segmentation into sentences and tokens were preserved. To explore the effect of clause boundaries, in one version of the files, the clause boundary markers were deleted, but in another version, they were replaced with commas unless immediately preceded or followed by punctuation.

Automatic Morphological Analysis of the Corpus. The corpus was morphologically analysed with an open-source Estonian morphological analyzer Vabamorf[2] which also performs initial disambiguation. It has been reported to find the correct analysis for over 99% of tokens in standard written Estonian [10]. On our new media corpus, it achieved the precision of 87.71% and recall of 82.37% with gold clause boundaries. Without clause boundaries, the results were 0.3% lower.

As the Vabamorf disambiguator still leaves several cases ambiguous but Malt-Parser needs conll format with only one analysis for each token, the disambiguator that is part of the constraint grammar syntactic analysis workflow described

---

[2] https://github.com/Filosoft/vabamorf.

in [9] was used. This disambiguator solves most ambiguities; for the remaining ones, the first analysis was chosen.

The most common errors in morphological analysis in terms of part-of-speech tags are shown in Table 2. As the corpus contains a lot of unknown words for the morphological analyser, it guesses them as nouns, resulting in many incorrect noun tags. Verbs are confused with adjectives in cases of participles - in those cases, human annotators often have problems as well choosing between an adjective and a verb analysis. Discourse particles get the analysis of interjection because the morphological analyser developed for standard written Estonian does not distinguish them.

Table 2. Errors in automatic part-of-speech tags without clause boundaries

Correct POS-tag	Automatic POS-tag	Count
Adverb	Noun	60
Verb	Noun	41
Verb	Adjective	39
Adjective	Noun	37
Adverb	Conjunction	29
Discourse particle	Interjection	27
Noun	Abbreviation	26
Conjunction	Adverb	23

In terms of morphological cases, the most common mistake is assigning a nominative case tag to a word that is unknown to the morphological analyser and has been incorrectly identified as a noun: out-of-vocabulary words most often receive morphological analysis of a noun in nominative case, unless they have some distinctive inflectional ending (or a suffix that is homonymous to this ending) of some other word class. This is illustrated by Example 1 where the incorrect compound word "niiet" ('so that') receives a nominative case tag while it actually should be an indeclinable adverb.

Another common source of errors is the homonymy of the forms of three most frequent cases: nominative, genitive, and partitive. These errors are significant because the case information helps to determine the syntactic function tags and distinguish between subjects, objects, predicatives (i.e. subject complements), and nominal modifiers. In Example 2, the names "Lasnamäe" and "Kopli" are in nominative case and should receive the syntactic function tag of a subject, but as the analyser assigns them genitive tags, they also do not receive the correct tag.

(1) niiet   see on väga väga kaua olnud seal   juba
     sothat-*NOUN.SG.NOM it   has very very long been   there already
     'So that it has already been there for a very long time'

(2)   Tallinnas Lasnamäe   ja   Kopli
      Tallinn   Lasnamäe-*NOUN.SG.GEN and Kopli-*NOUN.SG.GEN
      'Lasnamäe and Kopli in Tallinn'

Parsing. After conversion into conll format, both the corpus with manual clause boundaries and the corpus without were parsed with MaltParser. The results presented in Table 3 show that the difference between parsing with clause boundaries and without remains: the LAS and UAS are both 0.6% higher with clause boundaries than without. However, the difference in LA is less than 0.1%, meaning there is no improvement in the syntactic function tagging.

Table 3. Parsing results of corpus with automatic morphological tags

	LAS	UAS	LA
+CB	72.37	77.86	82.22
–CB	71.76	77.29	82.14

With automatic morphological analysis, as expected, the parsing accuracy is considerably lower than with gold morphological tags: UAS is 5.6–5.7% lower with clause boundaries than without, LA is 6.4–6.6% lower and LAS 7.7% lower.

Looking at the corpus files separately, we can see that LAS with clause boundaries varied from 67.2% to 81.8%, without clause boundaries from 64.9% to 81.8%. This time, all the files either benefitted from the clause boundaries or remained the same.

3.3   Parsing Errors Analysis

The most common errors in parsing with gold and automatic morphological analysis both with and without clause boundaries are shown in Table 4. The first five rows in Table 4 regard syntactically ambiguous cases that pose problems also for expert human annotators. E.g. while deciding between the syntactic labels of an adverbial and an adverbial modifier, both annotators often said that they find both analyses equally correct. The hierarchical relations between clauses also needed often to be discussed before reaching an agreement. It means that there was a lot of disagreement in which finite verb should be labelled as root.

In addition, deciding which constituent should be the subject and which one the predicative turned out to be a common point of discussion. This confusion arises when there are two nouns in an equational clause. In case of neutral word order, the noun preceding the copular verb is the subject, and the noun after the copula is the predicative. But word order in Estonian is rather free and, to great extent, determined by information structure. So, from the syntactic point of view, in a copular clause, both word orders are equally possible.

The error counts in Table 4 show that gold morphology was not helpful in the cases that were difficult for humans: the counts with and without clause boundaries in both experiments are quite similar. The only bigger difference is that the identification of a root is significantly better if parsing is done with gold morphological analysis and clause boundaries.

The last three rows of Table 4 present the errors that probably happen due to a faulty morphological analysis: the error counts with gold morphology are significantly smaller than with automatic morphology. Most of the errors result from an incorrectly assigned case - as described in Sect. 3.2, the grammatical cases in Estonian often look identical and therefore cause errors first in morphological analysis and then in parsing.

Table 4. Syntactic function tag error counts in parse with gold morphological analysis

Assigned tag	Correct tag	Gold morphology		Automatic morphology	
		+CB	−CB	+CB	−CB
Adverbial modifier	Adverbial	79	74	78	74
Subject	Predicative	64	60	65	60
Adverbial	Adverbial modifier	56	56	56	56
Root	Finite verb	54	70	65	70
Finite verb	Root	52	51	51	50
Subject	Object	42	42	73	77
Nominal modifier	Adverbial	42	46	65	68
Nominal modifier	Subject	34	37	60	67

3.4 Comparison with Standard Written Estonian

Our results in parsing Estonian new media texts with gold morphological analysis and without gold clause boundaries are slightly lower than those achieved on standard written Estonian by Muischnek et al. [1]: LAS received on standard texts is 0.8% higher at 80.3%, UAS is 0.5% higher at 83.4% and LA 0.1% higher at 88.6%. However, if we were able to compare the results with automatic morphological analysis, they would be lower on new media language because of the non-standard nature of those texts.

Most of the common parsing errors in syntactic function tags described in Sect. 3.2 were present in standard texts as well, except for the predicative-subject confusion. The fact that this error is so common in our corpus could be due to the non-standard word order. In standard Estonian, a common source of errors was the incorrect identification of a postpositional nominal modifier. Postpositional nominal modifier commonly occurs in nominalisations and thus is characteristic to more formal and compressed language.

Therefore, we can say the biggest problems in addition to identifying relations between clauses in both standard and non-standard written Estonian are the distinction between adverbials and modifiers as well as subjects and objects.

4   Conclusion and Future Work

This paper explored the dependency parsing of new media language with a parser trained for standard written Estonian. The aim was to find out on which aspect we should concentrate our efforts to be able to parse a non-standard language variety with high accuracy.

Our hypothesis was that the correct identification of clause boundaries would be useful in both morphological analysis and parsing. This was partially confirmed - the accuracy of morphological analysis was about 0.3% better and the parsing results about 0.6% higher with manually annotated clause boundaries than without. On the other hand, gold morphological analysis proved to increase the parsing accuracy significantly - with gold morphological tags, the parser trained on standard written Estonian performed on our new media data as accurately as on standard Estonian texts.

This means that, first and foremost, better morphological analysis would be the most useful factor to increase the accuracy of parsing new media texts. As there are many out-of-vocabulary word forms in new media texts, normalization would be needed for that.

In addition, the parsing quality could benefit from choosing a different morphological disambiguator, different syntactic models or syntactic annotation schemes, for example, the Universal Dependencies scheme for which the preconditions are fulfilled: there is an UD corpus of standard written Estonian and a semi-automatic transition system. Increasing the corpus size and adding new media texts to the parser training data would also be useful, especially for discourse particles which did not exist in the current parsing model's training data.

References

1. Müürisep, K., Nigol, H.: Disfluency detection and parsing of transcribed speech of Estonian. In: Proceedings of 3rd LTC, pp. 483–487 (2007)
2. Kong, L., Schneider, N., Swayamdipta, S., Bhatia, A., Dyer, C., Smith, N.A.: A dependency parser for tweets. In: Proceedings of EMNLP 2014, pp. 1001–1012 (2014)
3. Foster, J., et al.: From news to comment: resources and benchmarks for parsing the language of web 2.0. In: Proceedings of the 5th IJCNLP, pp. 893–901 (2011)
4. Seddah, D., Sagot, B., Candito, M., Mouilleron, V., Combet, V. The French social media bank: a treebank of noisy user generated content. In: Proceedings of COLING 2012, Technical Papers, pp. 2441–2458 (2012)
5. Särg, D.: Adapting constraint grammar for parsing Estonian chatroom Texts. In: Proceedings of TLT 14, pp. 300–307 (2015)
6. Kallas, J., Koppel, K., Tuulik, M.: Korpusleksikograafia uued võimalused eesti keele kollokatsioonisõnastiku näitel. In: Eesti Rakenduslingvistika Ühingu aastaraamat, vol. 11, pp. 75–94 (2015)
7. Muischnek, K., Müürisep, K., Puolakainen, T., Aedmaa, E., Kirt, R., Särg, D.: Estonian dependency treebank and its annotation scheme. In: Proceedings of TLT 13, pp. 285–291 (2014)

8. Nivre, J., Hall, J., Nilsson, J.: Malt-parser: a data-driven parser-generator for dependency parsing. In: Proceedings of the 5th LREC, pp. 2216–2219 (2006)

9. Muischnek, K., Müürisep, K., Puolakainen, T.: Parsing and beyond. Tools and resources for Estonian. Acta Linguist. Acad. **64**(3), 347–367 (2017)

10. Kaalep, H.-K., Vaino, T.: Complete morphological analysis in the linguist's toolbox. In: Congressus Nonus Internationalis FennoUgristarum Pars V, pp. 9–17 (2000)

Morphological Aanalyzer for the Tunisian Dialect

Roua Torjmen[1]([✉]) and Kais Haddar[2]

[1] Faculty of Economic science and Management of Sfax, Miracl Laboratory,
University of Sfax, Sfax, Tunisia
`rouatorjmen@gmail.com`
[2] Faculty of Sciences of Sfax, Miracl Laboratory, University of Sfax, Sfax, Tunisia
`kais.haddar@yahoo.fr`

Abstract. The morphological analysis is an important task for the
Tunisian dialect processing because the dialect does not respect any standard and it is different for modern standard Arabic. In order to propose
a method allowing the morphological analysis, we study many Tunisian
dialect texts to identify different forms of written words. The proposed
method is based on a self-constructed dictionary extracted from a corpus
and a set of morphological local grammars implemented in the NooJ linguistic platform. Indeed, the morphological grammars are transformed
into finite transducers while using NooJ's new technologies. To test and
evaluate the designed analyzer, we applied it on a Tunisian test corpus
containing over 18,000 words. The obtained results are ambitious.

Keywords: Tunisian dialect word · Morphological grammar
NooJ transducer

1 Introduction

The morphological analysis is an essential step in the automatic analysis of
natural languages. This analysis permits to recognize the words that are written
in several forms or written from an inflected form. Moreover, this analysis helps
in the creation of several applications like word normalization and parsing.

Unfortunately, Tunisian Dialect (TD) is not studied in Tunisian schools. This
fact misses the existence of a standard spelling. Also, the different TD pronunciation of one region to another complicates the situation. In addition, using
morphological analyzer designed for the Modern Standard Arabic (MSA) on TD
corpora can give poor results because TD is not only a derivative of Arabic but
also a mixture of several languages (Spanish, Italian, French and Amazigh).

In this paper, we are interested in treating TD and especially in creating linguistic resources for TD. Therefore, the purpose of this work was the construction of morphological analyzer applying to TD corpora. This analyzer is based
on a dictionary extracted from a study corpus and a set of morphological local
grammars. These resources are implemented in the NooJ linguistic platform that
provides several technologies allowing the proposed method implementation.

© Springer Nature Switzerland AG 2018
P. Sojka et al. (Eds.): TSD 2018, LNAI 11107, pp. 180–187, 2018.
https://doi.org/10.1007/978-3-030-00794-2_19

The paper is structured in six sections. In the second section, we present previous works dealing with the dialect processing. In the third section, we perform a linguistic study on the TD word forms. In the fourth section, we propose a method for TD morphological analysis. In the fifth section, we experiment and evaluate our analyzer on TD test corpus. Finally, this paper is closed by a conclusion and some perspectives.

2   Related Work

The TD works are not numerous and even if they exist, they essentially concern the speech recognition. These works are interested in the construction of the lexical resources for the Maghrebi dialects and especially for TD. In addition, they are based on statistical approaches. Among the works on the Maghrebi dialects, we quote [1]. The authors created firstly an annotated corpus. Secondly, from the created corpus, they elaborated a Moroccan morphological analyzer. For the Algerian dialect, the authors in [4] created their morphological analyzer based on BAMA and Al-Khalil analyzers. This work deals with the dialects of Alger and Annaba and neglects the other regions.

Turning now to the TD works, in [7,8], the authors built their morphological analyzer through the conversion of MSA patterns to TD ones. The output results are improved by Al-Khalil analyzer and then used machine learning algorithms. There is also another work for TD [2], the authors used a set of mapping rules and the ATB corpus to identify TD verb concepts in order to generate an extensional lexicon. Also, they constructed a tool called TD Translator (TDT) to generate TD corpora and to enrich semi-automatically their dictionaries.

Among the works using NooJ, we cite [3,5]. The authors created a dictionary containing 24,732 names, 10,375 verbs and 1,234 particles for MSA. They also built a set of morphological grammars containing 113 infected verb patterns, 10 broken plural patterns and a pattern of agglutination phenomenon.

In conclusion, the TD works do not have a large coverage especially with non-Arabic origin words. Also, we found that the studied work rule systems are not well developed. For this reason, they give bad results by applying them to a larger TD corpus. In addition, NooJ does not have lexical resources for TD.

3   Linguistic Study on the TD Word Form

In this linguistic study, our goal is the identification of words that have the same sense but that are written in a different way in TD. In the following subsections, we detail this specificity for three word types: adverbs, demonstrative pronouns and verbs. Nouns and adjectives will be treated later.

3.1   Adverbs

According to our linguistic study, we found that in TD there are adverbs having 2 writing forms presented in Table 1 and others having 4 writing forms presented in Table 2. Other types of adverbs can exist.

Adverbs of Table 1 can be terminated either by the letter "ش" 'ch' or by the letter "ه" 'h'. While adverbs of Table 2 can be terminated either by the letter "ا" 'a', by the letter "ه" 'h', by the letter "ة" 'h', or by the elimination of the last letter. Note that, there are other adverbs that have a unique writing form like the word "ياسر" 'yasir' (many).

Table 1. Adverbs with 2 writing forms

Adverb	Alternative writing form
كِيفَاش 'kiifaach' (how)	كِيفَاه 'kiifaah'
غْلَاش "alaach' (why)	غْلَاه "alaah'

Table 2. Adverbs with 4 writing forms

Adverb	Form 1	Form 2	Form 3
شْنِيَّة 'chniyyah' (what)	شْنِيَّا 'chniyyaa'	شْنِيَّه 'chniyyaah'	شْنِيّ 'chniyya'
زَعْمَة 'za'amah' (is it)	زَعْمَا 'za'amaa'	زَعْمَه 'za'amah'	زَعْمَ 'za'ama'

3.2 Demonstrative Pronouns

Turning now to demonstrative pronouns, this category has different words for the same meaning. In other words, all words in this category are synonyms. Some demonstrative pronouns used in TD are shown in Table 3. Other demonstrative pronouns can exist.

Table 3. Demonstrative pronouns

Demonstrative pronoun	Translation in English
أَوْكَا "awkaa'	this
هَاكَا 'haakaa'	this
هَذَا 'hadhaa'	this
هَاذَاكَا 'haadhaakaa'	this
هَذِيكَا 'hadhika'	this

Besides, demonstrative pronouns can be masculine singular, feminine singular, plural, or standard form. We notice for each gender and number, the demonstrative pronouns possess different writing forms. In Table 4, we present an example of the demonstrative pronoun "هَذَا" 'hadhaa'.

The demonstrative pronouns have specificities that must be respected during the construction of different lexical resources.

3.3   Verbs

The verbs in TD can be conjugated only in the past, the present and the imperative. We notice the absence of the third person plural in the feminine and the dual in both genders. Moreover, there is no difference between the second person singular feminine and the second person singular masculine.

Table 4. Example of demonstrative pronouns

Demonstrative pronoun with alternative writing	Gender and number
هَذَا 'hadhaa', هَذَة 'hadhah', هَذَيَا 'hadhaya', هَذِه 'hadhah'	Masculine singular
هَذِي 'hadhii', هَذِيَا 'hadhiya'	Feminine singular
هَذُم 'hadhum', هَذُمَا 'hadhumaa'	Plural
هَذَ 'hadh'	Standard

Besides, for the tense past, the first and the second person singular will be conjugated in the same way. For the tense present, the second person singular and the third person singular feminine also will be conjugated in the same way. In other words, in both cases, these verbs are written and pronounced in the same way. In the following example, Table 5 illustrates these conjugation cases through the regular verb "قْتِل" 'ktil' (To kill).

Table 5. Example of regular verb conjugation

	Past	Present	Imperative
1st person singular	قْتِلْت 'ktilt'	نِقْتِل [U+0646] 'niktil'	
1st person plural	قْتِلْنَا 'ktilna'	نِقْتِلُوا 'niktluu'	
2nd person singular	قْتِلْت 'ktilt'	تِقْتِل 'tiktil'	اقْتِل 'iktil'
2nd person plural	قْتِلْتُوا 'ktiltuu'	تِقْتِلُوا 'tiktiluu'	اقْتِلُوا 'iktiluu'
3rd person singular masculine	قْتِل 'ktil'	يِقْتِل 'yiktil'	
3rd person singular feminine	قْتِلْت 'ktilit'	تِقْتِل 'tiktil'	
3rd person plural	قْتِلُوا 'ktiluu'	يِقْتِلُوا 'yiktlou'	

Like MSA, regular and irregular verbs do not have the same conjugation. For this raison, we classify verbs under twelve different categories according to the nature of each verb.

This linguistic study will help us to construct a TD dictionary and to elaborate different rules to recognize different forms.

4   TD Morphological Analysis

The morphological analysis process that we propose recognizes words from a TD corpus and classifies these words according to their grammatical categories. This process is based on two stages which are the construction of a dictionary and the establishment of morphological grammars.

4.1   Dictionary

The dictionary that we created is considered as a set of entries having grammatical categories with the possibility of having inflectional and derivational paradigms [6]. In our created dictionary, we have added entries for the interrogative adverbs ADV+INTERR with the inflection paradigm called QUESTION, entries for adverbs ADV other types with the inflection paradigm called BARCHA, entries for demonstrative pronouns DEM with the inflection paradigm called DEM and entries for verbs V with the inflection paradigm VERBE. The fragment illustrated in Fig. 1 is an example of the dictionary entries that concern four types.

غْلاشْ, ADV+INTERR+FLX=QUESTION

تَـوْة, ADV+FLX=BARCHA

هَذَ ا, DEM+FLX=DEM

فْـبِـل, V+FLX=VERBE

Fig. 1. Entries example

In fact, the self-established dictionary has 29 interrogative adverbs, 21 adverbs other types, 9 demonstrative pronouns and more than 1,000 verbs.

4.2   Morphological Local Grammars

The morphological grammars that we propose are specified with a set of 15 finite transducers. Consequently, they will help us to recognize the TD words written in different forms and having the same grammatical categories.

In order to recognize the adverbs having two writing forms, we construct a transducer called QUESTION presented in Fig. 2. In this transducer, we make two paths to produce two word forms and we use the following operators: $<B2>$ deletes two last characters of the concerning word and $<E>$ designing the empty string keeps the word in its initial form. Now, we illustrate the transducer operation through an example of the word "غْلاشْ" "alaach' (why) which is already a dictionary entry. So, the last two characters are deleted through the operator $<B2>$. Thereby, the word "غَلَ" "alaa' is procured. Finally, the two characters "ه" and "آ" are added and the word "غَلَةآ" "alaah' (why) is obtained.

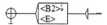

Fig. 2. Transducer QUESTION

Concerning the adverbs having four different writing forms, we create a transducer called BARCHA presented in Fig. 3. This transducer has four paths. In the first, second and third path, the last character is deleted though <B>. In the first and second path, either the character "ا" or the character "ه" are added. In the fourth path, we keep the entry without modification by <E>.

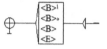

Fig. 3. Transducer BARCHA

For demonstrative pronouns, we create the transducer DEM with three operators <E>, <B> and <B2> shown in Fig. 4. With this flexional grammar, all demonstrative pronouns with all numbers (singular s and plural p) and genres (masculine m and feminine f) are recognized.

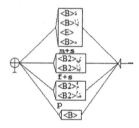

Fig. 4. Transducer DEM

In order to conjugate the TD verbs, we elaborate twelve transducers. In Fig. 5, we present the transducer VERBE which allows the conjugation of regular verbs by using the following operators: <E>, <B>, <LW> to beginning of word and <RW> to end of word. In this transducer, we offer different conjugation cases in all times (past I and present P) and modes (imperative Y) with all person (first person 1, second person 2 and third person 3) and all number (singular s and plural p).

In conclusion, we construct 15 transducers: 12 transducers for verbs, 1 for demonstrative pronouns, 1 for interrogative adverbs and 1 for adverbs other types. Then, all writing forms of treated categories can be detected.

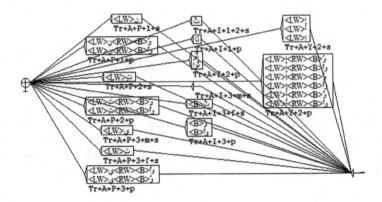

Fig. 5. Transducer VERBE

5   Experimentation and Evaluation

As we said previously, the self-established dictionary and the set of finite transducers are created in NooJ linguistic platform. The file barcha.nod is considered as the extensional version of our dictionary existed in file barcha.dic. Besides, barcha.nod uses the set of transducers in order to generate all flexional and derivational forms. Subsequently, the corpus is collected from social networks and from Tunisian novels. Note that, 1/3 of the corpus is for the study and 2/3 of the corpus is for the test and evaluation. Furthermore, our test corpus contains 18,134 words and his size is 522 Ko.

The experimentation of our morphological analyzer is based on the recognition of words having a grammatical category conceived in our dictionary. The linguistic analysis of the test corpus took only 2 seconds. In Table 6, we presented the number of Interrogative adverbs, adverbs, demonstrative pronouns and verbs in our test corpus and their number of recognized word.

Table 6. Obtained results for grammatical categories

	Verb	Interrogative adverb	Adverb other type	Demonstrative pronoun
Corpus 18134	3523 (100%)	206 (100%)	762 (100%)	139 (100%)
Recognized word	3121 (88%)	206 (100%)	762 (100%)	136 (97%)

We notice that our morphological analyzer detects all interrogative adverbs and all those of other types. Moreover, we observe that 0.03 demonstrative pronouns and 0.12 verbs are not recognized. This is due to agglutination. Therefore, this problem must be resolved. In our lexical resource, we have treated in our dictionaries words of different origin as "نجّم" 'najjim' (to can) which is a word of Amazigh origin. Also, we quote the word "مكيّج" 'makiyij' (to make up) which is word of French origin. The unrecognized words have other categories that are personal pronouns, nouns, adjectives and prepositions. Also, the unrecognized words are attached to the other words. The obtained results are ambitious for the treated categories and can be improved by increasing the coverage of our dictionary and by treating agglutinated forms, nouns and adverbs.

6 Conclusion

In the present paper, we develop a morphological analyzer for Tunisian dialect in NooJ linguistic platform based on a deep linguistic study. This morphological analyzer recognizes different writing word forms having specific grammatical categories. Also, it is based on self-established dictionary extracted from a test corpus and some morphological grammars. Thereby, the morphological grammars are specified by a set of finite transducers and by adopting the NooJ's new technologies. Thus, the evaluation is performed on a set of sentences belonging to a TD corpus. The obtained results are ambitious and show that our analyzer can treat efficiently different TD sentences despite the different origins of Tunisian words. As perspectives, we will increase the coverage of our designed dictionaries by treating other grammatical categories like nouns and adjectives. We will also treat the agglutinated phenomenon.

References

1. Al-Shargi, F., Kaplan, A., Eskander, R., Habash, N., Rambow, O.: Morphologically annotated corpora and morphological analyzers for Moroccan and Sanaani Yemeni Arabic. In: 10th Language Resources and Evaluation Conference, (LREC 2016), Portoroz, Slovenia, May 2016, pp. 1300–1306 (2016)
2. Boujelbane, R., Khemekhem, M. E., Belguith, L. H.: Mapping rules for building a Tunisian dialect Lexicon and generating corpora. In: Proceedings of the Sixth International Joint Conference on Natural Language Processing, Nagoya, Japan, 14–18 October 2013, pp. 419–428 (2013)
3. Hammouda, N.G., Haddar, K.: Parsing Arabic nominal sentences with transducers to annotate corpora. Comput. Sist. Adv. Hum. Lang. Technol. 21(4), 647–656 (2017). (Guest Ed. Gelbukh, A.)
4. Harrat, S., Meftouh, K., Abbas, M., Smaili, K.: Building resources for Algerian Arabic dialects. In: Fifteenth Annual Conference of the International Speech Communication Association, Singapore, 14–18 September 2014, pp. 2123–2127 (2014)
5. Mesfar, S.: Analyse morpho-syntaxique et reconnaissance des entités nommées en arabe standard. Doctoral dissertation, thèse, Université de franche-comté, France (2008)
6. Silberztein, M.: NooJ's dictionaries. In: Proceedings of LTC, Poland, 21–23 April 2005, vol. 5, pp. 291–295 (2005)
7. Zribi, I., Ellouze, M., Belguith, L.H., Blache, P.: Morphological disambiguation of Tunisian dialect. J. King Saud Univ.-Comput. Inf. Sci. 29(2), 147–155 (2017)
8. Zribi, I., Khemakhem, M.E., Belguith, L.H.: Morphological analysis of Tunisian dialect. In: Proceedings of the Sixth International Joint Conference on Natural Language Processing, Nagoya, Japan, 14–18 October 2013, pp. 992–996 (2013)

Morphosyntactic Disambiguation and Segmentation for Historical Polish with Graph-Based Conditional Random Fields

Jakub Waszczuk[1(✉)], Witold Kieraś[2], and Marcin Woliński[2]

[1] Heinrich Heine University Düsseldorf, Düsseldorf, Germany
waszczuk@phil.hhu.de
[2] Institute of Computer Science, Polish Academy of Sciences, Warsaw, Poland
{wkieras,wolinski}@ipipan.waw.pl

Abstract. The paper presents a system for joint morphosyntactic disambiguation and segmentation of Polish based on conditional random fields (CRFs). The system is coupled with Morfeusz, a morphosyntactic analyzer for Polish, which represents both morphosyntactic and segmentation ambiguities in the form of a directed acyclic graph (DAG). We rely on constrained linear-chain CRFs generalized to work directly on DAGs, which allows us to perform segmentation as a by-product of morphosyntactic disambiguation. This is in contrast with other existing taggers for Polish, which either neglect the problem of segmentation or rely on heuristics to perform it in a pre-processing stage. We evaluate our system on historical corpora of Polish, where segmentation ambiguities are more prominent than in contemporary Polish, and show that our system significantly outperforms several baseline segmentation methods.

Keywords: Word segmentation · Morphosyntactic tagging
Historical Polish · Conditional random fields

1 Introduction and Related Work

Despite the arguments raised in favor of performing end-to-end evaluation of Polish taggers rather than evaluating their disambiguation components only [14], the problem of word-level segmentation in Polish received little attention to this day. This is clearly due to relatively low frequency of segmentation ambiguities in Polish and, consequently, low influence of the phenomenon on tagging accuracy.

Several techniques of morphosyntactic tagging for Polish have been explored over the years, including trigrams [4], transformation-based methods[1] (TaKIPI [12]; Pantera [1]), conditional random fields (WCRFT [13]; Concraft [20]), and neural networks (Toygger [9]; KRNNT [22]; MorphoDiTa-pl [19]). The latter now obtain state-of-the-art results[2] in the task of morphosyntactic

---

[1] Based on algorithms involving automatic extraction of rules.
[2] See: http://poleval.pl/index.php/results/.

© Springer Nature Switzerland AG 2018
P. Sojka et al. (Eds.): TSD 2018, LNAI 11107, pp. 188–196, 2018.
https://doi.org/10.1007/978-3-030-00794-2_20

tagging for Polish [7]. All these taggers adopt a pipeline architecture, where morphosyntactic disambiguation (including guessing) is preceded by sentence segmentation, word segmentation, and morphosyntactic analysis (not necessarily in this order).[3] For instance, WMBT, WCRFT, Concraft, and KRNNT all relegate the three "subsidiary" preprocessing tasks to Maca [15]. For word segmentation, Maca relies on ad-hoc conversion rules, which transform and simplify the graph. If segmentation ambiguities persist, simple heuristics – e.g. choosing the shortest path among the remaining segmentation paths – are employed in the end. Another solution is used in MorphoDiTa-pl, which encodes all segmentation ambiguities as morphosyntactic ambiguities. More precisely, it relies on an expanded tagset and conversion routines which allow to encode a given segmentation DAG as a sequence over the expanded tagset. Other Polish taggers seem to neglect the problem of ambiguous segmentation altogether. Toygger, for instance, simply requires that the input text is already segmented and analyzed.

The issue with the existing solutions for Polish is that they assume that word segmentation is performed in preprocessing to morphosyntactic disambiguation. However, neither ad-hoc conversion rules nor simple heuristics are sufficient to deal with segmentation ambiguities, as the latter can require contextual information to be correctly dealt with. The method used in MorphoDiTa-pl actually avoids this pitfall to a certain extent, since it represents segmentation ambiguities in terms of morphosyntactic ambiguities. However, it relies on rather ad-hoc conversion routines which do not seem easily generalizable. One might want to enrich segmentation graphs to account for spelling errors, or to represent several segmentation hypotheses arising in a speech processing system, and it is hard to imagine how conversion routines could account for that.

The problem of word segmentation naturally received more attention for languages where it is more prevalent, such as Chinese or Japanese. Within the context of Chinese, segmentation is often regarded as a labeling task over sequences, where one of two labels – `Start` or `NonStart` – is assigned to each character in the sequence. CRFs, neural networks, and other labeling methods can be then used to discriminate between the possible `Start`/`NonStart` sequences for a given sentence, each sequence uniquely representing the corresponding segmentation [3,11].

The idea of modeling morphological segmentation graphs directly with CRFs was proposed by [10] for Japanese, where a DAG-based CRF assigns a probability to each path in a given segmentation DAG, thus allowing to discriminate between different segmentations and the corresponding morphological descriptions at the same time.

In this work, we use a method similar to [10] and apply it to Polish by extending an existing CRF-based tagger, Concraft, to handle ambiguous segmentation graphs (see Sect. 3). The system is coupled with Morfeusz [21], a morphosyntactic analyzer for Polish, which represents both morphosyntactic and segmentation ambiguities in the form of a DAG (see Sect. 2). Finally, we evaluate our system on historical Polish, where segmentation ambiguities are more prominent than in

---

[3] By extension, this holds true also for ensemble taggers, e.g. PoliTa [8].

the contemporary language, and show that our system significantly outperforms several baseline segmentation methods (see Sect. 4).

2   Morfeusz

Similarly to other systems listed in Sect. 1, we assume that morphological disambiguation is preceded by dictionary lookup providing all possible interpretations of the input text. This task is performed by the morphological analyser Morfeusz 2 [21], which is well suited to processing historical texts. Namely, Morfeusz allows to customize all linguistically sensitive parts of the analysis: inflectional dictionary, rules of segmentation and the tagset. Appropriate adaptation of Morfeusz to 19th century and Baroque Polish was done by the authors of the corpora we use, see [5,6].

Morfeusz accepts the text as a stream of characters, which it splits into tokens and describes each of them as an inflectional form by assigning a lemma and a morphosyntactic tag containing grammatical features of the form, starting with the part of speech. The tokens generated by Morfeusz are words or parts of words (they do not contain spaces). Segmentation in Morfeusz may be ambiguous. For that reason Morfeusz does not represent its output as a flat list, but as a DAG (directed acyclic graph) of morphological interpretations of tokens.

The past tense of Polish verbs has two variants, e.g. *czytałem* and *(e)m czytał* (1st person singular of 'to read'). The latter variant is interpreted by Morfeusz as consisting of two separate inflectional forms, *(e)m* being an auxiliary form of the verb BYĆ 'to be', which is written together with a preceding token. This variant of past tense was readily used in historical Polish, while in the contemporary language it is present only in specific constructions. The auxiliary form takes part in systematic homonymy with historical forms of numerous adjectives ending in -*em*, e.g. *waszem* ('yours' in instrumental or locative case of masculine or neuter gender). This word may be interpreted in ambiguous ways represented by the graph shown in Fig. 1. The first token on each path is a form of the adjective WASZ 'your' in various cases and genders (denoted with simplified Morfeusz tags).

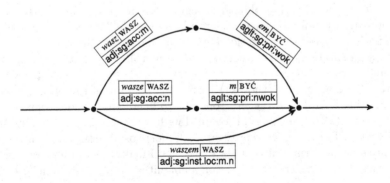

Fig. 1. Ambiguous segmentation of the word *waszem*

The second token is the auxiliary form of the verb BYĆ 'to be' used by the past tense. Depending on the context, each of the three alternative segmentation paths may constitute the correct interpretation.

Historical Polish provides also examples of accidental ambiguities in segmentation, e.g. the word *potym* can be interpreted as the preposition *po* 'after' written together with the form *tym* of the pronoun TO 'that' or as the form *poty* of the noun POT 'sweat' and an auxiliary *m*.

3   Graph-Based CRFs

A sequential CRF [16] defines the conditional probability of a sequence of labels $y \in Y^n$ given a sentence $x \in X^n$ of length n as:

$$p_\theta(y|x) = \frac{\Phi_\theta(y,x)}{Z_\theta(x)} \quad \text{with} \quad Z_\theta(x) = \sum_{y' \in Y^n} \Phi_\theta(y',x) \tag{1}$$

Intuitively, the *potential function* $\Phi_\theta(y,x)$ represents the plausibility of sequence of labels y given sentence x – the higher $\Phi_\theta(y,x)$ is, the more probable y w.r.t. x is – while the normalization factor $Z_\theta(x)$ ensures that the probabilities of the individual label sequences sum up to 1. In the particular case of 1-order sequential CRFs, the potential is defined as:

$$\Phi_{\theta(y,x)} = \exp\Big(\sum_{i=1...n} \sum_k \theta_k f_k(y_{i-1}, y_i, x)\Big), \tag{2}$$

where θ is a parameter vector and $f_k(y_{i-1}, y_i, x)$ is a binary feature function determining if the k-th feature holds within the context of (y_{i-1}, y_i, x).[4] Defining the exact form of feature functions is a part of the feature engineering process and will depend on the particular application. In our experiments (see Sect. 4), we relied on the Concraft's default feature templates.

Constrained CRFs. Concraft relies on a *constrained* version of sequential CRFs, in which to each position i in the input sequence a set of possible labels $r_i \subseteq Y$ is assigned[5]. When the sets of the potential morphosyntactic interpretations of the individual words in the sentence are available, such position-wise constraints can be successfully applied to both speed up processing and improve the tagging accuracy [20]. Formally, for a given sequence $y \in \prod_i r_i$:

$$p_\theta(y|x,r) = \frac{\Phi_\theta(y,x)}{Z_\theta(x,r)} \quad \text{with} \quad Z_\theta(x,r) = \sum_{y' \in \prod_i r_i} \Phi_\theta(y',x). \tag{3}$$

The probability of sequences not respecting the constraints is equal to 0. Note that such sequences are also not accounted for in $Z_\theta(x,r)$.

---

[4] Intuitively, f_k has a positive influence on the modeled probability if $\theta_k > 0$, negative influence if $\theta_k < 0$, and no influence whatsoever if $\theta_k = 0$.

[5] With $r_i = Y$ for out-of-vocabulary words.

Constrained DAG-Based CRFs. In this work, we rely on a further extension of the constrained model, where the structure of input is a DAG rather than a sequence. Let $D = (N_D, E_D)$ be a segmentation DAG of a given sentence, where N_D and E_D is the set of DAG nodes and edges, respectively. Let also $x_i \in X$ be the word assigned to $i \in E_D$ and $r_i \subseteq Y$ be the set of i's possible labels. We adapt the model to discriminate between the possible paths $y \in P(D, r)$, where $P(D, r)$ denotes the set of labeled paths encoded in D.

$$p_\theta(y|x, r, D) = \frac{\Phi_\theta(y, x, D)}{Z_\theta(x, r, D)} \quad \text{with} \quad Z_\theta(x, r, D) = \sum_{y' \in P(D, r)} \Phi_\theta(y', x, D). \quad (4)$$

The potential, in turn, is defined as:

$$\Phi_\theta(y, x, D) = \exp\left(\sum_{i \in Dom(y)} \sum_k \theta_k f_k(y_{i-1}, y_i, x, D)\right), \quad (5)$$

where $Dom(y) \subset E_D$ is the set of edges on the path, y_i denotes the label assigned to edge $i \in E_D$, and y_{i-1} denotes the label assigned to the preceding edge.

Within the context of morphosyntactic tagging, the above model assigns a probability to each DAG-licensed segmentation of the input sentence with a particular morphosyntactic description assigned to each segment on the path. Hence, maximizing $p_\theta(y|x, r, D)$ over all the labeled paths in D jointly performs segmentation and disambiguation, as desired.

Inference. The standard algorithms for sequential CRFs can be straightfor-wardly adapted to DAG-based CRFs. This includes the *max-product* algorithm used for Viterbi decoding (i.e., finding the most probable labeled path for a given DAG and constraints) and *sum-product* algorithm used for computing the forward and backward sums [18]. These two algorithms, in turn, allow to com-pute the posterior marginal probabilities of the individual segments and labels in the graph, the expected counts of CRF features per sentence, and to perform the maximum likelihood-based parameter estimation process, neither of which is particularly dependent on the underlying structure (sequence vs. DAG). We refer interested readers to [10] for more information on extending CRFs to DAGs.

Observations. Concraft relies on two types of features: 2-order *transition* fea-tures (t_{i-2}, t_{i-1}, t_i), and *observation* features (o_i, t_i), where o_i is an observation (wordform, suffix, prefix, shape, etc.) related to word i. Observations can include information about the preceding and following words – e.g., "the wordform of the segment on position $i-2$" – straightforward to obtain with sequential CRFs. However, in the case of DAGs position $i - 2$ may not be uniquely defined.

To overcome this issue, [10] limit the scope of features to two adjacent words, directly accessible in their 1-order model. We adopt a different solution, where the predecessor $i - 1$ (successor $i + 1$, respectively) of edge $i \in E_D$ is defined as the shortest (in terms of wordform length) edge preceding (following, respec-tively) i. This allows to define observations in terms of words arbitrarily distant from the current edge, which enables us to use Concraft's feature templates. Note that this does not mean that the model will prefer shorter paths, it simply means that observations are defined at a lower level of granularity. We believe this approach to be reasonable, as long as it is consistently used for both training and tagging.

4 Experimental Evaluation

Dataset. Our dataset consists of two separate gold-standard historical corpora of Polish. The first is a manually annotated subcorpus of the Baroque Corpus of Polish [5] which is still under development at the time of writing. It is currently ca. 430,000 tokens large and consists of samples (ca. 200 words each) excerpted from over 700 documents of various genres published between 1601 and 1772. The other dataset is a 625,000 tokens large manually annotated corpus of Polish texts published between 1830 and 1918 [6]. The corpus consists of samples (ca. 160 words each) excerpted from 1000 documents divided between five genres: fiction, drama, popular science, essays and short newspaper texts. The corpus is balanced according to genre and publication date.

The tagset of the 1830–1918 corpus consists of 1449 possible tags, from which 1292 were chosen at least once by human annotators. The Baroque tagset is much larger: it consists of 2212 possible tags and 1940 of them were used by annotators. The size of the Baroque tagset reflects the extensive time span covered by the corpus as well as significant grammatical changes which took place in that period, such as the grammaticalisation of masculine personal gender. It is assumed that since the turn of the 18th and 19th centuries Polish morphosyntactic system was not subject to major changes.

Table 1. Evaluation (our system on the left, segmentation baselines on the right)

	Baroque	1830-1918	Segm. baselines	Baroque	1830-1918
Tagging:			shortest path:		
precision	0.882724	0.903176	precision	0.712871	0.694111
recall	0.88303	0.903335	recall	0.503595	0.517577
Guessing:			longest path:		
precision	0.60125	0.610493	precision	0.264848	0.294253
recall	0.601214	0.609796	recall	0.41452	0.47628
Segmentation:			freq. based:		
precision	0.937455	0.951261	precision	0.838571	0.911858
recall	0.948684	0.965946	recall	0.724294	0.823025

Evaluation. The results of 10-fold cross-validation of our system on both historical corpora are presented in Table 1. We measured the quality of morphosyntactic tagging[6] and segmentation in terms of *precision* and *recall*. If several tags

---

[6] Note that these results abstract from the potential morphosyntactic analysis errors.

were assigned to a segment in gold data, we considered the choice of our system as correct if it belonged to this set. In case of segmentation, the choices of morphosyntactic tags were not accounted for.

We compared our system with three baseline segmentation methods. The first and the second one systematically chooses the shortest and the longest possible segmentation path, respectively. The third system is based on frequencies with which ambiguous segments are marked as chosen in gold data. Namely, we define the probability $p(x)$ of a segment x as #($x$chosen in gold + 1)/#(x present in gold + 2),[7] and the probability of a given segmentation path as a product of the probabilities of its component segments. Our system outperforms all three baseline methods significantly. Best among the baselines, the frequency-based method suffers from the *length bias* problem, as revealed by the differences between its precision and recall.

5   Conclusions and Future Work

The existing taggers for Polish either neglect the problem of ambiguous segmentation, or adopt ad-hoc approaches to solve it. By extending an existing CRF-based tagger for Polish, Concraft, to work directly on segmentation graphs provided by Morfeusz, we designed a system which addresses this deficiency by performing disambiguation and word-level segmentation jointly. Evaluation of our system on two historical datasets, both containing a non-negligible amount of segmentation ambiguities, showed that it significantly outperforms several baseline segmentation methods, including a frequency-based method.

The advantages of neural methods, now state-of-the-art in the domain, over CRFs include their ability to capture long-distance dependencies and to incorporate dense vector representations of words. For future work, we would like to explore the possibility of alleviating these weaknesses of CRFs, and the possibility of adapting neural methods to DAG-based ambiguity graphs. Following our claim that contextual information is required to properly deal with segmentation ambiguities, it seems clear that the principal way of improving segmentation accuracy is to focus on the quality of the subsequent NLP modules – disambiguation, parsing – as long as they are able to handle ambiguous segmentations.

Acknowledgements. The work being reported was partially supported by a National Science Centre, Poland grant DEC-2014/15/B/HS2/03119.

---

[7] Increasing all counts by 1 makes the probability of unseed segments equal to 1/2.

References

1. Acedański, S.: A morphosyntactic brill tagger for inflectional languages. In: Loftsson, H., Rögnvaldsson, E., Helgadóttir, S. (eds.) NLP 2010. LNCS (LNAI), vol. 6233, pp. 3–14. Springer, Heidelberg (2010). https://doi.org/10.1007/978-3-642-14770-8_3
2. Calzolari, N., et al., (eds.): Proceedings of the Ninth International Conference on Language Resources and Evaluation, LREC 2014. ELRA, Reykjavík, Iceland (2014). http://www.lrec-conf.org/proceedings/lrec2014/index.html
3. Chen, X., Qiu, X., Zhu, C., Liu, P., Huang, X.: Long short-term memory neural networks for Chinese word segmentation. In: Proceedings of the 2015 Conference on Empirical Methods in Natural Language Processing, pp. 1197–1206. ACL (2015). http://www.aclweb.org/anthology/D15-1141
4. Dębowski, L.: Trigram morphosyntactic tagger for Polish. In: Kłopotek, M.A., Wierzchoń, S.T., Trojanowski, K. (eds.) Intelligent Information Processing and Web Mining, pp. 409–413. Springer, Heidelberg (2004). https://doi.org/10.1007/978-3-540-39985-8_43
5. Kieraś, W., Komosińska, D., Modrzejewski, E., Woliński, M.: Morphosyntactic annotation of historical texts. The making of the baroque corpus of Polish. In: Ekštein, K., Matoušek, V. (eds.) TSD 2017. LNCS (LNAI), vol. 10415, pp. 308–316. Springer, Cham (2017). https://doi.org/10.1007/978-3-319-64206-2_35
6. Kieraś, W., Woliński, M.: Manually annotated corpus of Polish texts published between 1830 and 1918. In: Proceedings of the Ninth International Conference on Language Resources and Evaluation, LREC 2018. ELRA, Miyazaki, Japan (2018)
7. Kobyliński, Ł., Ogrodniczuk, M.: Results of the PolEval 2017 competition: part-of-speech tagging shared task. In: Vetulani and Paroubek [17], pp. 362–366
8. Kobyliński, Ł.: PoliTa: A multitagger for Polish. In: Calzolari et al. [2], pp. 2949–2954. http://www.lrec-conf.org/proceedings/lrec2014/index.html
9. Krasnowska-Kieraś, K.: Morphosyntactic disambiguation for Polish with bi-LSTM neural networks. In: Vetulani and Paroubek [17], pp. 367–371
10. Kudo, T., Yamamoto, K., Matsumoto, Y.: Applying conditional random fields to Japanese morphological analysis. In: Proceedings of the 2004 Conference on Empirical Methods in Natural Language Processing (2004). http://www.aclweb.org/anthology/W04-3230
11. Peng, F., Feng, F., McCallum, A.: Chinese segmentation and new word detection using conditional random fields. In: COLING 2004: Proceedings of the 20th International Conference on Computational Linguistics (2004). http://www.aclweb.org/anthology/C04-1081
12. Piasecki, M., Wardyński, A.: Multiclassifier approach to tagging of Polish. In: Proceedings of the International Multiconference on ISSN, vol. 1896, p. 7094
13. Radziszewski, A.: A tiered CRF tagger for Polish. In: Bembenik, R., Skonieczny, L., Rybinski, H., Kryszkiewicz, M., Niezgodka, M. (eds.) Intelligent Tools for Building a Scientific Information Platform, pp. 215–230. Springer, Heidelberg (2013). https://doi.org/10.1007/978-3-642-35647-6_16
14. Radziszewski, A., Acedański, S.: Taggers gonna tag: an argument against evaluating disambiguation capacities of morphosyntactic taggers. In: Sojka, P., Horák, A., Kopeček, I., Pala, K. (eds.) TSD 2012. LNCS (LNAI), vol. 7499, pp. 81–87. Springer, Heidelberg (2012). https://doi.org/10.1007/978-3-642-32790-2_9
15. Radziszewski, A., Śniatowski, T.: Maca-a configurable tool to integrate Polish morphological data. In: Proceedings of the Second International Workshop on Free/Open-Source Rule-Based Machine Translation (2011)

16. Sutton, C., McCallum, A.: An introduction to conditional random fields. Found. Trends® Mach. Learn. **4**(4), 267–373 (2012)
17. Vetulani, Z., Paroubek, P. (eds.): Proceedings of the 8th Language & Technology Conference: Human Language Technologies as a Challenge for Computer Science and Linguistics. Fundacja Uniwersytetu im. Adama Mickiewicza w Poznaniu, Poznań, Poland (2017)
18. Wainwright, M.J., Jordan, M.I., et al.: Graphical models, exponential families, and variational inference. Found. Trends® Mach. Learn. **1**(1–2), 1–305 (2008)
19. Walentynowicz, W.: MorphoDiTa-based tagger for Polish language (2017), CLARIN-PL digital repository. http://hdl.handle.net/11321/425
20. Waszczuk, J.: Harnessing the CRF complexity with domain-specific constraints. The case of morphosyntactic tagging of a highly inflected language. In: Proceedings of COLING 2012, pp. 2789–2804 (2012). http://www.aclweb.org/anthology/C12-1170
21. Woliński, M.: Morfeusz reloaded. In: Calzolari et al. [2], pp. 1106–1111. http://www.lrec-conf.org/proceedings/lrec2014/index.html
22. Wróbel, K.: KRNNT: Polish recurrent neural network tagger. In: Vetulani, Paroubek [17], pp. 386–391

Do We Need Word Sense Disambiguation for LCM Tagging?

Aleksander Wawer[1]([⊠]) and Justyna Sarzyńska[2]

[1] Institute of Computer Science, Polish Academy of Sciences,
Jana Kazimierza 5, 01-248 Warszawa, Poland
axw@ipipan.waw.pl
[2] Institute of Psychology, Polish Academy of Sciences,
Jaracza 1, 00-378 Warszawa, Poland
jsarzynska@psych.pan.pl

Abstract. Observing the current state of natural language processing, especially in the Polish language, one notices that sense-level dictionaries are becoming increasingly popular. For instance, the largest manually annotated sentiment dictionary for Polish is now based on plWordNet (the Polish WordNet) [13], also the Polish Linguistic Category Model (LCM-PL) [10] dictionary has its significant part annotated on sense level. Our paper addresses the important question: what is the influence of word sense disambiguation in real-world scenarios and how it compares to the simpler baseline of labeling using just the tag of the most frequent sense. We evaluate both approaches on data sets compiled for studies on fake opinion detection and predicting levels of self-esteem in the area of social psychology. Our conclusion is that the baseline method vastly outperforms its competitor.

Keywords: Linguistic Category Model · LCM · LCM-PL · Polish
Word sense disambiguation · Sense-level tagging

1 Introduction

This paper deals with the issue of practical design and usage of the Linguistic Category Model (LCM) dictionary, described in more detail in the next section, with regard to word senses.

The two opposing views on this matter are as follows.

The first, simple idea is to have a dictionary annotated on the level of lemmas. While it misses the nuances of contextual variations of word meaning, it does not require word sense disambiguation tools. In practical usage, the accuracy of annotations using such dictionary over a set of ambiguous words is determined by frequency of the most frequent sense of a word and its related labels (e.g. LCM tags).

The second, more elaborate method, is to have our dictionaries annotated on the level of word senses. Only this approach may, at least in theory, yield error-free annotations. However, its quality may be downgraded by the quality of word

© Springer Nature Switzerland AG 2018
P. Sojka et al. (Eds.): TSD 2018, LNAI 11107, pp. 197–204, 2018.
https://doi.org/10.1007/978-3-030-00794-2_21

sense disambiguation, needed to determine senses of words in their actual use in a sentence.

In this paper we compare both views, benchmarking them on two data sets, typical for LCM usage.

The paper is organized as follows: Sect. 2 describes the basics of Linguistic Category Model (LCM), Sect. 3 discusses how word senses influence LCM labels, Sect. 4 presents the current state of the Polish LCM dictionary. Sections 5 and 6 describe the word sense disambiguation (WSD) algorithm and data sets, used in our experiments. The results are summarized in Sect. 7.

2   Linguistic Category Model (LCM)

The LCM typology [7] is a well-established tool to measure language abstraction, applicable to multiple problems in psychology (for example [1,6,11]), psycholinguistics and more broadly, text analysis.

Its core idea is the categorization of verbs into classes reflecting their abstraction.

The most general, top level distinction of the Linguistic Category Model is the one between state verbs (SV) and action verbs. As LCM authors put it, *state verbs (SV) refer to mental and emotional states or changes therein. SVs refer to either a cognitive (to think, to understand, etc.) or an affective state (to hate, to admire, etc.).* This verb category is the most abstract one and also present in Levin's typology.

The other more concrete type of verbs in the LCM are action verbs. This type is always instantiated as one of its two sub-types, descriptive and interpretative action verbs (DAV and IAV) that all refer to specific actions (e.g., to hit, to help, to gossip, etc.) with a clearly defined beginning and end. SVs, in contrast, represent enduring states that don't have a clearly defined beginning and end.

The distinction between DAVs and IAVs is based on double criteria. The first states that DAVs have at least one physically invariant feature (e.g. to kick - leg, to kiss - mouth), whereas IAVs do not (therefore, are more abstract than DAVs). The second criterion, sentiment, states that IAVs have a pronounced evaluative component (e.g., positive IAVs such as to help, to encourage vs. negative IAVs such as to cheat, to bully), whereas DAVs do not (e.g., to phone, to talk). Descriptive action verbs (DAVs) are neutral in themselves (e.g. to push) but can gain an evaluative aspect dependent on the context (to push someone in front of a bus vs. to push someone away from an approaching bus).

In practice, the criteria sometimes overlap. Some verbs have physical invariants but also have clear evaluative orientation. For instance, "to cry" always involves tears (an invariant physical feature), but carries negative sentiment.

3   LCM and Word Senses

Ideally, LCM labels should be assigned to verb senses, not verb lemmas. To illustrate why, let us focus on one verb, picked from our dictionary: *dzielić* (eng. divide or share) in Table 1. In the table, column called 'domain' contains plWordNet verb domain of each specific sense https://en.wikipedia.org/wiki/PlWordNet.

Table 1. LCM tags for the senses of Polish verb *dzielić* (eng. to divide)

LCM	Lexical id	Synset id	Domain	Synsets/gloss
SV	81612	56818	State	Sharing
DAV	89828	63657	Ownership	Share sth.
IAV	89829	63654	Social life	Divide, separate
DAV	89826	63655	Change	Divide
DAV	89827	56841	Ownership	Separate
DAV	81339	56584	Thinking	Determine the quotient of two numbers

The verb has multiple senses that illustrate its various meanings, spanning across all possible LCM labels. The example proves that one verb lemma may have multiple LCM labels assigned to its senses. Let us look at it more closely.

The most abstract sense has the LCM label SV and its corresponding plWord-Net domain is state. It's English equivalent is 'to share'. In Polish, it refers to an abstract property of something being shared between multiple objects or people. For instance, in programming, an object reference may be shared between multiple class instances. A point of view may be shared between multiple people. No physical correlates are involved and the meaning is clearly an abstract one too, therefore SV label is the most appropriate.

In its sense related to social life domain, the verb becomes interpretative (IAV). It's English equivalent in this case means 'to divide'. An example of meaning reflected here may refer to groups of people divided by their opposite opinions, often linked to strong sentiments. There are no physical correlates and no objects are involved, therefore IAV tag is appropriate.

Finally, the verb may give a description of an observable event in a situational context. For example, a separation of ownership (e.g. ownership of something is divided between multiple owners). This situation usually refers to some owned entity, therefore in this meaning the verb becomes a DAV.

Generally, the principles behind LCM labels make the distinction between IAV and DAV sometimes vague. If a verb refers to observable events in a situational context, but requires additional interpretation and evaluation, it is an IAV. Otherwise, we assumed it's a DAV, especially if some physical correlates may be found. As for the verb 'to share', some of its meanings rely on context, whereas other meanings possess an autonomous, context-independent meaning.

Once attached to WordNet's senses, LCM labels could be used to tag texts. Word sense disambiguation routines could help distinguish whether particular verb occurrence in a sentence should be an SV, IAV or DAV (depending on LCM tags assigned to verb's synsets). In this article we examine the feasibility of this approach contrasted to the baseline of using an LCM tag of the most frequent sense of each verb.

4   The Polish LCM (LCM-PL): Current State

LCM-PL is a Polish language LCM dictionary [10]. It consists of multiple parts: manual sense-level annotations, lexeme-level manual annotations and automated annotations. The most recent version of the dictionary and its components are maintained at http://zil.ipipan.waw.pl/LCM-PL.

In our experiments we focused only on the sense-level annotated part. Sense-level annotation covers the most frequently used Polish verbs. As of March 2018, the number of senses with LCM labels exceeds 8000.

5   Word Sense Disambiguation (WSD)

There have been a few small-scale and experimental approaches to word sense disambiguation (WSD) in the Polish language. However, the only WSD method that promises universal applicability on any input domain is Page Rank WSD based on plWordNet sense repository [3]. The algorithm explores relations between units (synsets) in the plWordNet graph and textual contexts of word occurrences. It assumes that word senses that are semantically related occur more likely together in text than non-related.

In our experiments we used the implementation of this method available via REST API service described at http://nlp.pwr.wroc.pl/redmine/projects/ nlprest2/wiki. Input texts are loaded as strings. Users download .ccl (xml) files that contain plWordNet sense numbers as well as scores of each possible sense, assigned to each ambiguous token[1].

6   Data Sets

In this section we describe two Polish language data sets used to test the WSD method as well as the most frequent sense (MFS) method, our baseline. Both of the data sets are realistic examples of scenarios where LCM is normally applied.

6.1   Self-esteem

The first data set contains 427 textual notes on self-esteem (21,200 tokens).

Self-esteem is the positive or negative evaluations of the self, as in how we feel about it [8]. It might be measured on explicit as well as implicit level and

---

[1] Our data sets have been processed in March 2018. We have no information about version of the WSD module available at that time, including no information on potential open bugs that might influence sense annotation.

is assumed to be an important factor for understanding various psychological issues. Among 427 participants of the study 245 were woman. Measures of implicit (name-letter task) [2] and explicit (Single-Item Measure) [4] self-esteem were taken, accompanied with predictions of life-satisfaction based on electronic traces from social-media activity (Facebook LikeIDs) [12] and self-description. Subjects of the study were asked to write about themselves and their self-esteem for about ten minutes. To make the task easier they were provided with questions taken from an online study which is currently being conducted by James Pennebaker at http://www.utpsyc.org/Write/. It was hypothesized that higher levels of self-description abstractness is negatively associated with explicit self-esteem and positively with implicit self-esteem[2].

6.2   Fake Opinions

The other data set we used is 500 reviews of cosmetics (perfumes) available from http://zil.ipipan.waw.pl/Korpus%20Szczerosci. It contains 36,200 tokens.

Half of the reviews is fake: they have been written by professional fake opinion writers, hired for this task. The other half contains reviews written by well-established but moderately active members of one of the largest online communities interested in cosmetics. This part is almost certainly not fake, although it can not be fully ensured.

The corpus was collected for experiments on automated detection of fake reviews [5] using machine learning methods. LCM as a measure of language abstraction is one of the tools commonly applied to such tasks. It has been observed that untrue, fabricated texts are more likely to contain more abstract language than true utterances.

7   Results

From both data sets we selected 45 verbs detected by the WSD algorithm as occurring with more than one LCM tag. We did not include verbs that have more than one LCM tag in the dictionary, but were detected with just one LCM tag in both tested data sets.

For those 45 verbs, we asked a human annotator to provide LCM tags of their most frequent senses (MFS). In the tables below, MFS denotes the results of applying these LCM tags to every occurrence of each verb in both data sets. We also created a reference LCM labeling by manually examining all 466 verb occurrences in both corpora and assigning correct senses.

Table 2 illustrates accuracies of the two tested methods on both data sets. Surprisingly, the MFS method is overwhelmingly superior. It is over two times better on self esteem data and nearly two times better on fake opinions data.

Self-esteem data set appears to be better suited for the MFS heuristic than the more difficult language of fake opinions corpus, as they contain significant

---

[2] The study was conducted by members of the Warsaw Evaluative Learning Lab headed by professor Robert Balas.

Table 2. Accuracy of WSD and MFS approaches on both data sets.

		Accuracy	Support
Self Esteem	WSD	0.39	205
	MFS	0.92	
Fake Opinions	WSD	0.47	261
	MFS	0.77	

Table 3. Confusion matrix of WSD; self-esteem

actual

		iav	dav	sv
predicted	iav	17	20	4
	dav	0	24	32
	sv	14	53	40

Table 4. Confusion matrix of WSD; fake opinions

actual

		iav	dav	sv
predicted	iav	12	8	33
	dav	6	57	46
	sv	14	36	58

Table 5. Confusion matrix of MFS; self-esteem

actual

		iav	dav	sv
predicted	iav	37	0	6
	dav	0	55	0
	sv	2	9	96

Table 6. Confusion matrix of MFS; fake opinions

actual

		iav	dav	sv
predicted	iav	39	6	4
	dav	0	104	0
	sv	3	47	58

amounts of figurative language used to describe fragrances. Interestingly, the WSD method turns out to perform relatively better on the fake opinions corpus.

7.1 Error Analysis

Tables 3, 4, 5 and 6 present confusion matrices of both methods, WSD and MFS. The results reveal several notable observations. In MFS, the most frequent type of errors was confusing actual 'dav' and 'sv', yet actual 'sv' verbs were rarely mistaken for 'dav'. In the case of WSD, all types of errors occurred with relatively high frequency.

8   Conclusions and Future Work

We have benchmarked two approaches to LCM tagging of corpora in the Polish language. The first approach starts with a sense-level LCM dictionary followed by an application of WSD software to annotate sense occurrences and their LCM tags, the second is based on lemma-level annotation assuming LCM tags of the most frequent sense (MFS) of a verb.

Our experiments clearly point at the MFS approach as the results obtained by WSD method are far from satisfactory. Potential benefits of sense-level sensitivity are not realized due to low performance of the word sense disambiguation algorithm.

The main open issue worth investigating is the influence of existing WSD on the sentiment annotation. We plan to conduct a similar study using the reference sentiment data set from PolEval 2017 shared task [9] and apply the sense-level sentiment dictionary described in [13].

If the conclusions of our study apply also to sentiment detection, it might be worthwhile to reconsider using plWordNet as a base of annotations such as sentiment and LCM. Instead, it may be advisable to switch to a less granular dictionary framework: perhaps lemma level, but preferably lower granularity sense-level dictionary backed by high quality WSD, with a potential to outperform MFS.

References

1. Beukeboom, C., Tanis, M., Vermeulen, I.: The language of extraversion: extraverted people talk more abstractly, introverts are more concrete. J. Lang. Soc. Psychol. **32**(2), 191–201 (2013)
2. Hoorens, V.: What's really in a name-letter effect? Name-letter preferences as indirect measures of self-esteem. Eur. Rev. Soc. Psychol. **25**(1), 228–262 (2014)
3. Kędzia, P., Piasecki, M., Orlińska, M.: Word sense disambiguation based on large scale Polish Clarin heterogeneous lexical resources. Cogn. Stud. **15**, 269–292 (2015)
4. Robins, R.W., Hendin, H.M., Trzesniewski, K.H.: Measuring global self-esteem: construct validation of a single-item measure and the rosenberg self-esteem scale. Pers. Soc. Psychol. Bull. **27**(2), 151–161 (2001)
5. Rubikowski, M., Wawer, A.: The scent of deception: recognizing fake perfume reviews in polish. In: Kłopotek, M.A., Koronacki, J., Marciniak, M., Mykowiecka, A., Wierzchoń, S.T. (eds.) IIS 2013. LNCS, vol. 7912, pp. 45–49. Springer, Heidelberg (2013). https://doi.org/10.1007/978-3-642-38634-3_6
6. Rubini, M., Sigall, H.: Taking the edge off of disagreement: linguistic abstractness and self-presentation to a heterogeneous audience. Eur. J. Soc. Psychol. **32**(3), 343–351 (2002)
7. Semin, G.R., Fiedler, K.: The cognitive functions of linguistic categories in describing persons: social cognition and language. J. Pers. Soc. Psychol. **54**(4), 558 (1988)
8. Smith, E.R., Mackie, D.M., Claypool, H.M.: Social Psychology. Psychology Press, Hove (2014)
9. Wawer, A., Ogrodniczuk, M.: Results of the PolEval 2017 competition: sentiment analysis shared task. In: 8th Language and Technology Conference: Human Language Technologies as a Challenge for Computer Science and Linguistics (2017)
10. Wawer, A., Sarzyńska, J.: The linguistic category model in polish (LCM-PL). In: Chair, N.C.C., et al. (eds.) Proceedings of the Eleventh International Conference on Language Resources and Evaluation (LREC 2018). European Language Resources Association (ELRA), Paris, France, May 2018
11. Wigboldus, D.H., Semin, G.R., Spears, R.: How do we communicate stereotypes? Linguistic bases and inferential consequences. J. Pers. Soc. Psychol. **78**(1), 5 (2000)

12. Youyou, W., Kosinski, M., Stillwell, D.: Computer-based personality judgments are more accurate than those made by humans. Proc. Natl. Acad. Sci. **112**(4), 1036–1040 (2015)
13. Zaśko-Zielińska, M., Piasecki, M., Szpakowicz, S.: A large wordnet-based sentiment lexicon for Polish. In: Proceedings of the International Conference Recent Advances in Natural Language Processing, pp. 721–730. INCOMA Ltd., Shoumen, BULGARIA, Hissar, Bulgaria, September 2015. http://www.aclweb.org/anthology/R15-1092

Generation of Arabic Broken Plural Within LKB

Samia Ben Ismail[1]([✉]), Sirine Boukedi[2], and Kais Haddar[3]

[1] ISITCom Hammam Sousse, Miracl Laboratory, Sousse University, Sfax, Tunisia
samia_benismail@yahoo.fr
[2] National Engineering School, Miracl Laboratory, Gabes University, Sfax, Tunisia
sirine.boukedi@gmail.com
[3] Faculty of Sciences of Sfax, Miracl Laboratory, University of Sfax, Sfax, Tunisia
kais.haddar@yahoo.fr

Abstract. The treatment of Broken Plural (BP) for Arabic noun using a unification grammar is an important task in Natural Language Processing (NLP). This treatment contributes to construct extensional lexicons with a large coverage. In this context, the main objective of this work is to develop a morphological analyzer for Arabic treating BP with Head-driven Phrase Structure Grammar (HPSG). Therefore, after a linguistic study, we start by identifying different patterns of BP and representing them with HPSG. The designed grammar was specified in Type Description Language (TDL) and then was experimented with LKB system. The obtained results were encouraged and satisfactory because our system can generates all BP forms that can have an Arabic singular noun.

Keywords: Arabic broken plural · Morphological HPSG grammar
TDL specification · Linguistic Knowledge Builder (LKB)

1 Introduction

Morphology study has always been a center of interest for many researchers, especially for Arabic language. Among the most important morphological structures, we find the Arabic plural. In fact, Arabic language has two types of plurals: regular and Broken Plural. The Arabic plural essentially BP is very frequent in Arabic corpora. The treatment of such forms allows in the construction of extensional lexicon with a wide coverage, especially using a unification grammar. This greatly reduces the ambiguities and execution time of parser. Indeed, such formalism (i.e. unification grammar) offers complete representation with a minimum number of rules. However, research works on Arabic BP especially with HPSG are very missing and virtually absent. Indeed, the treatment of Arabic BP is very delicate. It must cover various forms. Also, there exist several classification criteria like the noun letter number, the noun nature, the noun type and the schema.

In this context, we propose a method based on HPSG treating Arabic BP. To do that, we begin our work by a large study on the different BP forms then

© Springer Nature Switzerland AG 2018
P. Sojka et al. (Eds.): TSD 2018, LNAI 11107, pp. 205–212, 2018.
https://doi.org/10.1007/978-3-030-00794-2_22

we propose an adequate classification. The identified paradigms are represented with HPSG formalism, specified in TDL language and experimented with LKB (Linguistic Knowledge Builder) system. The result is a morphological tool recognizing Arabic BPs. The originality of the present work appears in the use of a unification grammar (HPSG) and parsers generator (LKB). This kind of system is based on experimented algorithms. It is conceived for grammars specified in TDL. The use of such language represents another novelty. Indeed, TDL offers portability of the conceived grammar and an object oriented paradigm to construct the type hierarchy. This shows the great interaction between computer science and linguistics.

In this paper, we start by describing some previous works about morphological analyzers. Then, we present the proposed type hierarchy for Arabic BP. Based on this classification; we present the elaborated HPSG grammar for Arabic BP and its TDL specification. After that, we give the experimentation with LKB system and we evaluate the obtained results. Finally, we close the present paper by a conclusion and some perspectives.

2   Previous Works

The literature showed that there exist two main approaches treating the morphological analysis: statistical and symbolic ones. In this paper, we focus on works based on the symbolic approach.

The research work of [1] implements an Arabic algorithm applying two types of stemming: light and heavy, to extract the triliteral roots of words. This kind of algorithm is not based on a dictionary to detect the stemmer. It removes prefixes and suffixes, compares the output to standard word sources and correct the extracted root. As result, the accuracy of this work attains 75.03%. Contrariwise, in [5], the authors propose a method that detects the Arabic BP form the vowel or not vowel texts. This method was implemented within Nooj platform, based on a dictionary, a morphological and disambiguation grammar. For 3,158 words, the precision attains 80% and the recall attains 78.37%. The obtained results are good but not optimal. Indeed, some kinds of forms were not detected due to the absence of the vowel in the text.

The work of [2] implements an HPSG grammar generating some concatenative forms (i.e. verb conjugation and noun regular plural). As measure result, this work attains 87%, the total of performance percent, this work was implemented within the LKB platform. Furthermore, other works delight some morphological aspect such the declination in [8] and definiteness in [9]. All these works were focused on regular forms for an Arabic word. In fact, we remark that irregular forms such as the BP were always neglected and not well specified in an appropriate formalism.

3   Arabic Broken Plural

Referring to [4], an Arabic BP is an irregular plural noun obtained from a singular noun according to various schemas. As shown in Table 1, the change in structure is manifested either by adding or deleting one or more initial letters, either by modification of vocalization, or by the combination of two or three cases of change. In addition, in some cases, the BP can keep the same form of singular noun.

Table 1. Applied operation to obtain a BP.

Applied operation	Singular noun	BP
Add one or more letters	sahmun (arrow)	sihaāmun (arrows)
Delete one or more letters	rasuūlun (prophet)	Rusulun (prophets)
Modification of vocalization	ùasadun (lion)	ùusudun (lions)

Table 1 shows tree kinds of operations in the internal structure of the singular noun during its transformation into the BP. In the first one, the letter "ﺍ" ($\bar{a}$) is added to the singular noun "sahmun". In the second, the letter "ﻭ" (w) is removed from the singular noun "rasuwlun". While, in the third, the vocalization of the letters "ﺱ" (s) and "ﺍ" (ù) are modified to "ﺍ" (u). In addition, the BP transforms the defective letter "ﺍ" ($\bar{a}$) in the singular noun to its original way "ﻭ" (w) or "ﻱ" (y) such as "baābun (door) −> ùabwaabun (doors)". Besides, if a singular noun contains a Hamzated-letter (i.e. "ﺍ" (ù) or "ﺉ" (ỳ), it should be transformed respectively to "ﻭﺅ" (ùw) and "ﺉ" (ỳ/ā). So, the various changes lead to the appearance of patterns generating the BP easily and with a complete lexical category. Then, when we obtain the BP, we find two features (type and plural schema). We give in Fig. 1 the BP type hierarchy.

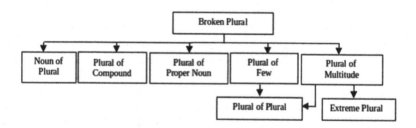

Fig. 1. BP type hierarchy

As shown in Fig. 1, the BP has five types. The most popular types are the Plural of Multitude and the Plural of Few. The last type of Plural is used when we have a number superior than three and less than ten. This type has four famous schemas that are: ùaf´ulu/ùaf´aālun/fi´laä/ùaf`ilaäun. However, the Plural of Multitude is used when we have a number superior than ten. This type has different famous schemas (i.e. fu´lun/fu´uūlun). Besides, the Plural of Multitude can be composed from Extreme Plural. This type represents plural that cannot be transformed another time in the plural. In certain cases, the plural of multitude or the plural of few can be transformed to regular plural or to BP one more time that is called Plural of Plural such as the singular noun "silaāḥun (weapon)" becomes "ủaslihaäun" then "ủasaālihun" unlike the Extreme Plural. Furthermore, we can transform the BP to duel such as "rimaāḥun (spears)" become "rimaāḥaāni (two spears)".

For the Arabic noun, the study showed that there exist plural nouns without singulars. This type of plural called Noun of Plural that can be transformed to Plural (i.e. qawmun (people) to ủaqwaāmun). In addition, this specific type can be treated syntactically as singular number or plural number. Furthermore, we find the plural of compound depending on the type of a compound. An Arabic compound can have various grammatical structures such as annexation. In fact, for annexation compound beginning with "ibn (son)", if it is used for a human noun, then it can have either a regular male plural or a BP like the compound word "ibn ´abbaās" has the plural "banū ´abbaās" or "ủabnā´ abbās". In the case of non-human noun, the noun "ibn" becomes "banāt". However, the annexation compound that begins with "dū" is always treated with a regular plural. Concerning the plural of proper Noun, male or female, we can treat them either to the regular plural or to the BP. In the case of the BP, the plural of Multitude and the Plural of Few will be applied according to the characteristics of the proper noun: if the proper noun is a masculine gender, it can be transformed into two types of plural (i.e. regular, BP) while if it is a female gender, it can be just transformed to regular plural.

After the linguistic, we can detect a set of patterns representing the proposed properties to generate there different forms. In the next section, we describe the elaborated Arabic HPSG representing the BP patterns.

4   Elaborated Arabic HPSG grammar for Broken Plural

Referring to Pollard and Sag, HPSG [7] represents the different linguistic structures (i.e. types, lexicon and morphological/syntactical rules) based on typed feature structure called AVM (Attribute Value Matrix). Besides, this grammar is based on a set of constraints and inheritance principle modeling the different grammatical phenomena. Inspired from some previous works such as [2] and based on our linguistic study [4], we adapted the HPSG representation of Arabic noun to elaborate the Arabic BP. According to [4], just the entire variable noun can have in-flected forms such as BP. This category is the first constraint added in our patterns to represent an Arabic BP. Moreover, it should be taken into consideration the type (NTYPE) and the nature of noun (NAT). So, we identify a set of 24 patterns that can be used to construct BPs. Then, each pattern has a set of schema. According to the pattern schema, the type and the nature of noun; we can deduce the possible change to obtain a BP. The following table illustrates some examples of Arabic BP.

As shown in Table 2, the noun of the structure "ḥml" has two forms of BP but the distinction is possible only by adding the schema vowels of the singular noun. In fact, if the schema is "fi`lun", the BP will be "ủḥmāl" while the

Table 2. Example of Arabic BP pattern.

Class	Plural schema	Example	
		Singular	Plural
$(C_3C_2C_1)$f'l	$(C_3 | C_2C_1$)ùf āl	hml(pregnency)	ùhmāl(pregnencies)
	$(C_3 | C_2C_1)$ f āl	Ǧml(camel)	Ǧmāl(camels)
	$(C_3{}_{,}C_2C_1)$f wl	hml (load)	hmūl (loads)

schema is "fa`lun", the BP will be (hmūl). Besides, according to Table 2 and our linguistic study, we deduce that we must to add two features when we represent the Arabic BP. These two features are "PluSCHEME" and "PluTYPE". For example, PluSCHEME can be "ùf āl" as shown in Table 2. Furthermore, for example, PluTYPE can be Plural of Few or an Extreme Plural Fig. 2, we give the BP "riǧāl/men", after application of a specified BP rule.

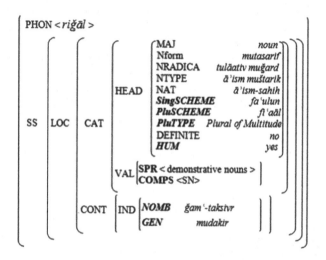

Fig. 2. AVM of the noun "riǧ āl" after application of a BP rule

As represented in Fig. 2 the used morphological BP rule inherits the features from its singular noun such as SingSCHEME and adds its both proper features. These two features were added at the level of the HEAD feature. Moreover, we specify another feature called IND, representing the gender and the number of the BP. Besides, this type of BP can be transformed to regular plural according to our type hierarchy of BP. For this case, we modify the value of feature PluTYPE to Plural of plural and the value of the feature NOMB. This last feature can be sound male plural or sound female plural, however, we can't find a regular rule to treat the value of this feature because the distinction can be just by the used context. In the next section, we present the TDL specification of the elaborated Arabic HPSG grammar for Broken Plural.

5   TDL Specification

To implement the proposed HPSG BP grammar within LKB system, it is necessary to specify it in TDL. Indeed, TDL [6] syntax is very similar to HPSG. Figure 3 illustrates the type specification of the Arabic BP in TDL.

tete := valeur & [MAJ string, DEC dec]. nom := tete & [MAJ "nom", NTYPE ntype, NFORM nform].	nom_variable:= nom & [NFORM متصرف, NGENRE ngenre, NRADICAL nradical, NAT nat, NRACINE string, SingSCHEME nscheme, ADJ boolean, DEFINI boolean].	nom_variable_pluBrise:= nom_variable & [PluType plutype, PluScheme pluscheme].

Fig. 3. Type specification of BP

In Fig. 3 shows that the specification of Arabic BP is inherited from the variable noun that is inherited itself from the noun type. As well as the Arabic noun is inherited from the base sign "tete". Moreover, for each level of inheritance, we add also the specific constraints and features of this type. So, each type has its own specifications developed in the type files (i.e. type.tdl and lex-type.tdl). Then, each entry of noun must be specified by features and constraints. Besides, this specification of lexical noun is treated in the file "lexicon.tdl". Moreover, the TDL specification for a noun represents the canonical form (i.e. singular and indefinite). The other forms are generated automatically by applying the elaborated rules. We give in Fig. 4 an example of these morphological rules.

```
%(letter-set (!f   ط ب ن ق ي د ك ج ح ع ج ر ف ب س م ل))
nom-pluriel-brisé1 :=
%suffix (!f !!f)
l2m-flex &
[SS[LOC[CAT[TETEnom_variable_pluBrise&        [NAT        اسم صحيح,NRADICAL
اسم مجرد.اسم تكثي,SingSCHEME #P1,PluScheme فعال, PluType جمع الكثرة, NGENRE #ngenre, NRACINE
#string, ADJ #boolean, NTYPE #ntype]],
CONT. IND[NOMB جمع تكبير, GEN اذكر]]],
ARGS⊲[SS.LOC.CAT.TETE[NAT  اسم صحيح,NRADICAL  مجرد,اسم تكثي,SingSCHEME  #P1,
NGENRE #ngenre, NRACINE #string, ADJ #boolean, NTYPE #ntype]] >].
```

Fig. 4. Example of morphological rule applied to generate a BP

The rule of Fig. 4 is used to generate the BP of the schema "fi`aāl". In fact, we add one letter "<ا>" to the suffix that belongs to the set of letter called "!f". This rule is applied to the nature of the noun ism-sahih (sound) and to a set of singular schema regrouped in Pattern called "P1" such as "fa´ulun" and "fa´lun". In addition, the role of symbol '#' attached to P1 or any features is allowing the inheritance of features values. Moreover, this singular noun must be triliteral. During the steps of specification, we create five TDL files; three for the type specification, one for the lexicon and one for the morphological rules specification that contains 33 rules to generate the Arabic BP. These files are added to the LKB platform. Therefore, in the following, we present our obtained results with LKB.

6   Experimentation and Evaluation

To generate the BP of a singular Arabic noun, we can use the linguistic platform LKB. This platform is developed by Copestake [3]. LKB can generate automatically different robust analyzers based on a set of reliable algorithms. It is composed of two types of files: LISP files representing the systems files and TDL files where we specified the elaborated Arabic HPSG. After application of the added rules, LKB generates an adequate derivation tree if we give it a noun in BP form. Thus, Fig. 5 shows an example of an obtained BP with LKB.

Fig. 5. Example of BP generated with LKB

As shown in Fig. 5, our LKB generated tool gives the BP form "riǧāl/men" from the canonical noun "raǧul/man". Also, the given forms contain all the added and the inherited morphological features. Indeed, when we specify a canonical noun, we add its vocalized schema. So, the elaborated LKB tool generates all BP forms for each type of singular noun. However, for some cases of BP, two derivation trees can be generated for BP. In fact, the singular schema "fa'laāäun" have the same treatment to obtain its BP that has various plural schemas. In fact, for example, the both of noun "marmaāäun/crosshair" and "sa'laāäun/cough" have the same type, nature and singular schema but we cannot distinguish between the correct plural schemas. In the following, we calculate the obtained average of the derivation tree number of each pattern. This average of performance (P) is defined by (1).

$$P = \frac{1}{\text{number of the generated derivation tree}}. \tag{1}$$

The performance percent is equal to 100% for 31 patterns and 50% for two patterns, giving as total performance percent equal to 97%. These ambiguities appear because of the not vocalization or writing some forms of Arabic BP with the same manner such as (maraāmin(crosshairs)/sa` aālin(cough)).

7   Conclusion

In this paper, we have elaborated a tool allowing the possessing of BP for Arabic singular nouns within LKB. This tool is based on a linguistic study, an established Arabic HPSG grammar treating BP and a TDL specification. The experimentation is performed by testing a set of BP nouns. The obtained results are encouraging and effectiveness showed by the reduction of ambiguity cases.

As perspectives, we intend to improve the obtained results by adding other lexical and morphological rules. Moreover, we aim to treat other irregular morphological phenomena such as gerund and agglutination resolutions. In addition, we plan to extend our Arabic HPSG grammar generates in order to treat all types of morphological phenomena.

References

1. Al-Kabi, M.N., Kazakzeh, S.A., Abu Ata, B.M., Al-Rababah, S.A., Alsmadi, I.M.: A novel root based Arabic stemmer. King Univ. Comput. Inf. Sci. **27**, 94–103 (2015)
2. Ben Ismail, S., Boukedi, S., Haddar, K.: LKB generation of HPSG extensional Lexicon. In: 14th ACS/IEEE International Conference on Computer Systems and Applications (AICCSA 2017), pp. 944–950, Hammamet-Tunisia (2017)
3. Copestake, A.: Implementing Typed Feature Structure Grammars. Cambridge University Press, Cambridge (2002)
4. Dahdah, A.: m′ĝm qwā′d āllĝtā′rbyt fy ĝdāwl w lwāt. 5th edn. Lebenon Library, Lebanon (1992)
5. Ellouze, S., Haddar, K., Abdelwahed, A.: NooJ disambiguation local grammars for Arabic broken plurals. In: Proceedings of the NooJ 2010 International Conference, pp. 62–72, Greece (2011)
6. Krieger, H., Schäfer, U.: TDL: a type description language for HPSG. Part 2: user guide. Technical reports, Deutsches Forschungszentrum für Künstliche Intelligenz, Saarbrücken, Germany (1994)
7. Pollard, C., Sag, I.: Head-Driven Phrase Structure Grammars. Chicago University Press, Chicago (1994)
8. Mahmudul Hasan, M., Muhammad Sadiqul, I., Sohel Rahman, M., Reaz, A.: HPSG analysis of type-based Arabic nominal declension. In: 13th International Conference on Arab Conference on Information Technology (ACIT 2012), pp. 10–13, Jordan (2012)
9. Mammeri, M.F., Bouhassain, N.: Implémentation d'un fragment de grammaire HPSG de l'arabe sur la plateforme LKB. In: 3rd International Conference on Arabic Language Processing (CITALA 2009), Rabat-Morocco (2009)

Czech Dataset for Semantic Textual Similarity

Lukáš Svoboda[(⊠)] and Tomáš Brychcín

University of West Bohemia, Univerzitní 22, 30100 Pilsen, Czech Republic
{svobikl,brychcin}@kiv.zcu.cz
http://www.zcu.cz/en/

Abstract. Semantic textual similarity is the core shared task at the International Workshop on Semantic Evaluation (SemEval). It focuses on sentence meaning comparison. So far, most of the research has been devoted to English.

In this paper we present first Czech dataset for semantic textual similarity. The dataset contains 1425 manually annotated pairs. Czech is highly inflected language and is considered challenging for many natural language processing tasks. The dataset is publicly available for the research community.

In 2016 we participated at SemEval competition and our UWB system were ranked as second among 113 submitted systems in monolingual subtask and first among 26 systems in cross-lingual subtask.

We adapt the UWB system for Czech (originally for English) and experiment with new Czech dataset. Our system achieves very promising results and can serve as a strong baseline for future research.

Keywords: Czech dataset · Semantic · Textual similarity

1 Introduction

Representing the meaning of the text is a key discipline in natural language processing (NLP). The Semantic textual similarity (STS) [1] - task assumes we have two textual fragments (word phrases, sentences, paragraphs, or full documents), the goal is to estimate the degree of their semantic similarity. STS systems are usually compared with the manually annotated data.

The authors of [2] explore the behavior of state-of-the-art word embedding methods on Czech, which is a representative of Slavic languages, characterized by a rich morphology. These languages are highly inflected and have a relatively free word order. Czech has seven cases and three genders. The word order is very variable from the syntactic point of view: words in a sentence can usually be ordered in several ways, each carrying a slightly different meaning. All these properties complicate building STS systems.

We experiment with techniques exploiting syntactic, morphosyntactic, and semantic properties of Czech sentence pairs. We discuss the results on new corpus

© Springer Nature Switzerland AG 2018
P. Sojka et al. (Eds.): TSD 2018, LNAI 11107, pp. 213–221, 2018.
https://doi.org/10.1007/978-3-030-00794-2_23

and give recommendations for further development. While state-of-the-art STS approaches for English achieve Pearson correlation of 89%, we show that the performance drops down to 80% when Czech data with the same domain are used.

In Sect. 2, we describe the properties of a new Czech dataset and in Sects. 3, 4 and 5, we describe applying Lexical and Semantic features to the system. In Sect. 6 we discuss the experiments and results of the system.

2   Czech STS Dataset

For Czech there are curpuses for measuring of the individual words embeddings properties, such as: RG-65 [3], WS-353 [4] and Czech Word analogy [2] corpora.

We introduce a new Czech dataset for the STS task. We have chosen the data with relatively simple and short sentence structure (Article Headlines and Image captions). With such data, we will create a better matching sentences to the original English data - as Czech has a free word order. Dataset has been divided into 925 training and 500 testing pairs (see Table 1) translated to Czech by four native speakers from previous SemEval years. In SemEval competition the data consist of pairs of sentences with a score between 0 and 5 (higher number means higher semantic similarity). For example, Czech pair:

Černobílý pes se dívá do kamery[1]
Černobílý býk se dívá do kamery[2]

has a score of 2, sharing information about camera, but it is about different animal. We kept annotated similarities unchanged.

Table 1. Dataset with STS gold sentences in Czech.

Dataset	Pairs
SemEval 2014–15 Images CZ – Train	550
SemEval 2013–15 Headlines CZ – Train	375
SemEval 2014–15 Images CZ – Test	300
SemEval 2013–15 Headlines CZ – Test	200

3   Data Preprocessing

To deal with Czech rich morphology, we use lemmatization [5] and stemming [6] to preprocess the training data. Stemming and lemmatization are two related

---

[1] A black and white dog looking at the camera.
[2] The black and white bull is looking at the camera.

fields and are among the basic preprocessing techniques in NLP. Both the methods are often used for similar purposes: to reduce the inflectional word forms in a text. Stemming usually refers to a crude heuristic process. Stemming removes the ends of words in the hope of achieving this goal correctly most of the time, and often includes the removal of derivational affixes. Product of lemmatization is a lemma which is a valid linguistic unit (the base or dictionary form of a word).

4 Semantic Textual Similarity

This section describes techniques used for estimating the text similarity.

4.1 Lexical and Syntactic Similarity

The authors of [7] address implementation of basic lexical and syntactic similarity features. Following techniques benefit from the weighing of words in a sentence using *Term Frequency - Inverse Document Frequency* (TF-IDF) [8].

We summarize these basic features as follows:

- **IDF weighted lemma n-gram overlapping**, measured with *Jaccard Similarity Coefficient* (JSC) [9].
- **IDF weighted POS n-gram overlapping**, measured with JSC.
- **Character n-gram overlapping**, measured with JSC.
- **TF-IDF** as standalone feature.
- **String features**, such as a longest common subsequence, longest common substring, where similarity is computed as fraction of longest common subsequence/substring divided by the length of both sentences.

4.2 Semantic Similarity

The Semantically oriented vector methods we use are based on the *Distributional Hypothesis* [10]. We employ semantic composition approach based on *Frege's principle of compositionality* [11]. We estimate the meaning of the text as a linear combination of word vectors with *TF-IDF* weights (more information can be found at [7]).

We use state-of-the-art word embedding methods, CBOW and SkipGram [12], and compare their semantic composition properties with FastText [13] method that enriches word vectors with subword information. This method promises significant improvement of word embeddings quality especially for languages with rich word morphology.

We train CBOW,SkipGram and FastText methods on Czech Wikipedia and provide experiments on standard dataset for word similarity (WS-353 [4] and RG-65 [3]) and word analogy [2]. Results are shown in Table 2.

Table 2. Word similarity and word analogy results on Czech Wikipedia.

Model	Word similarity		Word analogy
	WS-353	RG-65	
FastText - SG 300d wiki	67.04	67.07	71.72
FastText - CBOW 300d wiki	40.46	58.35	73.23
CBOW 300d wiki	54.31	47.03	58.69
SkipGram 300d wiki	65.93	68.09	53.74

5 STS Model

The combination of STS techniques mentioned in Sects. 4.1 and 4.2 is a regression problem. The goal is to find the mapping from input space $x_i \in \mathbb{R}^d$ of d-dimensional real-valued vectors (each value $x_{i,a}$, where $1 \leq a \leq d$ represents the single STS technique) to an output space $y_i \in \mathbb{R}$ of real-valued targets (desired semantic similarity). These mapping are learned from the training data $\{x_i, y_i\}_{i=1}^N$ of size N. System has been trained on 925 pairs and further tested on 500 pairs. We experiment with three regression methods (see Table 5):

- **Linear Regression**
- **Gaussian Process**
- **Support Vector Machines** (SVM) with Sequential Minimal Optimization (SMO) algorithm [14].

We use algorithms for the meaning representation in the same manner as we have used them for English at SemEval 2016. Methods benefit from various sources of information, such as lexical, syntactic, and semantic.

This section describes all measured settings and their reasons. The former is a traditional STS task with paired monolingual sentences originally translated from English data sources to Czech followed by crosslingual test. Gold data were evaluated as described in following subsections.

5.1 Lexical, Syntactic and Semantic Features

We evaluated each feature from three categories individually to see influence of particular feature (see Table 6).

5.2 Preprocessing Tests

Most of our STS models (apart from word alignment and POS n-gram overlaps) work with lemmas instead of word forms - this leads to better performance. We tested all features with three techniques of representing individual tokens in sentence - word, stemming and lemma (see Table 3).

Table 3. Pearson correlations on Czech evaluation data and comparison with the second best system from SemEval 2016 on English data. Test made with linear regression.

Model features\dataset	Correlation
ngram features (word)	0.6140
ngram features (lemma)	0.6959
ngram features (stem)	0.7319
ngram + string features (word)	0.7732
ngram + string features (lemma)	0.7897
ngram + string features (stem)	0.7829
all previous + syntactic (word)	0.7704
all previous + syntactic (lemma)	0.7860
all previous + syntactic (stem)	0.7865
all + CBOW composition (word)	0.7796
all + SkipGram composition (word)	0.7814
all + FastText - SG composition (word)	0.7774
all + CBOW composition (lemma)	0.7917
all + SkipGram composition (lemma)	0.7924
all + CBOW composition (stem)	0.7910
all + SkipGram composition (stem)	**0.7939**

5.3   Crosslingual Test

Cross-lingual STS involves assessing paired English and Czech sentences. Cross-lingual STS measure enables an alternative way to comparing text. Due to lack of the supervised training data in the particular language, cross-lingual task is getting still higher attention during last years.

We handled with the cross-lingual STS task with Czech-English bilingual sentence pairs in two steps:

1. Firstly, we translated original Czech sentences to English via *Google translator*. Why we have done this, if we do already have an original English data? We did not use the original-matching EN sentences, because we did not want to involve the manual translation by the native speaker. The Google translator, especially between EN and Czech is not accurate. That was also in most cases the way, how cross-lingual task was evaluated on SemEval2016. However, the situation is changing with new bilingual word embeddings methods coming up in recent years [15,16]. The Czech sentences were left untouched.
2. Secondly, we used the same STS system as for monolingual task. Because we have much bigger training set for English sentences, we wanted to see if such data-set will help us in performance on Czech, results can be seen in a Table 4.

Table 4. Comparison of Pearson correlations on monolingual STS task versus crosslingual STS task with automatic translation to English. Crosslingual model is trained on data from SemEval 2014 and 2015.

Model\dataset	Headlines	Images
Monolingual test	0.7999	**0.7887**
Czech-English crossling. (850 pairs)	0.8060	0.7583
Czech-English crossling. (3000 pairs)	**0.8198**	0.7649

Some of our word vector techniques are based on unsupervised learning and thus they need large unannotated dataset to train. We trained CBOW, Skip-gram and FastText models on Czech Wikipedia. Widipedia dump comes from 05/10/2016 with 847 milion token. Resulting models has vocabulary size of 773,952 words. This dump has been cleaned from any Wiki Markup and HTML tags. Dimension of vector for all these models was set to 300. All regression methods mentioned in Sect. 5 are implemented in WEKA [17].

6   Results and Discussion

Based on learning curve (see Fig. 1), system needs at least 170 pairs to set weights of individual features, therefore we can state that our system has reasonable amount of training data for learning - this theory is also supported by larger amount of training data thanks to cross-lingual test (see Table 4).

Fig. 1. Pearson correlation achieved by linear regression with different training data size (ranging between 50 and 850 pairs).

The best score of 78.87% on short *Images labels* we have achieved with simple Linear regression. Together with such short sentences we will not benefit from larger corpus as it can be seen in Table 4 from our evaluation of cross-lingual test with much larger corpus base (3000 pairs) that we have for English. From larger dataset we benefit on longer *Headlines* sentences, where we have achieved a score of 81.98%.

Table 5. Pearson correlations on Czech evaluation data and comparison with the second best system from SemEval 2016 on English data.

Model\dataset	Headlines	Images
Our best at SemEval 2016 (EN)	0.8398	0.8776
Linear regression	0.7918	**0.7887**
Gaussian processes regression	0.7986	0.7829
SVM regression	**0.7999**	0.7856

Interesting results can be seen in Table 6 for standalone vector composition. Standard Skipgram model seems to be more suitable to carry the meaning of a sentence as a simple linear combination of word vectors, despite the fact that it has lower score on similarity measurements of individual words (see Table 2).

Czech is a language with rich morphology, as it can be seen from Table 6, string features plays important role, especially *Greedy String Tiling*. The more matches is found in words endings, the higher is success of reasoning about two sentences. Results of testing lemma versus stemming techniques give a similar score. Of course without preprocessing, we get slightly lower score, this can be seen on n-gram features, where stemming is performing the best (see Table 3). When the model is covered by syntactic features, the situation for lemma and stemming techniques is nearly equal.

Table 6. Linear regression test of individual features, word base is lemma.

Model\dataset	Images	Headlines
Longest common subsequence	0.6586	0.6993
Longest common substring	0.4998	0.5886
Greedy string tiling	0.7005	**0.7983**
All string features	**0.7379**	0.7932
IDF weighted word n-grams	0.5979	0.6432
IDF weighted character n-grams	0.6885	0.7869
POS n-grams	0.5331	0.5618
TF-IDF	0.5785	0.5892
CBOW composition	0.6774	0.6355
SkipGram composition	0.6299	0.6785
FastText - SG composition	0.5966	0.6396
FastText - CBOW composition	0.4958	0.5102

Together with presented Czech dataset we have original matching sentences in English, so our dataset can be used for new STS crosslingual task without manual translating the sentences to English and can be evaluated directly with bilingual word embeddings methods [18,19] in future. These methods are getting popular in recent years and takes a key part in the recent SemEval 2017 competition.

The authors in [7] showed that use of syntactic parse tree and training with tree-based LSTM [20] does not help on English. Classic bag-of-words semantic approaches do a better job, however this situation might change on highly inflected languages like Czech and might be worth testing.

7   Conclusion

In this paper we introduced a new corpus for semantic textual similarity of Czech sentences. We created strong baseline based on state-of-the-art methods. Our Czech baseline achieved Pearson correlation of 80% (compared to 89% achieved on English data [7]).

Based on our system tests, we are getting lower accuracy with more complex sentences. As the translated sentences are relatively short and simple (Image captions and Headlines) with already significantly lower score, this paper shows a room for the potential future research focus.

The Czech STS corpus with its original matching sentences in English is available for free at following link: https://github.com/Svobikl/sts-czech.

Acknowledgements. This work was supported by the project LO1506 of the Czech Ministry of Education, Youth and Sports and by Grant No. SGS-2016-018 Data and Software Engineering for Advanced Applications. Computational resources were provided by the CESNET LM2015042 and the CERIT Scientific Cloud LM2015085, provided under the programme "Projects of Large Research, Development, and Innovations Infrastructures".

References

1. Agirre, E., et al.: Semeval-2016 task 1: semantic textual similarity, monolingual and cross-lingual evaluation. In: Proceedings of the 10th International Workshop on Semantic Evaluation (SemEval-2016), San Diego, California, pp. 497–511. Association for Computational Linguistics, June 2016
2. Svoboda, L., Brychcín, T.: New word analogy corpus for exploring embeddings of Czech words. arXiv preprint arXiv:1608.00789 (2016)
3. Krčmář, L., Konopík, M., Ježek, K.: Exploration of semantic spaces obtained from Czech corpora. In: Proceedings of the Dateso 2011: Annual International Workshop on DAtabases, TExts, Specifications and Objects, Pisek, Czech Republic, 20 April 2011, pp. 97–107 (2011)
4. Cinková, S.: WordSim353 for Czech. In: Sojka, P., Horák, A., Kopeček, I., Pala, K. (eds.) TSD 2016. LNCS (LNAI), vol. 9924, pp. 190–197. Springer, Cham (2016). https://doi.org/10.1007/978-3-319-45510-5_22
5. Straková, J., Straka, M., Hajic, J.: Open-source tools for morphology, lemmatization, POS tagging and named entity recognition. In: ACL (System Demonstrations), pp. 13–18 (2014)

6. Brychcín, T., Konopík, M.: HPS: high precision stemmer. Inf. Process. Manage. **51**(1), 68–91 (2015)
7. Brychcín, T., Svoboda, L.: UWB at SemEval-2016 task 1: Semantic textual similarity using lexical, syntactic, and semantic information. In: Proceedings of SemEval, pp. 588–594 (2016)
8. Manning, C.D., Schütze, H.: Foundations of Statistical Natural Language Processing. MIT Press, Cambridge (1999)
9. Niwattanakul, S., Singthongchai, J., Naenudorn, E., Wanapu, S.: Using of Jaccard coefficient for keywords similarity. In: Proceedings of the International MultiConference of Engineers and Computer Scientists, vol. 1 (2013)
10. Harris, Z.S.: Distributional structure. Word **10**(2–3), 146–162 (1954)
11. Pelletier, F.J.: The principle of semantic compositionality. Topoi **13**(1), 11–24 (1994)
12. Mikolov, T., Chen, K., Corrado, G., Dean, J.: Efficient estimation of word representations in vector space (2013)
13. Bojanowski, P., Grave, E., Joulin, A., Mikolov, T.: Enriching word vectors with subword information. Trans. Assoc. Comput. Linguist. **5**, 135–146 (2017)
14. Platt, J.: Fast training of support vector machines using sequential minimal optimization. Advances in Kernel Methods - Support Vector Learning. MIT Press, Cambridge (1998)
15. Levy, O., Søgaard, A., Goldberg, Y.: A strong baseline for learning cross-lingual word embeddings from sentence alignments. In: Proceedings of the 15th Conference of the European Chapter of the Association for Computational Linguistics, Volume 1: Long Papers, vol. 1, pp. 765–774 (2017)
16. Zou, W.Y., Socher, R., Cer, D., Manning, C.D.: Bilingual word embeddings for phrase-based machine translation. In: Proceedings of the 2013 Conference on Empirical Methods in Natural Language Processing, pp. 1393–1398 (2013)
17. Hall, M., Frank, E., Holmes, G., Pfahringer, B., Reutemann, P., Witten, I.H.: The WEKA data mining software: an update. ACM SIGKDD Explor. Newslett. **11**(1), 10–18 (2009)
18. Vulić, I., Moens, M.F.: Monolingual and cross-lingual information retrieval models based on (bilingual) word embeddings. In: Proceedings of the 38th International ACM SIGIR Conference on Research and Development in Information Retrieval, ACM, pp. 363–372 (2015)
19. Gouws, S., Søgaard, A.: Simple task-specific bilingual word embeddings. In: HLT-NAACL, pp. 1386–1390 (2015)
20. Tai, K.S., Socher, R., Manning, C.D.: Improved semantic representations from tree-structured long short-term memory networks. arXiv preprint arXiv:1503.00075 (2015)

A Dataset and a Novel Neural Approach for Optical Gregg Shorthand Recognition

Fangzhou Zhai$^{(\boxtimes)}$, Yue Fan, Tejaswani Verma, Rupali Sinha, and Dietrich Klakow

Spoken Language Systems, Saarland Informatics Campus, 66123 Saarbrücken, Saarland, Germany
thearkforyou@gmail.com, fanyue596893540@gmail.com, tejaswani.verma@gmail.com, rupalisinha23@gmail.com, dietrich.klakow@lsv.uni-saarland.de

Abstract. Gregg shorthand is the most popular form of pen stenography in the United States. It has been adapted for many other languages. In order to substantially explore the potentialities of performing optical recognition of Gregg shorthand, we develop and present Gregg-1916, a dataset that comprises Gregg shorthand scripts of about 16 thousand common English words. In addition, we present a novel architecture for shorthand recognition which exhibits promising performance and opens up the path for various further directions.

Keywords: Optical Gregg shorthand recognition
Character recognition · Convolutional neural networks
Recurrent neural networks

1 Introduction

Shorthand is an abbreviated symbolic writing system designed to increase speed and brevity of writing. Using it, writing speed can reach 200 words per minute. Shorthand scripts are designed to write down the pronunciations of words (e.g. 'through' would be written as if it were 'thru'). They are very concise, and rely on ovals and lines that bisect them to encode information (see Fig. 1). Gregg Shorthand, first published in 1888 in the US by John Robert Gregg, is the most prevalent form of shorthand. It has been adapted to numerous languages, including French, German, Russian, etc. Even with the invention of various electronic devices, Gregg shorthand has its advantages and is still in use today. It is therefore interesting to explore the possibilities of recognizing shorthand scripts.

Our key insight is, the characters in the words or their combinations could be seen as a kind of label, corresponding to different regions of the image. For example, an 'a' may correspond to an oval. Therefore, we could learn representations of the image regions and characters in a shared embedding space. The idea is very much inspired by the image captioning framework (see, e.g. [9]). However, our task is different from image captioning: we need to reconstruct

© Springer Nature Switzerland AG 2018
P. Sojka et al. (Eds.): TSD 2018, LNAI 11107, pp. 222–230, 2018.
https://doi.org/10.1007/978-3-030-00794-2_24

Fig. 1. A sentence in Gregg Shorthand from Encyclopedia Britannica [2]. Note how much more concise its corresponding image is compared to plain English text.

Fig. 2. Examples of Gregg shorthand. Note how different word endings are encoded in a subtle way.

a *ground truth* word, instead of a caption that is merely plausible. We will exploit an additional word retrieval module to accomplish this task.

Concretely, our contributions are twofold:

- We construct and present Gregg-1916, which is, to our knowledge, the first dataset for optical Gregg shorthand recognition of a considerable scale[1].
- We develop a novel deep neural network model that recognizes words from their Gregg shorthand versions. The architecture is also an attempt to conduct optical character recognition at word level when character segmentation is hardly possible.

2   Related Work

Shorthand Recognition with Auxiliary Devices. Exploiting auxiliary devices grants access to extra, informative, features. Leedham et al. conducted series of research on Pitman shorthand recognition with the help of a sensor that keeps track of the nib pressure. On a dataset of a couple thousands outlines, they were able to recognize Pitman consonants, vowels and diphthong outlines to accuracies of 75.33%, 96.86% and 91.86%, respectively, using dynamic programming techniques (see, e.g. [13]).

Optical Shorthand Recognition. To our knowledge, most existing research on optical shorthand recognition formulate the task as multi-class classifications. [19] used a neural network model to segment Pitman script and recognize the consonants, and reported an accuracy of 89.6% on a small set of 68 English words; [11] was able to get 94% accuracy recognizing from scans of Pitman scripts of 9 English words. [17] achieved perfect accuracy recognizing as many as 24 Gregg letters.

We observe two things. Firstly, it does not appear promising to directly formulate word-level shorthand recognition as a classification task. The subtlety of the scripts and the extremely large number of possible categories (up to the size

---

[1] The dataset, together with our code, is made publicly available at https://github. com/anonymously/Gregg1916-Recognition.

of the English vocabulary) make the task very challenging. Secondly, due to the absence of a dataset of considerable scale, the outcomes of most previous word-level shorthand recognition are not conclusive.

3   Method

3.1   The Dataset and What Makes Shorthand Recognition Challenging

In order to thoroughly study optical recognition of Gregg shorthand, we built **Gregg-1916**, a medium-sized dataset from the **Gregg Shorthand Dictionary** [6], which provides written forms for Gregg shorthand scripts. We acquired a scan of the book at 150 ppi, which yielded 15,711 **images** of shorthand scripts with their corresponding words as **labels**.

For the following reasons, Gregg shorthand recognition is inherently special and challenging.

* Designed for maximum brevity, shorthand inherently incorporates a trade-off between brevity and recognizability. The information in the scripts is hence encoded in a very concise way. A morpheme could be encoded in a tiny area, e.g. by a sheer shift in the strike directions, packed within a few dozen pixels. A recognizer would struggle painfully to locate and decode the information. See Fig. 2.
* Shorthand relies on both the sizes and the directions of small ovals and strikes to encode information. That means, the images are in general **not** invariant under scaling and rotation. For most convolutional network based methods, the possibility of data augmentation through scaling and rotation is limited to a minimum level.
* The images exhibit great variance in size and shape (the smallest being around 1% the size of the largest image; correlation between image height and width is −0.30). It is thus very difficult to apply attention-based image captioning methods (see, e.g. [16]), which would have helped a recognizer to locate the information.

3.2   Our Method

Our model goes as follows: a CNN based **feature extractor** generates a feature vector from the input image; the feature vector is then used to initialize the **sequence generator**, an array of recurrent neurons trained to be a generative model of labels. As the generated **hypotheses** are rarely completely correct English words themselves, we exploit a further **word retrieval** module to determine which label the sequence generator was trying to generate. This is done by ranking the words in the vocabulary according to some **retrieval criteria** that basically measure string similarities (see Fig. 3).

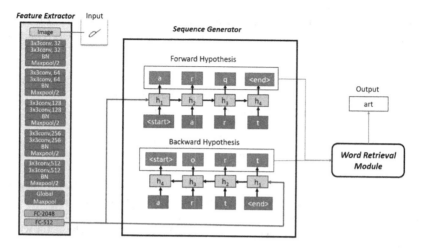

Fig. 3. The architecture. It exploits a forward decoder and a backward decoder. Both decoders are shown in the figure.

3.3 Feature Extractor

The feature extractor consists of 10 convolutional layers, with batch normalization layers (see, e.g. [8]) and max-pooling layers in between. The stack of convolutional layers is followed by two fully connected layers (See Fig. 3).

We tested the feature vector through a binary classification task, where given a label, the classifier needs to determine from the feature whether each specific letter exists in the label. In a few exploratory testing sessions, the feature extractor outperformed a Xeption net [5] and yields an average accuracy of 90% (with average chance level being 75%). Thus it should qualify as a baseline feature extractor for further research.[2]

3.4 Sequence Generator

The sequence generator consists of an array of GRU [3] or LSTM [7] cells, initialized with the feature vector. As one may expect, the information encoded in the feature vector may become noisy when the sequence gets longer. Consequently, single directional recurrent decoders often witness a decay in performance from the beginning of the sequence towards the end. Our model is no exception (see Fig. 4). To partially tackle this issue, we trained a **backward decoder** that generates sequences from the end to the beginning. Therefore, the sequence generator yields two hypotheses: the **forward hypothesis**, which is more accurate

---

[2] As Gregg shorthand is designed based on the word pronunciations instead of word spellings and that English word spellings are notoriously famous for the mismatch between the two, each letter may have multiple possible representations in the shorthand, which is actually designed according to the word pronunciations. Therefore, the binary classification task is highly non-trivial.

at the beginning; and the **backward hypothesis**, which is more accurate at the end. Both hypotheses would be taken into consideration by the word retrieval module.

Label	Hypothesis_f	Hypothesis_b
spiritualism	spernnii	pprrolalism
necromancy	necramaa	nnorancy
story	strrn	tarry
congeniality	conventalittl	coovintality

Fig. 4. The output of the sequence generator. The columns show the labels, the forward hypothesis and the backward hypothesis, respectively. Note how each hypothesis is making better predictions of the label on the corresponding ends. Quantitatively, the character-wise accuracy of the forward generator is 0.953 at the first character whereas 0.642 on average.

3.5   The Word Retrieval Module

Now we retrieve the word in the vocabulary that fits the hypotheses the most. The words were ranked by their similarities to the hypotheses, which were evaluated by a weighted sum of the following metrics.

Levenshtein Distance. Levenshtein distance (originally from [12]) counts the minimal number of editorial operations that converting one string to the other. Given the reference string **ref** and the hypothesis string **hyp**, we project the Levenshtein distance $e_{dist}(ref, hyp)$ between them onto $[0, 1]$ to become **editorial similarity** $e_{sim}(ref, hyp)$, thus normalizing the metric:

$$e_{sim}(ref, hyp) = 1 - \frac{e_{dist}(ref, hyp)}{max(|ref|, |hyp|)} \tag{1}$$

here $|\cdot|$ takes the length of a string.

Sentence BLEU. Originally proposed to evaluate the quality of machine translation, BLEU score (see [14]) focuses on the overlap between the n-grams in the reference and the hypothesis. More precisely, it takes the geometric average of the precision of hypothesis n-grams:

$$BLEU(ref, hyp) = BP \cdot exp(\sum_{i \leq N} w_i \log p_i) \tag{2}$$

here BP is a penalty term applied to hypothesis length; p_i is the n-gram precision, i.e. the proportion of hypothesis i-grams that is seen in the reference. Here we apply the sentence BLEU to a sequence of characters, i.e. the evaluation is based on character n-grams instead of word n-grams.

Bi-directional BLEU. As the forward hypothesis is more reliable at the beginning, while the backward hypothesis is more reliable towards the end, to give the better halves of both hypotheses more influence over the similarity metric, we developed a variant of the original BLEU, the **bi-directional BLEU**. Given the hypotheses $\mathbf{h_f}, \mathbf{h_b}$, we evaluate n-gram precision as

$$p_n = \sum_{x \in C_f} \psi_f(x, h_f) + \sum_{x \in C_b} \psi_b(x, h_b) \tag{3}$$

where C_f and C_b are the set of correct n-grams in h_f and h_b, respectively. The **forward positional weight**

$$\psi_f(x, h) := \frac{h.index(x)}{|h| - |x| + 1} \tag{4}$$

where $h.index(x)$ denotes the *index* of x in h. Basically, the bi-directional BLEU utilizes a new criterion of evaluating the n-gram precision p_n: instead of giving all n-grams equal significance, the n-grams that appear closer to the beginning of the forward hypothesis (i.e. the 'better' parts), would receive larger weights. The **backward positional weight** is defined analogously as:

$$\psi_b(x, h) := 1 - \psi_f(x, h) \tag{5}$$

4 Experimental Setup

Our model was implemented with Python 3.5. The neural network-based modules of our model (feature extraction and sequence generator) were implemented with Keras 2.1.2 [4].

Data Preprocessing and Augmentation. 5% of the data is randomly selected as the validation set, another 5% as the test set. All images are padded with zeros to be of maximum possible size. Conservative data augmentation operations were applied, including shifting, scaling to 96% and rotating by 2°.

Optimization. We used adam [10] optimizer to perform a two stage coarse-to-fine random hyper-parameter search (see [1]). Important hyper-parameters include learning rate, dropout [18], gradient clipping [15] threshold and mini-batch size (see Table 1).

5 Results

We considered the following metrics for evaluation.

- BLEU-1 to BLEU-4.
- *accuracy*@1 and *accuracy*@5.

Table 1. Hyper-parameters that yielded best character-wise accuracy. The same set of hyper-parameters unexpectedly yielded best performance for both decoders. Very noticeably, the backward model is significantly worse than its forward counterpart. This is because silent letters appear more often in the later part of the words (e.g. a silenced 'e' at the end of a word), and are not reflected by Gregg shorthand scripts, which is designed to reflect pronunciations.

	Dropout	Clip norm	Lr	Batch size	Accuracy	Perplexity	Neurons
Forward decoder	0.29	12	4.5e-5	64	0.642	3.16	LSTM
Backward decoder	0.29	12	4.5e-5	64	0.430	8.25	GRU

- Editorial similarity, as defined earlier. It could also be seen as a soft version of $accuracy@1$: it is 1 when the retrieval is correct; when the retrieval is incorrect, it considers the similarity between the hypothesis and the reference, instead of evaluating the retrieval with 0.

We tested various ways of formulating the retrieval criterion (see Table 2). Figure 5 shows a few sample retrievals according to criterion 3. Most noticeably, the word retrieval module is necessary: without it the sequence generator rarely outputs correct labels ($accuracy@1$ for raw forward/backward hypotheses was only 2.7%, 0.39%, respectively), and cannot make use of both hypotheses. The best performances were achieved by criterion 3, which averaged Levenshtein distances, and criterion 4, which further considered bi-directional BLEU. Although the original sentence BLEU yielded poor performance (probably due to the hypotheses being messy on the 'wrong' end), our bi-directional BLEU contributed to one of the best criteria.

Table 2. Results for some criteria. f and b stands for forward and backward hypotheses, respectively. bb stands for bi-directional BLEU.

Index	Retrieval criteria	BLEU-1 to BLEU-4	e_{sim}	acc@1	acc@5
1	Raw forward hypothesis	.581, .465, .395, .338	.574	.027	n/a
2	Raw backward hypothesis	.524, .369, .292, .238	.447	.0039	n/a
3	$0.5edit_{sim}^{f} + 0.5edit_{sim}^{b}$	.707, .600, .546, **.508**	.644	**.349**	**.580**
4	$0.5bb + 0.25edit_{sim}^{f} + 0.25edit_{sim}^{b}$	**.708, .604, .548,** .507	**.662**	.330	.576
5	$0.5bleu_f + 0.5bleu_b$	.596, .486, .427, .384	.539	.164	.301

Label	Hypothesis_f	Hypothesis_b	Hypothesis
necromancy	necramaa	nnorancy	necromancy
aeration	arration	aeration	aeration
schedule	scandll	jandll	scandal
circuitous	circumsutt	ssttious	circuitous

Fig. 5. Sample outputs. Labels together with the forward hypotheses, the backward hypotheses and the final hypotheses. Note how both hypotheses contributed to the retrieval of different halves of the label.

6 Conclusion

6.1 Further Directions

There are many possibilities to improve the system:

- **Pronunciations.** Since Gregg shorthand is based on the pronunciations of the words, it would be easier to retrieve the correct pronunciations than to retrieve the correct spellings. Besides, retrieving word given its pronunciation would be almost trivial given the availability of various dictionaries.
- **Character-level or Word-level Language Models.** A character-level or word-level language model can encode prior knowledge of English spelling, morphology, grammar, etc. to improve the sequence generator.
- **Better Retrieval Criteria.** Recall that the sequence generator saw a sharp performance drop from the prediction of the first letter to that of the end. The original Levenshtein distance cannot consider the cost of editorial operations in a manner related to the relative positions of the edition, which neutralizes a large amount of prior knowledge. With a more accordingly-designed retrieval criterion, we could surely expect better performance.

6.2 Summary

We developed a medium-size dataset for word level optical Gregg shorthand recognition, which allows us to explore shorthand recognition on an unprecedented amount of data. Different from previous works, we formulated the problem as an information retrieval task with a novel architecture. Achieving promising performance and showing many possibilities for further explorations, the architecture would make a nice baseline for future research.

References

1. Bergstra, J., Bengio, Y.: Random search for hyper-parameter optimization. J. Mach. Learn. Res. **13**(Feb), 281–305 (2012)
2. Encyclopaedia Britannica: Encyclopædia Britannica. Common Law, Chicago (2009)

3. Cho, K., et al.: Learning phrase representations using RNN encoder-decoder for statistical machine translation. arXiv preprint arXiv:1406.1078 (2014)
4. Chollet, F., et al.: Keras (2015). https://keras.io
5. Chollet, F.: Xception: Deep learning with depthwise separable convolutions. arXiv preprint (2016)
6. Gregg, J.R.: Gregg Shorthand Dictionary. Gregg Publishing Company, Upper Saddle River (1916)
7. Hochreiter, S., Schmidhuber, J.: Long short-term memory. Neural Comput. 9(8), 1735–1780 (1997)
8. Ioffe, S., Szegedy, C.: Batch normalization: Accelerating deep network training by reducing internal covariate shift. arXiv preprint arXiv:1502.03167 (2015)
9. Karpathy, A., Fei-Fei, L.: Deep visual-semantic alignments for generating image descriptions. In: Proceedings of the IEEE conference on computer vision and pattern recognition, pp. 3128–3137 (2015)
10. Kingma, D.P., Ba, J.: Adam: A method for stochastic optimization. arXiv preprint arXiv:1412.6980 (2014)
11. KumarMishra, J., Alam, K.: A neural network based method for recognition of handwritten English Pitmans shorthand 102, 31–35 (2014)
12. Levenshtein, V.I.: Binary codes capable of correcting deletions, insertions, and reversals. Soviet physics doklady, vol. 10, pp. 707–710 (1966)
13. Ma, Y., Leedham, G., Higgins, C., Myo Htwe, S.: Segmentation and recognition of phonetic features in handwritten Pitman shorthand 41, 1280–1294 (2008)
14. Papineni, K., Roukos, S., Ward, T., Zhu, W.J.: BLEU: a method for automatic evaluation of machine translation. In: Proceedings of the 40th Annual Meeting On Association For Computational Linguistics, pp. 311–318. Association for Computational Linguistics (2002)
15. Pascanu, R., Mikolov, T., Bengio, Y.: On the difficulty of training recurrent neural networks. In: International Conference on Machine Learning, pp. 1310–1318 (2013)
16. Pedersoli, M., Lucas, T., Schmid, C., Verbeek, J.: Areas of attention for image captioning. In: ICCV-International Conference on Computer Vision (2017)
17. Rajasekaran, R., Ramar, K.: Handwritten Gregg shorthand recognition 41, 31–38 (2012)
18. Srivastava, N., Hinton, G., Krizhevsky, A., Sutskever, I., Salakhutdinov, R.: Dropout: a simple way to prevent neural networks from overfitting. J. Mach. Learn. Res. 15(1), 1929–1958 (2014)
19. Zhu, M., Chi, Z., Wang, X.: Segmentation and recognition of on-line Pitman shorthand outlines using neural networks, vol. 37, no. 5, pp. 2454–2458, December 2002

A Lattice Based Algebraic Model for Verb Centered Constructions

Bálint Sass[(✉)]

Research Institute for Linguistics, Hungarian Academy of Sciences,
Budapest, Hungary
sass.balint@nytud.mta.hu

Abstract. In this paper we present a new, abstract, mathematical model for verb centered constructions (VCCs). After defining the concept of VCC we introduce proper VCCs which are roughly the ones to be included in dictionaries. First, we build a simple model for one VCC utilizing lattice theory, and then a more complex model for all the VCCs of a whole corpus combining representations of single VCCs in a certain way. We hope that this model will stimulate a new way of thinking about VCCs and will also be a solid foundation for developing new algorithms handling them.

Keywords: Verb centered construction · Proper VCC · Double cube
Corpus lattice

1 Verb Centered Constructions

What is a *verb centered construction (VCC)*? We will use this term for a broad class of expressions which have a verb in the center. In addition to the verb, a VCC consists of some (zero or more) other linguistic elements which are or can occur around the verb. In this paper, the latter will be PP and NP dependents of the verb, including the subject as well. The definition is rather permissive because our aim is to cover as many types of VCCs as we can, and provide a unified framework for them.

Sayings (*the ball is in your court*) meet this definition just as verbal idioms (*sweep under the rug*), compound verbs/complex predicates (*take a nap*), prepositional phrasal verbs (*believe in*) or simple transitive (*see*) or even intransitive verbs (*happen*). The first example above shows that it is useful to include the subject, as the concrete subject can be an inherent part of a VCC.

As elements of a VCC, we introduce the notion of *bottom, place* and *filler*. The bottom is the verb, there are places for PP/NP dependents around the verb, and fillers are words which occur at these places. Using this terminology, in *sweep under the rug* there is a place marked by the preposition *under* and filled by the word *rug*. Similarly, in *take a nap* there is a place for the object (designated by word order in English) filled by *nap*. The VCC *believe in* demonstrates the notion of a *free place*, marked by the preposition *in* and not filled by anything.

P. Sojka et al. (Eds.): TSD 2018, LNAI 11107, pp. 231–238, 2018.
https://doi.org/10.1007/978-3-030-00794-2_25

We can talk about different classes of VCCs – fully free, partly free, fully filled – according to how many places and fillers they have. *Take part in* has one filled place (object) and one free place (*in*), showing that a VCC can be a compound verb and a prepositional phrasal verb at the same time.

Have a closer look at *sweep under the rug*. We find that this VCC is in a certain sense not complete. In fact, it should have two additional (free) places: one for the subject and another for the object. Let us use the following notation for VCCs: [sweep + subj + obj + under ⌒ rug]. First element is the verb, places are attached by +, and fillers are attached to the corresponding place by ⌒. (This representation does not indicate word order: places are taken as a set.) If we narrowed down our focus to a certain kind of VCCs, we would obtain expressions which are incomplete in the above sense. For example, in their classical paper Evert [5] search for proposition+noun+verb triplets, they obtain for example *zur Verfügung stellen* which is clearly incomplete: it lacks free subject and object places.

Table 1. Illustrating the notion of *proper VCCs*. Clearly, the proper VCC is transitive *read* in the first sentence and *take part in* in the second (together with the free subject place). Other VCCs of these sentences are evidently not proper.

John reads the book.	
VCC	Proper?
[read + subj + obj]	+
[read + subj + obj ⌒ book]	−
John takes part in the conversation.	
VCC	Proper?
[take + subj + obj]	−
[take + subj + obj ⌒ part]	−
[take + subj + obj ⌒ part + in]	+
[take + subj + obj ⌒ part + in ⌒ conv.]	−

As we see, not all VCCs are multi-word, but most of them are multi-unit at least. Whether a certain VCC is multi-word or not depends on the language: the counterpart of a separate word (e.g. a preposition) in a language can be an affix (e.g. a case marker) in an other. Our target is the whole class of VCCs, so we do not lay down a requirement that a VCC must be multi-word.

All sentences (which contains a verb) contain several VCCs which are substructures of each other. From these, one VCC is of special importance: the *proper VCC*. This notion is essential for the following. The proper VCC is complete, that means it contains all necessary elements, and clean, that means it does not contain any unnecessary element. It contains free places constituting complements and does contain free places constituting adjuncts. It contains fillers which are idiomatic (or at least institutionalized [8]) and does not contain fillers

Table 2. A Hungarian example for interfering places in VCCs. Places follow their fillers in this table because places are cases in Hungarian: -t is a case marker for object, and -rA for something like English preposition *onto*.

Verb	Filler	Place	Filler	Place
vet	*pillantás*	-t		-rA
cast	glance	obj		onto

Cast a glance onto something

= look at something

Verb	Filler	Place	Filler	Place
vet		-t	*szem*	-rA
cast		obj	eye	onto

Cast something onto (somebody's) eye

= reproach somebody for something

which are compositional. In fact, we look for a combination of elements beside the verb that, together with the verb, form a *unit of meaning* [10]. In short, the proper VCC is the verbal expression from a given sentence which can be included in a dictionary as an entry (Table 1).

Notice that we have one certain set of linguistic tools for expressing places of VCCs in a language: word order and prepositions in English, prepositions and case markers in German, postpositions and case markers in Hungarian etc. Since we use them both for free and filled places, they can interfere with each other beside a verb: place A can be free and place B filled beside verb V in a VCC, while place B can be free and place A filled beside the same verb in another (Table 2).

We present an algebraic model for VCCs in the next sections.

2   Model for One VCC: The Double Cube Lattice

Let us take a cube, and generalize it for n dimensions. The 1-dimensional cube is a line segment. The 2-dimensional cube is a square. The 3-dimensional cube is the usual cube. The 4-dimensional cube is the tesseract. Now, let us create the so called *double cube* by adding another cube in every dimension to make a larger cube whose side is twice as long. The 1- and 2-dimensional double cube can be seen in Fig. 1, the 3-dimensional double cube can be seen in Fig. 2. The n-dimensional double cube consists of 2^n pieces of n-dimensional cubes.

Double cubes should always be depicted with one vertex at the bottom and one at the top. Edges of double cubes are directed towards the top. Notice that supplemented with these properties double coubes are in fact *bounded lattices* [7, part 11.2], and the ordering of the lattice is defined by the directed edges. Epstein [4] calls these structures n-dimensional Post lattices of order 3. Post lattices are a generalization of Boolean lattices ('simple cubes') as Boolean lattices are the same as Post lattices of order 2.

Now, we relate VCCs to these kind of general structures. A double cube will represent the fully filled VCC of a verbal clause taken from a corpus together

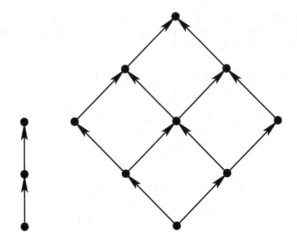

Fig. 1. The 1-dimensional double cube which consists of two line segments, and the 2-dimensional double cube which consists of four squares.

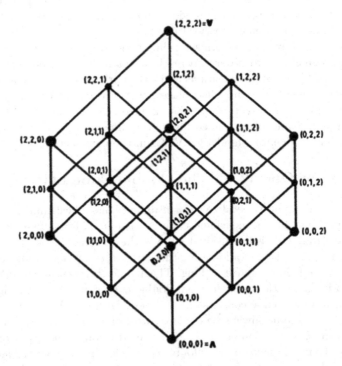

Fig. 2. The 3-dimensional double cube made from 8 usual cubes. This figure is taken from [4, p. 104] or [3, p. 309].

with all of its sub-VCCs. The dimension of the double cube equals with how many places are there in the fully filled VCC. All vertices are sub-VCCs of the fully filled VCC, while edges are *VCC building operations*. There are two such operations: *place addition* (represented by +) and *place filling* (represented by ⌒). The model is based on the idea that places and fillers are both kinds of elements, so place addition and place filling are treated alike as VCC building operations working with elements.

Of course, place addition must precede place filling with respect to a specific place. This very property is what determines the cubic form of the lattice. The bottom of the lattice is the bare verb, the top of the lattice is the fully filled VCC: it contains all fillers which present in the clause regardless whether they are part of the proper VCC or not. The proper VCC itself is one of the vertices (cf. Table 1). Figure 3 shows the first sentence of Table 1 as an example. Representing the second sentence is left to the reader. This would require a 3-dimensional double cube and the proper VCC would be the vertex marked by (1, 2, 1) in Fig. 2 if we define the order of places according to Table 1.

Our approach follows the traditional theory of valency [9] in some aspects, as we talk about slots beside the verb and take the subject as a complement as well, but we deviate from it in other aspects, as we do not care what kind of complements a verb can have in theory, but take all dependents we find in the

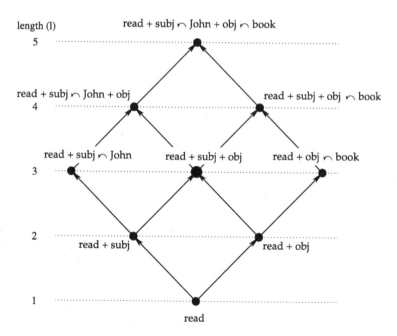

Fig. 3. The double cube representation of the clause *John reads the book*. The fully filled VCC is at the top of the lattice. The proper VCC is [read + subj + obj] which can be seen in the center marked with a larger dot. Length of a VCC (see left side) is defined as how many elements it consists of.

corpus instead, dealing complements and adjuncts in a uniform way, following the full valency approach [1] essentially.

As we see, the double cube serves two purposes at the same time: on the one hand, it is a representation of a verbal clause, on the other hand, one vertex of the double cube marks the proper VCC of this clause.

It is important to see that a double cube is not simply a graphical form of a power set. Unlike the power set where two fundamental possibilities exist (namely being or not being an element of a set), we have three possibilities here concerning a place of a VCC: the place does not exist, the place exists and it is free, the place exists and is filled by a given filler. Obviously, it is important to discriminate between no object (*happen*), free object (*see*) and filled object (*take part*). Graphical form of a power set would be a Boolean lattice, Post lattices are a generalization of them (from 2 to 3) as we mentioned earlier. The mere fact that a place occurs in a clause does not mean that this place will be a part of the proper VCC of the clause. The model must give some opportunity to omit certain places from the original clauses if necessary. The double cube model meets this requirement appropriately.

Some grammatically incorrect expressions can be noticed in Fig. 3 (e.g. *read+obj*). As the double cube is a formal decomposition of the original clause, it is not a problem to have some vertices representing ungrammatical expressions, the only thing which should be ensured that the chosen proper VCC is grammatically correct at the end.

3   Model for a Whole Corpus: The Corpus Lattice

Using lattice structures defined above, a complex model can be built which represents all VCCs occurring in a corpus. So far, we have built double cubes from elements, now the double cubes themselves will be the building blocks for assembling the corpus lattice. Having double cubes introduced explicitly is what allows us to build this larger lattice. The corpus lattice is considered as one of the main contributions of this paper. As it represents the distribution of all free and filled places beside verbs, we think that it is a representation which can be the basis for discovering typical proper VCCs of the corpus.

We define the *lattice combination* ($\oplus$) operation for lattices having the same bottom. Let $L_1 \oplus L_2 = K$ so that K a minimal $\wedge$-semilattice which is correctly labeled and into which both lattices can be embedded. In other words: let it be that $L_1 \subseteq K$ (with correct labels) and $L_2 \subseteq K$ (with correct labels) and K has the minimum number of vertices and edges, and labeled edges of $L_1 \cup L_2$ occurs only once in K (Fig. 4).

We build the corpus lattice this way: we go through the corpus, take the verbal clauses one by one, and combine the double cube of the actual clause to the corpus lattice being prepared using the $\oplus$ operation defined above. (As only lattices having the same bottom can be combined, we will obtain a separate $\wedge$-semilattice for every verb. One such $\wedge$-semilattice and the set of all both can be called corpus lattice).

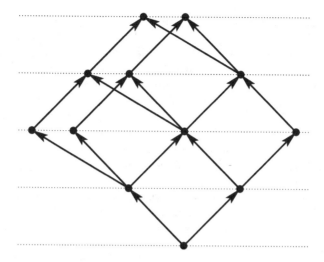

Fig. 4. An illustration of the lattice combination operation. This structure is the corpus lattice representation of a small example corpus containing only two sentences: *John reads the book* and *Mary reads the book*. It is a ∧-semilattice: the bottom (the verb) is unique, the top is clearly not.

Remark: in category theory, the lattice combination operation corresponds to the *coproduct* [6, pp. 62–63] defined on the lattice of corpus lattices in which the ordering is defined by the above embedding.

If we compare our model to other approaches of verbal relations (e.g. verb subcategorization, TAG or FrameNet), main difference can be phrased as follows. Firstly, our model puts great emphasis on filled places, and accordingly on complex proper VCCs (which have filled places and possibly free places as well), connecting our approach to multiword expression processing [2]. Secondly, the aim of our model is to represent not just one VCC but all VCCs of a corpus together including their relationships to each other, in order to be able to tackle proper VCCs based on this combined model. The corpus lattice is the tool which realizes this aim projecting VCCs onto each other in a sense, the double cubes can be considered as an aid for creating the corpus lattice.

4   Summary and Future Work

In this paper, we presented a model for VCCs. Hopefully, this model will allow us to talk about this type of constructions in a new way and it will also be a suitable basis for developing algorithms handling them. The model provides a unified representation for all kinds of VCCs being multi-word or not, regardless of the language they are in, and also regardless whether they have free or filled places opening up an opportunity to solve the interference problem exemplified in Table 2.

Our main future aim is to discover proper VCCs. To achieve this, new methods are needed which collect all the required places and determine whether a place is free or filled by a certain filler, in order to make the VCCs complete and clean. The corpus lattice – equipped with corpus frequencies at every vertex – is an appropriate starting point for developing this kind of new algorithms. We think that proper VCCs are at some kind of thickening points of the corpus lattice. The prospective algorithm would move through the corpus lattice (top-down or bottom-up) vertex by vertex, until it reach proper VCCs at such points. For this type of algorithms it is needed to be able to effectively advance from one vertex to another differing only in one element. Our representation is suitable exactly for this purpose.

Another future direction can be discovering parallel proper VCCs. They may be useful for tasks where multiple languages are involved (e.g. machine translation). On the one hand, proper VCCs are not to be translated element by element, they need to be known and interpreted as one unit. On the other hand, being complete, they can be corresponded to each other if not element by element, then at least free place by free place. For example Dutch *nemen deel aan* and French *participer à* have completely different structure (multi-word vs. single-word), but sharing the same meaning they both have one free place (beside the subject). We think that parallel proper VCCs can be discovered applying our model to parallel corpora in a way.

References

1. Čech, R., Pajas, P., Mačutek, J.: Full valency. Verb valency without distinguishing complements and adjuncts. J. Quant. Linguist. **17**(4), 291–302 (2010)
2. Constant, M., et al.: Multiword expression processing: a survey. Comput. Linguist. **43**(4), 837–892 (2017)
3. Epstein, G.: The lattice theory of Post algebras. Trans. Am. Math. Soc. **95**(2), 300–317 (1960)
4. Epstein, G.: Multiple-valued Logic Design: An Introduction. IOP Publishing, Bristol (1993)
5. Evert, S., Krenn, B.: Methods for the qualitative evaluation of lexical association measures. In: Proceedings of the 39th Meeting of the Association for Computational Linguistics, pp. 188–195. Toulouse, France (2001)
6. Mac Lane, S.: Categories for the Working Mathematician. GTM, vol. 5, 2nd edn. Springer, New York (1978). https://doi.org/10.1007/978-1-4757-4721-8
7. Partee, B.H., Ter Meulen, A., Wall, R.E.: Mathematical Methods in Linguistics. Kluwer Academic Publishers, Dordrecht (1990)
8. Sag, I.A., Baldwin, T., Bond, F., Copestake, A., Flickinger, D.: Multiword expressions: a pain in the neck for NLP. In: Gelbukh, A. (ed.) CICLing 2002. LNCS, vol. 2276, pp. 1–15. Springer, Heidelberg (2002). https://doi.org/10.1007/3-540-45715-1_1
9. Tesnière, L.: Elements of Structural Syntax. John Benjamins, Amsterdam (2015)
10. Teubert, W.: My version of corpus linguistics. Int. J. Corpus Linguist. **10**(1), 1–13 (2005)

Annotated Corpus of Czech Case Law for Reference Recognition Tasks

Jakub Harašta[1]([✉]), Jaromír Šavelka[2], František Kasl[1], Adéla Kotková[1],
Pavel Loutocký[1], Jakub Míšek[1], Daniela Procházková[1], Helena Pullmannová[1],
Petr Semenišín[1], Tamara Šejnová[1], Nikola Šimková[1,3], Michal Vosinek[1],
Lucie Zavadilová[1], and Jan Zibner[1]

[1] Faculty of Law, Masaryk University, Brno, Czech Republic
jakub.harasta@law.muni.cz
[2] Intelligent Systems Program, University of Pittsburgh, Pittsburgh, USA
[3] Faculty of Informatics, Masaryk University, Brno, Czech Republic

Abstract. We describe an annotated corpus of 350 decisions of Czech top-tier courts which was gathered for a project assessing the relevance of court decisions in Czech law. We describe two layers of processing of the corpus; every decision was annotated by two trained annotators and then manually adjudicated by one trained curator to solve possible disagreements between annotators. This corpus was developed as training and testing material for reference recognition tasks which will be further used for research on assessment of legal importance. However, the overall shortage of available research corpora of annotated legal texts, particularly in Czech language, leads us to believe that other research teams may find it useful.

Keywords: Reference recognition · Dataset · Legal texts
Manual annotation

1 Introduction

Identification and extraction of relevant information from unstructured documents is classical a NLP task. Legal documents are heavily interconnected by references. This property can be leveraged for more efficient legal information retrieval. In this paper, we present a dataset of 350 annotated court decisions. The documents vary greatly in length (from 4,746 characters to 537,470 characters with an average length of 36,148 characters). Annotations cover references to other court decisions and to literature. We have intentionally omitted references to legal acts as irrelevant for our broader inquiry into the importance of references to court decisions in Czech case-law. The research team used its strong domain knowledge of law to define the annotation task in a way most useful from a lawyers' point of view. Also, we present our approach to issues typically associated with reference recognition, such as missing citation standards etc. In Sect. 2 we explain the related work from which we drew inspiration for defining our

© Springer Nature Switzerland AG 2018
P. Sojka et al. (Eds.): TSD 2018, LNAI 11107, pp. 239–250, 2018.
https://doi.org/10.1007/978-3-030-00794-2_26

annotation task. In Sect. 3 we describe our approach to the task, presenting the annotation scheme and the basic assumptions behind it. Section 4 presents the process of manual annotations and adjudication of annotations to resolve their ambiguities. Section 5 presents basic statistics of the resulting dataset. Section 6 concludes the paper by outlining possible future use of the provided dataset.

2 Related Work

2.1 Reference Recognition in Legal Documents

One of the general applications of NLP is the pursuit of approaches that allow for identification and extraction of relevant information from unstructured documents. This allows for the establishment of references which further enrich the information available about the document through ties with other information sources. Reference recognition in legal texts helps us understand ties between court decisions, unravel the structure of legal acts and directly establish additional legal rules [9]. Even outside the common law system, court documents are often referring to each other to ensure coherence and legal certainty of the judicial decision-making.

There is a growing body of literature focusing on recognition of references in legal documents. Palmirani et al. [13] reported about the extraction of references from Italian legal acts. Recognition was based on a set of regular expressions. This work aimed to solve one of the basic issues and bring references under a set of common standards to ensure interoperability between legal information systems. They reported finding 85% of references and parsing 35% to referred documents.

Maat et al. [11] focused on automated detection of references to legal acts in the Dutch language. Their parser was based on grammar consisting of increasingly complex citation patterns. Research achieved accuracy of more than 95%.

Opijnen [12] aimed for reference recognition and reference standardization using regular expressions. These accounted for multiple variants of the same reference and recognition of multiple vendor-specific identifiers. Unlike previous research, Opijnen focused on the processing of court decisions. Reference recognition in court decisions is more difficult, given more diverse origin of court decisions compared to the legal acts.

Language specific work of Kríž et al. [8] focused on the detection and classification of references to other court decisions and legal acts. Authors reported F-measure of over 90% averaged over all entities. They implemented statistical recognition, specifically HMM and Perceptron algorithms. It is the state-of-the-art in automatic recognition of references in Czech court decisions. Resulting JTagger is a system capable of recognizing parts of references to legal acts, file identifiers of courts decisions, dates of effectiveness of legal acts, and names of certain public bodies.

Zhang and Koppaka [24] added a semantic layer to their work on reference recognition which allowed them to distinguish not only between different decisions being referred to, but also between different reasons for the reference.

Even when two references are parsed as referring to the same decision, these references can be made for different reasons altogether. Zhang and Koppaka used data obtained by [16], which allows to explore the reason for citing to create semantic-based network. Their approach thereby allows to determine which sentences near a reference are the best ones to represent the reason for citing.

Liu et al. [10] took inspiration from US legal information systems and distinguished between different depths of discussion of referred document. This allowed to further distinguish between more and less substantial references. Panagis and Šadl [15] used manual annotation with subsequent automated reference recognition using GATE framework [1]. Unlike previous works, the authors resolved references below the document level to create a paragraph-to-paragraph citation network.

Wyner [23] presented a methodology concerned with extracting textual elements of legal cases using the GATE framework. In the following research, Wyner et al. [22] presented a gold standard of case annotation indexing the specific case, assigning legal roles, establishing the facts of the case and reasoning outcomes. This work predominantly focuses on the common law system, which employs precedents. Such analysis of court decisions, in this legal setting, is therefore relatively more important in comparison with the continental civil law tradition (e.g. Austria, Germany, Czech Republic).

Dozier et al. [3] reported a hybrid system combining lookup and statistical methods for recognizing and resolving different entities in court documents. These entities were not concerned with references to other court decisions. However, this take suggests it may be easier to focus on a named entity recognition compared to complex references.

Our annotation scheme is formulated based on the abovementioned previous work. Specific annotation categories and approaches are further explained in Sect. 3.

2.2 Available Corpora of Legal Texts

As in every other field involved in linguistics and data processing of any kind, publicly available corpora of legal documents are essential. Similarly, these datasets have to be created through transparent methods. Walker [21] states, that in order to make progress in automated processing of legal documents, we need manually annotated corpora and evidence that these corpora are accurate. Similarly, Vogel et al. [20] emphasize that the availability of corpora of legal texts can yield benefits to adjudication, education, research, and legislation. Corpora inherently present in existing legal information systems are not sufficient, because the access to them is restricted by predesigned user interface [20].

The number of available datasets from the legal domain grows. However, these are mostly available as unannotated corpora of legal texts or corpora annotated for grammatical features. A non-exhaustive list of legal corpora includes HOLJ corpus for summarization of legal texts [4], The British Law Report Corpus [14], The Corpus of US Supreme Court Opinions [2], Corpus of Historical

English Law Reports [18], Juristisches Referenzkorpus [6], DS21 corpus [7], Corpus de Sentencias Penales [17] or JRC-Acquis [19]. These corpora vary in size, language and their overall purpose.

The statistics of our corpus are available below in Sect. 5. Our corpus is not annotated for grammatical features, but is expertly annotated for references to case law and to scholarly literature. It is comprised solely out of court decisions of Czech top-tier courts.

3   Annotation Scheme

One of the typical issues associated with reference recognition is the presence of non-standardized citations. Our approach to annotations is formed of issues reported by [13] and mentioned by [11]. Non-standard citations hamper reference recognition. Inspired by [3,8], we approach the references through annotation of more basic entities. These basic units have a higher degree of uniformity and we believe that these therefore are in some regards better suited for automatic detection. Specific references (even non-standard) are formed through various patterns of basic units. Some of the basic units are not present in specific references or are present in the text of the annotated decision text in a non-typical order. The overview of the specific basic units used as annotation types in our approach to annotations is presented in Table 1.

Table 1. Annotation types for basic units forming references.

c:id	Identifier of the referred court decision
c:court	Court issuing the referred decision
c:date	Date on which the decision was issued
c:type	Procedural type of the decision (e.g. decision, decree, opinion)
l:title	Title of the referred literature
l:author	Author of the referred literature
l:other	Any other possible information of interest, such as place or year of publication, publisher etc.
POI	Pointer to a specific place in the decision (e.g. paragraph) or scholarly work (e.g. page or chapter). Inspired by [15]
Content	Content associated with the reference as to why the court referred to a decision or a work. Inspired by [24]
Implicit	Text span referring to previous reference

We have also identified several value types that could be assigned to given reference or specific annotation type. Unlike annotation, value does not have corresponding text span, but is assigned to an annotation or reference (group of annotations) to provide further information. These value types are presented in Table 2.

Table 2. Value types in the annotation scheme.

Content type	Assigned to any annotation of type content. It segments the nature of the semantic layer into three values – **claim** where the court claims something based on the referred document, **citation** where the court directly cites a piece of text from a referred document, and **paraphrase** where the court repeats the content of a referred document in its own words
Polarity	Assigned to every reference. It specifies the overall sentiment expressed by the court towards referred documents. It has three values – **polarity+** where the court explicitly stated that the referred document is correct, **polarity-** where the court explicitly stated that the referred document is incorrect, and **polarity0** where the court did not state the referred document neither as correct nor as incorrect
Depth of discussion	Assigned to every reference. It specifies the depth of discussion of the referred document by the court. It has two values – **cited** for reference, which mentions the document and its argumentation, and **discussed** for cases where the court discussed the scope of the argumentation contained in the referred document. Inspired by [10]

We divide references into two main groups – references to court decisions (c:ref) and references to literature (l:ref). These two main groups can appear either as explicit references (c:ref(expl), l:ref(expl)) or implicit references to previously cited documents (c:ref(impl), l:ref(impl)). These types of references appear comprised of a specific set of basic units found in the text and annotated. A set of basic units creating every type of reference is indicated in Table 3.

Extremely important for our annotation scheme is the fact that only so called argumentative references were annotated. Courts often refer to other court decisions, because they are forced to do so. Court decisions often contain references as part of procedural history, where the court summarizes decisions made by lower-level bodies. Also, court decisions contain references to court decisions brought in by parties in their argumentation. These references appear, when the court summarizes the petitions of the parties. We have annotated only the references made by the court in the case reasoning.

The constituents of case-law references are comprised of an identification of the court that issued the referred decision (c:court), the issuance date (c:date), an unique court decision identifier or identifiers (c:id) and an identification of the formal category of the decision (c:type). Despite a largely standardized approach to the identification of various court instances it is to be noted that such references also occurred in an indirect form, referring instead to 'local court' or merely to 'court', whereas the specific identity of the court followed only from the broader context of the reference. The date constituent was the most uniform,

Table 3. Annotation types and value types grouped into references.

c:ref (expl)	c:id (may be multiple)
	c:court
	c:date
	c:type
	Content (may be multiple) & content type
	POI
	Polarity
	Depth of discussion
l:ref (expl)	l:title
	l:author (may be multiple)
	l:other (may be multiple)
	Content (may be multiple) & content type
	POI
	polarity
	depth of discussion
c:ref (impl)	Implicit
	Content (may be multiple) & content type
	POI
	Polarity
	Depth of discussion
l:ref (impl)	Implicit
	Content (may be multiple) & content type
	POI
	Polarity
	Depth of discussion

but months were inconsistently referred to either with a number or with a word. Unique court decision identifiers included file numbers, popular names of decisions as well as references to collections of court decisions where a given decision was published. The type then allowed differentiating between formal categories of decisions, particularly judgments, decrees, orders and opinions.

The references to scholarly literature consist of different basic units, which was reflected by targeting primary constituents identifying the title of the referred work (l:title), one or more authors (l:author) and also a variety of additional information about the referred work (l:other). The last mentioned constituent was aimed at information of interest, such as publishing house, year of publication or identifier like ISBN, as these were very often also included in the reference and the variety impeded further division into separate constituents.

Common for both groups of references were components representing eventual pointers to a specific part or place in the source document, especially the page or paragraph (POI), and also the particular text segment of the annotated decision that was deemed to be associated with the reference by the annotators (content). Three values were further assessed with respect to this link between reference and related segment of the annotated decision. These were aimed to provide better insight into the role of the given reference for the decision.

The values were assigned by the annotators based on specific instructions and the disparities between assessments were unified by the curator. In this regard we expect the presence of residual subjective bias pursuant to instructional classification and also to the limited number of annotators. The first assigned value was the polarity of the reference in relation to the argument in the decision. A positive reference was understood as apparent adoption of the opinion in the reference by the court. If the court obviously limited itself against the referred opinion, the polarity was deemed negative. Other references contained neutral value of polarity. The second value considered was the depth of discussion. The annotators had to assess, if the source is merely quoted as part of an argument, or if it is discussed through more elaborated commentary, polemic or counterargument. The final type of value, content type, was considered to aim at the specificity of the reference. The segment related to the reference could be direct quotation of the source, indirect paraphrase or derived claim based on the source, but not closely linked to its wording. The value of polarity and depth of discussion were assigned to the reference as a whole, whereas the value of content type is related to each specific segment of the annotated decision linked to the reference. This approach was chosen due to finding that one reference often connected to a larger segment of the text, which had a changing degree of specificity.

4   Annotation and Adjudication of References

Decisions in the dataset were annotated by thirteen annotators who were remunerated for their work. Annotations took place between April, 3 and May, 31 2017. Seven annotators were pursuing their Master degree in Law, five annotators were pursuing their Ph.D. degree in Law and one of the annotators pursued her Ph.D. in Economy.

Annotations proceeded in several phases. In the first phase, four annotators representing all three groups were involved in dummy runs testing the annotation manual. Any differences between annotators were brought up during discussions and the annotation manual was refined to solve these ambiguities. Surprisingly, the initial dummy runs showed that despite a rather minimalistic annotation manual of 13 pages, even vague concepts were surprisingly clear to all the annotators participating in this phase. The necessity to annotate only argumentative references or difficult concepts such as polarity or depth of discussion brought forth a surprisingly small amount of issues.

After clearing ambiguities in the annotation manual, we initiated dummy runs for all annotators to teach them the scheme. After that we moved to a round of double annotations. In this phase, different pairs of annotators independently annotated a batch of decisions for low-level constituents of references to case law and scholarly literature, combined these references into objects representing individual references and assigned these references respective values in terms of their sentiment and depth of discussion. The pair of annotators were assigned to decisions randomly.

To ensure a high quality of the resulting gold dataset the three most knowledgeable annotators were appointed as curators of the dataset. Each document was then further processed by one of the curators. A curator could not be assigned a decision that he himself annotated. The goal of the curators was to adjudicate the differences between the two annotations and modify the annotations to precisely follow the annotation manual. Curation took place between July, 29 and September, 5 2017.

The result of their work is the presented dataset.

5   Dataset Statistics

The presented dataset contains 350 double-annotated and curated court decisions. The decisions are distributed over three top-tier courts in the Czech Republic: Constitutional Court, Supreme Court and Supreme Administrative Court. The dataset contains 75 decisions of the Constitutional Court, 160 decisions of the Supreme Court and 115 decisions of the Supreme Administrative Court. The documents vary in length – the shortest decision has 4,746 characters, while the longest has 537,470 characters. The average length is 36,148 characters.

The presented dataset contains 54,240 individual annotations. Their distribution among individual annotation types is in the top row of Table 4. The dataset contains 11,831 content annotations with assigned content type value (Table 7). Annotations are clustered into 15,004 references (top line of Table 5) and assigned values of polarity and depth of discussion (Table 6).

Table 4. Statistics describing the individual annotation types.

	c:id	c:type	c:date	c:court	l:author	l:title	l:other	POI	Implicit	Content
Overall count	12427	6044	5479	4428	3565	2577	2747	3847	1295	11831
Maximum	885	345	271	164	231	247	194	322	75	822
Average	38.2	18.9	18.8	14.0	13.7	9.2	9.9	13.4	7.0	33.9
Median	18	10	10	9	7	4	5	5	3	19
Agreement (strict)	70.62	80.77	85.91	73.61	73.82	50.17	44.91	73.51	27.57	26.38
Agreement (overlap)	78.97	84.11	88.11	77.04	80.61	69.01	64.35	80.36	38.54	53.00

We report two types of inter-annotator agreement for pair of annotators annotating the same decision in the bottom two rows of Table 4. The strict

agreement is the percentage of annotations where the annotators agree exactly
– text span is the same as is the used annotation type. The overlap agreement
relaxes this condition. It is sufficient if the two annotations partially overlap as
long as the same annotation type is used.

Table 5. Statistics of references in the dataset.

	c:ref (expl)	c:ref (impl)	l:ref (expl)	l:ref (impl)
Overall count	9872	1418	3318	396
Maximum	673	119	305	44
Average	30.4	9.0	11.7	5.1
Median	16	4	6	3
Agreement (strict)	42.15	12.43	30.61	19.53
Agreement (overlap)	89.04	35.71	86.24	51.53

Also, we report two types of inter-annotator agreement for references in the
bottom two rows of Table 5. The strict agreement is the percentage of references
where annotators agreed completely on assigned annotations and values. The
overlap agreement relaxes this condition – it counts every assigned attribute
independently construing partially correct references.

Table 6. Statistics describing individual value types for references.

	Polarity+	Polarity−	Polarity0	Discussed	Cited
c:ref (expl)	649	224	8946	2433	7386
c:ref (impl)	76	56	1285	438	978
l:ref (expl)	147	55	3094	748	2548
l:ref (impl)	9	10	376	111	285

The curation of data cost significant resources, however it improved the overall quality of the dataset and also served as a thorough error analysis. It showed
where most of the disagreements in different annotation types stemmed from.

In c:id, most disagreements stemmed from annotating a text span containing multiple identifiers as one annotation. Most disagreements in c:type and
c:court stemmed from an improperly followed annotation manual. The annotators partially annotated specific bodies within court as c:court (e.g. *Fourth Sen-
ate of Constitutional Court* instead of *Constitutional Court* and specific bodies
as part of c:type (e.g. *Decree of Criminal Colegium* instead of *Decree*). Disagreements in l:author stemmed from annotating multiple authors as a single
text span, contrary to the manual requiring annotations spanning individual
authors. Czech courts often use literature in other languages than Czech. This

proved difficult when annotating l:title. Other issues with l:title and l:other were caused by annotators disregarding the annotation manual while annotating journal articles. Title of article and title of journal belonged to different annotation types and there were mistakes in following this rule. Annotating an implicit type proved to be challenging because some references to previously referred works were not apparent. Disagreements in this category were quite significant. This transposed into poor agreement in c:ref(impl) and l:ref(impl). Some of the issues with implicit references were solved by curation, but definitely not all of them. These disagreements on individual annotations and values passed on to references. Mistakes in references stemming from e.g. attribution of annotation containing additional whitespace were easy to fix.

Table 7. Statistics describing individual value types assigned to content annotations.

	Claim	Citation	Paraphrase
Content types	5806	1805	3005

6   Conclusions

We have presented a new corpus of Czech case law annotated for reference recognition tasks. Every decision in this dataset was manually annotated by two annotators and later curated by a single editor to ensure high quality of data. No data were construed automatically.

Manual annotations included low level constituents which group into complex references. The Resulting references are assigned with additional values, such as sentiment and depth of discussion to provide other characteristics.

The Dataset is intended for training and evaluation of reference recognition tasks. It is constructed in such a way that allowsus or others to undergo the challenging task of automatic recognition of references in the way we intended in our annotation scheme. However, it also permits to focus on individual annotation types (such as c:id) or values (depth of discussion) without much effort.

The Annotation scheme was not created with full automation in mind. Some parts of the scheme get in a way of automation using simple methods, such as CRF [5] and will require significant further effort. However, we believe the dataset can be used to train for individual selected tasks and as such provides a valuable addition to available language resources. Dataset and documentation is available at http://hdl.handle.net/11234/1-2647.

Contribution Statement and Acknowledgment. J.H. developed the annotation scheme, prepared the annotation manual, and selected the court decisions included in the dataset. J.H., T.Š., N.Š., and J.Z. participated in dummy runs and evaluation of annotation manual. J.H., F.K., A.K., P.L., J.M., D.P., H.P., P.S., T.Š., N.Š., M.V., L.Z., and J.Z. annotated the decisions. J.H., P.L., and J.M. curated/edited the decisions.

J.Š. programmed the annotation environment, prepared dataset for publication, and prepared dataset statistics. J.H., J.Š., F.K. wrote the paper with input from all authors. J.H., F.K., A.K., P.L., J.M., D.P., H.P., P.S., T.Š., M.V., L.Z., and J.Z. gratefully acknowledge the support from the Czech Science Foundation under grant no. GA17-20645S.

References

1. Cunningham, H., Maynard, D., Bontcheva, K., Tablan, V.: GATE: an architecture for development of Robust HLT applications. In: Proceedings of the 40th Annual ACL meeting, pp. 168–175 (2002)
2. Davies, M.: Corpus of US Supreme Court Opinions. https://corpus.byu.edu/scotus/
3. Dozier, C., Kondadadi, R., Light, M., Vachher, A., Veeramachaneni, S., Wudali, R.: Named entity recognition and resolution in legal text. In: Francesconi, E., Montemagni, S., Peters, W., Tiscornia, D. (eds.) Semantic Processing of Legal Texts. LNCS (LNAI), vol. 6036, pp. 27–43. Springer, Heidelberg (2010). https://doi.org/10.1007/978-3-642-12837-0_2
4. Grover, C., Hachey, B., Hughson, I.: The HOLJ corpus: supporting summarisation of legal texts. In: Proceedings of the 5th International Workshop on Linguistically Interpreted Corpora, pp. 47–53 (2004)
5. Harašta, J., Šavelka, J.: Toward linking heterogenous references in Czech court decisions to content. In: Proceedings of JURIX, pp. 177–182 (2017)
6. Hamann, H., Vogel, F., Gauer, I.: Computer assisted legal linguistics (CAL^2). In: Proceedings of JURIX, pp. 195–198 (2016)
7. Höfler, S., Piotrowski, M.: Building corpora for the philological study of Swiss legal texts. J. Lang. Technol. Comput. Linguist. 26(2), 77–89 (2011)
8. Kríž, V., Hladká, B., Dědek, J., Nečaský, M.: Statistical recognition of references in Czech court decisions. In: Gelbukh, A., Espinoza, F.C., Galicia-Haro, S.N. (eds.) MICAI 2014. LNCS (LNAI), vol. 8856, pp. 51–61. Springer, Cham (2014). https://doi.org/10.1007/978-3-319-13647-9_6
9. Landthaler, J., Waltl, B., Matthes, F.: Unveiling references in legal texts: implicit versus explicit network structures. In: Proceedings of IRIS, pp. 71–78 (2016)
10. Liu, J.S., Chen, H.-H., Ho, M.H.-C., Li, Y.-C.: Citations with different levels of relevancy: tracing the main paths of legal opinions. J. Assoc. Inf. Sci. Technol. 65(12), 2479–2488 (2014)
11. de Maat, E., Winkels, R., Van Engers, T.: Automated detection of reference structures in law. In: Proceedings of JURIX, pp. 41–50 (2006)
12. van Opijnen, M.: Canonicalizing complex case law citations. In: Proceedings of JURIX, pp. 97–106 (2010)
13. Palmirani, M., Brighi, R., Massini, M.: Automated extraction of normative references in legal texts. In: Proceedings of ICAIL, pp. 105–106 (2003)
14. Pęréz, J.M., Rizzo, C.R.: Structure and design of the british law report corpus (BLRC): a legal corpus of judicial decisions from the UK. J. Engl. Stud. 10, 131–145 (2012)
15. Panagis, Y., Šadl, U.: The force of EU case law: a multidimensional study of case citations. In: Proceedings of JURIX, pp. 71–80 (2015)
16. Automated System and Method for Generating Reasons that a Court Case is Cited. Patent US6856988

17. Pontrandolfo, G.: Investigating judicial phraseology with COSPE: a contrastive corpus-based study. In: Fantinuoli, C., Zanettin, F. (eds.) New Directions in Corpus-Based Translation Studies, pp. 137–159 (2015)
18. Rodríguez-Puente, P.: Introducing the corpus of historical english law reports: structure and compilation techniques. Revistas de Lenguas para Fines Específicos **17**, 99–120 (2011)
19. Steinberger, R. et al.: The JRC-Acquis: a multilingual aligned parallel corpus with 20+ languages. In: Proceedings of LREC, pp. 2142–2147 (2006)
20. Vogel, F., Hamann, H., Gauer, I.: Computer-assisted legal linguistics: corpus analysis as a new tool for legal studies. In: Law & Social Inquiry, Early View (2017)
21. Walker, V.R.: The need for annotated corpora from legal documents, and for (Human) protocols for creating them: the attribution problem. In: Cabrio, E., Graeme, H., Villata, S., Wyner, A. (eds.) Natural Language Argumentation: Mining, Processing, and Reasoning over Textual Arguments (Dagstuhl Seminar 16161) (2016)
22. Wyner, A.Z., Peters, W., Katz, D.: A case study on legal case annotations. In: Proceedings of JURIX, pp. 165–174 (2013)
23. Wyner, A.: Towards annotating and extracting textual legal case elements. Informatica e diritto **XIX**(1–2), 173–183 (2010)
24. Zhang, P., Koppaka, L.: Semantics-based legal citation network. In: Proceedings of ICAIL, pp. 123–130 (2007)

Recognition of the Logical Structure of Arabic Newspaper Pages

Hassina Bouressace$^{(\boxtimes)}$ and Janos Csirik$^{(\boxtimes)}$

University of Szeged, 13 Dugonics square, Szeged 6720, Hungary
bouressacehassina@hotmail.fr, jcsirik@gmail.com

Abstract. In document analysis and recognition, we seek to apply methods of automatic document identification. The main goal is to go from a simple image to a structured set of information exploitable by machine. Here, we present a system for recognizing the logical structure (hierarchical organization) of Arabic newspapers pages. These are characterized by a rich and variable structure. They may contain several articles composed of titles, figures, author's names and figure captions. However, the logical structure recognition of a newspaper page is preceded by the extraction of its physical structure. This extraction is performed in our system using a combined method which is essentially based on the RLSA (Run Length Smearing/Smoothing Algorithm) [1], projections profile analysis, and connected components labeling. Logical structure extraction is then performed based on certain rules of sizes and positions of the physical elements extracted earlier, and also on an a priori knowledge of certain properties of logical entities (titles, figures, authors, captions, etc.). Lastly, the hierarchical organization of the document is represented as an XML file generated automatically. To evaluate the performance of our system, we tested it on a set of images and the results are encouraging.

Keywords: Arabic language · Document recognition
Physical structure · Logical structure · Document processing
Segmentation

1 Introduction

In the area of analysis and document recognition, we apply methods of automatic identification. The intention is to turn a raw image into a set of structured information exploitable by the machine. After the revolution in the systematic recognition of writing over the last two decades, the document analysis and recognition procedure is moving towards the recognition of the logical structure of documents. The latter is a high-level representation in the form of a structured document of the components contained in the document image. The purpose of extracting the logical structure of a document is to understand the hierarchical organization of its elements and the relationships among them. In this study, we

© Springer Nature Switzerland AG 2018
P. Sojka et al. (Eds.): TSD 2018, LNAI 11107, pp. 251–258, 2018.
https://doi.org/10.1007/978-3-030-00794-2_27

are interested in recognizing the logical structure of the hierarchical organization of a category of documents with a complex structure, namely newspaper pages. This paper is organized as follows. In Sect. 2, we provide an overview of existing newspaper recognition methods, then we present our recognition approach in Sect. 3. In Sect. 4 we present our experimental results on Arabic newspaper page segmentation, and lastly in Sect. 5 we draw some conclusions and make suggestions for future study.

2   Related Work

We describe the main works and the different methods proposed in the literature for the recognition of document structures, while focusing on the documents with complex structures. This constitutes our area of interest in the present study.

2.1   Recognition of Physical Structure

There are a number of significant challenges that segmentation algorithms must overcome. Among these challenges are the quality deterioration of the scanned newspaper due to time and the complex layout of the newspaper pages [2]. The method of Liu et al. [3] used a bottom-up [4] method based on the clustering of related components. To merge text lines into blocks, neighboring connected components are examined and only the most valuable pair of connected components is chosen for the merge. Another bottom-up approach was proposed by Mitchell and Yan [5] as part of the complex structured document segmentation competition in 2001. Hadjar and Ingold [6] proposed an algorithm using a bottom-up approach based on the related components. The only difference is at the level of extracting the blocks, which is performed by merging the related components into large areas. Antonacopoulos et al. [7] analyzed text indentations, spaces between nearby lines and text line features in order to split regions into paragraphs which are merged together if they overlap significantly. In the 2009 ICDAR Complex Documents Segmentation Competition, the winning method was the Fraunhofer Newspaper Segmented method. The technique employed includes an ascending step guided by descending information in the form of the layout of the logical columns of the page. Text regions are separated from non-text by using statistical properties of a text (characters aligned on baselines, etc.) [8].

2.2   Recognition of the Logical Structure

The author in [9] proposed a model called 2 (CREM) for the recognition of log pages based on two-dimensional patterns. Methods based on relevance feedback can be applied to refine learning models. In [10], in conjunction with the extraction of the logical structure of the journal pages, the authors propose labeling the extracted blocks in figures, titles and texts. The figures are separated from the text in the first step of the extraction of the physical structure, and rules relating to the dominant height of characters and the average distance between the lines of text are used to perform the logical labeling of the text blocks into titles and texts. In [11], the authors proposed a method for the logical segmentation of articles in old newspapers. The purpose of the segmentation was to extract metadata from the digitized images by using a method of pixel sequence classification based on conditional random fields, associated with a set of rules that defines the very notion of article within a newspaper copy.

3 The Proposed System

Our system was designed to handle Arabic newspaper pages, and we chose the daily newspaper called Echorouk for our test corpus. The pages of this newspaper have a great variability in their structure and this makes their treatment and analysis very difficult. Our approach includes two parts, namely the extraction of the physical structure and recognition of the logical structure, where the first part seeks to analyze the document image in order to recognize its physical structure. It combines two phases: pre-processing to improve the quality of the input image, and segmentation to separate the physical entities contained in the document. The second part also has several phases: labeling by logical labels, physical entities previously extracted, and generating a structured XML file that represents the logical organization of the document, and generation of a dynamic tree, representing the hierarchical organization of the document.

3.1 Segmentation

Before segmenation we must do pre-processing, which contains two steps, namely a transformation into grayscale and thresholding. The aim of these transformations is to construct an image for the labeling of related components.

Image segmentation involves partitioning the image of newspaper page into several related regions. The three approaches for document segmentation are the bottom-up approach, the top-down approach, and the mixed approach. In our study, we perform a mixed segmentation. We commence with an upward segmentation that starts from the pixels of the image and merges them into related components. Then the related component information is used to separate the graphic components of the page (figures, bands, rectangles, and straight lines). Next, to divide the text of the newspaper page into articles, we use mixed segmentation based on the analysis of projection profiles, the RLSA smoothing algorithm, and the labeling of related components. Lastly, we apply a descending segmentation to divide the articles of the page into blocks, the blocks into lines and lines into words.

Labeling of Related Components. The labeling of the related components involves merging the neighboring black pixels into a separate unit, using the pixel aggregation method. The result of the labeling of the related components is a color image where each related component is displayed by a different color.

Detecting and Removing Graphics. Taking into account the fact that the header and the footer are always delimited at the top or bottom by a horizontal straight line, the detection of the header and the footer relies on the detection of these dividing lines. In order to detect the separating line of the header (or foot), the widest connected component is extracted from the top (or bottom) part of the page (1/6 of the height of the page). If the width of this component is greater than (the width of the image/2), then this component is treated as the dividing line of the header (or the foot of the page). Lastly, all connected components above the line separating the header are considered components of the header and all components below the foot dividing line are considered components of the footer. The same procedure is used for extracting the footer. The header is highlighted in yellow in Fig. 1. Separating the graphic components and the text is an important step before decomposing the text of the page, and it regroups several stages. It commences with the detection of the header/footer, then the detection of the figures, then the detection of the bands/rectangles/black threads/lines, and finally the elimination of all the detected components.

After separating the text and graphics, the next step is to divide a text into articles. The decomposition of the text into an article is carried out in our system based on the RLSA algorithm.

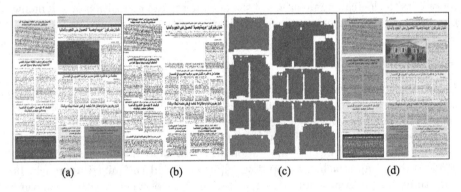

(a)   (b)   (c)   (d)

Fig. 1. Article segmentation: (a) graph detection, (b) graph elimination, (c) RLSA algorithm, (d) article division.

Segmentation of Articles into Blocks. After separating the articles, the next step is the decomposition of each article into blocks. Thus, two types of text blocks are distinguished, namely the header block representing the headings, and the text columns. The decomposition of the article into blocks is carried out in our system based on the profiles of horizontal and vertical projections. This method consists of calculating the number of black pixels accumulated in the horizontal or vertical directions in order to identify the separation locations.

Segmentation of Blocks in Lines. The next step in extracting the physical structure is the decomposition of each block into lines. To do this, we used the technique of line segmentation implemented in [12]. This technique relies on the application of a horizontal projection on each block separately in order to extract the lines that compose it. It consists of:

(1) The calculation of the histogram of the horizontal projections of the block.
(2) Extraction of local minima: if we treat the histogram of the projections as a discrete function $f(x)$, for k ranging from 1 to the size of the histogram-1, k is considered a local minimum if $f(k-1) > f(k)$ and $f(k+1) > f(k)$.
(3) Local minima filtering in two passes.
    In the first pass, local minima having a width greater than a given threshold are eliminated. The threshold is chosen as the width of the longest local minimum/2. The space between two successive minima corresponds to the height of a line of the text. In the second pass, one of the two very similar minima is removed because the line height of the text is almost the same throughout the block. To do this, one first calculates the average distance ($Mediumdistance$) between two successive minima. If the distance between two successive minima is $<2 * (Mediumdistance)/3$, the longer of the two will be removed. The remaining minima correspond to the areas of separation between the lines of the text.
(4) Conflict resolution. This is done by assigning the existing black pixels in the separator zones to the nearest line of text by proximity analysis.

Segmentation of Lines in Words. Related component labeling and RLSA smoothing are applied on each row separately to extract the words that make it up. Thus the segmentation of a line into words is done as follows:

(1) Vertical RLSA smoothing to interconnect diacritical points to words. The smoothing threshold is set to 30.
(2) Smoothing RLSA horizontal with a threshold equal to 2 to connect the sub-words (Part of Arabic Word) of the same word.
(3) Labeling related components for each line.
(4) Filtering to improve the separation between successive words. Spaces with a width smaller than a given threshold are eliminated. The threshold is found using the width and the height of each line.

3.2   Extraction of the Logical Structure

Labeling consists of recognizing all the components of the log page. The logical components considered in our system are: articles, page header, footer, columns (with their number), titles (with their level), and the author's name of each article, the figures, the lines of texts, and the legends of the figures. We chose the XML (eXtensible Markup Language) format because it is widely used in the area of Electronic Document Management and it also allows the exchange of results. We match each log page with a corresponding XML annotation file (Fig. 2).

Fig. 2. Example of a segmentation page and its logical structure.

Furthermore, the application makes it possible to build a dynamic tree that is enriched and well organized at each stage of the processing in our system. This tree provides all the information of the page in a dynamic and well-organized, structured and hierarchical form, and it can be treated as a navigation tool inside the page. However, the component tree of the page allows one to easily locate in a single click any physical or logical element of the page (titles, articles, lines, authors, captions, figures, etc.).

4   Experimental Study

In order to validate our system, we used all the images in the corpus, more than 100 pages being taken from the website of the newspaper called "Echourouk". The evaluation was performed on JPG images generated from PDF files, in order to evaluate the methods on noise-free images, and because the new method may be used to perform a layout analysis of encrypted documents. The viewing of the results of the experiments was carried out using the Java Runtime Environment. In fact, the user can detect any part of the page: articles, authors, pictures, header, footer, columns, lines, words, titles, and get all the information about the whole page, number of articles, number of (columns, lines, words) present in each article, and the user can also toggle the view of the different layers: image text separation, threads, text line extraction and line merging into blocks, and word extraction. The logical labels of the various elements of the test images were established manually in all the steps of newspaper recognition. Then we applied our system to all test images to label them automatically. The automatic labeling results of each image are compared with the actual labels (manually set) to determine the recognition rate. In order to verify the generality of our system, we attempted to vary the test images, so that they would contain a different number of articles, with different dispositions, and also contain straight lines, strips, figures; etc. Table 1 below summarizes the average recognition rate for each logical entity. In this table, we can see that the system has managed to recognize most of the existing logical entities, which means a recognition rate of 91.90%. When comparing the recognition performance across the two types of structures (physical and logical), it can be seen that the identification and verification results are different from other tests of the proposed methods due to the diversity of the structured pages. Hadjar and Ingold [6] got the best results in physical segmentation with 50.53% in thread segmentation, 97.59% in figure segmentation, 96.51% in line segmentation and 95.21% in block segmentation on a set of pages from ANNAHAR in an identification task. We solved many problems, such as the text line segmentation when diacritics of the first line and those of the second line are close to each other or merged deficient detection for a certain type of frame (a non-closed rectangle).

Table 1. Test results.

Label	Recognition score
Page header	99.23%
Footer	89.45%
Figures	90.20%
Black stripes	95.70%
Borders	88.55%
Straight horizontal lines	98.87%
Articles	90.03%
Blocks	90.32%
Lines	99.85%
Words	75.08%
Columns	90.28%
Legends	93.10%
Authors	94.16%
Average	91.90%

We got a logical recognition with a score of 93.10% for the legend, 94.16% for the author and a detection score at 75.08% in word segmentation, by using the run-length features, 90.03% in articles and we increased the segmentation score to 99.85% in line segmentation. However, problems were encountered with pages containing: very small sizes, in which case the system treats these titles as simple text lines; irregular spaces between words; segmentation between successive words on the same line that becomes problematic when the space between the title and the following paragraph is small and equal to the space between the lines; when the author is an abbreviation (the first two letters); a very luminous figure, or one that contains writing. The shape of the black bands is non-rectangular. Or the footer is not delimited by a dividing line. In general, our application provides good results. These results are, in our opinion, very encouraging considering the rich and complex structure of the newspaper pages that we examined.

5   Conclusions and Future Works

Here, we presented a system that converts a raw image of a newspaper page into a set of structured information that can be used to represent the logical organization of a document. The extraction of the logical structure is carried out by labeling the different physical elements extracted. Several tests were conducted to evaluate the performance of our system and the results obtained are encouraging. In the current state of our project, we have an application that satisfactorily met the goals set at the beginning, but our prototype version like any other application needs improving. Our ideas for future research include overcoming the resolution problem, refining our technique and applying it to different newspaper formats.

References

1. Wong, K.Y., Casey, R.G., Wahl, F.M.: Document analysis system. IBM J. Res. Dev. **26**, 647–656 (1982)
2. Gatos, B., Mantzarisl, S., Antonacopoulos, A.: First international newspaper segmentation contest. In: Proceedings of the 6th International Conference on Document Analysis and Recognition, pp. 1190–1194 (2001)
3. Liu, F., Luo, Y., Yoshikawa, M., Hu, D.: A new component based algorithm for newspaper layout analysis. In: Proceedings of the 6th International Conference on Document Analysis and Recognition (ICDAR), pp. 1176–1179. IEEE Computer Society (2001)
4. Jain, A.K., Yu, B.: Document representation and its application to page decomposition. IEEE Trans. Pattern Anal. Mach. Intell. J. **20**, 294–308 (1998)
5. Mitchell, P.E., Yan, H.: Newspaper document analysis featuring connected line segmentation. In: 6th International Conference on Document Analysis and Recognition, pp. 1181–1185 (2001)
6. Hadjar, K., Ingold, R.: Arabic newspaper page segmentation. In: 7th International Conference on Document Analysis and Recognition, pp. 895–899 (2003)
7. Antonacopoulos, C., Clausner, C., Papadopoulos, S., Pletschacher, S.: Historical document layout analysis competition. In: Proceedings of the 11th International Conference on Document Analysis and Recognition, pp. 1516–1520 (2011)
8. Antonacopoulos, A., Pletschacher, S., Bridson, D., Papadopoulos, C.: ICDAR 2009 page segmentation competition. In: 10th International Conference on Document Analysis and Recognition, University of Salford, pp. 1370–1374 (2009)
9. Robadey, L.: 2 (CREM): Une méthode de reconnaissance structurelle de documents complexes basée sur des patterns bidimensionnels, Doctoral thesis, University of Friborg-Suisse (2001)
10. Hadjar, K., Hitz, O., Ingold, R.: Newspaper page decomposition using a split and merge approach. In: 6th International Conference on Document Analysis and Recognition (ICDAR), pp. 1186–1189 (2001)
11. Palfray, T., Hébert, D., Tranouez, P., Nicolas, S., Paquet, T.: Segmentation logique d'images de journaux anciens, Francophone International Conference on Writing and Document, p. 317 (2012)
12. Boufersaoui, H., Frihi, I.: Extraction of the logical structure of documents, Master's thesis of Media Engineering, University, 08 May 1945-Guelma (2015)

A Cross-Lingual Approach for Building Multilingual Sentiment Lexicons

Behzad Naderalvojoud[1]([✉]) [iD], Behrang Qasemizadeh[2], Laura Kallmeyer[2],
and Ebru Akcapinar Sezer[1]

[1] Hacettepe University, 06800 Beytepe, Ankara, Turkey
{n.behzad,ebru}@hacettepe.edu.tr
[2] DFG SFB 991, Universität Düsseldorf, Düsseldorf, Germany
{zadeh,kallmeyer}@phil.hhu.de

Abstract. We propose a cross-lingual distributional model to build sentiment lexicons in many languages from resources available in English. We evaluate this method for two languages, German and Turkish, and on several datasets. We show that the sentiment lexicons built using our method remarkably improve the performance of a state-of-the-art lexicon-based BiLSTM sentiment classifier.

1 Introduction

Sentiment lexicons are important language resources for sentiment classification systems. The manual construction of these lexicons, however, is resource-intensive and thus expensive. When sentiment lexicons are not available for a language, one solution is to build them using automatic translation from available resources in other languages [19] such as the English *SentiWordNet* lexicon [1]. To this end, we propose a new cross-lingual distributional model to create a mapping between a pair of source–target languages so that the sentiment information about lexical items already known in the source language can be transferred to the target language.

We propose an extrinsic evaluation method to show the effectiveness of our method. We apply a stat-of-the-art neural-network *lexicon-based* sentiment classification method to a number of evaluation datasets in German and Turkish using off-the-shelf sentiment lexicons. We then augment/replace these sentiment lexicons with lexicons that are built using our method and redo the sentiment classification tasks. We interpret the gain in the performance of the sentiment classifier in these tasks as the quality of our constructed lexicons, thus the effectiveness of our method. In the remainder of this paper, Sect. 2 describes related work. Section 3 details our method. We report results from our experiments in Sect. 4 and conclude in Sect. 5.

This work was supported by TÜBİTAK Grant No. EEEAG-115E440. First author was supported by SFB991 as a SToRE visiting fellow. We acknowledge the support of NVIDIA Corporation with the donation of a GPU used for this research.

© Springer Nature Switzerland AG 2018
P. Sojka et al. (Eds.): TSD 2018, LNAI 11107, pp. 259–266, 2018.
https://doi.org/10.1007/978-3-030-00794-2_28

2   Related Work

The dominant approach in sentiment analysis is to specify and use sentiment polarity (or so-called subjectivity/objectivity) lexical databases, in which a word or phrase is often assigned to one or more quantities to describe its connotation (e.g., negative or positive) out-of-context. E.g., [10,11] assume that the polarity of words is defined independently of their domain of usage and they assign a *general* polarity to each word. In this respect, most lexicons available for sentiment classification are built using a method similar to [9,21].

Enlarging sentiment lexicons built through a manual annotation effort is a popular topic in sentiment analysis. For example, a sentiment dictionary is built simply by assigning positive and negative sentiment values to a small set of seed words; and then, it is expanded using semantic relations that are available in other lexical databases such as WordNet. Similarly, distributional similarities can be used. An example is [17], in which two sets of seed words are collected manually and then expanded by finding words that are *most similar* to them using statistical measures such as Pointwise Mutual Information. [8] combines these ideas and expand the lists of positive and negative seed words using semantic relations asserted in WordNet and predicts the polarity of unseen words using two probabilistic models. A similar idea can be found in [5]. Provided that a resource such as WordNet is available, [1] shows that it is possible to build a high quality sentiment lexicon such as SentiWordNet using automatic methods. Simply put, SentiWordNet (SWN) assigns polarity values to the WordNet synsets. However, machine-readable lexical knowledge bases such as WordNet are not available for many languages, and except for English, their coverage is often limited. Hence, these methods are not applicable to several languages, e.g. Turkish.

At the absence of high quality lexical knowledge bases, some studies have attempted to translate English resources such as SentiWordNet to other languages using machine translation techniques. E.g., [3] has generated sentiment lexicons from SentiWordNet for three Indian languages ('Bengali', 'Hindi' and 'Telugu') using a *word-level synset transfer technique*. Two sentiment lexicons have been proposed by [19] for German using a semi-automatic translation method from the *Subjectivity Clue List* [20] and *SentiSpin* [13]. Similarly, [18] compares methods for translating subjective terms in SentiWordNet to Turkish.

Lastly, although a few studies (e.g., [1]) take into account word senses and assign more than one polarity to terms (depending on the employed inventory of senses), most work focuses on assigning polarities based on term usages in context and provide contextualized/domain-specific sentiment lexicons [6,7].

3   Cross-Lingual Method for Building Sentiment Lexicon

We use English SWN (i.e., a sentiment lexicon organized around synsets) as input to our method. To use polarity values assigned to English synsets in a target language other than English, we must create a mapping between the target language words and WordNet's synsets. Hence, we build a model in which meanings of words in the target language are represented by WordNet synsets. Subsequently, we use this model to extract polarity values for words in the target language. Steps for deriving this model for a target language are described below:

Building a Cross-Lingual Distributional Model. First, we generate a co-occurrence matrix from a sentence-aligned parallel corpus. From the input corpus, we extract a vocabulary $S = \{w_1 \ldots w_n\}$ for the source language and another one $T = \{w'_1 \ldots w'_m\}$ for the target language. We instantiate a matrix $\mathbf{M}_{n \times m}$ and use it to keep track of the counts of w_is and w'_js that co-occur in the aligned sentences. Note that S contains both words and multiword expressions of maximum length of 3 tokens. For the source language, we distinguish between words of different part-of-speech categories (limited to nouns, verbs, adjectives and adverbs), e.g. instead of simply asserting the word-form *book* in S, we assert two entries *book-n* (i.e., the word book with the part-of-speech category noun) and *book-v* (with the part-of-speech category verb).

The obtained co-occurrence counts in matrix $\mathbf{M}$ are smoothed using a log-entropy transformation (similar to the one proposed in [16]). Each component m_{ij} of $\mathbf{M}$ is weighted using $m_{ij} = w_j \log(m_{ij} + 1)$, in which $w_j = 1 - \frac{H_j}{\log(n)}$ and H_j is the entropy of the column j of $\mathbf{M}$. That is, $H_j = -\sum_{i=1}^{n} p_{ij} \log(p_{ij})$, in which $p_{i,j} = \frac{m_{ij}}{\sum_{k=1}^{n} m_{kj}}$.

Synset Representation. The weighted $\mathbf{M}$ is used to represent the subjective synsets of SWN. In SWN, the subjectivity of each synset is shown using three sentiment scores, p (positive polarity), n (negative polarity), and u (neutrality) for which $p + n + u = 1$. We assign a single subjectivity value s to each synset by subtracting the negative polarity score from the positive one (i.e. $s = p - n$). The sign of s indicates the overall sentiment of its synset (i.e., positive or negative). A synset in WordNet can be interpreted and understood using (a) its *gloss* which is a textual description that describes the meaning of the synset, and/or (b) by looking at the *synset terms*, i.e., the collection of terms/words that share the same meaning represented by the synset. Here, we exploit the latter. Accordingly, each synset x is represented by one vector $\vec{x}$; $\vec{x}$ is the sum of the row vectors in $\mathbf{M}$ that represent the terms that belong to x. We call these $\vec{x}$s synset vectors. We replace row vectors of $\mathbf{M}$ with these synset vectors to form a synset-based co-occurance matrix $\mathbf{M}'_{|\vec{x}| \times m}$, where $|\vec{x}|$ is the number of synset vectors.

Synset Mapping. In this step, we build a mapping between target language words and synsets. Each synset i is mapped to k target words: for synset i ($1 \leq i \leq |\vec{x}|$), we sort $m'_{ij} \in M'$ for $1 \leq j \leq m$ in descending order and choose top k target words. The polarity values of these k words are set as the polarity of the synset i. Note that these top k words can appear in sorted lists of more than one synset. This is the major difference of our method and the previous translation-based method for building sentiment lexicon: instead of using a word-by-word translation, we use a synset-to-word translation strategy which allows a target word to express several meanings of different sentiment polarities.

4   Evaluation and Empirical Experiments

To assess the effectiveness of our method and in order to show its impact on sentiment classification tasks, we report results from a number of empirical investigations. To build our distributional model, we use Open-Subtitle corpora [15], a set of sentence-aligned parallel corpora built from movie subtitles. As mentioned earlier, our source language is English. As target language we choose Turkish and German and report result based on models for the pairs of English-German and English-Turkish; details regarding the construction of these models and respectively the lexicon induced from them are given in Sect. 4.1.

To evaluate our method for building sentiment lexicons, we employ a lexicon-based deep learning method based on BiLSTM proposed in [14] for sentiment classification. In this approach, the sentiment score of a sentence is computed based on the weighted sum (an interpolation) of the polarity values of the subjective words obtained from the lexicon. Simply put, these weights are learned from training samples to modify the prior polarity values of words with respect to their usage context.

We conduct our experiments on two datasets for German—i.e., German twitter data (SB10K) [2] and German customer feedback (GermEval2017) [22]—and three datasets for Turkish—i.e., hotel, movie, and product reviews. SB10K consists of 9949 tweets that are labelled as *Positive*, *Negative* and *Neutral*. The original train and test sets are used in the experiment; we choose randomly 10% of the train data and use it as the development set. GermEval2017, which is used as a benchmark in a GermEval 2017 shared task, is accompanied by two types of test sets (synchronic and diachronic). For GermEval2017 dataset, as an additional baseline, we report the best-obtained result (*Best-GermEval*) from the shared task results. For Turkish, we use hotel review dataset by [18] with its original split for train and test. The movie and product review datasets are proposed in [4]; we use (80% : 10% : 10%) splits as train, dev and test sets, respectively. In Turkish datasets, documents are labelled either as *Positive* or *Negative* class.

4.1   Lexicons

We created German and Turkish lexicons using the method proposed in Sect. 3 from roughly two million aliened sentences in OPUS. In our experiments, we choose $k = 10$. Since each target word has more than one polarity score (based on the synset mappings), we propose four ways to produce a single polarity: (1) we use the average of all polarity scores (*avg*), (2) we sum all the scores (*sum*), (3) the score is obtained by calculating the percentage of the assigned positive and negative polarities to the word and the polarity with the majority of votes is used as the polarity of the word (*major*) and (4) the score is obtained by subtracting the percentage of the negative synsets from the percentage of the positive ones (*subMajor*). Note that (1) and (2) are calculated based on the polarity scores, whereas (3) and (4) are obtained by counting the number of positive or negative synsets[1].

To build baselines, we repeat sentiment classification tasks using sentiment lexicons other than ones built by our method. For German, we employ the German sentiment lexicon proposed in [19] which uses the translation of English subjectivity clues [20]. The translation results from three online English-to-German translation systems have been used to construct this German lexicon. It is worth noting that another German lexicon is also available [19], however we selected the German subjectivity clue lexicon since the polarity values in this lexicon are assigned manually. For Turkish, we employ the sentiment lexicon proposed in [18]. This lexicon is built using a word-by-word translation of the subjective terms of SWN. Because this lexicon has been built using a 'parallel' translation method, we call it *parallel* in our experiments.

4.2   Result

GermEval2017 Test Sets: Tables 1 and 2 show the obtained results for the two test sets of the GermEval dataset. For the synchronic test set (Tables 1), without using a sentiment lexicon, our BiLSTM classifier yields a weak F-measure for both *Positive* and *Negative* classes. However, using a sentiment lexicon improves results noticeably; despite the lack of positive and negative instances in the train set (6% and 26%, respectively), the lexicon-based BiLSTM model achieves better results than the standard BiLSTM. Namely, BiLSTM achieves F-measure values of 24.11 and 65.56 by using *sum* and *major* lexicons, respectively. Moreover, we observe that lexicons built using our method outperform the model that uses the German subjectivity clue (subj.clue-BiLSTM). Similarly, we outperform the best system of the GermEval2017 share task. Both *sum* and *major* lexicons yield high macro and micro F-measure values.

Table 1. Results on GermEval2017 (synchronic test set)

Lexicon-model	Pos-F1	Neg-F1	Neu-F1	Macro-F1	Micro-F1
BiLSTM	00.00	23.60	80.20	34.60	68.30
Best-GermEval	-	-	-	48.06	74.94
subj.clue-BiLSTM	12.33	62.77	81.65	52.25	74.20
avg-BiLSTM	13.63	63.51	82.32	53.15	75.13
major-BiLSTM	14.91	**65.56**	**82.68**	54.38	**75.72**
majorSub-BiLSTM	14.04	65.23	81.51	53.59	74.39
sum-BiLSTM	**24.11**	63.99	82.03	**56.71**	75.10

---

[1] The constructed lexicons are available at https://github.com/nbehzad/CLSL.

Table 2. Results on GermEval2017 (diachronic test set)

Lexicon-model	Pos-F1	Neg-F1	Neu-F1	Macro-F1	Micro-F1
BiLSTM	00.00	25.20	81.60	35.60	70.00
Best-GermEval	-	-	-	51.65	73.62
subj.clue-BiLSTM	1.75	57.08	82.29	47.04	73.18
avg-BiLSTM	6.45	59.32	82.05	49.27	73.34
major-BiLSTM	26.67	59.07	81.83	55.86	73.13
majorSub-BiLSTM	15.07	58.86	82.16	52.03	73.18
sum-BiLSTM	**32.14**	**60.18**	**82.75**	**58.35**	**74.21**

Table 3. Results on the test set of SB10K dataset

Lexicon-model	Pos-F1	Neg-F1	Neu-F1	Macro-F1	Micro-F1
BiLSTM	46.70	23.10	77.50	49.10	66.20
subj.clue-BiLSTM	62.90	42.99	79.34	61.74	69.91
avg-BiLSTM	63.74	39.30	79.01	60.68	69.82
major-BiLSTM	62.15	39.62	76.04	59.27	67.59
majorSub-BiLSTM	62.87	42.94	77.44	61.09	68.31
sum-BiLSTM	**66.23**	**44.52**	**80.03**	**63.59**	**71.63**

We observe similar results for the diachronic test set of the GermEval dataset. From Table 2, sum-BiLSTM gives the best result and it achieves the best macro and micro F-measure values of 58.35 and 74.21, respectively. Similar to the synchronic test, we observe the positive effect of using lexicon-based sentiment data during classification. In this test, however, the subj.clue-BiLSTM model (i.e., our baseline) does not perform as well as it does in the synchronic test.

SB10K Test Set: Table 3 shows the result. We observe that the proposed German sentiment lexicons, likewise previous tests, yield the best results; particularly, the sum method yields the best macro and micro F-measure values of 63.59 and 71.63, respectively.

Turkish Hotel and Movie Reviews: All the Turkish datasets have a balanced distribution of positive and negative instances, hence we report results only using micro F-measure and F-measures for the positive and negative classes. Table 4 reports the obtained results. We observe that using our method, the micro F-measure value increases from 78.00 to 90.07 in hotel reviews, and from 84.50 to 89.31 in movie reviews. Although, the sum-BiLSTM again produces more consistent results than the other methods, all the lexicon-based models perform better than the standard BiLSTM (as well as when using *parallel* lexicon) in both hotel and movie reviews.

Table 4. Results on Turkish hotel and movie reviews

Lexicon-model	Hotel review			Movie review		
	Pos-F1	Neg-F1	Micro-F1	Pos-F1	Neg-F1	Micro-F1
BiLSTM	73.30	81.30	78.00	85.10	83.90	84.50
parallel-BiLSTM	79.97	85.42	83.12	87.51	87.90	87.71
avg-BiLSTM	85.79	88.60	87.34	88.44	89.21	88.84
major-BiLSTM	76.21	83.63	80.60	88.41	88.32	88.37
majorSub-BiLSTM	**89.35**	**90.70**	**90.07**	88.31	88.79	88.56
sum-BiLSTM	88.37	89.99	89.24	**89.06**	**89.54**	**89.31**

Turkish Product Reviews: To investigate the quality of Turkish sentiment lexicon built using our method in a cross-domain setting (e.g., as proposed in [12]), we repeat experiments over the product review dataset of 4 different domains (*books*, *DVD*, *electronics* and *kitchen* appliances). Table 5 reports the obtained results. As shown, all the sentiment lexicons consistently improve the performance of the base BiLSTM method in both classes with an exception for the F-measure value for the positive class in the *kitchen* domain. However, the gain in the performance using our method is higher than using the parallel lexicon.

Table 5. Results on Turkish product reviews

Lexicon-Model	Books		DVD		Electronics		Kitchen	
	Pos-F1	Neg-F1	Pos-F1	Neg-F1	Pos-F1	Neg-F1	Pos-F1	Neg-F1
BiLSTM	58.30	68.80	59.90	61.50	60.20	63.90	50.70	52.10
parallel-BiLSTM	**63.86**	73.29	63.70	66.20	79.69	81.63	37.25	64.04
avg-BiLSTM	62.90	70.51	**73.61**	72.06	85.71	**88.31**	44.04	64.33
major-BiLSTM	60.00	74.12	68.80	74.84	81.81	83.78	42.99	64.74
majorSub-BiLSTM	58.18	72.94	64.35	**75.15**	84.13	87.01	**58.33**	55.88
sum-BiLSTM	63.64	**76.47**	69.29	74.51	**86.57**	87.67	46.00	**70.00**

5 Conclusion

We proposed a cross-lingual method for building sentiment lexicons in a target language from sentiment lexicons available in another source language. We showed the effectiveness of our method and assessed the quality of the obtained lexicons through a number of experiments. Namely, we improved results from a state-of-the-art lexicon-based BiLSTM sentiment classification system for German and Turkish in several tasks. The obtained results verified that lexicons generated by our proposed method can boost the performance of sentiment analysis and outperform other translation-based methods for building sentiment lexicons.

References

1. Baccianella, S., Esuli, A., Sebastiani, F.: SentiWordNet 3.0: an enhanced lexical resource for sentiment analysis and opinion mining. In: LREC (2010)
2. Cieliebak, M., Deriu, J., Egger, D., Uzdilli, F.: A twitter corpus and benchmark resources for German sentiment analysis. In: SocialNLP 2017, p. 45 (2017)
3. Das, A., Bandyopadhyay, S.: SentiWordNet for Indian languages. In: ALR (2010)
4. Demirtas, E., Pechenizkiy, M.: Cross-lingual polarity detection with machine translation. In: WISDOM, p. 9. ACM (2013)
5. Esuli, A., Sebastiani, F.: Determining term subjectivity and term orientation for opinion mining. In: EACL, vol. 6, p. 2006 (2006)
6. Hung, C.: Word of mouth quality classification based on contextual sentiment lexicons. Inf. Process. Manag. **53**(4), 751–763 (2017)
7. Hung, C., Chen, S.J.: Word sense disambiguation based sentiment lexicons for sentiment classification. Knowl.-Based Syst. **110**, 224–232 (2016)
8. Kim, S.M., Hovy, E.: Determining the sentiment of opinions. In: COLING (2004)
9. Liu, B., Hu, M., Cheng, J.: Opinion observer: analyzing and comparing opinions on the web. In: WWW, pp. 342–351 (2005)
10. Mohammad, S., Dunne, C., Dorr, B.: Generating high-coverage semantic orientation lexicons from overtly marked words and a thesaurus. In: EMNLP. ACL (2009)
11. Strapparava, C., Valitutti, A., Stock, O.: The affective weight of lexicon. In: LREC (2006)
12. Taboada, M., Brooke, J., Tofiloski, M., Voll, K., Stede, M.: Lexicon-based splncs04 methods for sentiment analysis. Comput. Linguist. **37**(2), 267–307 (2011)
13. Takamura, H., Inui, T., Okumura, M.: Extracting semantic orientations of words using spin model. In: ACL, pp. 133–140. ACL (2005)
14. Teng, Z., Vo, D.T., Zhang, Y.: Context-sensitive lexicon features for neural sentiment analysis. In: EMNLP, pp. 1629–1638 (2016)
15. Tiedemann, J.: News from opus-a collection of multilingual parallel corpora with tools and interfaces. In: RNLP, vol. 5, pp. 237–248 (2009)
16. Turney, P.D.: Similarity of semantic relations. Comput. Linguist. **32**(3), 379–416 (2006)
17. Turney, P.D., Littman, M.L.: Measuring praise and criticism: inference of semantic orientation from association. TOIS **21**(4), 315–346 (2003)
18. Ucan, A., Naderalvojoud, B., Sezer, E.A., Sever, H.: SentiWordNet for new language: automatic translation approach. In: SITIS, pp. 308–315. IEEE (2016)
19. Waltinger, U.: GermanPolarityClues: a lexical resource for German sentiment analysis. In: LREC. Electronic Proceedings, Valletta, Malta, May 2010
20. Wilson, T., Wiebe, J., Hoffmann, P.: Recognizing contextual polarity in phrase-level sentiment analysis. In: EMNLP. pp. 347–354. ACL (2005)
21. Wilson, T., Wiebe, J., Hwa, R.: Just how mad are you? Finding strong and weak opinion clauses. In: AAAI, vol. 4, pp. 761–769 (2004)
22. Wojatzki, M., Ruppert, E., Holschneider, S., Zesch, T., Biemann, C.: Germeval 2017: shared task on aspect-based sentiment in social media customer feedback. In: GermEval (2017)

Semantic Question Matching in Data Constrained Environment

Anutosh Maitra[1], Shubhashis Sengupta[1], Abhisek Mukhopadhyay[1]([✉]),
Deepak Gupta[2], Rajkumar Pujari[2], Pushpak Bhattacharya[2], Asif Ekbal[2],
and Tom Geo Jain[1]

[1] Accenture Labs, Bangalore, India
{anutosh.maitra,shubhashis.sengupta,
abhisek.mukhopadhyay,tom.geo.jain}@accenture.com
[2] Indian Institute of Technology, Patna, India
{deepak.pcs16,asif}@iitp.ac.in, rajkumarsaikorian@gmail.com,
pb@cse.iitb.ac.in

Abstract. Machine comprehension of various forms of semantically similar questions with same or similar answers has been an ongoing challenge. Especially in many industrial domains with limited set of questions, it is hard to identify proper semantic match for a newly asked question having the same answer but presented in different lexical form. This paper proposes a linguistically motivated taxonomy for English questions and an effective approach for question matching by combining deep learning models for question representations with general taxonomy based features. Experiments performed on short datasets demonstrate the effectiveness of the proposed approach as better matching classification was observed by coupling the standard distributional features with knowledge-based methods.

Keywords: Question answering · Semantic matching · Taxonomy

1 Introduction

Question to question matching is significant in identifying previously answered similar questions for computationally lesser expensive, yet feasible, Question Answering systems [19]. This could also expand the question repositories with new information. Semantically similar questions with same answers appear in various lexical forms and machine comprehension of such forms is not easy. Unlike the information retrieval approaches that rely on a purely lexical metric of similarity between query and document, question matching needs a semantic knowledge base to improve its ability. Most similarity measures developed for document retrieval or the traditional sentence distance measures such as the Jaccard Coefficient and the Overlap Coefficient [1] work poorly where there is little word overlap. Question comprehension involves understanding of multiple dependency relationships amongst words to find a semantic match through the usage of

© Springer Nature Switzerland AG 2018
P. Sojka et al. (Eds.): TSD 2018, LNAI 11107, pp. 267–276, 2018.
https://doi.org/10.1007/978-3-030-00794-2_29

lexical resources. Since the advent of deep learning (DL) techniques, distributed representations of words and text have been used to solve NLP problems. Feedforward neural networks, recurrent neural networks (RNNs), long-short-term memory (LSTM), gated recurrent units (GRUs) etc. are being used to create useful representations of words, sentences, paragraphs and documents [21]. But DL models generally require voluminous training data, which is often not available in most industry practices. For semantic question matching in limited sized industrial datasets, augmenting generic DL representations with additional linguistic features is attempted to improve classifier performance. The Question to Question (QQ) matching framework proposed in this paper uses DL models trained on large generic datasets to generate dense semantic representation of questions. Further, a hierarchical taxonomy is created for questions having similar answers to belong to same taxonomy class. A rule-based algorithm is proposed to classify the new questions into appropriate taxonomy class(es) by understanding the focus of the questions. Information thus obtained from taxonomy is used along with the DL features to perform question matching. Empirical evidence establishes that the taxonomy, when used in conjunction with DL based similarity metrics, improves the performance of the system. A DL system trained on a large corpus generates meaningful semantic representation of a question and the taxonomy helps identify the intent of the question. The additional features can also be thought of as means to domain adaptation as the domain specific datasets are typically low volume.

2   Question Matching: State of the Art

With the rapid growth of community question and answer (cQA) forums, answer retrieval from a question corpus by identifying semantically similar question has drawn the attention of researchers in recent times [11,14]. The problem of finding the most similar match to a given question was addressed earlier too [3,12]. The authors in [19] have presented a syntactic tree based matching of semantically similar questions. Similar question retrieval has been modeled using various techniques such as topic modeling [10], knowledge graph representation [23] and machine translation [8]. Answer selection task, a related problem, has also been investigated with DL-based models [6,17,18]. Most of the existing works either focus on better representations for questions or linguistic information associated with the questions. In many practical cases, users often use specific terminology and pose rather technical questions that require very specific answers. Users more conversant with the domain will even pose similar questions with some very unconventional lexical forms. For example, the questions in software engineering domain, "How do I eliminate the error with code XY" and "What are the steps to correct XYZ errors" are essentially expecting the same answer; and it needs a complex mechanism to understand the similarity of the questions. Investigations on lexical disambiguation in natural language questions were also reported [15]; but these approaches were heavily dependent on contexts and ontology of domain concepts.

3   Proposed Question Matching Framework

When framed as an Information Retrieval (IR) problem, QQ matching becomes similarity score based ranking of the database questions to the user provided question. Existing state-of-the-art systems use either deep learning models [9] or traditional text similarity methods [8,19] to obtain the similarity scores. The present work proposes a mix classification model by combining distributed question representations and a linguistically motivated taxonomy. Algorithm 1 below details the method. Focus of a question is detailed later. A similarity score $s(\vec{p}, \vec{q})$ is defined to measure how close the candidate question $\vec{p}$ is to the question $\vec{q}$, where $\vec{p}$ and $\vec{q}$ are dense representations of questions p and q respectively, and $s(.)$ is the standard cosine similarity function. For the question pair $(\vec{p}, \vec{q})$, taxonomy based features of questions are augmented with $s(\vec{p}, \vec{q})$ to obtain a feature vector which is then passed through a binary classifier that outputs match/no-match classification along with a confidence score. A Support Vector Machine (SVM) [5] classifier was used for creating the model. The process is shown as a schematic in Fig. 1.

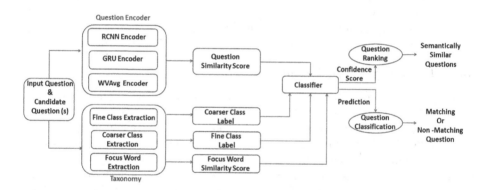

Fig. 1. Proposed model architecture for QQ matching.

3.1   Question Encoder Model

We use the output of deep learning based models trained on the open sourced Quora QQ Dataset question pairs to get a semantic similarity score of an arbitrary pair of questions. As the Quora data is diverse, we assume that the model learns in a domain agnostic manner and the similarity score given by the model is in a sense, generic. Investigations were made using two different DL models, namely Recurrent Convolutional Neural Network (RCNN) as proposed in [9] and Gated Recurrent Unit (GRU). Each of the models were trained with two sets of word embeddings (Word2vec and GloVe). For each of the models, annotated data from the Quora dataset $D = \{(q_i, p_i^+, p_i^-)\}$ is used to optimize $f(p.q.\phi)$, where $f(.)$ is a measure of similarity between the questions p and q, and ϕ is a parameter to be optimized. Here p_i^+ and p_i^- correspond to similar and non-similar question sets respectively for question q_i. The maximum margin approach was used to optimize the parameter ϕ. For a training example, where q_i is similar to p_i^+, the max-margin loss $L(\phi)$ is minimized. $L(\phi)$ is defined as:

$$\mathcal{L}(\phi) = \max_{p \in Q'(q_i)} \left\{ f(q_i, p; \phi) - f(q_i, p_i^+; \phi) + \lambda(p, p_i^+) \right\} \qquad (1)$$

where $Q'(q_i) = p_i^+ \cup p_i^-$, $\lambda(p, p_i^+)$ is a positive constant set to 1 when $p \neq p_i^+$, 0 otherwise.

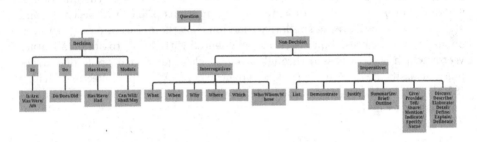

Fig. 2. Snapshot of a part of the proposed question taxonomy

3.2   Taxonomy Based Features

We assume that questions asked in a formal business setting are mostly syntactically well-formed and explicit. In the scope of this paper, a question is defined as, "A sentence asking for an explicit textual response". These questions can be broadly classified into two categories: 'decision questions' and 'non-decision questions'. Questions that expect a Boolean answer are Decision questions while Non-Decision questions demand specific answers, varying in length from possibly a single word to a few paragraphs. Analysis of question answer pairs of Stanford Question Answering Dataset (SQuAD) released in [16] revealed that Decision questions mainly appear in four different lexical constructs, namely, be-questions (is/are/was/were), do-questions (do/does/did), modal-questions (can/will/shall etc.) and has/ have/ had questions. Non-decision questions can further be classified into two sub-categories:

interrogatives (what/how/why/which/where etc.) and *imperatives* (describe/ provide/justify/list etc.). Again, each lexical construct can be sub-divided based on the answer types expected, such as time, person, location, descriptive, measure etc. Machine learning driven question classifier has been proposed in the seminal work of [20], but different kinds of words and syntactic structures used in data restrained industrial domains can make the classification complicated. Thus, a hierarchical structure in question taxonomy was conceived in comprehending and classifying any user query, a snapshot of which is shown in Fig. 2. Using this taxonomy, we design a mechanism for extracting the Coarse Class and Fine Class of a question as its representative features.

Coarse Classes: Questions contain both content words as well as function words. Content words provide pivotal information to understand semantic meaning of a question. A question word is a function word for asking questions, e.g. what, which, why, how etc. In the present work, the domain of a question word is extended to imperatives too. Question word for an imperative is defined as the verb that specifies the action expected from the answerer, such as mention, list, justify, describe etc. Question word of an interrogative can be obtained by extracting the first word in the question with PoS.tag $\in$ [WDT,WP,WPS,WRB]. For an imperative, it is the first word with PoS.tag $\in$ [VB,VBD,VBP,VBZ]. The Coarse Classes are based on the question word of a question, the cluster of question words forming different Coarse Classes is given in Table 1.

Focus of a Question: As per [13], focus of a question is a word or a phrase, which defines the question and disambiguate it to indicate the expected answer class. In the example, "Describe the customer service model for Talent and HR BPO", the word 'model' serves as the focus. [2] explained that focus of a question is contained within the noun phrases of a question. In the case of imperatives, the direct object (dobj) contains the focus word. Similarly, in case of interrogatives, there are certain dependencies that capture the relation between the question word and its focus. It has been shown that the "dobj" relation of the root verb or "det" relation of the question word for interrogatives contain the focus. Question word how has advmod relations that contain focus of the question. Priority order of the relations used to extract focus was obtained by observations on the SQuAD data and further corroborated against other datasets.

Fine Classes: Questions that differ syntactically might still be semantically equivalent as seen in the following example from a recruitment domain:

1. What is the number of new hires in 2017?
2. How many employees were recruited in 2017?

While Coarse Classes differentiate the questions syntactically, Fine Class, based on the nature of the expected answer, of a question attempts to provide that additional key information that captures the semantic sense of the question. Questions belonging to the same Fine Class need not always be semantically same; questions belonging to different Fine Classes rarely match. The focus words reveal the nature of the expected answer type. A rule-based approach is proposed to extract the focus of a question. Word Sense Disambiguation [22] is performed on the extracted focus and is followed by identification of its WordNet sense. Rule-based mappings of WordNet synsets to taxonomy classes, after considering potential synonyms and eliminating toponyms, was used to assign a Fine Class to a question. A total of 338 synsets were mapped to the described Fine Classes based on the observations on the training corpus. This set can be tuned to different domains. Table 2 below shows the set of proposed Fine Classes.

Coverage of the Taxonomy: Using the rules to generate the taxonomy features, it is observed that 85.23% of test data of a SQuAD duplicate dataset was assigned one of the Fine Classes. The remaining questions are classified as $<$ unknown $>$ Fine Class category. For a smaller practical dataset on Software Engineering, 82.17% of questions are classified to one of the taxonomy Fine Classes. A verification on the Quora dataset showed 99.99% of 537,930 questions being assigned to one of the known taxonomy Fine Classes; but the correctness of the assignments could not be ascertained.

Domain Adaptability: The Fine Classes defined above are mostly generic. For a new domain, they may be augmented with new domain specific classes, if found necessary. The rule-set mapping the Wordnet synsets to Fine Classes may have to be then appended as per the domain characteristics with human inputs.

Table 1. Set of proposed Coarse Classes

what	who/whom/whose	justify
where	is/are/was/were	demonstrate
when	has/have/had	list
why	do/does/did	brief/outline/summarize
which	provide/indicate/give/share/mention tell/specify/name/estimate	can/shall/will/may/could/ might/should/would
how	describe/delineate/elaborate/discuss /define/explain/detail	$<UNK>$

4   Datasets and Evaluation

The proposed technique is aimed at finding semantic match for questions where availability of training data is limited. Experiments are performed on following two different datasets:

SQuAD Duplicate Dataset is a semantically similar question-pair dataset built on a portion of SQuAD data. 6,000 question-answer pairs from SQuAD dataset were sampled and 12 human annotators were engaged to formulate semantically similar questions that ask for the same answer. Further, a set of 6,000 semantically dissimilar question pairs was constructed to train the classifier. A hold-out test set was created using 2,000 question pairs out of the 6,000 generated by the annotators. This dataset will be re-leased subsequently.

Software Engineering Dataset has been used in this work as a practical dataset and it involves client-organization interaction. 561 matched question pairs were used for experimentation. The dataset is divided into 432 training and 129 testing pairs. The average length of question body is 12 words. Non-matching pairs for training the SVM classifier were constructed in a fashion similar as described before. Typical IR metrics like Recall in top-k results (Recall@ k) for k = 1, 3 and 5 and Mean Reciprocal Rank (MRR) is used for evaluation.

5   Experimental Setup

The proposed question matching model uses a set of taxonomy based features as well as state-of-the-art deep linguistic feature based learning. The feature vector for a given question pair p and q consists of: <Coarse Class of p, Coarse Class of q, Fine Class of p, Fine Class of q, cosine similarity between p and q embeddings, FOCSIM(p.q)>, where FOC-SIM(p.q) is the cosine similarity of the word vectors of the focus words in p and q. The SVM classifier was trained on the fixed 6-dimensional representation of the question pairs. The libsvm implementation as in [4] was used with linear and polynomial kernels with degrees 2, 3 and 4. Best performance was obtained with linear kernel and results shown are with this setting. The McNemar's chi-squared test for each model was performed and the corresponding model results are presented. The results were found to be statistically significant with p-value < 0.001 for all cases. The experiments showcased the predictive power that the new hand-crafted features lend to any generic question matcher. Tests were also performed with different pre-trained word vectors and the Glove vectors trained on the Wikipedia corpus achieved better results. A baseline with HAL [7] embeddings is also presented which was significantly worse than Glove. Only the results of the SVM classifier with input from the best performing simple model are presented here, however the effect of the additional taxonomy features are shown with scores from each of the state-of-the-art algorithms presented previously.

Table 2. Set of proposed Fine Classes

location	person	organization	time	thing	attribute	mass
number	length	temperature	description	true/false	volume	$<UNK>$

5.1   Question Matching Score Formulation

The cosine similarity of the feature representations of the questions alongside other taxonomy based features were used as input to the SVM classifier. The MRR and Recall@k results are generated by using the cosine similarity based scoring of the representation of the test question pairs using multiple scoring mechanisms.

TF-IDF: The candidate questions are ranked using cosine similarity value obtained from the TF-IDF based vector representation.

Jaccard Similarity: The questions are ranked using Jaccard similarity calculated for each of the candidate questions with the input question.

BM-25: The candidate questions are ranked using BM-25 score, provided by Apache Lucene.

TF-IDF weighted Word Vector Composition: Question vectors were computed as:

$$\text{VEC}(q) = \frac{\sum_{t_i \in q} \text{VEC}(t_i) \times \text{tf-idf}_{t,Q}}{number\ of\ look\text{-}ups} \tag{2}$$

where q is question in interest, Q is set of candidate questions. *number of look-ups* represents the number of words in the question for which word embeddings are available. Experiments were conducted using four sets of pre-trained word embeddings: Google's Word2Vec embeddings, Glove embeddings, HAL and LSA embeddings, all of dimension 300.

6   Results and Discussion

In the Table 3 the term **R@k** denotes Recall@k and $k \in \{1,3,5\}$. **TF-IDF, Jaccard Similarity, BM-25** refer to results obtained using the corresponding methods, **WVAvg-XYZ** refers to the word vector averaging using XYZ embeddings $\in \{$HAL, GloVe$\}$, **RCNN-W2V** and **RCNN-GloVe** refer to the RCNN model trained using *word2vec* and *GloVe* embeddings respectively. Similarly, for **GRU**. **Tax** denotes augmentation of taxonomy features. Results of experiments on both SQuAD duplicate dataset(SQUAD) and Software Engineering dataset(SW) is shown in Table 3. It is evident from the feature based models that by nature the SQuAD Duplicate set was more challenging in terms of finding a semantic match of questions. Further observations are as follows: Firstly, simple models outperform complex DL models by a fair margin in case the dataset size is small. Since most practical use cases typically have constraints on the amount of available labelled training data, additional taxonomy features can help improve performance. Secondly, in a data-constrained environment, the addition of taxonomy features makes the feature representation more discriminative, thus improving the matching results. It is seen that that both recall

Table 3. Comparative results for software engineering and SQuaD duplicate dataset

Model	R@1 SQUAD	R@1 SW	R@3 SQUAD	R@3 SW	R@5 SQUAD	R@5 SW	MRR SQUAD	MRR SW
TF-IDF	54.75	69.76	66.15	82.94	70.25	85.27	61.28	76.23
Jaccard	48.95	48.06	62.8	64.34	67.4	66.67	57.26	58.92
BM-25	56.4	71.31	69.35	82.94	71.45	86.82	61.93	77.2
WVAvg-HAL	67.75	61.2	79.4	69	82.5	75.2	73.86	61.03
WVAvg-Glove	78.8	79.84	88.8	85.27	90.4	85.27	81.76	82.69
Tax+WVAvg-GloVe	**80.65**	79.84	**90.05**	86.82	**91.7**	89.14	**82.32**	83.28
GRU-GloVe	61.6	83.72	73.7	90.69	77.9	93.02	66.06	87.45
Tax+GRU-GloVe	62.45	84.49	75.05	**92.24**	79.45	**94.57**	66.89	**88.4**
RCNN-GloVe	63.7	84.49	77.6	91.47	81.4	91.47	68.48	87.38
Tax+RCNN-GloVe	65.1	**86.04**	78.9	91.47	82.15	92.24	69.21	88.35

and MRR numbers increase by a significant amount consistently across multiple base scoring methods with the addition of the taxonomy features. Thirdly, the features designed are general in the sense that the improvements are consistent across multiple datasets that differ in the nature and volume of data. Fourthly, the general improvement of results with the taxonomy based features indicates that the SVM Classifier can use these features in conjunction with any match score, thus making them good features that can be combined with any question similarity scoring engine for improving matching performance. Finally, the use of linear SVM, a linear classifier, denotes that the improvements are not due to powerful classification algorithm but due to the predictive nature of the carefully selected and hand-crafted features.

7 Conclusion

Existing works on semantic question matching are either based on DL models or traditional ML based approaches. This paper presents an effective hybrid model for question matching where DL models are combined with pivotal features obtained using linguistic analysis. It emerges that these features positively impact classifier performance by enriching the conventional DL models on limited sized industrial datasets. A taxonomy of questions is created and a two-layer architecture for both interrogatives and imperatives has been proposed. The coverage of the taxonomy, usefulness of the model and domain adaptability have been investigated by experimental analysis. Further improvement of the proposed model may be achieved by including selectorial preference for root verbs to classify questions with no explicit focus. Creation of a learning-based question classifier cascading the rule-based approach might be useful. The behavior of the model with respect to treatment of missing or non-computable features is yet to be established.

References

1. Achananuparp, P., Hu, X., Sheng, X.: The evaluation of sentence similarity measures. In: Proceedings of 10th International Conference on Data Warehousing and Knowledge Discovery, pp. 305–316 (2008)
2. Bunescu, R., Huang, Y.: Towards a general model of answer typing: question focus identification. In: CICLing (2010)
3. Burke, R.D.: Question answering from frequently asked question files: in experiences with the FAQ finder system. AI Mag. **18**(2), 57 (1997)
4. Chang, C.-C., Lin, C.-J.: LIBSVM: a library for support vector machines. ACM Trans. Intell. Syst. Technol. (TIST) **2**(3), 27 (2011)
5. Cortes, C., Vapnik, V.: Support-vector networks. Mach. Learn. **20**(3), 273–297 (1995)
6. Feng, M., Xiang, B., Glass, M.R., Wang, L., Zhou, B.: Applying deep learning to answer selection: a study and an open task. In: 2015 IEEE Workshop on Automatic Speech Recognition and Understanding (ASRU), pp. 813–820. IEEE (2015)

7. Gunther, F.: LSAfun - an R package for computations based on latent semantic analysis. Behav. Res. Methods **47**(4), 930–944 (2015)
8. Jeon, J., Croft, W.B., Lee, J.H.: Finding similar questions in large question and answer archives. In: Proceedings of the 14th ACM international conference on Information and Knowledge Management, pp. 84–90. ACM (2005)
9. Lei, T., et al.: Semi-supervised question retrieval with gated convolutions. In: Proceedings of the 2016 Conference of the North American Chapter of the Association for Computational Linguistics: Human Language Technologies, pp. 1279–1289. Association for Computational Linguistics, San Diego (2016)
10. Li, S., Manandhar, S.: Improving question recommendation by exploiting information need. In: Proceedings of the 49th Annual Meeting of the Association for Computational Linguistics: Human Language Technologies, vol. 1, pp. 1425–1434. Association for Computational Linguistics (2011)
11. M'arquez, L., Glass, J., Magdy, W., Moschitti, A., Nakov, P., Randeree, B.: Semeval-2015 task 3: answer selection in community question answering. In: Proceedings of the 9th International Workshop on Semantic Evaluation (SemEval 2015) (2015)
12. Mlynarczyk, S., Lytinen, S.: FAQFinder question answering improvements using question/answer matching. In: Proceedings of L&T-2005-Human Language Technologies as a Challenge for Computer Science and Linguistics (2005)
13. Moldovan, D., et al.: Lasso: a tool for surfing the answer net. In: Proceedings 8th Text Retrieval Conference (TREC-8) (2000)
14. Nakov, P., et al.: SemEval-2016 task 3: community question answering. In: Proceedings of the 10th International Workshop on Semantic Evaluation, vol. 16 (2016)
15. Al-Harbi, O., Jusoh, S., Norwawi, N.M.: Lexical disambiguation in natural language questions. Int. J. Comput. Sci. Issues **8**(4), 143–150 (2011)
16. Rajpurkar, P., Zhang, J., Lopyrev, K., Liang, P.: Squad: 100,000+ questions for machine comprehension of text. CoRR abs/1606.05250 (2016)
17. Severyn, A., Moschitti, A.: Learning to rank short text pairs with convolutional deep neural networks. In: Proceedings of the 38th International ACM SIGIR Conference on Research and Development in Information Retrieval, pp. 373–382. ACM (2015)
18. Wang, D., Nyberg, E.: A long short-term memory model for answer sentence selection in question answering. In: Proceedings of the 53rd Annual Meeting of the Association for Computational Linguistics, Beijing, China, pp. 707–712 (2015)
19. Wang, K., Ming, Z., Chua, T.-S.: A syntactic tree matching approach to finding similar questions in community-based QA services. In: Proceedings of the 32nd International ACM SIGIR Conference on Research and Development in Information Retrieval, pp. 187–194. ACM (2009)
20. Li, X., Roth, D.: Learning question classifiers. In: Proceedings of the 19th International Conference on Computational Linguistics, vol. 1 (2002)
21. LeCun, Y., Bengio, Y., Hinton, G.: Deep learning. In: Nature, vol. 521 (2015)
22. Zhong, Z., Ng, H.T.: It makes sense: a wide coverage word sense disambiguation system for free text. In: Proceedings of the ACL 2010 System Demonstrations, ACL Demos 10, pp. 78–83. ACM, Stroudsburg (2010)
23. Zhou, G., Liu, Y., Liu, F., Zeng, D., Zhao, J.: Improving question retrieval in community question answering using world knowledge. In: IJCAI, vol. 13, pp. 2239–2245 (2013)

Morphological and Language-Agnostic Word Segmentation for NMT

Dominik Macháček[ID], Jonáš Vidra[ID], and Ondřej Bojar[(⊠)][ID]

Faculty of Mathematics and Physics, Institute of Formal and Applied Linguistics,
Charles University, Malostranské náměstí 25, 118 00 Prague, Czech Republic
{machacek,vidra,bojar}@ufal.mff.cuni.cz
http://ufal.mff.cuni.cz

Abstract. The state of the art of handling rich morphology in neural machine translation (NMT) is to break word forms into subword units, so that the overall vocabulary size of these units fits the practical limits given by the NMT model and GPU memory capacity. In this paper, we compare two common but linguistically uninformed methods of subword construction (BPE and STE, the method implemented in Tensor2Tensor toolkit) and two linguistically-motivated methods: Morfessor and one novel method, based on a derivational dictionary. Our experiments with German-to-Czech translation, both morphologically rich, document that so far, the non-motivated methods perform better. Furthermore, we identify a critical difference between BPE and STE and show a simple pre-processing step for BPE that considerably increases translation quality as evaluated by automatic measures.

1 Introduction

One of the key steps that allowed to apply neural machine translation (NMT) in unrestricted setting was the move to subword units. While the natural (target) vocabulary size in a realistic parallel corpus exceeds the limits imposed by model size and GPU RAM, the vocabulary size of custom subwords can be kept small.

The current most common technique of subword construction is called byte-pair encoding (BPE) by Sennrich et al. [6] http://github.com/rsennrich/subword-nmt/. Its counterpart originating in the commercial field is wordpieces [10]. Yet another variant of the technique is implemented in Google's open-sourced toolkit Tensor2Tensor, http://github.com/tensorflow/tensor2tensor namely the SubwordTextEncoder class (abbreviated as STE below).

This work has been supported by the grants 18-24210S of the Czech Science Foundation, SVV 260 453 and "Progress" Q18+Q48 of Charles University, H2020-ICT-2014-1-645452 (QT21) of the EU, and using language resources distributed by the LINDAT/CLARIN project of the Ministry of Education, Youth and Sports of the Czech Republic (projects LM2015071 and OP VVV VI CZ.02.1.01/0.0/0.0/16 013/0001781). We thank Jaroslava Hlaváčová for digitizing excerpts of [7] used as gold-standard data for evaluating the segmentation methods.

© Springer Nature Switzerland AG 2018
P. Sojka et al. (Eds.): TSD 2018, LNAI 11107, pp. 277–284, 2018.
https://doi.org/10.1007/978-3-030-00794-2_30

The common property of these approaches is that they are trained in an unsupervised fashion, relying on the distribution of character sequences, but disregarding any morphological properties of the languages in question.

On the positive side, BPE and STE (when trained jointly for both the source and target languages) allow to identify and benefit from words that share the spelling in some of their part, e.g. the root of the English "*legalization*" and Czech "*legalizace*" (noun) or "*legalizační*" (adj). On the downside, the root of different word forms of one lemma can be split in several different ways and the neural network will not explicitly know about their relatedness. A morphologically motivated segmentation method could solve this issue by splitting words into their constituent semantics- and syntax-bearing parts.

In this paper, we experiment with two methods aimed at morphologically adequate splitting of words in a setting involving two morphologically rich languages: Czech and German. We also compare the performance of several variations of BPE and STE. Performance is analysed both by intrinsic evaluation of morphological adequateness, and extrinsically by evaluating the systems on a German-to-Czech translation task.

2   Morphological Segmentation

Huck et al. [2] benefit from linguistically aware separation of suffixes prior to BPE on the target side of medium-size English to German translation task (overall improvement about 0.8 BLEU). Pinnis et al. [5] show similar improvements with analogical prefix and suffix splitting on English to Latvian.

Since there are no publicly available morphological segmentation tools for Czech, we experimented with an unsupervised morpheme induction tool, Morfessor 2.0 [9], and we developed a simple supervised method based on derivational morphology.

2.1   Morfessor

Morfessor [9] is an unsupervised segmentation tool that utilizes a probabilistic model of word formation. The segmentation obtained often resembles a linguistic morpheme segmentation, especially in compounding languages, where Morfessor benefits from the uniqueness of the textual representation of morphs. It can be used to split compounds, but it is not designed to handle phonological and orthographical changes as in Czech words "*žeň*", "*žně*" ("*harvest*" in singular and plural). In Czech orthography, adding plural suffix "*e*" after "*ň*" results in "*ně*". This suffix also causes phonological change in this word, the first "*e*" is dropped. Thus, "*žeň*" and "*žn*" are two variants of the same morpheme, but Morfessor can't handle them appropriately.

2.2   DeriNet

Our novel segmentation method works by exploiting word-to-word relations extracted from DeriNet [11], a network of Czech lexical derivations, and MorfFlex [1], a Czech inflectional dictionary. DeriNet is a collection of directed trees of derivationally connected lemmas. MorfFlex is a list of lemmas with word forms and morphological tags. We unify the two resources by taking the trees from DeriNet as the basis and adding all word forms from MorfFlex as new nodes (leaves) connected with their lemmas.

The segmentation algorithm works in two steps: Stemming of words based on their neighbours and morph boundary propagation.

We approximate stemming by detecting the longest common substring of each pair of connected words. This segments both words connected by an edge into a (potentially empty) prefix, the common substring and a (potentially empty) suffix, using exactly two splits. For example, the edge *"mávat"* (to be waving)→*"mávnout"* (to wave) has the longest common substring of *"máv"*, introducing the splits *"máv-at"* and *"máv-nout"* into the two connected words.

Each word may get multiple such segmentations, because it may have more than one word connected to it by an edge. Therefore, the stemming phase itself can segment the word into its constituent morphs; but in the usual case, a multi-morph stem is left unsegmented. For example, the edge *"mávat"* (to be waving)→*"mávající"* (waving) has the longest common substring of *"máva"*, introducing the splits *"máva-t"* and *"máva-jící"*. The segmentation of *"mávat"* is therefore *"máv-a-t"*, the union of its splits based on all linked words.

To further split the stem, we propagate morph boundaries from connected words. If one word of a connected pair contains a split in their common substring the other word does not, the split is copied over. This way, boundaries are propagated through the entire tree. For example, we can split *"máva-jící"* further using the other split in *"máv-a-t"* thanks to it lying in the longest common substring *"máva"*. The segmentation of *"mávající"* is therefore *"máv-a-jící"*.

These examples also shows the limitations of this method: the words are often split too eagerly, resulting in many single-character splits. The boundaries between morphemes are fuzzy in Czech because connecting phonemes are often inserted and phonological changes occur. These cause spurious or misplaced splits. For example, the single-letter morph *a* in *máv-a-t* and *máv-a-jící* does not carry any information useful in machine translation and it would be better if we could detect it as a phonological detail and leave it connected to one of the neighboring morphs.

3   Data-Driven Segmentation

We experimentally compare BPE with STE. As we can see in the left side of Fig. 1, a distinct feature of STE seems to be an underscore as a zero suffix mark appended to every word before the subword splits are determined. This small trick allows to learn more adequate units compared to BPE. For example, the Czech word form *"tramvaj"* (*"a tram"*) can serve as a subword unit that, combined with zero suffix (*"_"*) corresponds to the nominative case or, combined with the suffix *"e"* to the genitive case *"tramvaje"*. In BPE, there can be either *"tramvaj"* as a standalone word or two subwords *"tramvaj@@"* and *"e"* (or possibly split further) with no vocabulary entry sharing possible.

Language agnostic		Linguistically motivated	
Tokenized	Blíží se k tobě tramvaj .	DeriNet	Bl@@ íž@@ í se k tobě tramvaj .
(*)	Z tramvaje nevystoupili .	(*)	Z tram@@ vaj@@ e nevyst@@ oup@@ ili .
STE	Blíží_ se_ k_ tobě_ tramvaj _ ..	DeriNet	Bl@@ íž@@ í_ se_ k_ tobě_ tramvaj_ ..
	Z_ tramvaj e_ nevysto upil i_ ..	+STE	Z_ tra m@@ vaj@@ e_ nevyst@@ oup@@ ili_ ..
BPE	Blíží se k tobě tram@@ vaj .	Morfessor	Blíží se k tobě tramvaj .
	Z tram@@ va@@ je nevy@@ stoupili .	(*)	Z tramvaj@@ e ne@@ vystoupil@@ i .
BPE und	Blíží_ se_ k_ tobě_ tram@@ vaj_ ..	Morfessor	Blíží_ se_ k_ tobě_ tramvaj_ ..
	Z_ tram@@ va@@ je_ nevy@@ stoupili_ ..	+STE	Z_ tramvaj@@ e_ ne@@ vystoupil@@ i_ ..
BPE und non-final	Blíží_ se_ k_ tobě_ tram@@ vaj_ .		
	Z_ tram@@ va@@ je_ nevy@@ stoupili_ .		

Fig. 1. Example of different kinds of segmentation of Czech sentences "*You're being approached by a tram. They didn't get out of a tram.*" Segmentations marked with (⋆) are preliminary, they cannot be used in MT directly alone because they do not restrict the total number of subwords to the vocabulary size limit.

To measure the benefit of this zero suffix feature, we modified BPE by appending an underscore prior to BPE training in two flavours: (1) to every word ("BPE und"), and (2) to every word except of the last word in the sentence ("BPE und non-final").

Another typical feature of STE is to share the vocabulary of the source and target sides. While there are almost no common words in Czech and German apart from digits, punctuation and some proper names, it turns out that around 30% of the STE shared German-Czech vocabulary still appears in both languages. This contrasts to only 7% of accidental overlap of separate BPE vocabularies.

4   Morphological Evaluation

4.1   Supervised Morphological Splits

We evaluate the segmentation quality in two ways: by looking at the data and finding typical errors and by comparing the outputs of individual systems with gold standard data from a printed dictionary of Czech morpheme segmentations [7]. We work with a sample of the book [7] containing 14 581 segmented verbs transliterated into modern Czech, measuring precision and recall on morphs and morph boundaries and accuracy of totally-correctly segmented words.

4.2   Results

Figure 1 shows example output on two Czech sentences. The biggest difference between our DeriNet-based approach and Morfessor is that Morfessor does not segment most stems at all, but in contrast to our system, it reliably segments inflectional endings and the most common affixes. The quality of our system depends on the quality of the underlying data. Unfortunately, trees in DeriNet are not always complete, some derivational links are missing. If a word belongs to such an incomplete tree, our system will not propose many splits. None of the

Table 1. Morph segmentation quality on Czech as measured on gold standard data.

Segmentation	Morph detection			Boundary detection			Word accuracy
	Precision	Recall	F1	Precision	Recall	F1	
BPE	21.24	12.74	15.93	77.38	52.44	62.52	0.77
BPE shared vocab	19.99	11.75	14.80	77.04	51.49	61.72	0.69
STE	13.03	7.79	9.75	77.08	51.77	61.93	0.23
STE+Morfessor	11.71	7.59	9.21	74.49	52.85	61.83	0.23
STE+DeriNet	13.89	10.44	11.92	70.76	55.00	61.89	0.35

methods handles phonological and orthographical changes, which also severely limits their performance on Czech.

The results against golden Czech morpheme segmentations are in Table 1.

The scores on boundary detection seem roughly comparable, with different systems making slightly different tradeoffs between precision and recall. Especially the DeriNet-enhanced STE ("DeriNet+STE") system sacrifices some precision for higher recall. The evaluation of morph detection varies more, with the best system being the standard BPE, followed by BPE with shared German and Czech vocab. This suggests that adding the German side to BPE decreases segmentation quality of Czech from the morphological point of view.

The scores on boundary detection are necessarily higher than on morph detection, because a correctly identified morph requires two correctly identified boundaries—one on each side.

Overall, the scores show that none of the methods presented here is linguistically adequate. Even the best setup reaches only 62% F1 in boundary detection which translates to meagre 0.77% of all words in our test set without a flaw.

5   Evaluation in Machine Translation

5.1   Data

Our training data consist of Europarl v7 [3] and OpenSubtitles2016 [8], after some further cleanup. Our final training corpus, processed with the Moses tokenizer [4], consists of 8.8M parallel sentences, 89M tokens on the source side, 78M on the target side. The vocabulary size is 807k and 953k on the source and target, respectively.

We use WMT http://www.statmt.org/wmt13 newstest2011 as the development set and newstest2013 as the test set, 3k sentence pairs each.

All experiments were carried out in Tensor2Tensor (abbreviated as T2T), version 1.2.9, http://github.com/tensorflow/tensor2tensor using the model transformer_big_single_gpu, batch size of 1500 and learning_rate_warmup_steps set to 30k or 60k if the learning diverged.

The desired vocabulary size of subword units is set to 100k when shared for both source and target and to 50k each with separate vocabularies.

Table 2. Data characteristics and automatic metrics after 300k steps of training.

de	cs	Tokens de	cs	Types de	cs	% shrd	BLEU	CharacTER	chrF3	BEER
STE	STE	97M	87M	54k	74k	29.89	18.78	61.27	47.82	50.34
STE	Morfessor+STE	95M	98M	63k	63k	26.42	18.22	62.27	47.30	50.00
Morfessor+STE	DeriNet+STE	138M	308M	63k	69k	36.82	16.99	64.26	45.64	49.04
Google translate							16.66	59.18	46.24	49.65
STE	DeriNet+STE	94M	138M	80k	56k	35.58	15.31	69.44	44.77	47.91
Morfessor+STE	STE	139M	86M	41k	84k	26.43	14.51	68.81	43.51	47.56
BPE shrd voc		95M	85M	56k	71k	26.78	13.79	97.94	46.44	42.49

Since T2T SubwordTextEncoder constructs the subword model only from a sample of the training data, we had to manually set the `file_byte_budget` variable in the code to 100M, otherwise not enough distinct wordforms were observed to fill up the intended 100k vocabulary size.

For data preprocessed by BPE, we used T2T TokenTextEncoder which allows to use a user-supplied vocabulary.

Final scores (BLEU, CharacTER, chrF3 and BEER) are measured after removing any subword splits and detokenizing with Moses detokenizer. Each of the metric implementation handles tokenization on its own.

Machine translation for German-to-Czech language pair is currently under-explored. We included Google Translate (as of May 2018, neural) into our evaluation and conclude the latest Transformer model has easily outperformed it on the given test dataset.

Due to a limited number of GPU cards, we cannot afford multiple training runs for estimating statistical significance. We at least report the average score of the test set as translated by several model checkpoints around the same number of training steps where the BLEU score has already flattened. This happens to be approximately after 40 h of training around 300k training steps.

5.2 Experiment 1: Motivated vs. Agnostic Splits

Table 2 presents several combinations of linguistically motivated and data-driven segmentation methods. Since the vocabulary size after Morfessor or DeriNet splitting alone often remains too high, we further split the corpus with BPE or STE. Unfortunately, none of the setups performs better than the STE baseline.

5.3 Experiment 2: Allowing Zero Ending

Table 3 empirically compares STE and variants of BPE. It turns out that STE performs almost 5(!) BLEU point better than the default BPE. The underscore feature allowing to model zero suffix almost closes the gap and shared vocabulary also helps a little.

As Fig. 2 indicates, the difference in performance is not a straightforward consequence of the number of splits generated. There is basically no difference

Table 3. BPE vs STE with/without underscore after every (non-final) token of a sentence and/or shared vocabulary. Reported scores are avg ± stddev of T2T checkpoints between 275k and 325k training steps. CharacTER, chrF3 and BEER are multiplied by 100.

Split	Underscore	Shared vocab	BLEU	CharacTER	chrF3	BEER
STE	After every token	✓	18.58 ± 0.06	61.43 ± 0.68	44.80 ± 0.29	50.23 ± 0.16
BPE	After non-final tokens	✓	18.24 ± 0.08	63.80 ± 0.88	44.37 ± 0.24	49.84 ± 0.15
BPE	After non-final tokens	-	18.07 ± 0.08	63.24 ± 1.98	44.21 ± 0.20	49.72 ± 0.11
BPE	After every token	✓	13.88 ± 0.18	81.84 ± 3.33	36.74 ± 0.51	42.46 ± 0.51
BPE	-	✓	13.69 ± 0.66	76.72 ± 4.03	36.60 ± 0.63	42.33 ± 0.60
BPE	-	-	13.66 ± 0.38	82.66 ± 3.54	36.73 ± 0.53	42.41 ± 0.56

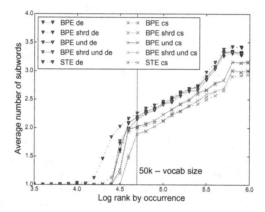

Fig. 2. Histogram of number of splits of words based on their frequency rank. The most common words (left) remain unsplit by all methods, rare words (esp. beyond the 50k vocabulary limit) are split to more and more subwords.

between BPE with and without underscore but shared vocabulary leads to a lower number of splits on the Czech target side. We can see that STE in both languages splits words to more parts than BPE but still performs better. We conclude that the STE splits allow to exploit morphological behaviour better.

6 Discussion

All our experiments show that our linguistically motivated techniques do not perform better in machine translation than current state-of-the-art agnostic methods. Actually, they do not even lead to linguistically adequate splits when evaluated against a dictionary of word segmentations. This can be caused by the fact that our new methods are not accurate enough in splitting words to morphs, maybe because of the limited size of DeriNet and small amount of training data for Morfessor, maybe because they don't handle the phonological and orthographical changes, so the amount of resulting morphs is still very high and most of them are rare in the data.

One new linguistically adequate feature, the zero suffix mark after all but final tokens in the sentence showed a big improvement, while adding the mark after every token did not. This suggests that the Tensor2Tensor NMT model benefits from explicit sentence ends perhaps more than from a better segmentation, but further investigation is needed.

7 Conclusion

We experimented with common linguistically non-informed word segmentation methods BPE and SubwordTextEncoder, and with two linguistically-motivated ones. Neither Morfessor nor our novel technique relying on DeriNet, a derivational dictionary for Czech, help. The uninformed methods thus remain the best choice.

Our analysis however shows an important difference in STE and BPE, which leads to considerably better performance. The same feature (support for zero suffix) can be utilized in BPE, giving similar gains.

References

1. Hajič, J., Hlaváčová, J.: MorfFlex CZ (2013). http://hdl.handle.net/11858/00-097C-0000-0015-A780-9, LINDAT/CLARIN dig. library, Charles University
2. Huck, M., Riess, S., Fraser, A.: Target-side word segmentation strategies for neural machine translation. In: WMT, pp. 56–67. ACL (2017)
3. Koehn, P.: Europarl: a parallel corpus for statistical machine translation. In: MT Summit, pp. 79–86. AAMT, Phuket (2005)
4. Koehn, P., et al.: Moses: open source toolkit for statistical machine translation. In: ACL Poster and Demonstration Sessions, pp. 177–180 (2007)
5. Pinnis, M., Krišlauks, R., Deksne, D., Miks, T.: Neural machine translation for morphologically rich languages with improved sub-word units and synthetic data. In: Ekštein, K., Matoušek, V. (eds.) TSD 2017. LNCS (LNAI), vol. 10415, pp. 237–245. Springer, Cham (2017). https://doi.org/10.1007/978-3-319-64206-2_27
6. Sennrich, R., Haddow, B., Birch, A.: Neural machine translation of rare words with subword units. In: ACL, pp. 1715–1725 (2016)
7. Slavíčková, E.: Retrográdní morfematický slovník češtiny. Academia (1975)
8. Tiedemann, J.: News from OPUS - a collection of multilingual parallel corpora with tools and interfaces. In: RANLP, vol. V, pp. 237–248 (2009)
9. Virpioja, S., Smit, P., Grönroos, S.A., Kurimo, M.: Morfessor 2.0: python implementation and extensions for Morfessor baseline. Technical report (2013). Aalto University publication series SCIENCE + TECHNOLOGY; 25/2013
10. Wu, Y., et al.: Google's neural machine translation system: bridging the gap between human and machine translation. CoRR abs/1609.08144 (2016)
11. Žabokrtský, Z., Ševčíková, M., Straka, M., Vidra, J., Limburská, A.: Merging data resources for inflectional and derivational morphology in Czech. In: LREC (2016)

Multi-task Projected Embedding for Igbo

Ignatius Ezeani[⊠], Mark Hepple, Ikechukwu Onyenwe, and Chioma Enemuo

Department of Computer Science, The University of Sheffield, Sheffield, UK
{ignatius.ezeani,m.r.hepple,i.onyenwe,clenemuo1}@sheffield.ac.uk
http://www.sheffield.ac.uk

Abstract. NLP research on low resource African languages is often impeded by the unavailability of basic resources: tools, techniques, annotated corpora, and datasets. Besides the lack of funding for the manual development of these resources, building from scratch will amount to the reinvention of the wheel. Therefore, adapting existing techniques and models from well-resourced languages is often an attractive option. One of the most generally applied NLP models is word embeddings. Embedding models often require large amounts of data to train which are not available for most African languages. In this work, we adopt an alignment based projection method to transfer trained English embeddings to the Igbo language. Various English embedding models were projected and evaluated on the *odd-word*, *analogy* and *word-similarity* tasks intrinsically, and also on the diacritic restoration task. Our results show that the projected embeddings performed very well across these tasks.

Keywords: Low-resource · Igbo · Diacritics · Embedding models
Transfer learning

1 Background

The core task in this paper is embedding-based diacritic restoration. Training embedding models requires large amounts of data which are unavailable in low resource languages. Web-scraped data are often relied upon but they are of poor quality. Languages with diacritics have most of the words wrongly written with missing diacritics. Diacritic restoration helps to improve the quality of corpora for NLP systems.

This work focuses on Igbo, mainly spoken in the south-eastern part of Nigeria and worldwide by about 30 million people. Igbo has diacritic characters (Table 1) which often determine the pronunciation and meaning of words with the same latinized spelling.

1.1 Previous Approaches

Key studies in diacritic restoration involve word-, grapheme-, and tag-based techniques [6]. Earlier examples include Yarowsky's works [17,18] which combined decision list with morphological and collocational information. POS-tags

© Springer Nature Switzerland AG 2018
P. Sojka et al. (Eds.): TSD 2018, LNAI 11107, pp. 285–294, 2018.
https://doi.org/10.1007/978-3-030-00794-2_31

Table 1. Igbo diacritic complexity

Char	Ortho	Tonal
a	–	à,á, ā
e	–	è,é, ē
i	ị	ì, í, ī, ị̀, ị́, ị̄
o	ọ	ò, ó, ō, ọ̀, ọ́, ọ̄
u	ụ	ù, ú, ū, ụ̀, ụ́, ụ̄
m	–	m̀,ḿ, m̄
n	ṅ	ǹ,ń, n̄

and language models have also been applied by Simard [15] to well resourced languages (French and Spanish). Hybrid of techniques are common with this task e.g. Yarowsky [18] used decision list, Bayesian classification and Viterbi decoding while Crandall [1] applied Bayesian- and HMM-based methods. Tufiş and Chiţu [13] combined the two approaches by backing off to character-based method when dealing with "unknown words".

However, these methods are mostly on well-resourced languages (French and Spanish) with comparatively limited diacritic complexity. Mihalcea *et al.* [8] proposed an approach that used character based instances with classification algorithms for Romania. This inspired the works of Wagacha *et al.* [16], De Pauw *et al* [2] and Scannell [14] on a variety of relatively low resourced languages. However, it is a common position that the word-based approach is superior to character-based approach for well resourced languages. Diacritic restoration can also be modelled as a classification task. For Maori, Cocks and Keegan [12] used naïve Bayes algorithms with word n-grams to improve on the character based approach by Scannell [14].

For Igbo, however, one major challenge to applying most of the techniques mentioned above that depend on annotated datasets is the lack of these datasets for Igbo e.g. tags, morph-segmented or dictionaries. This work aims to apply a resource-light approach that is based on a more generalisable state-of-the-art representation model like word-embeddings which could also be tested on other tasks.

1.2 Igbo Diacritic Restoration

Igbo was among the languages in a previous work [14] with 89.5% accuracy using a version of their *lexicon lookup* methods, *LL2*. This technique used the most frequent word and a bigram model to determine the right replacement. However, we could not directly compare their work to ours as the task definitions are slight different. While their accuracy is based on the restoration of every word in a sentence, our work focuses on only the ambiguous words. Besides, their training corpus was too little (31k tokens and 4.3k types) to be representative and there was no language speaker in their team to validate their results. However, we re-implemented a version of the *LL2* and bigram model as our baseline for the restoration task reported in this work.

Ezeani *et al.* [3] implemented a more complex set of *n*–gram models with similar techniques on a larger corpus but though they reported improved results, their evaluation method assumed a closed-world by training and testing on the same dataset. While a more standard evaluation method was used in [4], the data representation model was akin to *one-hot* encoding which is inefficient and could not easily handle large vocabulary sizes.

Another reason for using embedding models for Igbo is that diacritic restoration does not always eliminate the need for sense disambiguation. For example, the restored word *àkwà* could be referring to either *bed* or *bridge*. Ezeani *et al.* [4] had earlier shown that with proper diacritics on ambiguous wordkeys (e.g. *akwa*), a translation system like *Google Translate* may perform better at translating Igbo sentences to other languages. This strategy, therefore, could be more easily extended to sense disambiguation in future.

Table 2. Disambiguation challenge for *Google Translate*

Statement	Google Translate	Comment
O ji *egbe* ya gbuo *egbe*	He used his **gun** to kill *gun*	wrong
O ji **égbè** ya gbuo **égbé**	He used his **gun** to kill **kite**	correct
Akwa ya di n'elu *akwa* ya	It was on the **bed** in his room	fair
Ákwà ya di n'elu **àkwà** ya	his **clothes** on his **bed**	correct
Oke riri *oke* ya	Her addiction	confused
Òké riri **òkè** ya	**Mouse** ate his **share**	correct
O jiri *ugbo* ya bia	He came with his *farm*	wrong
O jiri **ugbọ** ya bia	He came with his **car**	correct

2 Experimental Setup

Our experimental pipeline follows four fundamental stages:

1. pre-processing of data (Sect. 2.1);
2. building embedding models (Sect. 2.2);
3. enhancing embedding models (Sect. 2.3);
4. evaluation of models (Sect. 2.4)

Models are intrinsically evaluated on the *word similarity, analogy* and *odd-word identification* tasks as well as the key process of diacritic evaluation.

2.1 Experimental Data

We used the Igbo-English parallel bible corpora, available from the *Jehova Witness* website[1], for our experiments. There are 32,416 aligned lines of text, bible

---

[1] jw.org.

verses, and chapter headings, from both languages. Total token sizes, without punctuations, are 902,429 and 881,771 while vocabulary lengths 16,084 and 15,000.

Over 50% of both the Igbo tokens (595,221) and vocabulary words (8,750) have at least one diacritic character. There are 550 ambiguous *wordkeys*[2]. Over 97 % of the ambiguous wordkeys have 2 or 3 variants.

2.2 Embedding Models

Inspired by the concept of the universality of meaning and representation (Fig. 1) in distributional semantics, we developed an embedding-based diacritic restoration technique. Embedding models are very generalisable and therefore will constitute essential resources for Igbo NLP work. We used both trained and projected embeddings, as defined below, for our tasks.

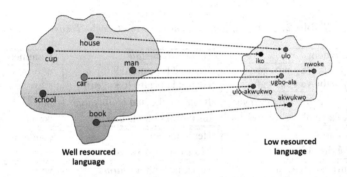

Fig. 1. Embedding projection

Embedding Training. We built the **igBbltrain** embedding from the data described in Sect. 2.1 using the Gensim *word2vec* Python libraries [11]. Default configurations were used apart from optimizing $dimension(default = 100)$ and $window\_size(default = 5)$ parameters to 140 and 2 respectively on the **Basic** restoration method described in Sect. 2.4[3].

Embedding Projection. We adopt an alignment-based projection method similar to the one described in [7]. It uses an Igbo-English alignment dictionary $A^{I|E}$ with a function $f(w_i^I)$ that maps each Igbo word w_i^I to all its co-aligned

---

[2] A *wordkey* is a word stripped of its diacritics if it has any. Wordkeys could have multiple diacritic variants, one of which could be the same as the wordkey itself.

[3] The pre-trained Igbo model from *fastText Wiki word vectors* project [19] was also tested but its performance was so bad that we had to drop it.

English words $w_{i,j}^E$ and their counts $c_{i,j}$ as defined in Eq. 1. $|V^I|$ is the vocabulary size of Igbo and n is number of co-aligned English words.

$$A^{I|E} = \{w_i^I, f(w_i^I)\}; i = 1..|V^I|$$
$$f(w_i^I) = \{w_{i,j}^E, c_{i,j}\}; j = 1..n \qquad (1)$$

The projection is formalised as assigning the weighted average of the embeddings of the co-aligned English words $w_{i,j}^E$ to the Igbo word embeddings $\mathbf{vec}(w_i^I)$ [7]:

$$\mathbf{vec}(w_i^I) \leftarrow \frac{1}{C} \sum_{w_{i,j}^E), c_{i,j} \in f(w_i^I)} vec(w_{i,j}^E) \cdot c_{i,j} \qquad (2)$$

where $C \leftarrow \sum\limits_{c_{i,j} \in f(w_i^I)} c_{i,j}$

Using this projection method, we built 5 additional embedding models for Igbo:

- **igBblproj** from a model we trained on the English bible.
- **igGNproj** from the pre-trained *Google News*[4]*word2vec* model.
- **igWkproj** from *fastText* Wikipedia 2017, UMBC webbase corpus and statmt.org news dataset.
- **igSwproj** from same as **igWkproj** but with subword information.
- **igCrlproj** from *fastText* Common Crawl dataset

Table 3 shows the vocabulary lengths ($Vocabs^L$), and the dimensions (*Dimension*) of each of the models used in our experiments. While the pre-trained models and their projections have vector sizes of 300, our trained **Igbo-Bible** performed best with vector size of 140 and so we trained the **IgboEnBbl** with the same dimension.

Table 3. Igbo and English models: vocabulary, vector and training data sizes

Model	Dimension	Vocabs^I	Vocabs^E	Data
igBbltrain	140	4,968	–	902.5k
igBblproj	140	4,057	6.3k	881.8k
igGNproj	300	3,046	3m	100bn
igWkproj	300	3,460	1m	16bn
igSwproj	300	3,460	1m	16bn
igCrlproj	300	3,510	2m	600bn

---

[4] https://code.google.com/archive/p/word2vec/.

2.3   Enhancing Embedding Models

For this experiment, our dataset consists of 29 ambiguous *wordkeys*[5] from our corpus. For each wordkey, we keep a list of sentences (excluding punctuations and numbers), each with a place-holder (see Table 4) to be filled with the correct variant of the wordkey.

Table 4. Instances of the wordkey *akwa* in context

Variant	Left context	Placeholder	Right context	Meaning
àkwá	ka okwa nke kpokotara	____	o na-eyighi eyi otu	egg
ákwà	a kpara akpa mee	____	ngebichi nke onye na-ekwe	cloth
ákwá	ozugbo m nuru mkpu	____	ha na ihe ndi a	cry

In both trained and projected embedding models, vectors are assigned to each word in the dictionary, and that includes each diacritic variant of a wordkey. The **Basic** restoration process (Sect. 2.4) uses this initial embedding model *as-is*. The models are then refined by "learning" new embeddings for each variant that correlate more with its context words embeddings.

For example, let mcw_v contain the top n (say $n = 20$) of the most co-occurring words of a certain variant, v and their counts, c. The diacritic embedding is derived by replacing each diacritic variant vector with the weighted average of the vectors of its most co-occurring words (see Eq. (3)).

$$\textbf{diac}_{\textbf{vec}} \leftarrow \frac{1}{|mcw_v|} \sum_{w \in mcw_v} w_{vec} * w_c \qquad (3)$$

where w_c is the 'weight' of w i.e. the count of w in mcw_v.

2.4   Model Evaluation

We evaluation the models on their performances on the following NLP tasks: *odd-words, analogy* and *word similarity* and diacritic restoration. As there are no standard datasets for these tasks in Igbo, we had auto-generate them from our data or transfer existing ones from English. Igbo native speakers were used to refine and validate instances of the dataset or methods used.

The *odd Word*. In this task, the model is used to identify the *odd word* from a list of words e.g. *breakfast, cereal, dinner, lunch* → *"cereal"*. We created four simple categories of words Igbo words (Table 5) that should naturally be mutually exclusive. Test instances were built by randomly selecting and shuffling three words from one category and one from another e.g. *okpara, nna, ogaranya, nwanne* → *ogaranya*.

---

[5] Highly dominant variants or very rarely occurring *wordkeys* were generally excluded from the datasets.

Analogy. This is based on the concept of analogy as defined by [9] which tries to find y_2 in the relationship: $x_1 : y_1$ as $x_2 : y_2$ using vector arithmetic e.g. $king - man + woman \approx queen$. We created pairs of opposites for some common nouns and adjectives (Table 6) and randomly combined them to build the analogy data e.g. di (husband) $-$ $nwoke$ (man) $+$ $nwaanyi$(woman) $\approx$ $nwunye$(wife)?

Table 5. Word categories for *odd word* dataset

category	Igbo words
nouns(family) *e.g. father, mother*	ada, ọkpara, nna, nne, nwanna, nwanne, di, nwunye
adjectives *e.g. tall, rich*	ọcha, ọgaranya, ogbenye, ogologo, oji, ọjọọ, okenye, ọma
nouns(humans) *e.g. man, woman*	nwaanyị, nwoke, nwata, nwatakịrị, agbọghọ, okorobịa
numbers *e.g. one, seven*	otu, abụọ, atọ, anọ, ise, isii, asaa, asatọ, itoolu, iri

Table 6. Word pair categories for *analogy* dataset

Category	Opposites
oppos-nouns	nwoke:nwaanyị, di:nwunye, okorobịa:agbọghọ, nna:nne, ọkpara:ada
oppos-adjs	agadi:nwata, ọcha:oji, ogologo:mkpụmkpụ, ọgaranya:ogbenye

Word Similarity. We created Igbo word similarity dataset by transferring the standard *wordsim353* dataset [5]. Our approach used *Google Translate* to translate the individual word pairs in the combined dataset and return their human similarity scores. We removed instances with words that could not be translated (e.g. cell $\rightarrow$ *cell* & phone$\rightarrow$*ekwenti*,7.81) and those with translations that yield compound words (e.g. situation $\rightarrow$ *ọnọdụ* & conclusion $\rightarrow$ *nkwubi okwu*,4.81)[6].

Diacritic Restoration Process. The restoration process computes the cosine similarity of the variant and context vectors and chooses the most similar candidate. For each wordkey, wk, candidate vectors, $D^{wk} = \{d_1, \ldots, d_n\}$, are extracted from the embedding model on-the-fly. C is defined as the list of the context words and vec_C is the context vector of C (Eq. (4)).

$$\mathbf{vec_C} \leftarrow \frac{1}{|C|} \sum_{w \in C} vec_w \tag{4}$$

$$\mathbf{diac_{best}} \leftarrow \operatorname*{argmax}_{d_i \in D^{wk}} sim(\mathbf{vec_C}, d_i) \tag{5}$$

---

[6] An alternative considered is to combine the word e.g. *nkwubi okwu* $\rightarrow$ **nkwubi-okwu** and update the model with a projected vector or a combination of the vectors of constituting words.

3 Results and Discussion

Our results on the odd-word, analogy and word-similarity tasks (Table 7, Fig. 2) indicate that the projected embedding models, in general, capture concepts and their relationships better. This is not surprising as the trained model, **igBible**, and the one from its parallel English data, **igEnBbl** are too little and cover only religious data. Although **igWkSbwd** includes subword information which should be good for an agglutinative language like Igbo, these subword patterns are different from the patterns in Igbo. Generally, the models from the news data, **igGNews, igWkNews**, did well on these tasks.

On the diacritic restoration task, the embedding based approaches, with semantic information, generally performed comparatively well with respect to the n-gram models that capture syntactic details better. **IgBible**'s performance is impressive especially as it outperformed the bigram model[7].

Expectedly, compared to other projected models, **IgBible** and its parallel, **IgEnBbl**, clearly did better on this task. **IgBible** was originally trained with the same dataset and language of the task and its vocabulary directly aligns

Table 7. Trained and Project Embeddings on odd-word prediction

| | Odd-word | Similarity | Analogy | |
Models	Accuracy	Correlation	nouns	adjectives
igBible	78.27	48.02	23.81	06.67
igGNews	**84.24**	60.00	64.29	**56.67**
igEnBbl	75.26	58.96	54.76	13.33
igWkSbwd	84.18	58.56	64.29	50.00
igWkCrl	80.72	**62.07**	78.57	21.37
igWkNews	81.51	59.69	**80.95**	50.00

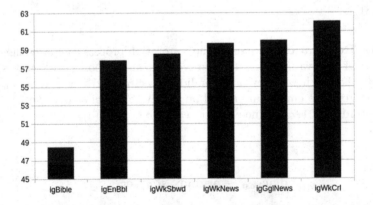

Fig. 2. Worst-to-Best word similarity correlation performance

---

[7] We intend to implement higher level n-gram models.

with that of **IgEnBbl**. Clearly, the enhanced diacritic embeddings improved the performances of all the models which is expected as each variant is pulled to the center of its most co-occurring words.

Table 8. Performances of Basic and Diacritic versions of the *Trained* and *Projected* embedding models on diacritic restoration tasks

Baselines: n-gram models								
	Unigram				*Bigram*			
	72.25%				80.84%			

Embedding models								
	Accuracy		Precision		Recall		F1	
	Basic	Diac	Basic	Diac	Basic	Diac	Basic	Diac
igBible	69.28	**82.26**	61.37	77.96	61.90	**82.28**	57.19	**76.16**
igEnBbl	64.72	78.71	59.60	75.18	59.65	79.52	50.51	72.93
igGNews	57.57	74.14	32.20	72.50	49.00	74.56	19.06	62.47
igWkSbwd	62.10	73.83	13.82	73.81	47.64	74.03	10.65	66.62
igWkCrl	60.78	73.30	40.07	**78.02**	49.16	76.24	25.36	68.62
igWkNews	61.07	72.97	14.16	76.04	46.10	75.14	8.31	65.20

4   Conclusion and Future Research Direction

This work contributes to the IgboNLP[8] [10] project. The goal of the project is to build a framework that can adapt, in an effective and efficient way, existing NLP tools to support the development of Igbo. In this paper, we demonstrated that projected embedding models can outperform the ones built with small language data on a variety of NLP tasks on low resource languages.

We also introduced a technique for learning diacritic embeddings which could be applied to the diacritic restoration task. Our next focus is to refine our techniques and datasets and train models with sub-word information as well as consider sense disambiguation task.

References

1. Crandall, D.: Automatic Accent Restoration in Spanish text (2005). http://www.cs.indiana.edu/~djcran/projects/674_final.pdf. Accessed 7 Jan 2016
2. De Pauw, G., De Schryver, G.M., Pretorius, L., Levin, L.: Introduction to the special issue on African language technology. Lang. Resour. Eval. **45**, 263–269 (2011)
3. Ezeani, I., Hepple, M., Onyenwe, I.: Automatic restoration of diacritics for Igbo language. In: Sojka, P., Horák, A., Kopeček, I., Pala, K. (eds.) TSD 2016. LNCS (LNAI), vol. 9924, pp. 198–205. Springer, Cham (2016). https://doi.org/10.1007/978-3-319-45510-5_23

---

[8] See igbonlp.org.

4. Ezeani, I., Hepple, M., Onyenwe, I.: Lexical disambiguation of Igbo using diacritic restoration. In: Proceedings of the 1st Workshop on Sense, Concept and Entity Representations and Their Applications, pp. 53–60 (2017)
5. Finkelstein, L., et al.: Placing search in context: the concept revisited. In: Proceedings of the 10th International Conference on World Wide Web, pp. 406–414. ACM (2001)
6. Francom, J., Hulden, M.: Diacritic error detection and restoration via POS tags. In: Proceedings of the 6th Language and Technology Conference (2013)
7. Guo, J., Che, W., Yarowsky, D., Wang, H., Liu, T.: Cross-lingual dependency parsing based on distributed representations. In: Proceedings of the 53rd Annual Meeting of the Association for Computational Linguistics and the 7th International Joint Conference on Natural Language Processing (Vol1: Long Papers), pp. 1234–1244 (2015)
8. Mihalcea, R.F.: Diacritics restoration: learning from letters versus learning from words. In: Gelbukh, A. (ed.) CICLing 2002. LNCS, vol. 2276, pp. 339–348. Springer, Heidelberg (2002). https://doi.org/10.1007/3-540-45715-1_35
9. Mikolov, T., Chen, K., Corrado, G., Dean, J.: Efficient Estimation of Word Representations in Vector Space, arXiv preprint arXiv:1301.3781 (2013)
10. Onyenwe, I.E., Hepple, M., Chinedu, U., Ezeani, I.: A basic language resource kit implementation for the IgboNLP project. ACM Trans. Asian Low-Resource Lang. Inf. Process. **17**(2), 101–1023 (2018)
11. Řehůřek, R., Sojka, P.: Software framework for topic modelling with large corpora. In: Proceedings of the LREC 2010 Workshop on New Challenges for NLP Frameworks, 22 May, pp. 45–50. ELRA, Valletta (2010). http://is.muni.cz/publication/884893/en
12. Cocks, J., Keegan, T.-T.: A word-based approach for diacritic restoration in Māori. In: Proceedings of the Australasian Language Technology Association Workshop 2011, Canberra, Australia, pp. 126–130, December 2011. http://www.aclweb.org/anthology/U/U11/U11-2016
13. Tufiş, D., Chiţu, A.: Automatic diacritics insertion in Romanian texts. In: Proceedings of the International Conference on Computational Lexicography, Pecs, Hungary, pp. 185–194 (1999)
14. Scannell, K.P.: Statistical unicodification of African languages. Lang. Resour. Eval. **45**(3), 375–386 (2011)
15. Simard, M.: Automatic insertion of accents in French text. In: Proceedings of the Third Conference on Empirical Methods for Natural Language Processing, pp. 27–35 (1998)
16. Wagacha, P.W., De Pauw, G., Githinji P.W.: A grapheme-based approach to accent restoration in Gĩkũyũ. In: Fifth International Conference on Language Resources and Evaluation (2006)
17. Yarowsky, D.: A comparison of corpus-based techniques for restoring accents in Spanish and French text. In: Proceedings of 2nd Annual Workshop on Very Large Corpora, Kyoto, pp. 19–32 (1994)
18. Yarowsky, D.: Corpus-Based Techniques for Restoring Accents in Spanish and French Text, Natural Language Processing Using Very Large Corpora. Kluwer Academic Publishers, pp. 99–120 (1999)
19. Bojanowski, P., Grave, E., Joulin, A., Mikolov, T.: Enriching Word Vectors with Subword Information, arXiv preprint arXiv:1607.04606 (2016)

Corpus Annotation Pipeline for Non-standard Texts

Zuzana Peliknov and Zuzana Nevilov[✉]

Natural Language Processing Centre, Masaryk University, Brno, Czech Republic
pelikanova@mail.muni.cz,xpopelk@fi.muni.cz

Abstract. According to some estimations (e.g. [9]), web corpora contain over 6% of foreign material (borrowings, language mixing, named entities). Since annotation pipelines are usually built upon standard and correct data, the resulting annotation of web corpora often contains serious errors.

We studied in depth annotation errors of the web corpus *czTenTen 12* and proposed an extension to the tagger desamb that had been used for czTenTen annotation. First, the subcorpus was made using the most problematic documents from czTenTen. Second, measures were established for the most frequent annotation errors. Third, we established several experiments in which we extended the annotation pipeline so it could annotate foreign material and multi-word expressions. Finally, we compared the new annotations of the subcorpus with the original ones.

Keywords: Non-standard language · Interlingual homographs Corpora annotation

1 Introduction

Web corpora are annotated using standard annotation pipelines that consist of text and encoding normalization, sentence splitting, tokenization, morphological analysis, lemmatization and tagging. In case of unknown words, the pipelines usually use guessers for the particular language. However, web corpora also contain non-standard language, language mixing, slang, and foreign named entities, thus the pipeline for standard language is sometimes insufficient. In Czech web corpus annotation, the guesser is overused, i.e. used on foreign material.

We modify the pipeline for annotation of Czech web corpus so that foreign material is identified *before* the possible use of the guesser. We took the corpus cztenten [4] and focused on two phenomena in the annotation: interlingual homographs and possible overuse of the guesser, mainly on foreign words.

First, we prepared a subcorpus that contains many occurrences of these two phenomena. The corpus contains 7,661 sentences and 256,922 tokens. Second, we set up several experiments in which we extended the vocabulary of morphological analysis with foreign single words or multi-word expressions (MWEs). In some

© Springer Nature Switzerland AG 2018
P. Sojka et al. (Eds.): TSD 2018, LNAI 11107, pp. 295–303, 2018.
https://doi.org/10.1007/978-3-030-00794-2_32

settings, we employed a foreign chunk detector. We compared the annotations of the subcorpus.

The work resulted in a corpus of non-standard Czech and an extension to the standard pipeline. Even though several pipelines exist for Czech, we believe our extension can be used in all of them, possibly with only a slight modification.

2   Related Work

One of the first works dealing with web corpora is [5]. WebCorp is a follow-up project[1]. These works deal with the web as a whole rather than web texts.

Web texts often make part of a new genre sometimes called *internet language*. Firstly defined by [3], internet language is characterized by the use of informal, non-standard style, sharing many features with the spoken language. The pragmatic features are often expressed by emojis, capital letters and punctuation. The style also comprises high use of foreign, mainly English words.

An example work concerning annotation of web texts, [10], uses Freeling for standard Spanish and employed manual post-processing. Another project of German web corpus [8] adapts Ucto for word and sentence tokenization of emojis and uses TreeTagger complemented with about 3,800 entries that occur in web corpora. In addition, named entity recognition was used. In case of Slovak corpus from the Aranea family [1], the input text is tagged by taggers for several languages and the final tag is estimated on the aggregated result.

3   Building Subcorpus of Foreign Material

Since Czech and English are not very close languages, the number of interlingual homographs is relatively low. Well-known examples comprise *copy* (meaning *plaits*) or a cross-POS *top* (meaning imperative of *to drown*).

Apart from interlingual homographs, Czech and English text share a considerable number of cognates (i.e. words of the same origin) with the same meaning (e.g. *sport*, *film*) that are not important from our point of view. Homographic cognates are used in the same syntactic situations in both languages except of gender assignment in Czech. This means that if a word is adopted in Czech, it has to have a gender and most frequently, it is assigned according to the word ending, for example *sport* is inanimate masculine noun.

From a total of 1,579 words appearing both in English and Czech texts (using a frequency threshold), 327 interlingual homographs were selected, while cognates with the exact same meaning such as *hamburger* or *badminton* were discarded. Sentences containing interlingual homographs from `cztenten` were selected, in total 250,000 tokens. The subcorpus was prepared based on manually sought English or mixed collocations frequently appearing in Czech text, e.g. *copy and paste*, *user friendly* or *top modelka* (meaning *top model*). We used Sketch Engine [6] to search for such collocations, as well as filtering and sampling

---

[1] http://www.webcorp.org.uk.

separate query results and compiling a final subcorpus. We named the corpus BULKY as a reference to one of the Czech-English interlingual homographs (meaning *buns* in Czech).

4   Extension of the Annotation Pipeline

The standard pipeline for `cztenten` annotation consists of tokenization, morphological analysis using `majka`, sentence splitting, guessing, and tagging using `desamb` as described in [11].

In our experiments, we used three resources independently: list of one-word named entities, list of multi-word named entities, script for English chunks detection. The lists of named entities were built from gazetteers of person names, place names from the Czech cadastre and from Geonames[2], Wikipedia articles, and from the list of multi-word expressions as described in Sect. 4.1.

We made different pipelines:

- EXTENDED: the dictionary of the morphological analyser `majka` was extended by one-word named entities, the pipeline stayed the same,
- MWE+EXTENDED: we detected multi-word expressions before lemmatization, the morphological analyser was extended by one-word named entities
- FOREIGN+EXTENDED: before lemmatization, we identified English chunks, afterwards, we used the extended dictionary for `majka`
- FOREIGN+MWE+EXTENDED: before lemmatization, we identified English chunks, afterwards, we identified MWEs, and finally, we used the extended dictionary for `majka`

In addition, we had two versions of the EXTENDED part of the pipeline. In the first version, EXTENDED1, we extend the morphological analyser `majka` by the complete list of one-word named entities. In the second version, EXTENDED2, we excluded words already present in `majka`, e.g. *York* is present in the `majka` dictionary as masculine inanimate (the city), while in EXTENDED1, it is also present as masculine animate and feminine (person names).

4.1   Multi-word Expressions

MWE processing consists of two steps: MWE discovery and MWE identification. MWE discovery results in datasets containing MWEs, MWE identification outputs a MWE annotation for particular texts. The problem is that many MWEs allow two readings: a MWE reading where the whole sequence is one unit, a non-MWE reading is a sequence of tokens arbitrarily appearing in the text. A non-MWE reading example is provided by [2]: *He passed by and large tractors passed him* where *by and large* is not a MWE.

We benefited from our previous work [7] in which we built a collection of 2,700 MWE lemmata. We focused on frozen and non-decomposable MWEs such

---

[2] http://www.geonames.org.

as *a priori* that contain out-of-vocabulary (OOV) tokens, interlingual homographs, or syntactic anomalies. We used a straightforward MWE identification: if a sequence of words appears in the MWE resource, it is annotated as MWE. In our case, this is relatively safe since we deal with continuous MWEs. The non-MWE readings are often possible e.g. in case of compound verbs where the components can appear in distant positions in the sentence. We are nevertheless aware that such cases possibly occur in our annotations as well.

4.2 Foreign Chunk Detection

Foreign chunk detection script compares relative frequencies of each token in two large web corpora: `cztenten` and `ententen`: if a token appears significantly more frequently in the English corpus or it is a Czech OOV, we annotate it as English. We also take the neighboring tokens into account. In cases of interlingual homographs, the script compares bigram frequencies.

5   Experiments

We set three measures in order to compare results of experiments. One of the most straightforward measures is the number of guessed tokens. As second measure, we selected the number of imperatives. The reason is that imperatives are the shortest word forms of Czech verbs and therefore are more likely homographic. Finally, we measured the number of infinitives because English words with the -t ending were often annotated as infinitives, for instance *next* or *assault*.

We also analyzed samples of the BULKY corpus annotation in order to explore the overall influence of each pipeline on the tagging. We consider errors in POS tagging severe, errors in other grammatical categories as less serious. The annotations are also compared to these created by the original pipeline.

5.1   Experiment 1: The EXTENDED1 Pipeline

Comparing by the original annotation to the EXTENDED1, our main focus was on guesser usage: Originally, the total number of words tagged by guesser was 30,640, with extended pipeline, the number dropped to 24,573.

Originally, the number of imperatives yielded a total of 2,038 forms and 1,948 within the EXTENDED1 annotation. Reduction of imperative tag is a promising indication as in current `czTenTen` corpus, imperative forms are noticeably overused. The number of infinitives dropped from 3,430 to 3,428 occurrences.

Foreign words and proper names were in some cases tagged and lemmatized more suitably; there was, however, not much consistency involved. The interlingual homograph *mine*, used as an English pronoun, for instance, was originally assigned an incorrect Czech lemma (*minout* meaning *to miss*) and was therefore tagged as a verb in all 106 occurrences, while the EXTENDED1 pipeline resulted in annotation of half of the occurrences tagged again as a Czech word (63 times), the other half given a more suitable tag and lemma (62 times assigned lemma *Mine* with a tag of a feminine noun in singular form).

While tagging of some of the proper names and English words slightly improved, the drawback to this extended annotation was oftentimes incorrect tagging of capitalized words that open a sentence as nouns (in contrast to the original corpus with correct tagging of such words). This was a common issue especially for prepositions and pronouns. For example, the preposition *Na* (meaning *on*) was (incorrectly) tagged as a proper noun.

5.2   Experiment 2: The EXTENDED2 Pipeline

Comparing the annotation of EXTENDED2 to EXTENDED1, the incorrect annotation of capitalized words as nouns was fixed. However, annotation of several homographs and proper names was impaired similarly to the original pipeline. For example, for a Czech-English homograph *Line*, used as an English word, we retrieved the correct lemma only within the EXTENDED1 pipeline.

Tagging another Czech-English homograph, *man*, with EXTENDED1 pipeline resulted in 134 cases lemmatized as *Man* or *man* and only 6 incorrectly with a Czech lemma *mana*. Correct tag of animate masculine noun was retrieved 60 times, while a more incorrect tag of a feminine noun 80 times (including tokens lemmatized as *mana*). With EXTENDED2 pipeline, in spite of more words being wrongly lemmatized as a Czech word *mana* (24 times), the tagging quality improved, as the rest 116 times, animate masculine noun tag was retrieved.

The EXTENDED2 pipeline resulted in 24,228 guesser uses, 2,095 imperatives, 3,442 infinitives, which is more than EXTENDED1. The annotation quality was overall objectively better than EXTENDED1.

5.3   Experiment 3: The MWE+EXTENDED1 Pipeline

We compared the original pipeline with the extended annotation together with MWE recognition. We did not filter out proper names that were recognized by `majka`. In this process, identified MWEs were simply treated as one word. In order to not change the pipeline too much, we replaced spaces in each MWE by underscores (not previously appearing in the data). Afterwards, a tag similar to one word expressions was assigned to each MWE. For example, *Top Gun*, that was initially tagged as two tokens – a verb in imperative form (lemma *topit* meaning *to drown*) and a noun, is now lemmatized as a single token *Top_Gun*, tagged as a noun. A total of 2,085 MWE's were newly identified in the corpus.

Guesser use again dropped significantly, from a total of 30,640 to 22,050 cases. Number of imperatives dropped from 2,038 to 1,706, infinitives decreased fairly slightly from 3,428 to 3,385. Annotating MWEs also resulted in improvement of one word proper noun tagging, e.g. words *Harry* or *Harryho* were newly assigned a correct lemma (*Harry*) and a tag of masculine noun in all 7 occurrences.

As with Experiment 1, some tokens newly received an incorrect annotation. In these cases, POS was preserved but grammatical features such as gender were altered.

5.4   Experiment 4: The FOREIGN+EXTENDED1 Pipeline

Foreign chunks (as described in Sect. 4.2) were detected after tokenization. Similarly to MWE identification, when a foreign multi-word sequence is found, it is treated as one word. We observed that not all foreign sequences were identified, e.g. *Baby Boom Baby Design Sprint*. Foreign sequences such as this one are difficult to identify because although composed from English words, they do not refer to objects appearing in English corpora. Particularly, in our example, the sequence refers to a Polish article (a baby coach).

Occasionally, secondary incorrect annotation occurred with words neighboring to foreign nouns, e.g. the preceding word was incorrectly annotated as adjective in the sequence *i kdy bude hostn ham and eggs* (*even though he will be served ham and eggs*). The original pipeline annotated *hostn* (*served*) as verb but this pipeline annotated it as an adjective.

In this experiment, the guesser was used 17,597 times, imperatives appeared 1,129 times, infinitives 3,324 times. Using EXTENDED2 instead of EXTENDED1 did not change the results significantly. With EXTENDED2, 17,598 words were guessed, imperative count was 1,131, infinitive count 3,325.

5.5   Experiment 5: The FOREIGN+MWE+EXTENDED1 Pipeline

In experiment 5, we used all possible tools: first, we detected foreign chunks, subsequently, we identified MWEs, and finally, we extended the morphological analysis and lemmatization with one-word proper nouns. Not surprisingly, the quantitative results were the best ones: only 17,416 words were guessed, 1,100 words were tagged as imperatives and 3,328 as infinitives. Replacing EXTENDED1 by EXTENDED2 did not change the results in any way.

6   Results

In the five experiments, we run different pipelines on the BULKY corpus. In all experiments, we used the same tokenizer, tagger, and guesser. However, we extended the dictionary of the morphological analyser. We counted the number of guesser uses since we had been aware that the guesser was overused. We also focused on imperatives since they are the shortest word forms of Czech verbs and thus are more likely to be homographic. We also observed the number of infinitives since the guesser assigned an infinitive tag to words with -t ending.

Table 1 shows that the pipeline with foreign chunks detection has better annotation results. Extension of the dictionary reduces the number of guesser uses. MWE identification contributes to the improvement as well. However, in

Table 1. Quantitative comparison of the experiments

Pipeline	Guesser	Imperatives	Infinitives
Original	30,640	2,038	3,428
EXTENDED1	24,573	1,948	3,430
EXTENDED2	24,228	2,095	3,442
MWE+EXTENDED1	22,050	1,706	3,385
FOREIGN+EXTENDED1	17,597	1,129	**3,324**
FOREIGN+EXTENDED2	17,598	1,131	3,325
FOREIGN+MWE+EXTENDED1	**17,416**	**1,100**	3,328

the qualitative analysis we made on samples, we discovered that the extension of the dictionary sometimes leads to new tagging errors.

We analysed a frequent interlingual homograph *top* more in detail (see Table 2). Lemmatization is a consensual process to some extent. For instance, some systems consider the lowercase word to be the base form (lemma), others preserve the case. The same applies for multi-word lemmata. We can see that the word *top* occurs with three lemmata, the first (meaning *to drown* in imperative) is incorrect in all cases. The other two (*top/Top*) are correct.

We also observed the tags of this word: annotating *top* as a verb is always incorrect, on the other hand, annotating *top* as either masculine or feminine noun or an indefinite noun or even an unknown POS is acceptable. The annotation improved in about one third of the cases. The remaining two third are expressions with a Czech word or a number (such as *top 10*). We annotated the interlingual homograph *top* only in the context of an English sequence or a MWE. In this case, the word seems to become an integral part of the Czech language.

Table 2. Interlingual homograph *top*: lemma and tag distribution. The asterisk means that the lemma is not necessarily complete, e.g. *Top* is part of *Top_Gun*

Pipeline	Lemma			Tag			
	Topit	Top*	Top*	Verb	Noun	m/f noun	Unknown
Original	147	0	0	147	0	0	0
EXT1	147	0	0	147	0	0	0
EXT2	147	0	0	147	0	0	0
MWE+EXT1	112	27	8	112	7	28	0
FOREIGN+EXT1	105	34	8	106	0	35	6
FOREIGN+EXT2	105	33	9	106	0	35	6
FOREIGN+EXT1+MWE	103	10	34	104	2	35	6

7   Conclusion and Future Work

This work concerns annotation of web corpora. Currently, web texts are annotated using the same pipeline as standard Czech texts which leads to systematic annotation errors. First, we made a corpus that contains texts typical for the web (non-standard Czech, frequent language mixing). Second, we modified the standard pipeline in order to reduce the number of known annotation errors. In the most successful setting, the guesser was used only in 57% cases compared to the original pipeline. Also, the number of incorrect annotations of interlingual homographs dropped: the number of imperatives decreased to 54% of the original number. The resulting corpus is available in the LINDAT/CLARIN repository[3].

Future work has to focus on the remaining errors. One possible solution is to use word sketches or word embeddings to discover the semantic role of a OOV and assign the tag of a similar word. In the future, we will annotate the whole web corpus cztenten. Currently, it is not clear whether to use the present tagger or to replace it by another one. In any case, foreign sequences detection and MWE processing can compensate the partial inappropriateness of the standard tools for the non-standard text.

References

1. Benko, V.: Language code switching in web corpora. In: The 11th Workshop on Recent Advances in Slavonic Natural Languages Processing, RASLAN 2017, Karlova Studanka, Czech Republic, 1–3 December 2017, pp. 97–105 (2017). http://nlp.fi.muni.cz/raslan/2017/paper11-Benko.pdf
2. Constant, M., Eryiit, G., Monti, J., Plas, L.V.D., Ramisch, C., Rosner, M., Todirascu, A.: Multiword expression processing: a survey. Comput. Linguist. 0(ja), 1–92 (2017). https://doi.org/10.1162/COLI_a_00302
3. Crystal, D.: Language and the Internet. Cambridge University Press (2006). https://books.google.cz/books?id=cnhnO0AO45AC
4. Jakubek, M., Kilgarriff, A., Kov, V., Rychl, P., Suchomel, V.: The TenTen Corpus Family. In: 7th International Corpus Linguistics Conference CL 2013, Lancaster, pp. 125–127 (2013). http://ucrel.lancs.ac.uk/cl2013/
5. Kilgarriff, A., Grefenstette, G.: Introduction to the special issue on the web as corpus. Comput. Linguist. 29(3), 333–347 (2003). https://doi.org/10.1162/089120103322711569
6. Kilgarriff, A., Rychl, P., Smr, P., Tugwell, D.: The sketch engine. In: Proceedings of the Eleventh EURALEX International Congress, pp. 105–116 (2004). http://www.fit.vutbr.cz/research/view_pub.php?id=7703
7. Nevilov, Z.: Annotation of multi-word expressions in czech texts. In: Hork, A., Rychl, P., Rambousek, A. (eds.) Ninth Workshop on Recent Advances in Slavonic Natural Language Processing, pp. 103–112. Tribun EU, Brno (2015)
8. Schfer, R.: Processing and querying large web corpora with the COW14 architecture. In: Baski, P., Biber, H., Breiteneder, E., Kupietz, M., Lngen, H., Witt, A. (eds.) Proceedings of Challenges in the Management of Large Corpora 3 (CMLC-3). IDS, Lancaster (2015). http://rolandschaefer.net/?p=749

[3] http://hdl.handle.net/11234/1-2822.

9. Schfer, R., Bildhauer, F.: Web corpus construction. Synth. Lect. Hum. Lang. Technol. (2013). https://doi.org/10.2200/S00508ED1V01Y201305HLT022

10. Taul, M., et al.: Spanish treebank annotation of informal non-standard web text. In: Daniel, F., Diaz, O. (eds.) Current Trends in Web Engineering, pp. 15–27. Springer International Publishing, Cham (2015). https://doi.org/10.1007/978-3-319-24800-4_2

11. Šmerk, P.: Unsupervised learning of rules for morphological disambiguation. In: Sojka, P., Kopeček, I., Pala, K. (eds.) TSD 2004. LNCS (LNAI), vol. 3206, pp. 211–216. Springer, Heidelberg (2004). https://doi.org/10.1007/978-3-540-30120-2_27

Recognition of OCR Invoice Metadata
Block Types

Hien T. Ha, Marek Medved'[(✉)], Zuzana Nevěřilová, and Aleš Horák

Natural Language Processing Centre, Faculty of Informatics, Masaryk University,
Botanická 68a, 602 00 Brno, Czech Republic
{xha1,xmedved1,xpopelk,hales}@fi.muni.cz

Abstract. Automatically cataloging of thousands of paper-based structured documents is a crucial fund-saving task for future document management systems. Current optical character recognition (OCR) systems process the tabular data with a sufficient level of character-level accuracy; however, the overall structure of the document metadata is still an open practical task.

In this paper, we introduce the OCRMiner system designed to extract the indexing metadata of structured documents obtained from an image scanning process and OCR. We present the details of the system modular architecture and evaluate the detection of text block types that appear within invoice documents. The system is based on text analysis in combination of layout features, and is developed and tested in cooperation with a renowned copy machine producer. The system uses an open source OCR and reaches the overall accuracy of 80.1%.

Keywords: OCR · Scanned documents · Document metadata
Invoice metadata extraction

1 Introduction

Nowadays, large companies deal with enormous numbers of both paper and digital-born documents that do not have a fixed predefined structure easily parsable by automatic techniques. Precise metadata annotation is thus an inevitable (and expensive[1]) prerequisite of further document processing by a standard information system. An example class of business documents that share a common, yet very varying, structure are financial statements and billing invoices.

In the following text, we present the design and development of the OCR-Miner project aiming at automatic processing of (semi-)structured business documents, such as contracts and invoices, based solely on the analysis of OCR processing of the document pages. We describe the modules used for feature

---

[1] A 2016 report by the Institute of Finance and Management [8] suggested that the average cost to process an invoice was $12.90.

© Springer Nature Switzerland AG 2018
P. Sojka et al. (Eds.): TSD 2018, LNAI 11107, pp. 304–312, 2018.
https://doi.org/10.1007/978-3-030-00794-2_33

extraction from the document layout and content properties. Then, we offer a detailed evaluation of the system on the task of detecting and annotating discovered text blocks with the corresponding informational type.

Previous works aiming at automatic processing of OCR documents rely on techniques based on layout graphs accompanied with several approaches to rule-based classification. One of the first systems [4] used a specific programming language with syntax-driven search for the description of the frame representation language for structured documents (FRESCO). The recognition rate for the analysed invoice blocks was between 40–60%.

In several works [7,13], a case-base reasoning (CBR) approach is used to extract invoice structure. The systems define similarity measures to compare two graphs based on weighted graph edit distance. The document invoice analysis here is composed of two phases: global solving and local solving. In the former, the system checks if a similar case (document graph consisting of tables and keyword structures) exists in the document database by using graph probing. In the local solving phase, the nearest structure's solution is applied adaptively to the given keyword structure. The recognition rate for both phases reaches 76–85%.

Bart and Sarkar [3] proposed a semi-automatic extraction method by applying given solutions for repeated structures in documents with the same or a similar format. Candidate fields of repeated structures are evaluated by the overall match quality between the candidate and a reference record in term of the perceptual coherence features (alignment, height, width and presence of overlaps, separation, gaps, etc.). In the evaluation, the system was able to identify 92% of invoice fields, which corresponds to 63% of testing invoices being processed successfully. This method can generalize well to different domains, however, it requires a large number of annotations.

In recent study [1], Aslan et al. apply the part-based modeling (PBM) approach to invoice processing based on deformable compositions of invoice parts candidates obtained from machine learning based classification techniques. The presented evaluation of invoice block detection ranges from 69% to 87% with the average of 71% accuracy.

The presented OCRMiner project is based on a combination of advantages of the above approaches. OCRMiner represents the documents as a graph of hierarchical text blocks with automatic modular feature annotations based on keywords, text structures, named entity processing and layout information. These features are then processed by both rule based and machine learning based classification to identify the appropriate document parts for the information extraction task.

2   The OCRMiner Pipeline

The OCRMiner system consists of a set of interconnected modules (a "pipeline") that allow to add any kind of partial annotations to the analysed document. An

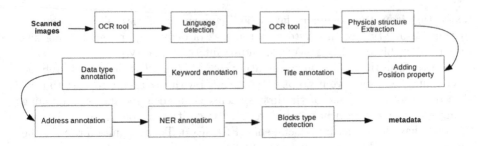

Fig. 1. The processing pipeline

overall schema of the pipeline is illustrated in Fig. 1. The invoice image is first processed by an OCR tool, then the language of document is detected based on text and other attributes.[2] In the next step, the basic document layout is analysed – from words received from OCR process, higher physical structures such as lines and blocks with their position properties are built. From now on, a series of annotations using different techniques is added in form of XML tagging, involving title, keywords, data type, addresses and name entities. Based on these annotations, the presented final module assigns the informational type to each text block using a rule-based model. In this stage, we focus on locating the most important groups of information in an invoice, including common information (invoice date, invoice number, order date, order number), seller/customer information (company name, address, contact, VAT number), payment term (payment method, dates, amount paid, balance due, and other terms), bank information, and delivery information.

2.1   Physical Structures and Their Position Properties

The document structure is separated into two categories: physical structure (or layout structure) and logical structure [9,12]. In the former, the document consists of pages, each page contains some blocks, each block has some lines, and lines are formed by words. In the latter, logical units are found. For example, in a scientific publication, they are the title, author, abstract, table, figure, form and so on. The layout structure is domain-independent while the other is domain-dependent. There are different methods to extract physical structure: bottom-up [4,5], top-down, and hybrid [11,15]. In this paper, we use the bottom-up approach.

The invoice image is first processed by the OCR engine[3] to get words and layout attributes such as bounding box, font name and font size. Words are grouped into lines based on three criteria: alignment, style, and distance. The distance between two words is the minimum distance between the east and the west edges of two bounding boxes for horizontal text. If two words have similar

---

[2] See [6] for details of the OCR setup.
[3] The open source Tesseract-OCR [14] is used now.

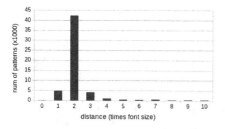

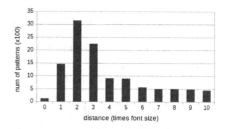

Fig. 2. Histogram of word distances in a line

Fig. 3. Histogram of line distances in a block

alignment and style, and the distance between them is less than a threshold (determined as a function of the font size), then they are in the same line. The threshold function was derived from the histogram of distances between each couple of adjacent words in 215 invoice images (see Fig. 2 for the chart) and currently corresponds to twice the font size of the first word in the line.

The process of combining lines into blocks is similar to the process of forming lines. However, while the distances between words in a line usually correspond to the space character, the distances between lines in a block vary a lot depending on the graphical format (see Fig. 3). We have chosen the block-line threshold as three times the font size of the previous line in the block.

After forming blocks, the position properties are added including the absolute position in the page and the relative position of the block to other blocks. The absolute position property separates a page into nine equal parts whereas the relative one looks for block's neighbors. If the block has the same alignment and there is no block between them, then the block id is added to top, bottom, left, or right properties respectively.

2.2   Annotation Modules

Further processing of the document is based on a modular series of task-oriented annotators. Each such module operates independently on the rest of the pipeline, however, some modules can employ information provided by previous annotators (e.g. the text structure or keyword annotation).

The first modules operate over the plain text of the recognized block lines to identify basic structures such as invoice specific keywords, dates, price, or VAT number. These modules can also cope with some character-level errors from the OCR process.

The subsequent modules provide higher-level information such as the presence of specific named entities (personal names, organizations, cities, etc.) and formatted address specifications.

Named Entity Recognition. The task of named entity recognition (NER) consists of two steps: named entity identification (including named entity boundaries detection in case of multi-word NER) and the entity classification (typically person name, place name, organization, sometimes a product name, artwork, date, time).

Currently, the best results in Czech NER are reported in [10,16]: the former report F-measure 74.08% on ConLL data, the latter report F-measure 82.82% on Czech NER Corpus CNEC.[4] The most efficient methods for NER are based on conditional random fields or maximum entropy.

We use Stanford NER with the standard MUC model for English, and the Czech model trained on CNEC for invoices detected as Czech ones. Since line breaks are very often important, we run NER on each line, not on larger chunks of text. The observations show that NER has (not surprisingly) plausible results for location detection (city and country names) and organization names. Nevertheless, it can also be helpful in case of street names since they are often detected as person names.

Recognition of named entities in invoices faces at least three problems that are not taken into consideration in the existing general models:

- text length
- use of uppercase text
- multilinguality

The existing models are suitable for larger text chunks, e.g. sentences. However, in invoice blocks, the text chunks are rather short. In addition, uppercase (which is an important feature in NER models for Czech and English) is used more frequently in this type of documents, for example for headings or company names. The last problem is multilinguality of invoices, e.g. in English invoices, the names of organization or street can be in different languages.

Location Names Recognition. Since location names are one the most important information we want to recognize, we have implemented two modules: one for detection of addresses, one for detection of locations in general. The former uses `libpostal` [2], a statistical model based on conditional random fields trained on Open Street Maps[5] and Open Addresses.[6] The latter uses Open Street Maps directly via the Nominatim API.[7] Even though both modules are based on the same data, they provide slightly different information.

Each of the two modules has positive matches on different data. For example, street names such as *Běly Pažoutové 680/4* are well recognized by the location names recognition module while the NER module recognizes *Běly Pažoutové* as person name. On the other hand, location names recognition matches *Konica Minolta Business* as an office building in the U.S. while the NER module annotates it as an organization name.

---

[4] http://hdl.handle.net/11858/00-097C-0000-0023-1B04-C.
[5] http://openstreetmap.org/.
[6] http://openaddresses.io/.
[7] http://nominatim.openstreetmap.org/.

3   Experiments and Evaluation

3.1   Dataset

The dataset is collected in cooperation with a renowned copy machine producer. It contains business documents written in several languages, mainly in Czech and English. We conduct the experiment on the English invoice set, which consists of 219 invoices from more than 50 suppliers all over the world. For developing and testing purposes, 60 invoices were randomly selected. Out of 60, 10 first invoices are used as the development set. They are from nine different suppliers in Austria, Poland, US, UK, The Netherlands, Germany, Italy. The other 50 invoices are used as evaluation set. Nearly a half of them are from suppliers seen in the development set and the rest are from 13 suppliers which have not appeared in the development set. The ground truth XML files have been manually annotated.

3.2   Block Type Detection

The main goal of the experiment is to recognise important text blocks within the invoice document and assign each block zero or more type labels.

The invoice text blocks are categorized into 9 main groups. First, the general information blocks contain the invoice date, invoice number, order date, and order number. They usually go with keywords. Next groups are the seller information and the buyer information, including company name, address, VAT number and other contacts such as person name, telephone number, email, fax, and website. Other groups are delivery information (delivery address, date, method, code, and cost), payment information (date, due date, method, terms), bank information (name, branch, address, account name, account number, swift code), invoice title, and page number. Blocks that do not belong to any of the previously mentioned categories are assigned an empty label.

The current block type detection technique is based on a set of logical rules that combine information obtained in preceding pipeline steps. The rules are in human readable and easy to edit form (see an example in Fig. 4). Each rule is applied to each block in the invoice document. If a block meets the rule's condition, then the label is added to the block type.

```
seller info = block_annot.data in3 [ORGANIZATION, CITY, COUNTRY,
                                     LOCATION, PHONE, PERSON]
              and abspos_y == bottom
```

Fig. 4. Block type rule example: if the intersection of all annotations in the block and the set {ORGANIZATION, CITY, COUNTRY, LOCATION, PHONE, PERSON} is more than three labels and the block is at the bottom of the page then the block type is seller info.

Table 1. Evaluation results with the testing set

	Blocks	In %
Match	1,189	80.01
Partial match	35	2.36
Mismatch	262	17.63
Total	1,486	100.00

3.3 Evaluation

The detection system was developed with the development set that consists of 10 invoice documents with 395 text blocks. The block type is correctly detected in 94.9% of blocks, 1.3% the (multiple) block type is partially correct and 3.8% of blocks are misclassified.

The evaluation set, which was not consulted during the development of the detection rules, consists of 50 invoice documents that have been manually annotated. The detection results of the evaluation set are presented in Table 1 reaching an average of **80.1%** correct block type detection. Analysis of the detection accuracy for each block type is presented in Table 2, where a match means correct identification of the block type (both positive and negative), and the precision and recall express the corresponding ratios of true positive results.

3.4 Error Analysis

A detailed analysis of the errors distinguishes 5 categories of errors. First, Tesseract OCR errors (6.1% of 262 mismatches in the test set) lead to missing important keywords. For examples: "sell date", "delivery address", "IBAN" are recognized as "se'' date", "deliveg address", "IBAN" respectively.

Table 2. Evaluation of individual block type categories

Block type	Match		Mismatch		Precision	Recall
	Blocks	In %	Blocks	In %	In %	In %
bank info	631	97.5	16	2.5	57.7	75.0
buyer info	715	89.4	85	10.6	70.7	69.2
company info	618	98.6	9	1.4	100.0	18.2
delivery info	624	97.3	17	2.7	100.0	32.0
general info	749	89.8	85	10.2	77.8	73.1
page no	636	100.0	0	0.0	100.0	100.0
payment term	819	94.3	50	5.7	93.6	84.9
seller info	658	85.7	110	14.3	63.64	32.8
title	748	98.9	7	1.1	88.9	91.4

Second, lines and blocks are formed from words and lines based on alignment, style and distance. However, because of wide variety in invoice formats, the threshold does not always cover all information into a block as it should be. The situation, where one block is split into multiple separate blocks causes a mismatch in 44 cases (16.8% of errors). It happens only in 8 invoices (7 out of 8 comes from the same vendor which does not appear in development set) but causes a big damage.

Third, as we mention in 3.2, in many invoices there is no keyword to determine if a block of company information is seller or buyer information. The seller information lies usually at the header or footer of the page, or the seller information usually appears before the buyer information, but there is a number of exceptions. Interchanging the buyer, seller and delivery information thus causes a mismatch in 34 cases (13% of errors).

Last but not least, keywords are not annotated (for a same item, there are various ways of using keywords) leads to 51 mismatches (19.5% of errors) leaving 117 cases (44.7%) for other mixed error causes.

4   Conclusions

Efficient information extraction of semi-structured OCR business documents relies on adaptable multilingual techniques allowing to transfer the task of document cataloging to an automatic document management system. In this paper, we have presented the current results of the OCRMiner system, which allows to combine text analysing techniques with positional layout features of the recognized document blocks. The results of the system evaluation confirm the flexibility of the combined approach reaching the overall accuracy of 80.1% (with open source OCR) which surpasses published state-of-the-art systems that use commercial OCR input.

In the future research, the system detection will concentrate on adaptation to various kinds of OCR errors (including layout), global rules for address assignment and on extending the range of language families covered to non-latin alphabet languages.

Acknowledgments. This work has been partly supported by Konica Minolta Business Solution Czech within the OCR Miner project and by the Masaryk University project MUNI/33/55939/2017.

References

1. Aslan, E., Karakaya, T., Unver, E., Akgül, Y.S.: A part based modeling approach for invoice parsing. In: Proceedings of the 11th Joint Conference on Computer Vision, Imaging and Computer Graphics Theory and Applications, VISIGRAPP 2016, pp. 392–399 (2016)
2. Barrentine, A.: Statistical NLP on OpenStreetMap: Part 2, Training Conditional Random Fields on 1 billion street addresses (2017). https://medium.com/@albarrentine/statistical-nlp-on-openstreetmap-part-2-80405b988718

3. Bart, E., Sarkar, P.: Information extraction by finding repeated structure. In: Proceedings of the 9th International Workshop on Document Analysis Systems, pp. 175–182. ACM (2010)
4. Bayer, T., Mogg-Schneider, H.: A generic system for processing invoices. In: Proceedings of the Fourth International Conference on Document Analysis and Recognition, vol. 2, pp. 740–744. IEEE (1997)
5. Chao, H., Fan, J.: Layout and content extraction for PDF documents. In: Marinai, S., Dengel, A.R. (eds.) DAS 2004. LNCS, vol. 3163, pp. 213–224. Springer, Heidelberg (2004). https://doi.org/10.1007/978-3-540-28640-0_20
6. Ha, H.T.: Recognition of invoices from scanned documents. In: Recent Advances in Slavonic Natural Language Processing, RASLAN 2017, pp. 71–78 (2017)
7. Hamza, H., Belaid, Y., Belaïd, A.: A case-based reasoning approach for invoice structure extraction. In: Ninth International Conference on Document Analysis and Recognition, vol. 1, pp. 327–331. IEEE (2007)
8. The Institute of Finance and Management (IOFM): Special Report: The True Costs of Paper-Based Invoice Processing and Disbursements. Diversified Communications (2016). https://www.concur.com/en-us/resources/true-costs-paper-based-invoice-processing-and-disbursements
9. Klink, S., Dengel, A., Kieninger, T.: Document structure analysis based on layout and textual features. In: Proceedings of International Workshop on Document Analysis Systems, DAS 2000, pp. 99–111. Citeseer (2000)
10. Konkol, M., Konopík, M.: CRF-based Czech named entity recognizer and consolidation of Czech NER research. In: Habernal, I., Matoušek, V. (eds.) TSD 2013. LNCS (LNAI), vol. 8082, pp. 153–160. Springer, Heidelberg (2013). https://doi.org/10.1007/978-3-642-40585-3_20
11. Liang, J., Ha, J., Haralick, R.M., Phillips, I.T.: Document layout structure extraction using bounding boxes of different entitles. In: Proceedings 3rd IEEE Workshop on Applications of Computer Vision, WACV 1996, pp. 278–283. IEEE (1996)
12. Mao, S., Rosenfeld, A., Kanungo, T.: Document structure analysis algorithms: a literature survey. In: Document Recognition and Retrieval X, vol. 5010, pp. 197–208. International Society for Optics and Photonics (2003)
13. Schulz, F., Ebbecke, M., Gillmann, M., Adrian, B., Agne, S., Dengel, A.: Seizing the treasure: transferring knowledge in invoice analysis. In: 10th International Conference on Document Analysis and Recognition, pp. 848–852. IEEE (2009)
14. Smith, R.: An overview of the Tesseract OCR engine. In: Ninth International Conference on Document Analysis and Recognition, vol. 2, pp. 629–633. IEEE (2007)
15. Smith, R.W.: Hybrid page layout analysis via tab-stop detection. In: 10th International Conference on Document Analysis and Recognition, pp. 241–245. IEEE (2009)
16. Straková, J., Straka, M., Hajič, J.: A new state-of-the-art Czech named entity recognizer. In: 16th International Conference on Text, Speech, and Dialogue, TSD 2013, pp. 68–75 (2013). https://doi.org/10.1007/978-3-642-40585-3_10

Speech

Automatic Evaluation of Synthetic Speech Quality by a System Based on Statistical Analysis

Jiří Přibil[1,2(✉)], Anna Přibilová[3], and Jindřich Matoušek[2]

[1] Institute of Measurement Science, SAS, Bratislava, Slovakia
Jiri.Pribil@savba.sk
[2] Faculty of Applied Sciences, Department of Cybernetics,
UWB, Pilsen, Czech Republic
jmatouse@kky.zcu.cz
[3] FEE & IT, Institute of Electronics and Photonics, SUT in Bratislava,
Bratislava, Slovakia
Anna.Pribilova@stuba.sk

Abstract. The paper describes a system for automatic evaluation of speech quality based on statistical analysis of differences in spectral properties, prosodic parameters, and time structuring within the speech signal. The proposed system was successfully tested in evaluation of sentences originating from male and female voices and produced by a speech synthesizer using the unit selection method with two different approaches to prosody manipulation. The experiments show necessity of all three types of speech features for obtaining correct, sharp, and stable results. A detailed analysis shows great influence of the number of statistical parameters on correctness and precision of the evaluated results. Larger size of the processed speech material has a positive impact on stability of the evaluation process. Final comparison documents basic correlation with the results obtained by the standard listening test.

Keywords: Listening test · Objective and subjective evaluation
Quality of synthetic speech · Statistical analysis

1   Introduction

At present, many objective and subjective criteria are used to evaluate quality of synthetic speech that can be produced by different synthesis methods implemented mainly in text-to-speech (TTS) systems. Practical representation of a subjective evaluation consists of a listener's choice from several alternatives (e.g.

The work was supported by the Czech Science Foundation GA16-04420S (J. Matoušek, J. Přibil), by the Grant Agency of the Slovak Academy of Sciences 2/0001/17 (J. Přibil), and by the Ministry of Education, Science, Research, and Sports of the Slovak Republic VEGA 1/0905/17 (A. Přibilová).

© Springer Nature Switzerland AG 2018
P. Sojka et al. (Eds.): TSD 2018, LNAI 11107, pp. 315–323, 2018.
https://doi.org/10.1007/978-3-030-00794-2_34

mean opinion score, recognition of emotion in speech, or age and gender recognition) or from two alternatives, speech corpus annotation, etc. [1]. Spectral as well as segmental features are mostly used in objective methods for evaluation of speech quality. Standard features for speaker identification or verification, as well as speaker age estimation, are mel frequency cepstral coefficients [2]. These segmental features usually form vectors fed to Gaussian mixture models [3,4] or support vector machines [5] or they can be evaluated by other statistical methods, e.g. analysis of variance (ANOVA) or hypothesis tests, etc. [6,7]. Deep neural networks can also be used for speech feature learning and classification [8]. However, they are not sufficient to render the way of phrase creation, prosody production by time-domain changes, speed of the utterance, etc. Consequently, supra-segmental features derived from time durations of voiced and unvoiced parts [9] must be included in the complex automatic system for evaluation of synthetic speech quality by comparison of two or more utterances synthesized by different TTS systems. Another application may be evaluation of degree of resemblance between the synthetic speech and the speech material of the corresponding original speaker whose voice the synthesis is based on.

The motivation of this work was to design, realize, and test the designed system for automatic evaluation of speech quality which could become a fully-fledged alternative to the standard subjective listening test. The function of the proposed system for automatic judgement of the synthetic speech signal quality in terms of its similarity with the original is described together with the experiments verifying its functionality and stability of the results. Finally, these results are compared with those of the listening tests performed in parallel.

2   Description of Proposed Automatic Evaluation System

The whole automatic evaluation process consists of two phases: at first, databases of spectral properties, prosodic parameters, and time duration relations (speech features – SPF) are built from the analysed male and female natural utterances and the synthetic ones generated by different methods of TTS synthesis, different synthesis parameters, etc. Then, separate calculations of the statistical parameters (STP) are made for each of the speakers and each of the types of speech features. The determined statistical parameters together with the SPF values are stored for next use in different databases depending on the used input signal ($DB_{ORIG}, DB_{SYNT1}, DB_{SYNT2}$) and the speaker (male/female). The second phase is represented by practical evaluation of the processed data: at first, the SPF values are analysed by the ANOVA statistics and the hypothesis probability assessment resulting from the Ansari-Bradley test (ASB) or the Wilcoxon test [10,11], and for each of their STPs the histogram of value occurrence is calculated. Subsequently, the root-mean-square (RMS) distances (D_{RMS}) between the histograms stemming from the natural speech signals and the synthesized ones are determined and used for further comparison by numerical matching. Applying the majority function on the partial results for each of SPF types and STP values, the final decision is got as shown in the block diagram in Fig. 1. It is given by the proximity of the tested synthetic speech produced by the TTS system to the sentence uttered by the original speaker (values "1" or "2" for two evaluated types of the speech synthesis). If differences between majority percentage results derived from the STPs are not statistically significant for any type of the tested synthesis, the final decision is set to a value of "0". This objective evaluation result corresponds to the subjective listening test choice "*A sounds similar to B*" [1] with small or indiscernible differences.

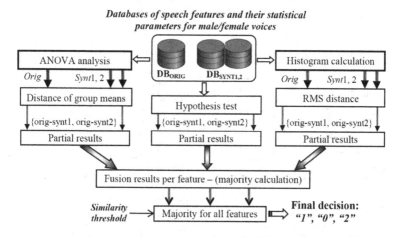

Fig. 1. Block diagram of the automatic evaluation system of the synthetic speech.

For building of SPF and STP databases, the speech signal is processed in weighted frames with the duration related to the speaker's mean fundamental frequency F0. Apart from the supra-segmental F0 and signal energy contours, the segmental parameters are determined in each frame of the input sentence. The smoothed spectral envelope and the power spectral density are computed for determination of the spectral features. The signal energy is calculated from the first cepstral coefficient c_0 (En_{c0}). Further, only voiced or unvoiced frames with the energy higher than the threshold En_{MIN} are processed to eliminate speech pauses in the starting and ending parts. It is very important for determination of the time duration features (TDUR). In general, three types of speech features are determined:

1. time durations of voiced/unvoiced parts in samples Lv, Lu for a speech signal with non-zero F0 and $En_{c0} \geq En_{MIN}$, their ratios Lv/u_L, Lv/u_R, Lv/u_{LR} calculated in the left context, right context, and both left and right contexts as $Lv_1/(Lu_1 + Lu_2), \ldots Lv_N/(Lu_{M-1} + Lu_M)$.
2. Prosodic (supra-segmental) parameters – F0, En_{c0}, differential F0 microintonation ($F0_{DIFF}$), jitter, shimmer, zero-crossing period, and zero-crossing frequency.

3. Basic and supplementary spectral features – first two formants (F_1, F_2), their ratio (F_1/F_2), spectral decrease (tilt), spectral centroid, spectral spread, spectral flatness, harmonics-to-noise ratio (HNR), spectral Shannon entropy (SHE).

Statistical analysis of these speech features yields various STPs: basic low-level statistics (mean, median, relative max/min, range, dispersion, standard deviation, etc.) and/or high-level statistics (flatness, skewness, kurtosis, covariance, etc.) for the subsequent evaluation process. The block diagram of creation of the speech feature databases can be seen in Fig. 2.

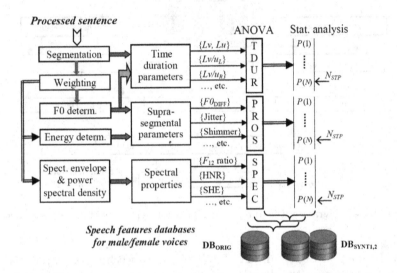

Fig. 2. Block diagram of speech feature databases creation from time durations, prosodic parameters, spectral properties, and their statistical parameters.

3   Material, Experiments and Results

The synthetic speech produced by the Czech TTS system based on the unit selection (USEL) synthesis method [12] and the sentences uttered by four professional speakers – 2 males (M1 and M2) and 2 females (F1 and F2) were used in this evaluation experiment. The main speech corpus was divided into three subsets: the first one consists of the original speech uttered by real speakers (further called as *Orig*), the second and third ones comprise synthesized speech signals produced by the TTS system with voices based on the corresponding original speaker using two different synthesis methods: with a rule-based prosody manipulation (TTSbase – *Synt1*) [13] and a modified version of the USEL method that reflects the final syllable status (TTSsyl – *Synt2*) [14]. The collected database consists of 50 sentences from each of four original speakers (200 in total), next sentences of two synthesis types giving 50 + 50 sentences from the male voice

M1 and 40 + 40 ones from the remaining speakers M2, F1, and F2. Speech signals of declarative and question sentences were sampled at 16 kHz and their duration was from 2.5 to 5 s. The main orientation of the performed experiments was to test functionality of the developed automatic evaluation system in every functional block of Fig. 1 – calculated histograms and statistical parameters are shown in demonstration examples in Figs. 3, 4 and 5. Three auxiliary comparison experiments were realized, too, with the aims to analyse:

1. effect of the number of used statistical parameters $N_{STP} = \{3, 5, 7, 10\}$ on the obtained evaluation results – see numerical comparison of values in Table 1 for the speakers M1 and F1,
2. influence of the used type of speech features (spectral, prosodic, time duration) on the accuracy and stability of the final evaluation results – see numerical results for speakers M1 and F1 in Table 2,
3. impact of the number of analysed speech signal frames on the accuracy and stability of the evaluation process – compare values for limited (15 + 15 + 15 sentences for every speaker), basic (25 + 25 + 25 sentences), and extended (50 + 40 + 40) testing sets in Table 3 for the speakers M1 and F1.

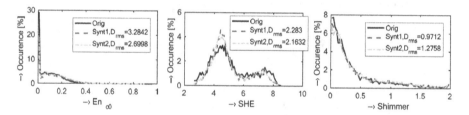

Fig. 3. Histograms of spectral and prosodic features En_{c0}, SHE, Shimmer together with calculated RMS distances between the original and the respective synthesis for the male speaker M1, using the basic testing set of 25 + 25 + 25 sentences.

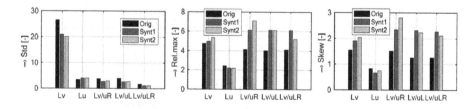

Fig. 4. Comparison of selected statistical parameters std, relative maximum, skewness calculated from values of five basic TDUR features, for the female speaker F1 and the basic testing set.

Finally, numerical comparison with the results obtained by the listening test was performed using the extended testing set. The maximum score using the determined STPs and the mixed feature types (spectral + prosodic + time duration) is evaluated for each of four speakers – see the values in Table 4.

Subjective quality of the same utterance generated by two different approaches to prosody manipulation in the same TTS synthesis system (TTS-base and TTSsyl) was evaluated by a preference listening test. Four different male and female voices were used, each to synthesize 25 pairs of randomly selected utterances, so that the whole testing set was made up of 100 sentences. The order of two synthesized versions of the same utterance was randomized too, to avoid bias in evaluation by recognition of the synthesis method. Twenty two evaluators (8 women and 14 men) within the age range from 20 to 55 years of age participated in the listening test experiment open from 7th to 20th March 2017. The listeners were allowed to play the audio stimuli as many times as they

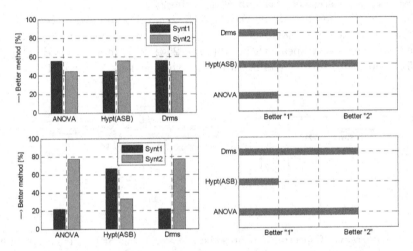

Fig. 5. Visualization of partial percentage results per three evaluation methods together with final decisions for speakers M1 (upper set of graphs) and F1 (bottom set), using only basic spectral properties from the basic set of sentences, $N_{STP} = 3$.

Table 1. Influence of the number of used statistical parameters on partial evaluation results for speakers M1 and F1, when spectral properties and prosodic parameters are used.

$N_{STP}[-]^{(A)}$	Male speaker M1		Female speaker F1	
	Partial	Final[B]	Partial	Final[B]
3	1 (65%), 2 (35%)	"**1**"	1 (60%), 2 (40%)	"**2**"
5	1 (67%), 2 (33%)	"**1**"	1 (48%), 2 (52%)	"**0**"
7	1 (71%), 2 (29%)	"**1**"	1 (44%), 2 (56%)	"**1**"
10	1 (73%), 2 (27%)	"**1**"	1 (37%), 2 (63%)	"**1**"

[A] used basic testing set (of 25+25+25 processed sentences),
[B] used "1"= TTSbase better, "0"= Similar, "2"= TTSsyl better.

Table 2. Influence of the used type of speech features (spectral, prosodic, time duration) on the accuracy and stability of the evaluation results for speakers M1 and F1.

Speech feature types[A]	Male speaker M1		Female speaker F1	
	Partial	Final[B]	Partial	Final[B]
Spectral only	1 (63%), 2 (37%)	"1"	1 (54%), 2 (46%)	'1"
Spectral+prosodic	1 (58%), 2 (42%)	"1"	1 (52%), 2 (48%)	"0"
Spectral+prosodic+ time duration	1 (46%), 2 (54%)	"2"	1 (44%), 1 (56%)	"2"

[A] used basic testing set (of 25+25+25 processed sentences), the maximum of determined STPs is applied.
[B] used "1"= TTSbase better, "0"= Similar, "2"= TTSsyl better.

Table 3. Partial evaluation results for different lengths of used speech databases for speakers M1 and F1 using only time duration features.

Speech corpus (No of sentences)[A]	Male speaker M1		Female speaker Fl	
	Partial	Final[B]	Partial	Final[B]
Limited (15+15+15)	1 (36%), 2 (64%)	"2"	1 (49%), 2 (51%)	"0"
Basic (25+25+25)	1 (29%), 2(71%)	"2"	1 (44%), 2 (56%)	"2"
Extended (50+40+40)	1 (22%), 2 (78%)	"2"	1 (37%), 1 (63%)	"2"

[A] per type of Orig+Syntl+ Synt2, the maximum of determined STPs is applied.
[B] used "1"= TTSbase better, "0"= Similar, "2"= TTSsyl better.

Table 4. Final comparison of objective and subjective evaluations for all four speakers.

Speaker	Automatic evaluation[A]		Listening test[B]		
	Partial	Final	"1"	"0"	"2"
Ml (AJ)	1 (40.7%), 2 (59.3%)	"2"	21.3%	20.0%	58.7%
M2 (JS)	1 (44.9%), 2 (55.1%)	"2"	16.5%	27.1%	56.4%
Fl (KI)	1 (44.4%), 2 (55.6%)	"2"	13.1%	21.8%	53.6%
F2 (SK)	1 (46.1%), 2 (54.9%)	"2"	17.1%	29.3%	58.5%

[A] used extended set of processed sentences, the maximum of determined STPs and all three types of speech features are applied.
[B] used evaluation as "1"= TTSbase better, "0"= Similar, "2"= TTSsyl better.

wished; low acoustic noise conditions and headphones were advised. Playing of the stimuli was followed by the choice between "*A sounds better*", "*A sounds similar to B*", or "*B sounds better*" [14]. The results obtained in this way were further compared with the objective results of the currently proposed system of automatic evaluation.

4   Discussion and Conclusion

The performed experiments have confirmed that the proposed evaluation system is functional and produces results comparable with the standard listening test method as documented by numerical values in Table 4. Basic analysis of the obtained results shows principal importance of application of all three types of speech features (spectral, supra-segmental, time-duration) for complex evaluation of synthetic speech. This is relevant especially when the compared synthesized speech signals differ only in their prosodic manipulation, as in the case of this speech corpus. Using only the spectral features brings non-stable or contradictory results, as shown in "Final" columns of Table 2. The detailed analysis showed principal dependence of the correctness of evaluation on the number of used statistical parameters – compare particularly the values for the female voice in Table 1. For $N_{STP} = 3$ the second synthesis type was evaluated as better and increase of the number of parameters to 5 resulted in considering both methods as similar. Further increase of the number of parameters to 7 and 10 gave stable results with preference of the first synthesis type. Additional analysis has shown that a minimum number of speech frames must be processed to achieve correct statistical evaluation and significant statistical differences between the original and tested STPs derived from the same speaker. If these were not fulfilled, the final decision of the whole evaluation system would not be stable and no useful information would be got by "0"category of the automatic evaluation system equivalent to "*A sounds similar to B*" in the subjective listening test. Tables 1, 2 and 3 show this effect for the female speaker F1. In general, the tested evaluation system detects and classifies male speakers better than female ones. It may be caused by higher variability of female voices and its effect to the supra-segmental area (changes of energy and F0), the spectral domain, and the changes in time duration relations.

In the near future, we will try to collect larger speech databases, including greater number of speakers. Next, in the databases, there will be incorporated more different methods of speech synthesis (HMM, PSOLA, etc.) produced by more TTS systems in other languages – English, German, etc. In this way, we will carry out complex testing of automatic evaluation with the final aim to substitute subjective evaluation based on the listening test method.

References

1. Grűber, M., Matoušek, J.: Listening-test-based annotation of communicative functions for expressive speech synthesis. In: Sojka, P., Horák, A., Kopeček, I., Pala, K. (eds.) TSD 2010. LNCS (LNAI), vol. 6231, pp. 283–290. Springer, Heidelberg (2010). https://doi.org/10.1007/978-3-642-15760-8_36
2. Monte-Moreno, E., Chetouani, M., Faundez-Zanuy, M., Sole-Casals, J.: Maximum likelihood linear programming data fusion for speaker recognition. Speech Commun. **51**(9), 820–830 (2009)
3. Reynolds, D.A., Rose, R.C.: Robust text-independent speaker identification using Gaussian mixture speaker models. IEEE Trans. Speech Audio Process. **3**, 72–83 (1995)

4. Xu, L., Yang, Z.: Speaker identification based on state space model. Int. J. Speech Technol. **19**(2), 407–414 (2016)
5. Campbell, W.M., Campbell, J.P., Reynolds, D.A., Singer, E., Torres-Carrasquillo, P.A.: Support vector machines for speaker and language recognition. Comput. Speech Lang. **20**(2–3), 210–229 (2006)
6. Lee, C.Y., Lee, Z.J.: A novel algorithm applied to classify unbalanced data. Appl. Soft Comput. **12**, 2481–2485 (2012)
7. Mizushima, T.: Multisample tests for scale based on kernel density estimation. Stat. Probab. Lett. **49**, 81–91 (2000)
8. Hussain, T., Siniscalchi, S.M., Lee, C.C., Wang, S.S., Tsao, Y., Liao, W.H.: Experimental study on extreme learning machine applications for speech enhancement. IEEE Accesss **5**, 25542 (2017)
9. van Santen, J.P.H.: Segmental duration and speech timing. In: Sagisaka, Y., Campbell, N., Higuchi, N. (eds.) Computing Prosody. Springer, New York (1997). https://doi.org/10.1007/978-1-4612-2258-3_15
10. Martinez, C.C., Cassol, M.: Measurement of voice quality, anxiety and depression symptoms after therapy. J. Voice **29**(4), 446–449 (2015)
11. Rietveld, T., van Hout, R.: The t test and beyond: recommendations for testing the central tendencies of two independent samples in research on speech, language and hiering pathology. J. Commun. Disord. **58**, 158–168 (2015)
12. Hunt, A.J., Black, A.W.: Unit selection in a concatenative speech synthesis system using a large speech database. In: Proceedings of the IEEE International Conference on Acoustics, Speech and Signal Processing (ICASSP). Atlanta (Georgia, USA), pp. 373–376 (1996)
13. Tihelka, D., Kala, J., Matoušek, J.: Enhancements of Viterbi search for fast unit selection synthesis. In: Proceedings of INTERSPEECH 2010, Makuhari, Japan, pp. 174–177 (2010)
14. Jůzová, M., Tihelka, D., Skarnitzl, R.: Last syllable unit penalization in unit selection TTS. In: Ekštein, K., Matoušek, V. (eds.) TSD 2017. LNCS (LNAI), vol. 10415, pp. 317–325. Springer, Cham (2017). https://doi.org/10.1007/978-3-319-64206-2_36

Robust Recognition of Conversational Telephone Speech via Multi-condition Training and Data Augmentation

Jiří Málek[✉], Jindřich Ždánský, and Petr Červa

Institute of Information Technologies and Electronics,
Technical University of Liberec, Studentská 2, Liberec 460 10, Czech Republic
{jiri.malek,jindrich.zdansky,petr.cerva}@tul.cz

Abstract. In this paper, we focus on automatic recognition of telephone conversational speech in scenario, when no amount of genuine telephone recordings is available for training. The training set contains only data from a significantly different domain, such as recording of broadcast news. Significant mismatch arises between training and test conditions, which leads to deteriorated performance of the resulting recognition system. We aim to diminish this mismatch using the data augmentation.

Speech compression and narrow-band spectrum are significant features of the telephone speech. We apply these effects to the training dataset artificially, in order to make it more similar to the desired test conditions. Using such augmented dataset, we subsequently train an acoustic model. Our experiments show that the augmented models achieve accuracy close to the results of a model trained on genuine telephone data. Moreover, when the augmentation is applied to the real-world telephone data, further accuracy gains are achieved.

Keywords: Compression · Data augmentation
Multi-conditional training · Conversational speech

1 Introduction

Nowadays, the research in Automatic Speech Recognition (ASR) is focused on robustness against detrimental distortions applied to the speech signal by the environment and the recording devices [24]. ASR is thus able to operate in the real-world scenarios, such as automatic subtitle production for audio-visual broadcast or transcription of telephone conversations (e.g. in telemarketing context). The later application is focus of this paper.

The telephone speech features many specific qualities. From the perspective of speaking style, it is highly spontaneous. The environment surrounding the speakers distorts the signal by effects such as background noise, concurrent speech or reverberation. During recording, some kind of compression may be applied to the recordings, for the purposes of storing or transmission.

© Springer Nature Switzerland AG 2018
P. Sojka et al. (Eds.): TSD 2018, LNAI 11107, pp. 324–333, 2018.
https://doi.org/10.1007/978-3-030-00794-2_35

To train a robust system for signals affected by these effects, a huge amount of speech from diverse conditions is required. For example, the state-of-the-art systems compared in [25] use 2000 h of speech for training. When significantly smaller datasets (tens of hours of recordings) are used, it is beneficial to utilize data, which originate from similar environments (acoustic conditions, means of recording/transmission etc.) as the potential test speech. However, it can be challenging to collect such dataset, e.g., for less resourced languages.

To mitigate the lack of suitable training data, several techniques were presented in the literature. Semi-supervised training [12,17] deals with small amount of precisely annotated acoustic data. Here, a large amount of imprecise automatic transcripts are directly used to train the acoustic model. In contrast, data selection [7,8] aims at selection of limited amount of manually annotated data, which is highly relevant for the training.

Another technique is the *augmentation*, which generates new speech signals/features artificially using existing ones. This is done to extend the speech variability and/or to introduce some specific environmental effect to the training dataset. The augmentation techniques often aim to preserve the labels assigned to the data (label-preserving transformations). The acoustic models are subsequently trained in a multi-condition manner [16] using both genuine and augmented data.

The augmentation through the vocal tract length perturbation (VTLP) [10] generates new training samples by scaling of the spectrum of the original samples along the frequency axis. The modification of speech time-rate (which is not label-preserving) was discussed in [15]. Both these types were simultaneously studied in [11] and referred to as elastic spectral distortion. Stochastic feature mapping [4] is a technique inspired by voice conversion. It seeks to transform parametrized utterance of a source speaker into parametrized utterance of another speaker using speaker dependent transformation. Addition of speech corrupted by various environmental noises to the training set was reported in [21] and generation of reverberated speech was discussed in [13].

In this paper, we focus on compression. It is an integral part of telephone speech recordings, due to the presence of broad range of transmission channels, such as cell phones, voip, landlines, computer-based clients etc. The compression was shown to be detrimental to a lot of speech-related tasks, such as ASR [2,3,20], speaker recognition [19] or emotion recognition [22]. The genuine diverse telephone dataset is difficult to collect, because many combinations of recording devices and transmitting channels should be considered. However, these conditions can be to some extend approximated by artificial distortion of training data by a broad set of encoding schemes.

Our paper aims at scenario, when an ASR system usable for telephone speech is trained using a database from a significantly different domain (recordings of broadcast news, in our case). Performance of such system deteriorates compared to a system trained on telephone speech, due to highly mismatched training-test conditions. Although the compression is not the sole reason for this accuracy loss, it significantly contributes to is. To mitigate, we augment the training data through various compression schemes, in order to make it more similar to the desired test conditions. Our goal is to approach performance of a model trained on genuine telephone recordings. We investigate the performance achieved on real-world telephone recordings, where we have no control, which specific codecs were applied.

Compared to the aforementioned recent papers investigating the compression in the ASR context, the present work differs in the following points. The papers [2,3] focus specifically on mp3 compression, which is not commonly applied to the telephone speech. The papers also do not aim at augmentation of the data. The paper [3] deals with acoustic model adaptation tailored specifically to the mp3 format and [2] analyzes application of dithering (addition of low-energy white noise) to speech compressed by mp3. The paper [20] investigates specifically the effect of compression by itself, using artificial/generated data only. The effects of significant domain difference between training and test datasets (e.g. speaking style, channel effects) are thus not present in [20]; the investigated systems are both trained and tested on TIMIT database [9]. In contrast, our current paper aims specifically on telephone speech and significant training-test set mismatch, which we aim to diminish by augmentation. We compare the augmented models to their counterparts trained on genuine real-world signals. Moreover, we investigate the benefits of augmentation for dataset of genuine telephone data.

The paper is organized as follows. Section 2 describes available speech databases, investigated codecs and our implementation of the augmentation process. Section 3 describes the utilized recognition system; its acoustic and linguistic part. Section 4 presents the achieved experimental results. Section 5 discusses the achieved results and concludes the paper.

2   Augmentation Process

2.1   Available Datasets

Due to the availability of training/test datasets, the recognition of Czech speech is discussed in this paper (without any loss of generality to the discussed topic). We utilize two training datasets. The so called *Broadcast dataset* (abbreviated in the experiments as "Broad") consists of 132 h of broadcast news and dictated speech (sampled at 16 KHz sampling rate). To this dataset we apply the augmentation and use it to train acoustic models in the multi-condition manner. The *Telephone dataset* (abbreviated as "Phone") consists of 116 h of telephone conversations (sampled at 8 KHz). The performance of a system trained on this dataset we consider as target in this paper and aim to approach it with systems trained on augmented Broadcast dataset. We also present the performance of a system trained on union of the Broadcast dataset (decimated to 8 KHz) and the Telephone dataset, which we denote as *Combined dataset* (abbreviated as "Combi").

We test the augmented acoustic models on real-world telephone conversations. The test dataset is comprised of about 3 h (12929 words) speech, which originates mostly from dialogs of customers with various call centers.

2.2 Considered Codecs and the Augmentation Details

In our study, we aim to transcribe narrow-band speech with 8 KHz sampling rate. We consider the following set of codecs, which are popular in the context of digital or cellular telephony and Voice-over-IP (VoIP) transmissions. The details about the encoding schemes can be found in Table 1. The codecs G.711A and G.726 use waveform encoding, whereas the six remaining are hybrid. With the exception of traditional G.711A, we consider very low bitrates, lower or equal to 24 kbps. As indicated for example in [20], such low bitrates severely deteriorate the ASR performance. The compression is performed using ffmpeg software [6]. The Speex encoding provides variable bitrate based on desired quality of the result. We apply two different options: 3 and 6.

The augmentation is performed by separate application of the considered codecs to the training speech dataset. This results into several instances of the training dataset, each corresponding to one of the codecs. Next, we consider also the augmentation by multiple encoders. When N codecs are utilized, the training dataset is split into $N + 1$ parts. One part is downsampled to 8 KHz without any compression, the other parts are compressed by the respective codecs.

Table 1. Overview of codecs used in our study. Abbreviation "vbr" denotes variable bitrate

Codec	Type	Context of utilization	Bitrate [kbps]	Notes
G.711A	Waveform	Digital telephony	64	
GSM	Hybrid	Cellular telephony	13	
AMR-NB	Hybrid	Cellular telephony	7.95	
G.723.1	Hybrid	VoIP	6.3	
G.726	Waveform	VoIP	24	
Speex	Hybrid	VoIP	vbr ($\approx$8/11)	Quality levels: 3/6
ILBC	Hybrid	VoIP	15.2	
Opus	Hybrid	VoIP	12	

3 Recognition System

We use our own ASR system; its core is formed by a one-pass speech decoder performing a time-synchronous Viterbi search. The system consists of the acoustic and language models. The acoustic models vary with respect to the augmented training datasets, which we investigate; the linguistic part remains the same for all investigated system variants.

3.1   Multi-condition Training of the Acoustic Models

The models are trained on augmented datasets described in the previous section. All models are based on Hidden Markov Model-Deep Neural Network (HMM-DNN) hybrid architecture [5]. Two underlying Gaussian Mixture Models (GMM) were trained, one for training sets derived from Broadcast dataset (3737 physical states) and one for training sets derived from Phone dataset (2638 physical states). Both models are context dependent, speaker independent.

The DNNs have feed-forward structure with five fully-connected hidden layers. Each hidden layer consists of 768 units. We employ the ReLU activation function as nonlinearity. The configuration of hyper-parameters for all acoustic models corresponds to the best performance in preliminary experiments with uncompressed data.

For feature extraction, 39 filter bank coefficients [26] are computed using 25-ms frames of signal and frame shift of 10 ms. The input of the DNNs consists of 11 consecutive feature vectors, 5 preceding and 5 following the current frame. Concerning the feature normalization, we employ the Mean Subtraction [18] with a floating window of 1 s.

The DNN parameters are trained by minimization of the negative log-likelihood criterion via the stochastic gradient descent method. The training procedure ends when the criterion does not improve anymore on a small validation dataset, which is not part of the training set or after 50 epochs over the data. The training is implemented in the Torch library [23].

3.2   Linguistic Part of the System

The linguistic part of the system consists of a lexicon and a language model. The lexicon contains 550 k entries (word forms and multi-word collocations) that were observed most frequently in a 10 GB large corpus covering newspaper texts and broadcast program transcripts. Some of the lexical entries have multiple pronunciation variants. Their total number is 580 k.

The employed Language Model (LM) is based on bigrams, due to very large vocabulary size. Our supplementary experiments showed that the bigram structure of the language model results in the best ASR performance with reasonable computational demands. In the training word corpus, 159 million unique word-pairs (1062 million in total) belonging to the items in the 550 k lexicon were observed. However, 20% of all word-pairs actually include sequences containing three or more words, as the lexicon contains 4 k multi-word collocations. The unseen bigrams are backed-off by the Kneser-Ney smoothing technique [14].

4   Experiments

We report the results of our experiments via recognition accuracy [%]; all improvements are stated as absolute. Throughout the experiments, we denote the considered acoustic models by convention "Dataset Abbreviation": "Augmentation codec(s)". The "Dataset Abbreviation" refers to dataset, which was subject to the augmentation and the "Augmentation codec(s)" describes which codec(s) were applied to this dataset prior acoustic model training.

All training data have 8 KHz sampling frequency, thus the extracted features use band 0–4 KHz. The sole exception is the system "Broad: None (16 KHz)", which is trained using the genuine wide-band broadcast data in the Broadcast Dataset, i.e., wide-band features exploiting band 0–8 KHz. Its performance serves as our baseline; the test data were upsampled in order to be transcribed via this system.

4.1 Models Using Augmentation by a Single Codec

The results in Fig. 1 indicate that the augmentation partly compensates the deterioration of the performance caused by mismatched training-test conditions. The baseline model "Broad: None (16 KHz)" achieves the lowest accuracy 58.1%. This is caused partly by the mentioned condition mismatch and partly by the fact that the wide-band features exploit for classification also the information from band 4–8 KHz, which is not present in narrow-band test data.

The accuracy is improved by 5.6% to 63.7%, when downsampling is applied to the training data and the narrow-band features exploiting band 0–4 KHz only are extracted (see "Broad: Decimation"). Nevertheless, the performance of "Phone: None", i.e. 66.9%, is still not achieved.

The augmentation by a single codec can improve but also degrade the performance compared to pure downsampling of wide-band recordings. The results depends on the similarity of the applied codec to the true but unknown compression present in the test data. For our specific test set, the best accuracy 65.2% is achieved by G.723.1 codec. This represents improvement of 7.1% over the baseline "Broad: None (16 KHz)" model. This accuracy is still lower by about 1.7% compared to "Phone: None". We argue, that this is caused by other differences between augmented train set and the test set, such as the speaking style.

4.2 Models Using Augmentation via Multiple Codecs

The augmentation by a single codec analyzed in the previous section is not practical, since the best codec must vary with the test data, in order to avoid mismatched training-test conditions. This section investigates the approach of splitting the train set and applying a different codec to each resulting subset, creating a multi-condition training-set.

The results in Fig. 2 indicate that this approach, with more general augmentation, is plausible and leads to comparable accuracy to specific augmentation by a single codec. We consider three different cases with various number of codecs, namely 2, 6 and 9 codecs; the details about the sets are provided in Table 2.

The highest accuracy is obtained using set "Broad: 6 codecs", which achieves a slight improvement of 0.4% compared to "Broad: G.723.1", which was found best in the previous section. The comparable performance of specific and general model is in accordance with findings in [20], where the general model is denoted as "cocktail model".

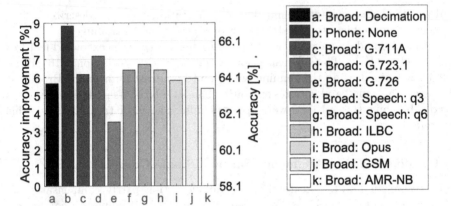

Fig. 1. Absolute improvement of accuracy (left axis) and accuracy (right axis) for models trained on broadcast dataset augmented by a single codec. The baseline accuracy of 58.1% is achieved by the model "Broad: None (16 KHz)".

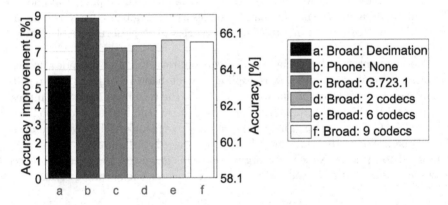

Fig. 2. Absolute improvement of accuracy (left axis) and accuracy (right axis) for models trained on broadcast dataset augmented by a set of codecs. The baseline accuracy of 58.1% is achieved by the model "Broad: None (16 KHz)".

4.3   Combination of Augmented Datasets

Next, we investigate the potential benefits of augmentation applied on the Telephone dataset and combination of both available datasets.

The results shown in Fig. 3 indicate that the augmentation increases the accuracy even for models trained on genuine telephone data; the accuracy of "Phone: 6 codecs" is increased by 1.2% compared to "Phone: none". The addition of more speech data is also beneficial, the accuracy of "Combi: None" is about 1.0% better compared to "Phone: None". This holds even though the added data come from the broadcast background and are not the telephone conversations. Finally, the best performance is achieved, when the augmentation is applied to

Table 2. Details about the codec sets utilized for augmentation via multiple codecs.

Set	Included codecs
2 codecs	G.711A, G.723.1
6 codecs	G.711A, G.723.1, G.726, GSM, Speex:q3, Speex:q6
9 codecs	G.711A, G.723.1, G.726, GSM, Speex:q3, Speex:q6, ILBC, AMR-NB, Opus

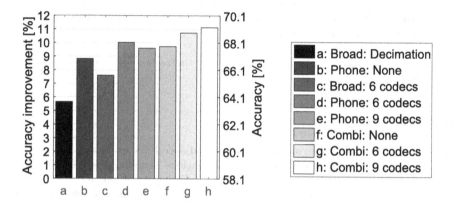

Fig. 3. Absolute improvement of accuracy (left axis) and accuracy (right axis) for models, trained using augmented telephone and combined datasets. The baseline accuracy of 58.1% is achieved by the model "Broad: None (16KHz)".

the combined dataset; model "Combi: 9 codecs" achieves 1.4% higher accuracy compared to "Combi: None".

5   Discussion and Conclusions

We investigated the benefits of data augmentation focused on compression (followed by multi-condition training) in the context of telephone conversational speech. From the results described above, we draw the following conclusions.

The augmentation is able to partially mitigate the deteriorated performance of the acoustic models trained on broadcast speech, when the models are applied to telephone speech (occurred mismatch in training-test conditions). The performance does not achieve the accuracy of model trained on real-world telephone data though, possibly due to very different speaking style. Therefore, as a future avenue of research, it may be beneficial to investigate joint augmentation focused on compression and various speaker styles. The latter is achieved through techniques such as VTLP [10] or elastic spectral distortion [11].

It is possible to create a general model trained in a multi-condition fashion using multiple codecs applied to different parts of the training dataset.

Such model seems robust to unknown/mismatched training-test conditions and achieves slightly better accuracy compared to a specialized model trained on dataset augmented via a single codec.

In accordance with literature (see, e.g., [1]), plain extension of telephone training set (116 h), even with recordings from mismatched broadcast context (132 h), improves the accuracy. However, this improvement is smaller compared to accuracy gained by augmentation of the telephone dataset alone. This means that augmentation is applicable even to matched training data, creating more diverse training dataset.

The best results are achieved by augmentation of the combined telephone and broadcast datasets; here, the benefits of extended training dataset and data augmentation have additive gains.

Acknowledgments. This work was supported by the Technology Agency of the Czech Republic (Project No. TH03010018).

References

1. Amodei, D., et al.: Deep speech 2: End-to-end speech recognition in English and Mandarin. In: International Conference on Machine Learning, pp. 173–182 (2016)
2. Borsky, M., Mizera, P., Pollak, P., Nouza, J.: Dithering techniques in automatic recognition of speech corrupted by MP3 compression: analysis, solutions and experiments. Speech Commun. **86**, 75–84 (2017)
3. Borsky, M., Pollak, P., Mizera, P.: Advanced acoustic modelling techniques in MP3 speech recognition. EURASIP J. Audio Speech Music Process. **2015**(1), 20 (2015)
4. Cui, X., Goel, V., Kingsbury, B.: Data augmentation for deep neural network acoustic modeling. IEEE/ACM Trans. Audio Speech Lang. Process. (TASLP) **23**(9), 1469–1477 (2015)
5. Dahl, G., Yu, D., Deng, L., Acero, A.: Context-dependent pre-trained deep neural networks for large-vocabulary speech recognition. IEEE Trans. Audio Speech Lang. Process. **20**(1), 30–42 (2012). https://doi.org/10.1109/TASL.2011.2134090
6. FFmpeg team: Ffmpeg - cross-platform solution to record, convert and stream audio and video. Software version: 20170525–b946bd8. https://www.ffmpeg.org/
7. Fraga-Silva, T., et al.: Active learning based data selection for limited resource STT and KWS. In: Sixteenth Annual Conference of the International Speech Communication Association (2015)
8. Fraga-Silva, T., et al.: Improving data selection for low-resource STT and KWS. In: 2015 IEEE Workshop on Automatic Speech Recognition and Understanding (ASRU), pp. 153–159. IEEE (2015)
9. Garofolo, J.S., et al.: TIMIT acoustic-phonetic continuous speech corpus. Linguist. Data Consortium, **10**(5) (1993)
10. Jaitly, N., Hinton, G.E.: Vocal tract length perturbation (VTLP) improves speech recognition. In: Proceeding of the ICML Workshop on Deep Learning for Audio, Speech and Language, pp. 625–660 (2013)
11. Kanda, N., Takeda, R., Obuchi, Y.: Elastic spectral distortion for low resource speech recognition with deep neural networks. In: 2013 IEEE Workshop on Automatic Speech Recognition and Understanding (ASRU), pp. 309–314. IEEE (2013)

12. Kemp, T., Waibel, A.: Unsupervised training of a speech recognizer: recent experiments. In: Eurospeech (1999)
13. Kinoshita, K., et al.: A summary of the REVERB challenge: state-of-the-art and remaining challenges in reverberant speech processing research. EURASIP J. Adv. Signal Process. **2016**(1), 7 (2016)
14. Kneser, R., Ney, H.: Improved backing-off for m-gram language modeling. In: 1995 International Conference on Acoustics, Speech, and Signal Processing 1995, ICASSP 1995, vol. 1, pp. 181–184. IEEE (1995)
15. Ko, T., Peddinti, V., Povey, D., Khudanpur, S.: Audio augmentation for speech recognition. In: INTERSPEECH, pp. 3586–3589 (2015)
16. Li, J., Deng, L., Gong, Y., Haeb-Umbach, R.: An overview of noise-robust automatic speech recognition. IEEE/ACM Trans. Audio Speech Lang. Process. **22**(4), 745–777 (2014)
17. Ma, J., Schwartz, R.: Unsupervised versus supervised training of acoustic models. In: Ninth Annual Conference of the International Speech Communication Association (2008)
18. Mammone, R.J., Zhang, X., Ramachandran, R.P.: Robust speaker recognition: a feature-based approach. IEEE Signal Process. Mag. **13**(5), 58 (1996)
19. Polacky, J., Jarina, R., Chmulik, M.: Assessment of automatic speaker verification on lossy transcoded speech. In: 2016 4th International Workshop on Biometrics and Forensics (IWBF), pp. 1–6. IEEE (2016)
20. Raghavan, S., et al.: A comparative study on the effect of different codecs on speech recognition accuracy using various acoustic modeling techniques. In: 2017 Twenty-third National Conference on Communications (NCC), pp. 1–6. IEEE (2017)
21. Seltzer, M.L., Yu, D., Wang, Y.: An investigation of deep neural networks for noise robust speech recognition. In: 2013 IEEE International Conference on Acoustics, Speech and Signal Processing (ICASSP), pp. 7398–7402. IEEE (2013)
22. Siegert, I., Lotz, A.F., Maruschke, M., Jokisch, O., Wendemuth, A.: Emotion intelligibility within codec-compressed and reduced bandwidth speech. In: ITG Symposium, Proceedings of Speech Communication, vol. 12, pp. 1–5. VDE (2016)
23. Torch team: Torch - a scientific computing framework for luajit. http://torch.ch
24. Vincent, E., Watanabe, S., Nugraha, A.A., Barker, J., Marxer, R.: An analysis of environment, microphone and data simulation mismatches in robust speech recognition. Comput. Speech Lang. **46**, 535–557 (2016)
25. Xiong, W., et al.: The Microsoft 2016 conversational speech recognition system. In: 2017 IEEE International Conference on Acoustics, Speech and Signal Processing (ICASSP), pp. 5255–5259. IEEE (2017)
26. Young, S., Young, S.: The HTK hidden Markov model toolkit: design and philosophy. Entrop. Cambridge Res. Lab. Ltd. **2**, 2–44 (1994)

Online LDA-Based Language Model Adaptation

Jan Lehečka[✉] and Aleš Pražák

Department of Cybernetics, Faculty of Applied Sciences,
University of West Bohemia, Univerzitní 8, 306 14 Plzeň, Czech Republic
{jlehecka,aprazak}@kky.zcu.cz
http://www.kky.zcu.cz

Abstract. In this paper, we present our improvements in online topic-based language model adaptation. Our aim is to enhance the automatic speech recognition of a multi-topic speech which is to be recognized in the real-time (online). Latent Dirichlet Allocation (LDA) is an unsupervised topic model designed to uncover hidden semantic relationships between words and documents in a text corpus and thus reveal latent topics automatically. We use LDA to cluster the text corpus and to predict topics online from partial hypotheses during the real-time speech recognition. Based on detected topic changes in the speech, we adapt the language model on-the-fly. We are demonstrating the improvement of our system on the task of online subtitling of TV news, where we achieved 18% relative reduction of perplexity and 3.52% relative reduction of WER over non-adapted system.

Keywords: Topic modeling · Language model adaptation

1 Introduction

Language model (LM) adaptation is a standard mechanism used to improve automatic speech recognition (ASR) in tasks, where the domain (specifically the topic, genre or style) is changeable, because different domains tend to involve relatively disjoint concepts with markedly different word sequence statistic [1].

A typical task to employ LM adaptation is broadcast news transcription, where each few-minutes-long report can be from a completely different topic. When topic labels are not available in the training text corpus, an unsupervised LM adaptation approach must be employed.

In the last two decades, there has been published a lot of studies dealing with unsupervised LM adaptation with an application on broadcast news transcription problem. An extensive survey of older approaches including also unsupervised methods, such as cache models, triggers or LSA, can be found in [1] and the experimental comparison in [2]. However, the majority of unsupervised LM adaptation approaches presented in the last 15 years for broadcast news transcription task is based on Latent Dirichlet Allocation (LDA) [3]. For example, [4]

© Springer Nature Switzerland AG 2018
P. Sojka et al. (Eds.): TSD 2018, LNAI 11107, pp. 334–341, 2018.
https://doi.org/10.1007/978-3-030-00794-2_36

implemented LM adaptation by interpolating the general LM with the dynamic unigram LM estimated by the LDA model, [5] extended LDA-based LM adaptation with a syntactic context-dependent state for each word in the corpus, [6] used efficient topic inference in LDA model, [7] used also named entity information for LDA-based topic modeling and LM adaptation, [8] computed adapted LMs using minimum discriminant information (MDI) and [9] used LDA-weights normalization to estimate topic mixture weights in adapted LMs.

All mentioned approaches are *offline* methods requiring more passes over the signal. In these approaches, usually a background, domain-independent LM is used to generate hypotheses in the first pass, then topics are inferred (predicted) using LDA model, adapted topic-mixture LM is prepared and used in the second pass to rescore or redecode the hypotheses.

In this paper, we propose a system using an unsupervised LDA-based LM adaptation scheme that can work in *online* mode in the real-time, i.e. the adaptation is done on-the-fly as soon as possible after the topic in the speech is changed. A typical task to employ online LM adaptation is subtitling of live TV shows, where topics are highly changeable, e.g. news, TV debates, sports summaries etc. We demonstrate the improvement of the proposed approach over a non-adapted baseline model on the task of live subtitling of Czech TV news. We also compare our online system with the state-of-the-art offline system.

2   LDA

Latent Dirichlet Allocation (LDA) [3] is the leading paradigm in unsupervised topic modeling. In LDA, documents are random mixtures over latent topics generated from Dirichlet distribution with parameter $\alpha = (\alpha_1, \alpha_2, ..., \alpha_K)$ where K is the number of latent topics, and latent topics are random mixtures over words generated from Dirichlet distribution with parameter $\beta = (\beta_1, \beta_2, ..., \beta_V)$, where V is the size of corpus vocabulary.

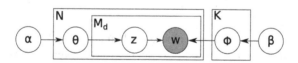

Fig. 1. Plate notation of LDA model. In this model, word w is the only observable variable, N is number of documents, M_d is number of words in document d, K is number of latent topics and a value in the upper left corner of each rectangle means repetition count of its content.

The plate notation of LDA is shown in Fig. 1. Each document is represented by a topic distribution $\theta \sim \text{Dirichlet}_K(\alpha)$, while each topic is represented by a word distribution $\Phi \sim \text{Dirichlet}_V(\beta)$. Then for each word position in each document, a latent topic z is drawn from θ, corresponding word distribution Φ_z is found in Φ and word w is drawn from Φ_z.

In order to predict the topic distribution for an unseen document, we use online variational inference [10] in our experiments.

3   Scheme of the System

The scheme of our solution is outlined in Fig. 2. The core of the system is online automatic speech recognizer (*Online ASR* block) developed on our department [11], which is processing the input audio stream in the real-time and generating partial hypotheses about the content of the current speech. The standard ASR system was extended by adding one more LM into the decoder resulting in a *Parallel Decoder* generating two partial hypotheses at each time step: one using a general LM and one using an adapted LM, which can be replaced on-the-fly.

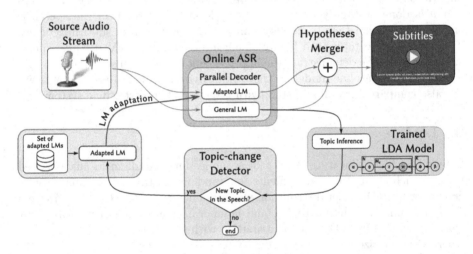

Fig. 2. The scheme of our system.

Since the adapted LM context is broken when changing LM during the decoding and there can be some gaps in decoded results, because LM changes may take several seconds, we improved this system by online merging both hypotheses together in a *Hypotheses Merger*, where the hypothesis from the adapted-LM decoder is always favored over hypothesis from the general-LM decoder. Only at the very beginning of the recognition (when the topic is not yet known) and when the hypothesis from the adapted-LM decoder is temporarily unavailable, the final output is backfilled with words recognized by the general-LM decoder. The outputs from *Hypotheses Merger* are desired subtitles which can be streamed online to TV viewers.

The 1-best hypothesis from the general-LM decoder is used to infer the current-speech topic using an LDA model. The inferred topic is checked if it is new in the speech (*Topic-change Detector* block) and if so, a corresponding pre-trained adapted LM is selected and used to replace the existing adapted LM in the ASR's decoder (*LM adaptation* arrow). After that, the ASR immediately starts to generate transcription using this new LM. In this way, the system is being adapted online based on the current (or very recent) topic in the speech.

3.1 System Settings

For experiments in this paper, we set our online ASR system to generate one partial result in the form of 1-best hypothesis per second from each decoder. To predict current topic, partial hypotheses are cropped to the last 50 words (representing approximately last 20 s of audio) and the most probable topic is inferred from the LDA model. Based on our experiments, predicting only the one best topic is sufficient in this task.

The system is not adapting the LM every time a different prediction from LDA model is observed. Instead, the system waits for 5 more predictions (seconds) to ensure, that the topic change has not been a false alarm. In order to compensate this 5-s delay behind the real topic change, we use five seconds *retro-recognition* when adapting the LM. It means that when a new adapted-LM decoder starts to recognize the input stream, it first redecodes the last five seconds. Based on our experiments, five seconds is enough to ensure the topic change is not a false alarm and at the same time it is short enough for online corrections of the last words using the new adapted LM.

4 Experimental Setup

To test online LM adaptation in the news domain, we chose TV show *Události*, which is the main daily broadcast TV news show in the Czech Republic. Each show is approximately 48 min long and contains about 23 individual reports related to many various topics. At each boundary of two consecutive reports, the topic is usually changing, which makes these data suitable to test the LM adaptation. In our experiments, we are evaluating perplexity of adapted LMs and word error rates (WER).

4.1 Audio Data

We selected 13 TV shows from November 2013 to April 2014. All test shows were transcribed by human annotators. In sum, our test data consists of 10.3 h of 16 KHz audio, 303 individual reports and 84k reference words in the transcripts.

We used our own speech recognition system optimized for low-latency real-time operation for experiments [11]. For acoustic modeling we used common three-state HMMs with output probabilities modeled by a Deep Neural Network (DNN).

4.2 Text Data

In order to train high-quality and robust topic-specific LMs, a large collection of in-domain text data must be accumulated. In our department, we have developed a framework solution for mining, processing and storing large amounts of electronic texts for language modeling purposes [12]. This system is periodically importing and processing news articles from many Czech news servers. We also supplemented the system with additional transcripts of selected TV and radio shows and a large amount of newspaper articles. Until now, we have accumulated Czech news-related text corpus amounting to almost 1.3 billion tokens in 3.7 million documents. All texts were preprocessed by text cleaning, tokenization, text normalization, true-casing and vocabulary-based text replacements to unify distinct word forms and expressions. We didn't use any word normalization like lemmatization or stemming.

4.3 LMs

From all available text data, we trained a general, topic-independent, trigram LM and used it to compute a baseline performance. Since the LM is trained from a lot of text data, we limited the minimal count of bigrams to 3 and the minimal count of trigrams to 6. To avoid misspelled words and other eccentricities to be present in subtitles, we checked all tokens in our text corpus against a list of known and correctly-spelled words, and marked all out-of-list tokens as unknown words, which reduced the vocabulary size from 4.4 to 1.2 million words. A general LM trained from all available texts contains 1.2 million unigrams, 27 million bigrams and 20 million trigrams.

Then we trained LDA model to distinguish between K topics in 5 passes over training data. Our stop-word list consisted of words which were contained in more than 40% of documents. When trained, we used the LDA model to separate our text corpus into topic clusters in order to train a topic-specific LM for each topic. According to our experiments, the best separation is to assign each text into topic clusters based on a probability threshold (half of the maximum predicted probability). To increase robustness, we trained an adapted LM for each topic as an interpolated mixture of topic-specific LM and general LM. Since topic-specific clusters are reasonably small, we do not have to constrain minimal counts of n-grams so rigorously as in the case of general LM.

In this paper, we are experimenting with various values of K, various interpolation weights in adapted LMs and various minimal counts of n-grams included in topic-specific LMs.

5 Experimental Results

In the first two experiments, we limited the minimal count of bigrams to 3 and the minimal count of trigrams to 6 for both general and topic-specific LM, therefore the LM adaptation does not bring any new n-gram into the recognizer, but it can assign more probability mass to topic-related n-grams existing in the general LM and hence accentuate the topic in the adapted LM. In the third experiment, we relaxed these constraints and include more n-grams into the topic-specific LMs, which allows the adaptation to fetch also new topic-related n-grams into the recognizer with only a reasonably small enlargement of adapted LMs.

We also compare our results with the stat-of-the-art *offline* system used for example in [9]. The offline system employs the same LDA models and LMs as in the online system, but in an offline 2-pass decoding scheme: the first pass is used to decode hypotheses with the general LM, then for each report, a specific LM is interpolated based on predicted topic distribution from the LDA model (we consider 5 best topics), it is mixed together with the general LM and the resulting adapted LM is used to redecode the report in the second pass. We do not expect our online system to perform better than offline (where the processing and mixing time is not an issue), but to be close enough to the state-of-the-art results with the benefit of the real-time usage.

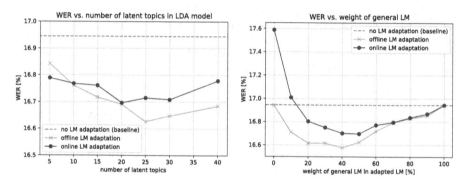

Fig. 3. Results of Exp. 1: number of latent topics (left) and results of Exp. 2: interpolation weights (right).

Experiment 1: Number of Latent Topics

In the first experiment, we fixed the interpolation weight of general LM to 0.5 and ran the recognition with varying number of latent topics. Results are shown in Fig. 3 (left subfigure).

We can see that all results with adapted LMs are slightly better than the baseline and the best result was achieved using LDA with 20 latent topics. Further increasing the number of latent topics didn't bring any improvement. As for the offline system, two results are very close to online system, one result is even worse and the best result was achieved using 25 latent topics.

Experiment 2: Interpolation Weights

In the second experiment, we fixed the number of latent topics to 20 (online), resp. 25 (offline) and experimented with various interpolation weights when mixing adapted LM. Results are shown in Fig. 3 (right subfigure).

The rightmost point of the adapted system equals to the baseline performance, because 100% of general LM was used in this case. The best interpolation ratio between general and topic-specific LM is roughly 50:50 (online), resp. 40:60 (offline), which stress the positive contribution of both LMs. An interesting result is that the performance can be also deteriorated compared to the non-adapted system if we put too small weight for the general LM.

Experiment 3: N-gram Count Limits

In the last experiment, we relaxed the constraints for minimal counts of n-gram when training topic-specific LM. Let $N_{123} = (N_1, N_2, N_3)$ be the minimal counts of unigrams, bigrams and trigrams when training a LM from a text. So far, all LMs were trained with $N_{123} = (1, 3, 6)$. In this experiment, we trained the general LM with $N_{123} = (1, 3, 6)$ and topic-specific LMs with $N_{123} = (1, 2, 3)$, $N_{123} = (1, 1, 2)$, resp. $N_{123} = (1, 1, 1)$. In this case, the adaptation can bring also new topic-related n-grams into the recognizer, which are not present in the general LM due to a low overall count in the text corpus.

Table 1. Results of Exp. 3: n-gram count limits. For different n-gram limitations of topic-specific LMs (the first column), we are showing the average size of used LMs during the recognition (LM size), real-time ratio on Intel Core i7-7800X machine (RT-ratio), LM perplexity (PPL) and word error rate (WER). We show also relative reduction of PPL and WER over the baseline system.

	LM size	RT-ratio	PPL	WER [%]
Baseline	1.25 GB	0.50	633.6	16.945
$N_{123} = (1, 3, 6)$/*online*/	1.25 GB	0.65	613.8 (−3.1%)	16.697 (−1.46%)
$N_{123} = (1, 3, 6)$/*offline*/	1.25 GB	1.23	562.9 (−11.2%)	16.578 (−2.17%)
$N_{123} = (1, 2, 3)$/*online*/	1.31 GB	0.66	592.1 (−6.6%)	16.632 (−1.85%)
$N_{123} = (1, 1, 2)$/*online*/	1.80 GB	0.67	561.3 (−11.4%)	16.478 (−2.76%)
$N_{123} = (1, 1, 1)$/*online*/	3.71 GB	0.70	519.9 (−18.0%)	16.394 (−3.52%)

Results are shown in Table 1. In the experiment, where we kept the same limits for general and topic-specific LM ($N_{123} = (1, 3, 6)$), we achieved 3.1% relative reduction of perplexity and 1.46% relative reduction of WER while preserving the exact same size of LM and slowing the recognition by 30% relatively (due to the parallel decoding and online topic inference). We achieved slightly better performance using the offline system, but at the expense of 146% relative slowdown of the recognition.

During the real-time recognition, there is no time for mixing suitable adapted LMs on-the-fly. Instead, we have to select one adapted LM from the pre-trained set. That is why our online results are slightly worse than state-of-the-art offline results, but our system has the benefit of the real-time usage.

In experiments, where also less-frequent topic-related n-grams were included in adapted LMs, we achieved up to 18.0% relative reduction of perplexity and 3.52% relative reduction of WER with only a modest further slowdown of the system.

6 Conclusion

In this paper, we presented a fully unsupervised way of topic-based LM adaptation in an online system capable of generating subtitles for multi-topic TV shows in the real-time. We achieved 3.1% relative reduction of perplexity and 1.46% relative reduction of WER with a fixed size of LM (only by accentuating topic-related n-grams) and 18.0% relative reduction of perplexity and 3.52% relative reduction of WER when including also less-frequent topic-related n-grams in adapted LMs.

Acknowledgments. This paper was supported by the project no. P103/12/G084 of the Grant Agency of the Czech Republic and by the grant of the University of West Bohemia, project no. SGS-2016-039.

References

1. Bellegarda, J.R.: Statistical language model adaptation: review and perspectives. Speech Commun. **42**(1), 93–108 (2004)
2. Chen, L., Lamel, L., Gauvain, J.L., Adda, G.: Dynamic language modeling for broadcast news. In: Eighth International Conference on Spoken Language Processing (2004)
3. Blei, D.M., Ng, A.Y., Jordan, M.I.: Latent Dirichlet allocation. J. Mach. Learn. Res. **3**, 993–1022 (2003)
4. Tam, Y.C., Schultz, T.: Dynamic language model adaptation using variational bayes inference. In: Ninth European Conference on Speech Communication and Technology (2005)
5. Hsu, B.J.P., Glass, J.: Style & topic language model adaptation using HMM-LDA. In: Proceedings of the 2006 Conference on Empirical Methods in Natural Language Processing, pp. 373–381. Association for Computational Linguistics (2006)
6. Heidel, A., Chang, H., Lee, L.: Language model adaptation using latent Dirichlet allocation and an efficient topic inference algorithm. In: Eighth Annual Conference of the International Speech Communication Association (2007)
7. Liu, Y., Liu, F.: Unsupervised language model adaptation via topic modeling based on named entity hypotheses. In: IEEE International Conference on Acoustics, Speech and Signal Processing 2008, ICASSP 2008, pp. 4921–4924. IEEE (2008)
8. Haidar, M.A., O'Shaughnessy, D.: Unsupervised language model adaptation using latent Dirichlet allocation and dynamic marginals. In: 2011 19th European Signal Processing Conference, pp. 1480–1484. IEEE (2011)
9. Jeon, H.B., Lee, S.Y.: Language model adaptation based on topic probability of latent dirichlet allocation. ETRI J. **38**(3), 487–493 (2016)
10. Hoffman, M., Bach, F.R., Blei, D.M.: Online learning for latent Dirichlet allocation. In: Advances in Neural Information Processing Systems, pp. 856–864 (2010)
11. Pražák, A., Loose, Z., Trmal, J., Psutka, J.V., Psutka, J.: Novel approach to live captioning through re-speaking: tailoring speech recognition to re-speaker's needs. In: INTERSPEECH (2012)
12. Švec, J., et al.: General framework for mining, processing and storing large amounts of electronic texts for language modeling purposes. Lang. Resour. Eval. **48**(2), 227–248 (2014)

Recurrent Neural Network Based Speaker Change Detection from Text Transcription Applied in Telephone Speaker Diarization System

Zbyněk Zajíc$^{(\boxtimes)}$ ⓘ, Daniel Soutner ⓘ, Marek Hrúz ⓘ, Luděk Müller ⓘ,
and Vlasta Radová ⓘ

Faculty of Applied Sciences, NTIS - New Technologies for the Information Society
and Department of Cybernetics, University of West Bohemia, Univerzitní 8,
306 14 Plzeň, Czech Republic
{zzajic,dsoutner,mhruz,muller,radova}@ntis.zcu.cz

Abstract. In this paper, we propose a speaker change detection system based on lexical information from the transcribed speech. For this purpose, we applied a recurrent neural network to decide if there is an end of an utterance at the end of a spoken word. Our motivation is to use the transcription of the conversation as an additional feature for a speaker diarization system to refine the segmentation step to achieve better accuracy of the whole diarization system. We compare the proposed speaker change detection system based on transcription (text) with our previous system based on information from spectrogram (audio) and combine these two modalities to improve the results of diarization. We cut the conversation into segments according to the detected changes and represent them by an i-vector. We conducted experiments on the English part of the CallHome corpus. The results indicate improvement in speaker change detection (by 0.5% relatively) and also in speaker diarization (by 1% relatively) when both modalities are used.

Keywords: Recurrent neural network
Convolutional Neural Network
Speaker change detection · Speaker diarization · I-vector

1 Introduction

The problem of Speaker Diarization (SD) is defined as a task of categorizing speakers in an unlabeled conversation. The Speaker Change Detection (SCD) is often applied to the signal to obtain segments which ideally contain a speech of a single speaker [1]. The telephone speech is a particular case where the speaker turns can be extremely short with negligible between-turn pauses and frequent

This research was supported by the Ministry of Culture Czech Republic, project No. DG16P02B009.

P. Sojka et al. (Eds.): TSD 2018, LNAI 11107, pp. 342–350, 2018.
https://doi.org/10.1007/978-3-030-00794-2_37

overlaps. SD systems for telephone conversations often omit the SCD process and use a simple constant length window segmentation of speech [2]. In our previous papers [3,4], we introduced the SD system with SCD based on Convolutional Neural Network (CNN) for segmentation of the acoustic signal. This SD system is based on i-vectors [5] that represent speech segments, as introduced in [6]. The i-vectors are clustered in order to determine which parts of the signal were produced by the same speaker and then the feature-wise resegmentation based on Gaussian Mixture Models is applied.

In all SD systems mentioned above, only the audio information is used to find the speaker change in the conversation. In this work we aimed to use the lexical information contained in the transcription of the conversation, which is a neglected modality in the SCD/SD task: The work [7] investigates whether the statistical information on the speaker sequence derived from their roles (using speaker roles n-gram language model) can be used in speaker diarization of meeting recordings. Using Automatic Speech Recognition (ASR) system transcription for diarization of a telephone conversation was used in [8] where only speech and non-speech regions were classified.

We can see the lexical information as an additive modality compared to the acoustic data. Also, both the SCD based on the linguistic and acoustic information can be combined to improve the accuracy of the SD system. A similar approach was recently published in [23].

2   Segmentation

2.1   Oracle Segmentation

We implemented oracle segmentation as in [9] for the purpose of comparison: the conversations are split according to the reference transcripts, each individual speaker turn from the transcript becomes a single segment.

2.2   CNN Based SCD on Spectrogram

In our previous work [10], we introduced the CNN for SCD task (Fig. 1). We trained the CNN as a regressor on spectrograms of the acoustic signal with a reference information L about the existing speaker changes, where L can be seen as a fuzzy labeling [3] with a triangular shape around the labeled speaker change time points produced by humans. The main idea behind it is to model the uncertainty of the annotation. The speaker changes are identified as peaks in the network's output signal P using non-maximum suppression with a suitable window size. The detected peaks are then thresholded to remove insignificant local maxima. We consider the signal between two detected speaker changes as one segment. The minimum duration of one segment is limited to one second, shorter parts are not used for clustering, and the decision about the speaker in them is waiting for the resegmentation step. This condition is made to avoid clustering segments containing an insignificant amount of data from the speaker to be modeled as an i-vector. It is also possible to use this system for SCD (with small modification) in the online SD system [9].

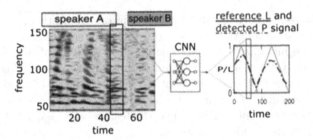

Fig. 1. The input speech as spectrogram is processed by the CNN into the output probability of change P (the dashed line). The reference speaker change L for the CNN training is depicted also (the solid line).

2.3 RNN Based SCD on Lexical Information

From the global point of view, a change of a speaker mainly occurs when the speaker ended a word as opposed to in the middle of pronunciation. The probability of change is even higher when he/she finished a sentence. This is the reason, why we decided to acquire extra information about the speaker change from text transcriptions using detection of utterance endings. This process might produce over-segmentation of the conversation. Although this means the coverage measure of the results will be lower, the purity of the segments will be high. Nevertheless, the over-segmentation of the conversation is not such a crucial problem, because our goal is to make the whole diarization process more accurate (not just SCD). If the segments are long enough to represent the speaker by an i-vector, the segmentation step of the SD system will assign the proper speakers to the segments. That's why we deduce that to find the end of an utterance is a reasonable requirement for segmentation.

We conducted two experiments. First with the reference transcriptions that were force-aligned with an acoustic model and the second with the recognized text from the ASR system. We followed this procedure: Obtain aligned text with time stamps (force aligned or from the ASR) from the recordings. Train a language model as Recurrent Neural Network [11] with Long Short-Term Memory (LSTM) layers [12]. Label every word from text with lexical probability, that the next word is the end of an utterance. The output from the RNN is the probability of speaker change in time (see Fig. 2).

2.4 Combination of both SCD Approaches

Both the approaches to the SCD problem can be combined to refine the information about the speaker change for segmentation step of SD system. Both systems output the probability of a speaker change in time. The combined system can decide about the speaker change considering two sources; CNN on a spectrogram (audio) and RNN on a transcription (text). The output of the combined system is also a probability of speaker change (a number between zero and one). We used a weighted sum of both speaker change probabilities

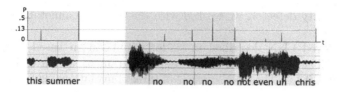

Fig. 2. The output of the RNN based SCD on lexical information. The probability P of the speaker/utterance change in time.

$P_{comb} = w * P_{spectr} + (1-w) * P_{transc}$ and normalized the results into an interval $\langle 0; 1 \rangle$. The value of the parameter w was found experimentally to be 0.5.

3   Segment Description

To describe a segment of conversation we first construct a supervector of accumulated statistics [13] and then the i-vectors are extracted using Factor Analysis [14]. In our work [4], we introduced an approach to the statics refinement using the probability of speaker change as a weighting factor into the accumulation of statistics. We also use this approach in this paper.

4   Experiments

We designed the experiment to investigate our proposed approach to SCD from RNN on transcription compared with CNN on spectrogram and with the combined system.

4.1   Corpus

The experiment was carried out on telephone conversations from the English part of CallHome corpus [15]. We mixed the original two channels into one and we selected only two speaker conversations so that the clustering can be limited to two clusters. This subset contains 109 conversations in total each has about 10 min duration in a single telephone channel sampled at 8 kHz. For training of the CNN, we used only 35 conversations, the rest we used for testing the SD system.

4.2   System

The SD system presented in our paper [4] uses the feature extraction based on Linear Frequency Cepstral Coefficients, Hamming window of length 25 ms with 10 ms shift of the window. We employ 25 triangular filter banks which are spread linearly across the frequency spectrum, and we extract 20 LFCCs. We add delta coefficients leading to a 40-dimensional feature vector ($D_f = 40$). Instead of the voice activity detector, we worked with the reference annotation about the missed speech.

We employed CNN described in [3] for segmentation based on the audio information. The input of the net is a spectrogram of speech of length 1.4 s, and the shift is 0.1 s. The CNN consists of three convolutional layers with ReLU activation functions and two fully connected layers with one output neuron. Note that for the purposes of this paper we reimplemented the network in Tensorflow[1], thus the results slightly differ from our previous work.

As our language model for computing lexical scores, we have chosen neural network model with two LSTM [12] layers with the size of hidden layer 640. We trained our model from Switchboard corpus [16], which is very near to our testing data. We split our data into two folds: train with 25433 utterances and development data with 10000 utterances. The vocabulary has the size of 29600 words (only from the training part of the corpus) plus the $\langle unk \rangle$ token for the unknown words. We used SGD as the optimizer. We employed dropout for regularization, and the batch size was 30 words. We evaluated our model on text data, and we achieved 72 in perplexity on development data and 70 on test data.

The ASR system setup, for automatic transcription of the data, was the same as the standard Kaldi [17] recipe s5c for Switchboard corpus; we used the "chain" model. We trained the acoustic model as Time Delayed Neural Network with seven hidden layers, each with an output of 625, the number of targets (states) was 6031. We set the inputs as MFCC features with a dimension of 40 and the i-vectors for adaptation purposes. We recognized all the recordings as one file, the Word Error Rate on tested data was 26.8%.

For the purpose of training the i-vector, we model the Universal Background Model as a Gaussian Mixture Model with 1024 components. We have set the dimension of the i-vector to 400. For clustering, we have used K-means algorithm with cosine distance to obtain the speaker clusters.

4.3   Results

The results as Purity [18] vs. Coverage [19] curve for SCD can be seen in Fig. 3 for all approaches to the segmentation, where dual evaluation metrics Purity and Coverage are used according to the work [20] to better evaluate the SCD process. The slightly modified Equal Error Rate (EER), where the Coverage and Purity have the same value, for each SCD method with the particular threshold T_{EER} can be seen in the first two columns of Table 1. The goal of the general SCD system is to get the best Purity and Coverage, but for our SD system, we want to get the best "Purity" of all segments with enough segments longer than 1 s. The 1-s threshold we set empirically as enough speech for training the i-vector to represent the speaker accurately in the segment for diarization of two-party conversation. For CallHome data with relatively long conversations (5–10 min), it is better for the SD system to leave some short segments out of clustering and wait for the re-segmentation step to decide about the speaker in these segments.

---

[1] Available on https://www.tensorflow.org.

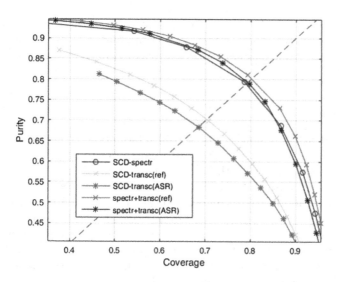

Fig. 3. Purity vs coverage curve for SCD system with CNN on spectrogram, RNN on transcripts and combined system.

Table 1. EER [%] for each the SCD system with particular threshold T_{EER} and DER [%] of the whole SD systems with the segmentation based on SCD. The SCD using spectrogram, reference transcription and transcription from ASR. Also, the results from combination using spectrogram with reference transcription and spectrogram with ASR transcription are reported. The experimentally chosen threshold T for segmentation is in the last column.

Segmantation	EER	T_{EER}	DER	T
SCD-oracle	0.00	-	6.76	-
SCD-spectr	0.21	0.75	6.93	0.70
SCD-transc(ref)	0.30	0.17	8.07	0.17
SCD-transc(ASR)	0.32	0.08	8.62	0.12
spectr+transc(ref)	**0.20**	0.50	**6.86**	0.45
spectr+transc(ASR)	0.21	0.49	7.06	0.45

We use the Diarization Error Rate (DER) for the evaluation of our SD system to be comparable to other methods tested on CallHome (e.g., [2,21]). DER has been described and used by NIST in the RT evaluations [22]. We use the standard 250 ms tolerance around the reference boundaries. DER is a combination of several types of errors (missed speech, mislabeled non-speech, incorrect speaker cluster). We assume the information about the silence in all testing recordings is available and correct. That means that our results represent only the error of incorrect speaker clusters. Contrary to a common practice in telephone speech diarization, we do not ignore overlapping segments during the evaluation. The last two columns of Table 1 shows the SD system using the SCD based on spectrogram, transcription, and combination of both and the experimentally chosen threshold T (to remove insignificant local maxima in SCD system outputs) for each method.

4.4 Discussion

The proposed SCD approach using lexical information from transcription performed worse than the SCD on the spectrogram. We think the main reason for this is due to the quality of the conversation: the sentences in the telephone recordings are not always finished due to the frequent crosstalks, so the SCD based on transcription has incomplete information about the speaker change. Nevertheless, this information brings an additive knowledge about the speaker change. The combined SCD system ("spectr+transc") improved the results of the SD system. When using the transcription from ASR we obtain slightly worse results due to the accuracy of the ASR system. More sophisticated classifier using both SCD from spectrogram and transcription can be trained. However, there is a problem with the training criterion because our goal is to get better results on the SD system, not only to find the precise boundaries of the speaker changes. Also, the mistakes in the reference annotations of CallHome corpus are limiting the performance (see the result of oracle segmentation). Authors of a similar approach [23] tried to find SCD using also both acoustics and lexical information combined together and propagated thru only one LSTM neural network. Unfortunately, the evaluation of their approach was made on different data.

5 Conclusions

In this paper, we proposed a new method for SCD using the lexical information from the transcribed conversation. For this purpose, we have trained RNN with LSTM layers to evaluate the transcription of the conversation and find the speaker changes in it. This approach brings new information about the speaker change and can be used in combination with SCD method based on the audio information to improve the diarization. Our future work is to train a complex classifier to improve the speaker change detection using both modalities (text and audio).

References

1. Rouvier, M., Dupuy, G., Gay, P., Khoury, E., Merlin, T., Meignier, S.: An open-source state-of-the-art toolbox for broadcast news diarization. In: Interspeech, Lyon, pp. 1477–1481 (2013)
2. Sell, G., Garcia-Romero, D.: Speaker diarization with PLDA I-vector scoring and unsupervised calibration. In: IEEE Spoken Language Technology Workshop, South Lake Tahoe, pp. 413–417 (2014)

3. Hrúz, M., Zajíc, Z.: Convolutional neural network for speaker change detection in telephone speaker diarization system. In: ICASSP, New Orleans, pp. 4945–4949 (2017)
4. Zajíc, Z., Hrúz, M., Müller, L.: Speaker diarization using convolutional neural network for statistics accumulation refinement. In: Interpeech, Stockholm, pp. 3562–3566 (2017)
5. Dehak, N., Kenny, P.J., Dehak, R., Dumouchel, P., Ouellet, P.: Front-end factor analysis for speaker verification. IEEE Trans. Audio Speech Lang. Process. **19**(4), 788–798 (2011)
6. Shum, S., Dehak, N., Chuangsuwanich, E., Reynolds, D., Glass, J.: Exploiting intra-conversation variability for speaker diarization. In: Interspeech, Florence, pp. 945–948 (2011)
7. Valente, F., Vijayasenan, D., Motlicek, P.: Speaker diarization of meetings based on speaker role n-gram models. In: ICASSP, pp. 4416–4419. IEEE, Prague (2011)
8. Tranter, S.E., Yu, K., Evermann, G., Woodland, P.C.: Generating and evaluating segmentations for automatic speech recognition of conversational telephone speech. In: ICASSP, pp. 753–756. IEEE, Montreal (2004)
9. Kunešová, M., Zajíc, Z., Radová, V.: Experiments with segmentation in an online speaker diarization system. In: Ekštein, K., Matoušek, V. (eds.) TSD 2017. LNCS (LNAI), vol. 10415, pp. 429–437. Springer, Cham (2017). https://doi.org/10.1007/978-3-319-64206-2_48
10. Hrúz, M., Kunešová, M.: Convolutional neural network in the task of speaker change detection. In: Ronzhin, A., Potapova, R., Németh, G. (eds.) SPECOM 2016. LNCS (LNAI), vol. 9811, pp. 191–198. Springer, Cham (2016). https://doi.org/10.1007/978-3-319-43958-7_22
11. Soutner, D., Müller, L.: Application of LSTM neural networks in language modelling. In: Habernal, I., Matoušek, V. (eds.) TSD 2013. LNCS (LNAI), vol. 8082, pp. 105–112. Springer, Heidelberg (2013). https://doi.org/10.1007/978-3-642-40585-3_14
12. Hochreiter, S., Urgen Schmidhuber, J.: Long short-term memory. Neural Comput. **9**(8), 1735–1780 (1997)
13. Zajíc, Z., Machlica, L., Müller, L.: Robust adaptation techniques dealing with small amount of data. In: Sojka, P., Horák, A., Kopeček, I., Pala, K. (eds.) TSD 2012. LNCS (LNAI), vol. 7499, pp. 480–487. Springer, Heidelberg (2012). https://doi.org/10.1007/978-3-642-32790-2_58
14. Kenny, P., Dumouchel, P.: Experiments in speaker verification using factor analysis likelihood ratios. In: Odyssey, Toledo, pp. 219–226 (2004)
15. Canavan, A., Graff, D., Zipperlen, G.: CALLHOME American English speech, LDC97S42. In: LDC Catalog. Linguistic Data Consortium, Philadelphia (1997)
16. Godfrey, J.J., Holliman, E.: Switchboard-1 release 2. In: LDC Catalog. Linguistics Data Consortium, Philadelphia (1997)
17. Daniel, P., et al.: Modelos animales de dolor neuropático. In: Workshop on Automatic Speech Recognition and Understanding, IEEE Catalog No.: CFP11SRW-USB (2011)
18. Harris, M., Aubert, X., Haeb-Umbach, R., Beyerlein, P.: A study of broadcast news audio stream segmentation and segment clustering. In: EUROSPEECH, Budapest, pp. 1027–1030 (1999)
19. Bredin, H.: TristouNet: triplet loss for speaker turn embedding. In: ICASSP, New Orleans, pp. 5430–5434 (2017)

20. Bredin, H.: pyannote.metrics: a toolkit for reproducible evaluation, diagnostic, and error analysis of speaker diarization systems. In: Interspeech, Stockholm, pp. 3587–3591 (2017)
21. Sell, G., Garcia-Romero, D., Mccree, A.: Speaker diarization with I-vectors from DNN senone posteriors. In: Interspeech, Dresden, pp. 3096–3099 (2015)
22. Fiscus, J.G., Radde, N., Garofolo, J.S., Le, A., Ajot, J., Laprun, C.: The rich transcription 2006 spring meeting recognition evaluation. Mach. Learn. Multimodal Interact. **4299**, 309–322 (2006)
23. India, M., Fonollosa, J., Hernando, J.: LSTM neural network-based speaker segmentation using acoustic and language modelling. In: Interspeech, Stockholm, pp. 2834–2838 (2017)

On the Extension of the Formal Prosody Model for TTS

Markéta Jůzová[1]([✉]), Daniel Tihelka[1], and Jan Volín[2]

[1] Faculty of Applied Sciences, New Technologies for the Information Society and Department of Cybernetics, University of West Bohemia, Pilsen, Czech Republic
{juzova,dtihelka}@kky.zcu.cz
[2] Faculty of Arts, Institute of Phonetics, Charles University, Prague, Czech Republic
jan.volin@ff.cuni.cz

Abstract. The formal prosody grammar used for TTS focuses mainly on the description of final prosodic words in phrases/sentences which characterize a special prosodic phenomenon representing a certain communication function within the language system. This paper introduces an extension of the prosody model which also takes into account the importance and distinction of the first prosodic words in the prosodic phrases. This phenomenon can not change the semantic interpretation of the phrase, but for higher naturalness, the beginnings of the prosodic phrases differ from subsequent words and should be, based on the phonetic background, dealt with separately.

Keywords: Unit selection · Formal prosody grammar · Prosodeme

1 Introduction

The formal prosody grammar [9], used for the description of prosody in our TTS system *ARTIC* [14], describes the required supra-segmental prosody features of an utterance to be synthesized on the deep semantic structure level [10]. In this way, there is no need to render any surface level prosody behaviour which prescribes the particular intonation pattern by means of energy+F_0 contours and phone durations. The reason we try to avoid the use of surface level prosody is the ambiguity between these two descriptions – there are multiple possible, completely valid and natural sounding, energy, F_0 and duration patterns which all respect the given deep level requirements. And vice versa, a particular surface representation can have more different deep level descriptions assigned, which in turn means that a particular segmental-level {energy, F_0, duration} pattern may be used within all (and possibly others) of the descriptions [16].

This research was supported by the Czech Science Foundation (GA CR), project No. GA16-04420S, and by the grant of the University of West Bohemia, project No. SGS-2016-039.

© Springer Nature Switzerland AG 2018
P. Sojka et al. (Eds.): TSD 2018, LNAI 11107, pp. 351–359, 2018.
https://doi.org/10.1007/978-3-030-00794-2_38

In the context of unit selection, our deep-level-based prosody description matches the *independent feature formulation* (IFF) [12]. The task of unit selection algorithm, and especially its target cost component, is to ensure that the selected sequence of candidates matches the deep level requirements, regardless which particular surface level prosody contours emerge in the sequence (matching the deep level ensures the intended perception). This is an opposite to statistical-parametric-synthesis [5] where the surface level (including spectral features as well) is synthesized from the trained models and forced to the vocoder where speech having these surface level prosody characteristics is created.

Having the prosody grammar, the text sentence is parsed in a derivation tree. The level on which the sequence of "null" and "functionally involved" prosodemes is generated by the grammar, is supposed to express the required semantic interpretation (what is intended to be expressed) of the given text the sequence was generated for. Still, there is no prescription for the particular duration, intensity or F_0 patters, but it must be ensured that there is no ambiguity in the interpretation of the utterance by the listener – e.g. there is no perception of question when an declarative phrase was expected. This, of course, is extremely important for languages without fixed sentence structure, Czech being one of them. Thus, the prosodic word labels are able to express particular semantic representations and they must not be mutually exchanged.

On the other hand, we can exchange the speech units (or surface level structure prosody holders [9]) to be concatenated unless the sequence renders the required understanding. The easiest way is to use only the units which appeared in the same prosodeme type as that required for the text to be rendered (synthesized). Such treatment, although working satisfactory in most of the situations, leads to the data sparsity as mentioned in [12].

From time to time in some synthesized utterances, we have perceived a noticeable increase of melody within phrases in place of *null* prosodeme. Further analysis showed that this pattern is related to the beginnings of utterances, but it does not have a certain communication function which would mean the need for another non-*null* prosodeme. Instead, it has been perceived as a special kind of *null* prosodeme (labeled $P_{0.1}$), i.e. it must not be exchanged with other functionally involved prosodemes (change of meaning), neither it should be placed to the "classic" P_0 *null* prosodeme (sometimes unnatural perception).

The present paper describes the changes in our prosody grammar required to avoid these unintended prosody renderings together with phonetic substantiation of this change. Both theoretical and empirical observations were confirmed by a listening test designed to focus on $P_{0.1} \leftrightarrow P_0$ exchanges.

2   Phonetic Background

One of the crucial roles of prosody is to divide the speech continuum, which can be conceived as a train of phones, into configurations that facilitate the recovery of the meaning (i.e. communication function). Although the terminology pertaining to prosodic units is quite diverse across the research community, some prosodic division can be found in every known language of the world. It is obvious that speech divided into typical prosodic units is easier to process in the brains of the listeners [1, 4, 7] and vice versa: the train of phones undivided into prosodic units is cerebrally demanding and occasionally even unintelligible.

The boundary signals between prosodic units vary across languages and across speaking styles and it is clear that language-specific prosodic cues lead to language specific prosodic strategies [2, 3]. However, it should not be overlooked that prosodic forms that are not typical for a language not only sound unnatural, but they also hinder the reception of the message by the listener.

When considering boundary cues at the level of the prosodic phrase and larger, researchers often focus on the end of the unit and notice two prosodic features. First, it is the phrase-final lengthening, which is considered a prosodic quasi-universal. Second, it is the occurrence of a melodic (F_0) pattern that signals finality, continuation and possibly other information about the ongoing linguistic structure. However, the phrase-final cues are often complemented by phrase-initial signals [11]. Phrase-initial acceleration was attested for Czech in [18] and the increase in phrase-initial F_0 is suggested by findings of [17].

The latter deserves a comment. The regular F_0 movements that mark prosodic word contours tend to decrease throughout prosodic phrases and larger units. This process is called declination and has been found in many languages. It can be demonstrated by regression lines fitted through the intonation contours. The trend itself, however, is not necessarily linear and the scale of the declination reset (pitch up-step at the beginning of a new prosodic unit) can be influenced by both phrase-final lowering and phrase-initial increase in F_0. The material in [17] suggests that, indeed, the first prosodic word often carries a more expanded pitch movement than the following words.

3 Formal Prosodic Structures

The authors of [9, 10, 16] introduced a new formal prosodic model to be used in text-to-speech system *ARTIC* [14] to control the appropriate usage of intonation schemes within the synthesized sentence – the idea was based on the Czech classical phonetic view described in [8]. The proposed prosodic grammar consists of the following alphabet [9]:

- *PS – prosodic sentence* – a prosodic manifestation of a sentence
- *PC – prosodic clause* – a segment of speech delimited by pauses
- *PP – prosodic phrase* – a segment of speech with a certain intonation scheme
- P_0, P_X – *prosodeme* – an abstract unit established in a certain communication function (explained below)
- *PW – prosodic word* – a group of words subordinated to one word accent (stress); it also establishes a rhytmic unit.

According to [10], every prosodic phrase *PP* consists of 2 prosodemes, a *null prosodeme* P_0 and one of *functional prosodemes* which are in our current TTS *ARTIC* [14] reduced to the following types:

- P_1 – prosodeme terminating satisfactorily (last PWs of declarative sent.)
- P_2 – prosodeme terminating unsatisfactorily (last PWs of questions)
- P_3 – prosodeme non-terminating (last PWs in intra-sentence PPs).

3.1 Formal Prosodic Grammar

Now, let us briefly summarize the formal prosodic grammar from [10]. The terminal symbol $ means an inter-sentence pause, # means an intra-sentence pause and w_i stands for a concrete word; Fig. 1a illustrates its usage:

$$PS \rightarrow PC\{1+\} \ \$\{1\} \tag{1}$$
$$PC \rightarrow PP\{1+\} \ \#\{1\} \tag{2}$$
$$PP \rightarrow P_0\{1\} \ P_X\{1\} \tag{3}$$
$$P_0 \rightarrow \emptyset \tag{4}$$
$$P_0 \rightarrow PW\{1+\} \tag{5}$$
$$P_X \rightarrow PW\{1\} \tag{6}$$
$$PW \rightarrow w_i\{1+\} \tag{7}$$

The number in {} parenthesis corresponds to the number of proceeding symbols generated; e.g. $PC\{1+\}$ means that at least one PC is generated.

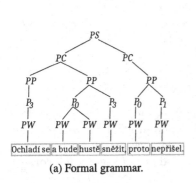

(a) Formal grammar.   (b) Extended formal grammar.

Fig. 1. The illustration of the tree built using the prosodic grammar for the Czech sentence "It will get colder and it will snow heavily, so he did not come".

3.2 Extension of Formal Prosodic Grammar

The grammar presented in Sect. 3.1 postulated that each prosodic phrase could be comprised of some (or none) prosodic words with prosodeme type P_0 and exactly one phrase-final prosodic word with a functional prosodeme P_X; let us note that the null prosodeme P_0 and any functional prosodeme P_X must not be interchangeable. However, it was pointed out in Sect. 2 that the first prosodic word in each prosodic phrase may differ from the other (non-final) prosodic words in the phrase in the sense of prosody. And so we decided to introduce a new prosodeme type $P_{0.1}$ as a special case of P_0, which will describe the first PWs in phrases. Based on that, we suggest to extend the grammar by changing the Eq. (5) by the following rule:

$$P_0 \rightarrow P_{0.1}\{1\}\ PW\{1+\} \tag{5}$$

and by adding 2 additional rules defined by Eqs. (8) and (9):

$$P_{0.1} \rightarrow \emptyset \tag{8}$$
$$P_{0.1} \rightarrow PW\{1\} \tag{9}$$

The application of the extended formal prosodic grammar is shown in Fig. 1b.

The authors decided to introduce the new prosodeme $P_{0.1}$ only as a "subset" of P_0 since this new type differs from the other non-*null* prosodemes. While P_a t the same level cannot be mutually exchanged without a change of sentence meaning, the $P_{0.1}$ prosodeme does not have any certain communication function, it is just related to the sentence beginnings.

4 Formal Prosodic Grammar in TTS

In the unit selection TTS [14], the prosody of the output sentence is partially controlled by *join cost* computation, which should ensure the smoothness in F_0 on the units' transitions, and also by the *target cost* which relies on the formal prosodic grammar and should ensure the keeping of the required meaning. In our TTS, the whole speech corpora (and thus all the units) are aligned with the prosodic structure derived from the grammar described in Sect. 3.1, and the synthesized sentences are described in the same way. During the optimal unit sequence search using Viterbi algorithm [15], the target cost penalizes units with mismatching prosodeme type assigned.

Based on the phonetic background described in Sect. 2, the usage of units newly assigned with $P_{0.1}$ (according to the extended grammar presented in Sect. 3.2), used in the P_0 context, are highly penalized in the modified version of our TTS, while in the opposite direction, the interchange is allowed, but slightly penalized.

4.1 Listening Tests

To verify the contribution of the new prosodeme type, we carried out a 3-scale preference listening test. Based on the methodology in [13], for 4 large professional synthetic voices [14], two male and two female (more than 13 hours each), we synthesized 10,000 shorter simple sentences (up to 5 words[1]) with the baseline system TTS_{base}. The output was analyzed to find sentences with more than

---

[1] Based on authors' knowledge, it is much easier for the listeners to be concentrated and to compare 2 short sentences in the listening test rather then compare 2 long compound sentences.

5 units (diphones) which were newly assigned with the prosodeme $P_{0.1}$ used in P_0 positions, which was the criterion for the selection of examples to evaluate. Sets of 15 sentences for all voices were then randomly selected for the listening test itself and all the sentences were synthesized by TTS_{base} and the modified TTS_{new} with the new prosodeme considered – in total, the test consisted of 60 pairs of samples.

15 listeners participated in the test, 6 of them being speech synthesis experts, 2 of them being phoneticians and the others were naive listeners. They were instructed to use earphones throughout the whole listening test and to judge the overall quality of the synthesized samples plus there were complementary questions. To sum up, for each pair of samples p in the listening test T, the listeners had to choose one of the following choices (*choice box area*):

- *Sample A sounds better.*
- *I can not decide which sample is better.*
- *Sample B sounds better.*

and they could also check any of the following complementary evaluation (*check box area*)

- *The sample contains an unnatural intonation pattern.*
- *The sample contains an unnatural stressed word.*

The answers of listeners in the *choice box area* were then normalized to $p = 1$ for that pairs where the modified version TTS_{new} was preferred, $p = -1$ where TTS_{base} was preferred and $p = 0$ otherwise. These values were used for the final computation of the listening test score s, defined by Eq. (10):

$$s = \frac{\sum_{p \in T} p}{\sum_{p \in T} 1} \tag{10}$$

Thus, the positive value of the score s indicates the improvement of the overall quality when using TTS_{new}.

In a similar way, we could define the formula for counting $s_{intonation}$ and s_{stress} to evaluate the *check box area* answers. These answers where normalized to $p_{base} = 1/p_{new} = 1$ for checked boxes box and $p_{base}/p_{new} = 0$ otherwise.

The scores $s_{intonation}$ and s_{stress} are defined by Eq. (11) as a measure of proportional improvement of the targetted characteristic (intonation or stress) in the output sentences when using TTS_{new} instead of TTS_{base}:

$$s_X = \frac{\sum_{p \in T} p_{base}}{\sum_{p \in T} 1} - \frac{\sum_{p \in T} p_{new}}{\sum_{p \in T} 1} \tag{11}$$

5   Results

The results of listeners' answers are shown in Table 1 which also contains the overall quality evaluation s defined by Eq. (10). In general, the listeners preferred

Table 1. The results on overall quality obtained from listening tests. The table contains the percentage of listener answers and the score values s.

	Male spkr 1	Male spkr 2	Female spkr 1	Female spkr 2	All speakers
TTS_{new} better	56.9%	43.6%	65.8%	55.1%	55.3%
Can not decide	24.9%	32.9%	16.9%	24.4%	24.8%
TTS_{base} better	18.2%	23.6%	17.3%	20.4%	19.9%
Score value s	0.387	0.200	0.484	0.347	0.354

the outputs of modified TTS_{new} and all these results are statistically significant which was confirmed by a *sign test*, similarly e.g. to [6].

The considerable number of *"can not decide"* answers in Table 1 indicates that some first prosodic words in phrases (newly assigned with the prosodeme $P_{0.1}$ instead of P_0) are probably not so different from P_0 words and their units used in non-first words do not cause any audible or distinct disturbing artefact in the synthesis. In these cases, the units used are in synonymy with respect to $P_{0.1}$ and P_0 [16], i.e. they can be used to render both $P_{0.1}$ and P_0 interchangeably.

Table 2 summarizes the results concerning unnatural intonation and stress in the synthesized sentences. Generally, the outputs of TTS_{new} were less often marked to contain something strange, even though the improvement is not so noticeable for all speakers. In any case, the score values are always positive and so they indicate the improvement.

Table 2. The evaluation of two special characteristics in synthesized sentences.

	Male spkr 1	Male spkr 2	Female spkr 1	Female spkr 2	All speakers
TTS_{new}–intonation	8.0%	8.4%	15.1%	13.8%	11.3%
TTS_{base}–intonation	26.2%	13.8%	27.6%	32.4%	25.0%
Score value $s_{intonation}$	0.182	0.053	0.124	0.187	0.137
TTS_{new}–stress	1.8%	5.3%	9.8%	8.4%	6.3%
TTS_{base}–stress	12.9%	11.6%	47.1%	12.0%	20.9%
Score value s_{stress}	0.111	0.062	0.373	0.036	0.146

6   Conclusion and Future Work

The extension of the grammar, based on the phonetic knowledge concerning the importance and specificity of the first prosodic words in phrases, brought encouraging results, as confirmed by listening tests performed on 4 large professional synthetic voices. In a short time, we are planning to release this modification into the publicly available version of our TTS system.

As the further step, we also want to verify the feasibility of $P_0 \rightarrow P_{0.1}$ exchanges, as these are supposed to be homonymous and the grammar has been modified with this regard.

References

1. Christophe, A., Gout, A., Peperkamp, S., Morgan, J.: The elastic phrase: modelling the dynamics of boundary-adjacent lengthening. J. Phon. **31**, 585–598 (2003)
2. Cutler, A., Dahan, D., Donselaar, W.V.: Prosody in the comprehension of spoken language: a literature review. Lang. Speech **40**, 141–201 (1997)
3. Cutler, A., Otake, T.: The elastic phrase: modelling the dynamics of boundary-adjacent lengthening. J. Mem. Lang. **33**, 824–844 (1994)
4. Gee, J., Grosjean, F.: Performance structures: a psycholinguistic appraisal. Cogn. Psychol. **15**, 411–458 (1983)
5. Hanzlíček, Z.: Czech HMM-based speech synthesis. In: Sojka, P., Horák, A., Kopeček, I., Pala, K. (eds.) TSD 2010. LNCS (LNAI), vol. 6231, pp. 291–298. Springer, Heidelberg (2010). https://doi.org/10.1007/978-3-642-15760-8_37
6. Jůzová, M., Tihelka, D., Skarnitzl, R.: Last syllable unit penalization in unit selection TTS. In: Ekštein, K., Matoušek, V. (eds.) TSD 2017. LNCS (LNAI), vol. 10415, pp. 317–325. Springer, Cham (2017). https://doi.org/10.1007/978-3-319-64206-2_36
7. Nooteboom, S.G.: Perceptual goals of speech production. In: Proceedings of the 12th International Congress of Phonetic Sciences, Aix-en-Provence, vol. 1, pp. 107–110 (1991)
8. Palková, Z.: Rytmická výstavba prozaického textu. Studia ČSAV; čís. 13/1974. Academia (1974)
9. Romportl, J.: Structural data-driven prosody model for TTS synthesis. In: Proceedings of the Speech Prosody 2006 Conference, pp. 549–552. TUD Press, Dresden (2006)
10. Romportl, J., Matoušek, J.: Formal prosodic structures and their application in NLP. In: Matoušek, V., Mautner, P., Pavelka, T. (eds.) TSD 2005. LNCS (LNAI), vol. 3658, pp. 371–378. Springer, Heidelberg (2005). https://doi.org/10.1007/11551874_48
11. Saltzman, E., Byrd, D.: The elastic phrase: modelling the dynamics of boundary-adjacent lengthening. J. Phon. **31**, 149–180 (2003)
12. Taylor, P.: Text-to-Speech Synthesis, 1st edn. Cambridge University Press, New York (2009)
13. Tihelka, D., Grůber, M., Hanzlíček, Z.: Robust methodology for TTS enhancement evaluation. In: Habernal, I., Matoušek, V. (eds.) TSD 2013. LNCS (LNAI), vol. 8082, pp. 442–449. Springer, Heidelberg (2013). https://doi.org/10.1007/978-3-642-40585-3_56
14. Tihelka, D., Hanzlíček, Z., Jůzová, M., Vít, J., Matoušek, J., Grůber, M.: Current state of text-to-speech system ARTIC: a decade of research on the field of speech technologies. In: Sojka, P. (ed.) TSD 2018. LNAI, vol. 11107, pp. 369–378. Springer, Cham (2018). https://doi.org/10.1007/978-3-030-00794-2_z

15. Tihelka, D., Kala, J., Matoušek, J.: Enhancements of Viterbi search for fast unit selection synthesis. In: Proceedings of Interspeech 2010, pp. 174–177. ISCA, Makuhari (2010)
16. Tihelka, D., Matoušek, J.: Unit selection and its relation to symbolic prosody: a new approach. In: Proceedings of Interspeech 2006, vol. 1, pp. 2042–2045. ISCA, Bonn (2006)
17. Volín, J.: Extrakce základní hlasové frekvence a intonační gravitace v češtině. Naše řeč **92**(5), 227–239 (2009)
18. Volín, J., Skarnitzl, R.: Temporal downtrends in Czech read speech. In: Proceedings of Interspeech 2007, pp. 442–445. ISCA (2007)

F_0 Post-Stress Rise Trends Consideration in Unit Selection TTS

Markéta Jůzová[1]([⊠]) and Jan Volín[2]

[1] New Technologies for the Information Society and Department of Cybernetics,
Faculty of Applied Sciences, University of West Bohemia, Pilsen, Czech Republic
juzova@kky.zcu.cz
[2] Institute of Phonetics, Faculty of Arts, Charles University,
Prague, Czech Republic
jan.volin@ff.cuni.cz

Abstract. In spoken Czech language, the stress and post-stress syllables in human speech are usually characterized by an increase in fundamental frequency F_0 (except for phrase-final stress groups). In unit selection text-to-speech systems, where no contour of F_0 is generated to be followed, however, the F_0 behaviour is usually tended very vaguely. The paper presents an experiment of making the unit selection TTS to follow the trends of fundamental frequency rise in synthesized speech to achieve higher naturalness and overall quality of speech synthesis itself.

Keywords: Unit selection · Stress and post-stress syllables · Fo rise

1 Introduction

In unit selection speech synthesis, the measurement of F_0 has traditionally been used in concatenation cost to measure the smoothness of units being joined together [23]. Contrary to a hybrid speech synthesis [14,20], where a contour of F_0 (among other characteristics) can directly be followed by the target cost, the behaviour of F_0 is driven only on a "phrase level" when symbolic prosody features (i.e. independent feature formulation – IFF [18]) are used [9,22]. This has the significant advantage of avoiding an artificiality potentially introduced by a F_0 generation model [19], while still keeping the required communication function (i.e. distinguishing e.g. questions from declarative phrases [5]). However, except the concatenation smoothness, there are no fine-grained limits of F_0 behaviour out of the main prosody descriptions, e.g. within P0 prosodeme [15,16]. The breaching of these limits is sometimes manifested in synthetic speech by unnatural dynamic melody or by inappropriate stress perception in a synthesized phrase.

This research was supported by the Czech Science Foundation (GA CR), project No. GA16-04420S, and by the grant of the University of West Bohemia, project No. SGS-2016-039.

P. Sojka et al. (Eds.): TSD 2018, LNAI 11107, pp. 360–368, 2018.
https://doi.org/10.1007/978-3-030-00794-2_39

It has been observed (as described in Sect. 2) that stress groups (i.e. prosodic words) in Czech speech in non-phrase-final position are typically represented by an increase in F_0 – the lower first (stress) syllable is followed by a higher second (post-stress) syllable (i.e., L*+H in ToBI transcription [17], see also below). In the presented paper, the authors suggest to use this knowledge and to control the F_0 behaviour in that way to ensure the F_0 post-stress rise. Hopefully, it should result in an increase of overall quality of the synthesized speech since the TTS will follow the common F_0 behaviour used in human speech, as described in [13,24].

In addition, we join this research with the experiment concerning a duration-related phenomenon – the last syllable units penalization [4] – since first, the former results were very encouraging and second, some outputs of the current study proved audible speech artefacts caused by placing a last syllable unit into a non-last syllable position. Moreover, we want to test the combination of the two studies since, based on our experience, it cannot be assumed that two promising experiments combined together result in an improvement of speech synthesis.

2 Phonetic Background

The Czech language belongs typologically to the so-called *fixed-stress* languages. This means that the lexical stress is consistently attached to a certain syllable whose position is considered with regard to the word boundary (initial, final, etc.) In the case of Czech, it is the first syllable of the word. Early accounts of the acoustic properties of the Czech stress syllables proposed an increased fundamental frequency on their nuclei [1,3]. These accounts were put forward by instrumental phoneticians so it would not be reasonable to doubt their validity.

However, informal observations of the current spoken Czech suggest that a post-stress rise is a preferred option for Czech speakers. An empirical evidence of this is provided, for instance, by [13] or [24]. The former study tested ambiguous syllable chains that differed in their F_0 contours and could have been perceived as one longer or two shorter stress-groups. The perceptual split into two units was caused by a drop of F_0 inside the chain followed by a rise. This means, that users of the Czech language tend to decode a low melodic target followed by an increase in F_0 as a beginning of a new stress-group or a prosodic word. The latter study reports an analysis of 402 stress-groups from continuous news-reading. Only 20% of these stress-groups actually contained the stressed syllable that was melodically higher than the second (i.e., post-stress) syllable and even these were actually quite often two-syllable words found in the phrase-final position in which they signalled a completed statement.

In the internationally known transcription ToBI [17], the melodic peak on a stressed syllable would be marked as H*, while the post-stress rise would be captured as and L*+H. The controversy between the accounts of former phoneticians and the current state deserves an additional comment. The older phoneticians actually admitted that the post-stress rise occurred in Czech intonation. The author of [1] even states that such situation is not rare. To the author of

[3], however, the post-stress rise was only used for emphasis. The recent findings then signal a change in the Czech melodic system or just focus on different speaking styles by the earlier and current phoneticians.

3   Analysis of F_0 Post-stress Rise in Speech Corpora

Based on the phonetic background in Sect. 2, we analysed our speech corpora [20] to verify that on large data. First, the glottal signal was used for the estimation of the fundamental frequency F_0 contour [7,8] for all the sentences. However, we discovered some miss-detected and wrongly detected glottal closure instances (GCI) which caused an incorrect F_0 contour generation. Thus we used *reaper* method [2] in this study to estimate the fundamental frequency. We plan to employ the algorithm in [12] to correct the GCI detection and then we will use the algorithm of F_0 estimation based on GCI computation. From the F_0 contours, mean F_0 values were computed for each speech unit in the corpora and after that, they were used during the analysis.

The authors set the condition for the F_0 post-stress rise of two subsequent speech units u_1, u_2 (phones, diphones) in first two syllables in non-final prosodic words (Fig. 1 in Sect. 4), defined by Eq. (1) (with a 5% tolerance band suggested by phoneticians):

$$meanF_0(u_2) - meanF_0(u_1) \geq -0.05 \cdot meanF_0(u_1) \tag{1}$$

For all corpora used, the analysis showed that more than 80% of all non-final prosodic words meet the defined condition (i.e. they follow the trend of F_0 post-stress rise) and thus confirmed the observation described in Sect. 2.

4   Handling F_0 Post-stress Rise Trends in Our TTS System

As described in Sect. 1, in TTS, the behaviour of F_0 is mainly guarded by join cost which ensures the smoothness of concatenated speech units. Furthermore, the formal prosodic grammar [16,22] and its derived symbolic prosody features are used to indirectly drive F_0 on a phrase level and guarantee the keeping of the required communication function at the phrase-final prosodic words. In non-final prosodic words, there are no other limits set for the F_0 contour, except for selection of units with the required prosodic features (prosodeme type P_0) – see [16] for more details.

The cost in unit selection systems, in general, consists of *target cost* and *join cost* computation, each of which is computed of more features [9]. Traditionally, the target cost ensures the selection of appropriate features and, in our TTS *ARTIC* [20], consists of the following:

– *type of prosodeme* – ensures the required prosody behaviour [16]
– *left and right context* – penalizes disagreements in phonetic contexts [6]

- *prosodic word position* – in our baseline TTS *ARTIC*, evaluates the difference in position within a prosodic word by a non-linear penalization; in [4], we experimented with binary feature associated with last syllable

The join cost ensures, in general, the smooth transition between units in the sequence selected by Viterbi search [21]. It consists of three sub-components listed below [19] and a new feature established to control the F$_0$ post-stress rise in the presented experiment:

- *difference in energy*
- *difference in* F$_0$
- *difference of* 12 *MFCC coefficients*
- newly: *penalization of* F$_0$ *decrease in defined prosodic words positions*

Based on the statements in Sect. 2, verified on our data in Sect. 3, we suggest to control the F$_0$ behaviour in first two syllables (stressed and post-stress) of the non-phrase-final prosodic words in the synthesized sentences – as described in Sect. 3 and illustrated in Fig. 1. The mean F$_0$ values counted for each speech unit in the corpora are used in the latter join cost feature computation: If the two speech units do not meet the condition defined in Eq. (1), their join in highly penalized in the modified TTS. The presented experiment has been carried out only in our scripting interface [20] but the results of the baseline system exactly correspond to the outputs of our commercial TTS *ARTIC*.

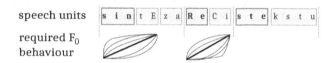

speech units

required F$_0$
behaviour

Fig. 1. The illustration of units (phones) which are controlled by join cost computation in unit selection to follow the F$_0$ rise, presented on the Czech phrase "syntéza řeči z textu" (EN: "text-to-speech synthesis"); the prosodic word boundaries are marked by dotted lines, the syllables are bordered by rectangles and the stressed syllables are highlighted; the F$_0$ contours are only illustrative.

5   Listening Test Overview

To verify the contribution of the compliance of the post-stress rise, we carried out a 3-scale preference listening test. For 4 large professional synthetic voices, two male and two female, we synthesized 10,000 shorter sentences with the baseline system TTS_{base} and analysed them to find sentences where the F$_0$ decreases in the first two syllables of prosodic words (according to Eq. (1)) and we also count the occurrences of last-syllable units in a non-last syllable position. Finally, for each voice, we randomly selected the following numbers of sentences for the listening test itself:

- 10 sentences with at least 5 occurrences of F$_0$ decrease

– 15 sentences with at least 5 occurrences of F_0 decrease and at least 2 occurrences of last-syllable units in other position

All the selected sentences were synthesized by TTS_{base} and the modified TTS_{F_0} or $TTS_{F_0+LastSyl}$ – in total, the test consisted of 100 pairs of samples. The order of samples in the pairs was randomized, so the listeners did not know which sample was synthesized by TTS_{base} and the modified version of TTS.

At this point, let us note that we intentionally combine the current experiment with the last syllable units penalization [4] since some outputs of TTS_{F_0} evince the problems with last syllable units used in a non-final context – which sometimes leads to unnatural lengthening within the synthesized sentence. And we are also verifying the effect of combining of more experiments whether the new added and the changed features do not affect each other negatively.

The participants of the listening test were instructed to use earphones throughout the whole listening test and to judge the overall quality of the synthesized samples – for each pair of samples in the listening test, they had to choose which sample sounded better or whether they could not decide which one was better (*choice box area*). They could also evaluate special characteristics by checking the check boxes if they think that the sample A/B contains an unnatural intonation pattern or an unnatural lenghtening (*check box area*).

Let p denotes the pair of samples and T represents the set of pairs comparing the same systems. The answers of listeners in the *choice box area* can then be normalized to $p = 1$ for that pairs where the modified version TTS_{F_0} or $TTS_{F_0+LastSyl}$ was preferred, $p = -1$ where TTS_{base} was preferred and $p = 0$ otherwise. These values are used for the final computation of the listening test score s, defined by Eq. (2):

$$s = \frac{\sum_{p \in T} p}{\sum_{p \in T} 1} \tag{2}$$

Thus, the positive value of the score s indicates the improvement of the overall quality when using TTS_{F_0}.

In a similar way, we could define the formula for counting $s_{intonation}$ and $s_{lenghtening}$ to evaluate the *check box area* answers. These where normalized to $p_{base} = 1/p_{new} = 1$ for checked boxes and $p_{base}/p_{new} = 0$ for non-checked boxes.

The scores $s_{characteristic}$ are defined by Eq. (3) as a measure of proportional improvement of the focused characteristic (unnatural intonation or lengthening) in the output sentences when using modified TTS instead of TTS_{base}:

$$s_{characteristic} = \frac{\sum_{p \in T} p_{base}}{\sum_{p \in T} 1} - \frac{\sum_{p \in T} p_{new}}{\sum_{p \in T} 1} \tag{3}$$

6   Results

15 listeners participated in the test, 6 of them being speech synthesis experts, 2 of them being phoneticians and the other were naive listeners. The results of listeners' answers are shown in Tables 1 and 2. Since some results were not so

convincing, we also decided to prove the statistical significance of all the results by a *sign test* (similarly to [4], the results are also listed in the tables).

Table 1. The results on overall quality obtained from the listening tests – the comparison of TTS_{base} and TTS_{F_0}.

	Male spkr 1	Male spkr 2	Female spkr 1	Female spkr 2	All speakers
TTS_{F_0}	40.7%	31.3%	39.3%	34.7%	36.5%
can not decide	32.7%	58.7%	32.7%	40.0%	41.0%
TTS_{base}	26.7%	10.0%	28.0%	25.3%	22.5%
score value s	**0.140**	**0.213**	**0.113**	**0.093**	**0.140**
p-value	0.046	<0.001	0.111	0.170	<0.001
better TTS	TTS_{F_0}	TTS_{F_0}	–	–	TTS_{F_0}

Table 2. The results on overall quality obtained from the listening tests – the comparison of TTS_{base} and $TTS_{F_0+LastSyl}$.

	Male spkr 1	Male spkr 2	Female spkr 1	Female spkr 2	All speakers
$TTS_{F_0+LastSyl}$	48.0%	48.4%	43.1%	48.9%	47.1%
can not decide	24.0%	34.7%	25.8%	38.7%	30.8%
TTS_{base}	28.0%	16.9%	31.1%	12.4%	22.1%
score value s	**0.200**	**0.316**	**0.120**	**0.364**	**0.250**
p-value	0.001	<0.001	0.044	<0.000	<0.001
better TTS	$TTS_{F_0+LastSyl}$	$TTS_{F_0+LastSyl}$	$TTS_{F_0+LastSyl}$	$TTS_{F_0+LastSyl}$	$TTS_{F_0+LastSyl}$

As can be seen from Tables 1 and 2, the usage of the modified version of TTS (TTS_{F_0} or $TTS_{F_0+lastSyl}$) results in an increase of the overall quality of synthesized samples compared to the outputs of TTS_{base}. But, although all the score values are positive, they are quite low for some voices in Table 1 and the sign test has also proved that these results are not statistically significant. However, the total numbers counted for all 4 voices in the last column confirm the quality improvement due to F₀ rise adherence. The contribution of the last syllable units penalization is unquestionable, even in the connection with F₀ experiment – the score values in Table 2 are much higher.

The analysis of the several sentences where TTS_{base} were preferred showed that the output of the modified system suffered from an annotation or segmentation error in the source sentences [10,11] – naturally, that sometimes appears also in our baseline TTS, regardless constant corrections of the speech corpora. And there were also some artefacts caused by breaching the last/non-last syllable issue in TTS_{F_0} sentences.

Table 3 summarizes the results concerning unnatural intonation and lengthening in the synthesized sentences. Generally, the outputs of TTS_{new} were less often marked to contain something strange compared to the outputs of TTS_{base}, even the improvement is not so noticeable for all speakers. In any case, the score values are always positive and so they indicate the improvement. Thus, the results confirm the authors' assumption that the following of the F_0 post-stress rising trend could lead to less unnatural and affected melody occurrences in the perception of speech synthesis. In addition, as presented also in [4], the penalizing last-syllable units in non-last positions eliminates the unnatural lengthening in synthesized sentences.

Table 3. The evaluation of two special characteristics in synthesized sentences – unnatural intonation and unnatural lengthening.

	Male spkr 1	Male spkr 2	Female spkr 1	Female spkr 2	All speakers
TTS$_{base}$ vs. **TTS$_{Fo}$**:					
TTS_{new} – intonation	15.3%	7.3%	7.3%	14.0%	11.0%
TTS_{base} – intonation	18.7%	15.3%	11.3%	14.7%	15.0%
$s_{intonation}$	0.033	0.080	0.040	0.007	0.040
TTS_{new} – lengthening	6.7%	5.3%	9.8%	8.4%	6.3%
TTS_{base} – lengthening	10.0%	14.6%	47.1%	12.0%	20.9%
$s_{lenghtening}$	0.089	0.076	0.013	0.040	0.146
TTS$_{base}$ vs. **TTS$_{Fo+LastSyl}$**:					
TTS_{new} – intonation	12.0%	11.6%	16.0%	12.9%	13.1%
TTS_{base} – intonation	20.9%	19.1%	17.3%	16.9%	18.6%
$s_{intonation}$	0.089	0.076	0.013	0.040	0.054
TTS_{new} – lengthening	7.6%	6.7%	22.7%	3.6%	10.1%
TTS_{base} – lengthening	10.7%	20.0%	30.7 %	22.7%	21.0%
$s_{lenghtening}$	0.031	0.133	0.080	0.191	0.109

7   Conclusion

The results of the listening test have confirmed the importance of following the F_0 post-stress rise trend, observed by phoneticians in Czech speech, in the speech synthesis. Also the combining of this experiment with another phenomenon (which contribution to the quality of synthesized sentences was proved recently) has evinced improvements on the overall quality and naturalness.

Facing the problem of excessive limitations to the unit selection TTS by adding a new feature, it is worth highlighting that the new feature which controls the keeping of F_0 rise post-stress trends does not limit the algorithm to search for units with a specific F_0 behaviour, it just penalizes the usage and concatenation of very inappropriate units – similarly to the new positional feature which avoids their use in positions where they are known to cause audible artefacts instead of forcing them into the expected position. By that, we follow the principle of units synonymy/homonymy established in [22].

References

1. Chlumský, J.: Česká kvantita, melodie a přízvuk. Studia ČSAV. Československá akademie věd, Praha (1928)
2. Google: Reaper github. https://github.com/google/REAPER
3. Hála, B.: Rytmická, výstavba prozaického textu. Studia ČSAV. Československá akademie věd, Praha (1962)
4. Jůzová, M., Tihelka, D., Skarnitzl, R.: Last syllable unit penalization in unit selection TTS. In: Ekštein, K., Matoušek, V. (eds.) TSD 2017. LNCS (LNAI), vol. 10415, pp. 317–325. Springer, Cham (2017). https://doi.org/10.1007/978-3-319-64206-2_36
5. Jůzová, M., Tihelka, D., Volín, J.: On the extension of the formal prosody model for TTS. In: Sojka, P., Horák, A., Kopeček, I., Pala, K. (eds.) TSD 2018. LNCS, vol. 11107, pp. 351–359. Springer, Cham (2018)
6. Legát, M.: Impact of phonetic context mismatches on quality of vowel concatenations. In: Proceedings of 2012 IEEE 11th International Conference on Signal Processing, Beijing, China, pp. 523–526 (2012)
7. Legát, M., Matoušek, J., Tihelka, D.: A robust multi-phase pitch-mark detection algorithm. In: Proceedings of Interspeech 2007, pp. 1641–1644 (2007)
8. Legát, M., Matoušek, J., Tihelka, D.: On the detection of pitch marks using a robust multi-phase algorithm. Speech Commun. **53**(4), 552–566 (2011)
9. Matoušek, J., Legát, M.: Is unit selection aware of audible artifacts? In: Proceedings of the 8th Speech Synthesis Workshop SSW 2013, pp. 267–271. ISCA, Barcelona (2013)
10. Matoušek, J., Tihelka, D.: Annotation errors detection in TTS corpora. In: Proceedings of INTERSPEECH 2013, Lyon, France, pp. 1511–1515 (2013)
11. Matoušek, J., Tihelka, D.: Anomaly-based annotation errors detection in TTS corpora. In: Proceedings of INTERSPEECH 2015, Dresden, Germany, pp. 314–318 (2015)
12. Matoušek, J., Tihelka, D.: Classification-based detection of glottal closure instants from speech signals. Proc. INTERSPEECH 2017, 3053–3057 (2017)
13. Palková, Z., Volín, J.: The role of F$_0$ contours in determining foot boundaries in Czech. In: Proceedings of the 15th International Congress of Phonetic Sciences, vol. 2, pp. 1783–1786. UAB & IPA, Barcelona (2011)
14. Qian, Y., Soong, F.K., Yan, Z.J.: A unified trajectory tiling approach to high quality speech rendering. IEEE Trans. Audio Speech Lang. Process. **21**(2), 280–290 (2013)
15. Romportl, J.: Structural data-driven prosody model for TTS synthesis. In: Proceedings of the Speech Prosody 2006 Conference, pp. 549–552. TUD Press, Dresden (2006)
16. Romportl, J., Matoušek, J.: Formal prosodic structures and their application in NLP. In: Matoušek, V., Mautner, P., Pavelka, T. (eds.) TSD 2005. LNCS (LNAI), vol. 3658, pp. 371–378. Springer, Heidelberg (2005). https://doi.org/10.1007/11551874_48
17. Silverman, K.E.A., et al.: ToBI: a standard for labeling English prosody. In: Proceedings of ICSLP 1992, pp. 867–870. ISCA, Banff (1992)
18. Taylor, P.: Text-to-Speech Synthesis, 1st edn. Cambridge University Press, New York (2009)
19. Tihelka, D.: Symbolic prosody driven unit selection for highly natural synthetic speech. In: Proceedings of INTERSPEECH 2005, pp. 2525–2528. ISCA, Bonn (2005)

20. Tihelka, D., Hanzlíček, Z., Jůzová, M., Vít, J., Matoušek, J., Grůber, M.: Current state of text-to-speech system ARTIC: A decade of research on the field of speech technologies. In: Sojka, P., Horák, A., Kopeček, I., Pala, K. (eds.) TSD 2018. LNCS, vol. 11107, pp. 369–378. Springer, Cham (2018)
21. Tihelka, D., Kala, J., Matoušek, J.: Enhancements of Viterbi search for fast unit selection synthesis. In: Proceedings of INTERSPEECH 2010, pp. 174–177. ISCA, Makuhari (2010)
22. Tihelka, D., Matoušek, J.: Unit selection and its relation to symbolic prosody: a new approach. In: Proceedings of INTERSPEECH 2006, vol. 1, pp. 2042–2045. ISCA, Bonn (2006)
23. Tihelka, D., Matoušek, J., Hanzlíček, Z.: Modelling F¡Subscript¿0¡/Subscript¿ dynamics in unit selection based speech synthesis. In: Sojka, P., Horák, A., Kopeček, I., Pala, K. (eds.) TSD 2014. LNCS (LNAI), vol. 8655, pp. 457–464. Springer, Cham (2014). https://doi.org/10.1007/978-3-319-10816-2_55
24. Volín, J.: Z intonace čtených zpravodajství: výška první slabiky v taktu. Čeština doma a ve světě **1–2**, 89–96 (2008)

Current State of Text-to-Speech System ARTIC: A Decade of Research on the Field of Speech Technologies

Daniel Tihelka[1(✉)], Zdeněk Hanzlíček[1], Markéta Jůzová[2], Jakub Vít[2], Jindřich Matoušek[1,2], and Martin Grůber[1]

[1] New Technologies for the Information Society, Faculty of Applied Sciences, University of West Bohemia, Pilsen, Czech Republic
{dtihelka,zhanzlic}@ntis.zcu.cz
[2] Department of Cybernetics, Faculty of Applied Sciences, University of West Bohemia, Pilsen, Czech Republic
{juzova,jvit,jmatouse}@kky.zcu.cz, gruber@ntis.zcu.cz

Abstract. This paper provides a survey of the current state of ARTIC – the modern Czech concatenative corpus-based text-to-speech system. Through more than a decade of research & development in the field of speech technologies and applications, the system was enriched with new languages (and, as a consequence, language-dependent NLP methods), and its speech generation capabilities were significantly improved when new progressive speech generation modules (SPS, DNN, HSS) were (and are still being to) designed and incorporated into it. Also, ARTIC has to deal with various requirements on data used to generate speech from, ranging in size, quality and domain of the output speech, while there always was the requirement to achieve the highest quality in terms of both naturalness and intelligibility. Thus, the paper summarizes some of the most significant achievements and demanding tasks which had to be tackled by the system, illustrating the universality and flexibility of this Czech TTS system.

Keywords: Speech synthesis · Unit selection
Statistical-parametric synthesis · DNN · WaveNet · Hybrid synthesis
Personalized speech synthesis · Voice banking

1 Introduction

The present paper provides a survey of the current state of the text-to-speech (TTS) system ARTIC (Artificial Talker in Czech), presenting the enhancements achieved through more than a decade of its research & development since [21]. About 10 years ago, ARTIC was mostly centered around single unit instance

This research was supported by the Technology Agency of the Czech Republic, project No. TH02010307.

© Springer Nature Switzerland AG 2018
P. Sojka et al. (Eds.): TSD 2018, LNAI 11107, pp. 369–378, 2018.
https://doi.org/10.1007/978-3-030-00794-2_40

(concatenative synthesis) and multiple unit instance (unit selection) synthesis methods, and both are still present within the system with various enhancements. While unit selection is still predominantly used in a wide range of applications, the former is not used very often now. However, new progressive methods such as HMM, hybrid speech synthesis and DNN-based synthesis were experimented with as a part of ARTIC's framework. In addition to new synthesis methods summarized in Sect. 2, there are also several new languages and many new voices added to the system, ranging in their size, purpose and origin, as described in more detail in Sects. 3 and 4.

All the voices and most of the synthesis methods are implemented in a platform-independent way, making it possible to run the system on a wide range of devices and operating systems ranging from Linux/Windows PCs, through smartphones with Android OS [37], to Raspberry Pi class of devices (Apple's OSs support is under the development).

2   Implemented Speech Synthesis Methods

Simultaniously with rapidly changing speech technologies research, we try to keep the TTS system up-to-date with the modern synthesis methods. The following sections present all the production and development methods incorporated into the system or built on the basis of it.

2.1   Single Unit Instance

The *single unit instance* (SUI, also referred as concatenative synthesis [31]) speech synthesis method has been implemented within ARTIC as the very first, and is in detail described in [21]. It uses the clustered-*triphone* as units, with the best candidate for each triphone then being selected. Having single unit instances, their signal needs to be OLA modified to meet the required prosodic characteristics generated by data [27] or HMM-driven prosody generator. Although this method is somewhat outdated in 2018, it is still kept in the synthesizer for its simplicity and low resource consumption (especially CPU power) as a backup solution for extremely resource limited devices.

2.2   Multiple Unit Instance

The method using *multiple unit instances* (MUI) is widely known as *unit selection*. In most of the ARTIC usages, it is the primary choice due to its highly natural-sounding output and relatively high performance on modern hardware. To achieve the high level of naturalness, the method usually needs large corpora [22] recorded by a skilled speaker, see Sect. 3, but we were able to achieve relatively high quality with significantly smaller non-professional speakers [11], as described in Sect. 4.

The selection procedure is driven exclusively by symbolic features [17], i.e. deep level prosody description [27,28] or independent feature formulation (IFF)

[31]. Although such feature treatment is viewed as suffering from data sparsity and being hard to tune, our very first experiments with driving the unit selection by prosody contours, i.e. prosody described on surface level in the same way as used in SUI module, also called acoustic space formulation (ASF) [31], showed lower naturalness than the use of symbolic target features [32]. This ASF features description, however, becomes more popular in case of hybrid speech synthesis, mentioned in Sect. 2.5.

Despite many years of unit selection research, the method still suffers from the occasional occurrences of unnatural glitches/artefacts [15,17,23,38]. However, it is crucial for commercial customers that the errors in acoustic unit inventory (AUI) can continuously be fixed [20,36] and thus the occurrence of the artefacts can be minimised to some extent. Our various comparison tests show that the speech of this method is still being perceived as the most natural [6]. Through the course of time, the method has also been (and still is being) tuned to provide the highest performance possible [4,13,33,35], which is another reason for its success within ARTIC.

2.3   Statistical-Parametrical Synthesis

The main disadvantage of the MUI synthesis is its need for relatively large corpora. Since 2010, the ARTIC was extended with HMM-based speech synthesis method [1], also generally called as *statistical-parametric synthesis* (SPS).

The quality of the speech produced by this method is naturally directly related to the amount of training data. To achieve a "usable" speech quality, SPS needs less training data than MUI. SPS can also work even in cases of very small voice corpus, when some units are missing for MUI and thus it cannot work properly (see Sect. 4). Moreover, it is not always necessary to build a new voice from scratch. Model adaptation techniques can be used to transform models to a new voice [2]. However, MUI generally produces more natural synthetic voices for large speech corpora, since there is the need of a vocoder which, as the last step, converts the trajectories of model parameters into output speech. Let us also note that the trajectories generated by the HMM can also be used in the SUI method (F_0, duration, energy), or we experiment with their use in hybrid speech synthesis (see Sect. 2.5), where we use them (+ spectral features) as ASF-based target features.

Our SPS implementation employs multi-stream hidden semi-Markov models, by default with 5 states. In [3], we have carried out experiments with a phone-dependent variable number of states which showed preference in listening tests and thus this modification is being transferred into the production-ready ARTIC version. In addition, a number of experiments with various analysis/synthesis methods were also performed, including e.g. standard mel cepstral analysis, STRAIGHT, GlottHMM, Ahocoder and WORLD vocoder. The conclusion, however, was that the preferences of particular methods differ, depending on the voice they are used for.

Regarding the system performance, the main advantage of SPS is that the size of units inventory is comparable to the SUI method (contains just statistical models). The overall CPU requirements depend on more factors, e.g. used type of parameters and vocoding, but in general, the runtime is definitely more CPU demanding than the SUI and roughly comparable with the MUI synthesis with the full units inventory.

Despite of continuous tuning of the SPS method, it starts to be obsolete today, since HMMs have been progressively replaced by neural networks [40].

2.4   Deep Neural Networks and WaveNet

WaveNet is a new state-of-art convolutional deep neural network [24]. It is capable to generate directly the raw audio samples, so no speech parameterization or vocoding are necessary in both training and synthesis stages.

Recently, we have performed initial experiments on WaveNet-based speech synthesis [39]. Even though we did not reach the quality produced by our MUI system, the initial results are very promising [6]. The WaveNet synthesis approach is generally recognised as the new-generation TTS method with the potential to outperform all other methods [24,25]. Its main disadvantage, however, is high computational demands – due to the sequential generation of single audio samples, it must run on powerful GPUs, and even though it is very difficult to achieve even close to real-time performance. However, this restriction seems to be overcome by the recently introduced parallel WaveNet [25], still needing very powerful GPUs, though.

Regarding the output speech quality, in [6] we have carried out a MUSHRA test [8] where the same utterances synthesized by MUI, SPS and WaveNet methods were mutually compared to the reference natural version of that utterance. The summary of the results is shown in Table 1.

Table 1. The results of the recent MUSHRA test comparing the overall quality of speech produced by the individual ARTIC modules. Taken from [6].

	Natural	MUI	SPS	WaveNet
Mean±Std	99.9±0.6	88.4±16.5	57.4±23.1	73.4±20.9
Median	100	93	62	80

2.5   Hybrid Speech Synthesis

As the MUI suffers from the artefacts, "raw" SPS synthesis from feature flattening and vocoder imperfection, and the DNN requires powerful hardware to run on, there is research interest in *hybrid speech synthesis* (HSS). This tries to combine the advantages of HMM or DNN acoustic features generation which can then drive the unit selection [26,34]. There is the expectation that the smooth ASF-based target specification will help to minimize the occurrences of unnatural artefacts, using the waveform concatenation to build the speech with high level of naturalness (avoiding any type of vocoder).

As the research on the field of hybrid synthesis is still ongoing, we do not have any definite conclusions to report now. However, very informally, the first results suggest that the output is not worse than is the output of the MUI module. Of course, deeper analysis is in progress.

2.6   Scripting Interface

The ARTIC TTS can be built with a special layer making it possible to override the behaviour of its individual modules in Python language. As the override of a particular module brings some, sometimes significant, overhead (the layer itself does not add almost any overhead at all), the primary purpose the this interface is to simplify the experimenting with the system allowing to re-implement only the part the researcher is interested in and rely on default implementations of the other parts of the system.

At the time of writing, the scripting is supported for unit selection module and sound builder module. The former allows us to override the computation of individual costs, replace the selection algorithm (e.g. in [13]) or to run-time exclude particular units from the selection, while the latter allows to modify the signal of the individual units before the output speech is created. As the scripting layer is relatively simple to add, the other modules can be extended by the layer when it will be required for them.

3   Commercial Voices in ARTIC

As the generic purpose TTS system, there are several large voices available in the ARTIC. All were recorded by professional or semi-professional speakers (radio

Table 2. Available commercial synthetic voices. Live demo for most of them is available at http://www.speechtech.cz/speechtech-text-to-speech/speechtech-tts-online-demo/.

Speaker ID	Lang.	Corpus		Methods used			
		Sentences	Time	MUI	SUI	HMM	DNN
♂ Jan	CZ	12,241	14.7 h	✓	✓	✓	✓
♂ Stanislav	CZ	12,150	15.7 h	✓	✓	✓	✓
♂ Jiří	CZ	12,487	14.1 h	✓	✓	✓	✓
♀ Iva	CZ	12,151	15.2 h	✓	✓	✓	✓
♀ Kateřina	CZ	12,707	13.0 h	✓	✓	✓	✓
♀ Radka	CZ	12,136	16.6 h	✓	✓	✓	✓
♀ Alena	CZ	4,999	10.3 h	✓			
♀ Melánie	SK	11,982	18.4 h	✓	✓	✓	
♀ Olga	RU	12,580	18.4 h	✓			
♂ Jeremi	EN_GB	10,924	13.6 h	✓			✓
♂ Chris	EN_US	6,669	5.5 h	✓			
♂	HY	10,254	20.7 h	*In processing*			

newsreaders or presenters) and are available for customers willing to use speech technologies in their products. All the voices are available for PC (Linux, Windows) and when usage-occurrence-based reduction [4] is carried out, for Android phones and other resource-limited devices as well (Table 2).

4   Non Commercial ARTIC Involvements

Since the primary purpose of the ARTIC synthesizer is speech technologies research, it has been involved in many beneficial projects, e.g. [7, 14, 16, 29, 41]. In most of them however, its role was only to support speech generation using the professional voices listed in Sect. 3. In this section, though, we emphasize only these with considerable impact on the system internals.

4.1   Personalized Speech Synthesis

Approximately since 2011, we have been recording people facing total laryngectomy or other fatal diseases able to cause the loss of their voices [5], in order to offer *personalised* speech synthesizer which these people can use in their portable devices as an alternative way of communication (and some of them really do).

This required the change of several paradigms, for example the way of text selection. As we do not know how many sentences patients will be able to, or be willing to, record, a special balanced text corpus was designed for this purpose [11]. Also, the recordings are usually of a lower quality since these people are not professional speakers and speaking for a long time is sometimes painful for them.

Nowadays we are able to prepare a SPS voice for everyone who records about 200 sentences. From about 500 recorded sentences we are able to employ the MUI module with reasonable output quality (given the quality of the source recordings); see Table 3 for the summary of personalized synthetic voices built so far. Naturally, the quality of the generated speech depends very much on the number of the recorded sentences and also on the quality of the patient's voice and the consistency of the recordings.

The ultimate goal of the ongoing LARYNGO-Voice project (see https:// www.certicon.cz/index.php/laryngo-voice/) is the full automation of the voice building process [12] since the preparation of voices still requires some non-negligible amount of manual work, not being error-prone [18,19]. At the end, however, the whole system will be available online for anyone who desires to make a "backup" of his/her voice together with personalized synthesizer speaking with his/her voice.

4.2   Foreign Accent Voices for Air-Traffic Control Training

During the project IT-BLP (see itblp.zcu.cz), we have faced the problem of synthesizing English voices with foreign accents. These have been used within both the automatic air-traffic control simulator and training tool [30] to speak

Table 3. Personalized speech synthetic voices with sizes (and quality) very dependent on speakers health status and speaking capabilities. The examples of the original and synthetic voices can be listened to at https://goo.gl/XoUPqA.

Speaker ID	Corpus		Methods used			Speaker ID	Corpus		Methods used		
	Sents.	Time	MUI	SUI	HMM		Sents.	Time	MUI	SUI	HMM
♀ P01	3,500	3.8 h	✓	✓	✓	♂ P14	700	1.2 h		✓	✓
♀ P02	3,480	2.3 h	✓		✓	♂ P15	684	38 min	✓	✓	✓
♂ P03	2,099	1.7 h	✓		✓	♀ P16	683	53 min	✓	✓	✓
♀ P04	2,016	2.3 h	✓		✓	♂ P17	652	39 min	✓		✓
♂ P05	2,014	1.9 h	✓	✓	✓	♂ P18	555	35 min	✓	✓	✓
♂ P06	1,895	2.1 h	✓	✓	✓	♂ P19	477	35 min	✓		✓
♂ P07	1,800	2.1 h	✓	✓	✓	♀ P20	473	32 min			✓
♂ P08	1,431	1.7 h	✓		✓	♂ P21	469	46 min	✓	✓	✓
♀ P09	1,101	1.1 h	✓	✓	✓	♂ P22	403	26 min	✓	✓	✓
♂ P10	1,049	2.5 h	✓		✓	♂ P23	350	16 min		✓	✓
♂ P11	1,038	1.0 h	✓		✓	♂ P24	300	17 min		✓	✓
♂ P12	856	1.3 h	✓	✓	✓	♂ P25	230	15 min		✓	✓
♂ P13	769	41 min	✓	✓	✓	♂ P26	210	15 min		✓	✓

as various pilots of simulated planes, replacing a person playing pilots of several planes simultaneously and usually speaking English with Czech accent only. In addition, a special noising module was added to the system, post-processing the output speech by adding various background noises, signal distortions, cracklings and other effects making the communication much more realistic (Table 4).

Table 4. IT-BLP synthetic voices recorded by skilled speakers; designed to work exclusively in an air traffic control domain, and are thus rather lower in size. Online examples can be listened to at http://itblp.zcu.cz/tts (*note the domain limitation!*)

Speaker ID	Accent	Corpus time	Speaker ID	Accent	Corpus time
♂ Jeremy	US	38.5 min	♂ Tsu-Keng	TW	57.1 min
♂ Bill	US	48.1 min	♂ John	CA	50.0 min
♀ Stephanie	DE	2.6 h	♂ Kumar	IN	54.6 min
♂ Thomasz	PL	2.3 h	♂ Milivoje	SR	1.9 h
♂ Adam	CZ	1.7 h	♂ Matthias	FR	2.0 h

Due to a somewhat special nature of the air-traffic communication, we have developed new algorithm for the building of text corpus used to record speech synthesis voice, which is tied to the domain limited in one or another way [9,10]. Moreover, also the letter-to-sound front-end (LTS) has to be adjusted to reflect voice-dependent pronunciation in order to convert common input texts into the given style of the voice chosen to render them.

5   Conclusion and Future Work

In our future work we will continuously aim at enhancing the quality of the synthetic speech produced by our TTS system, both by improving the existing methods and by implementing new progressive methods, such as the use of DNN.

In spite of the intensly researched modern methods with high theoretical potential, the highest quality is still being achieved with the MUI (unit selection) synthesis approach. Its other advantage is the relative simplicity of implementation, not requiring any special hardware to run on. In addition, even the MUI synthesis still has a potential for improvement, whether the hybrid approach or the consideration of units synonymy/homonymy principle (not handled satisfactory so far). One of the biggest research & development tasks is also the full automation of the voice build process, starting from the recordings and ending with the fully workable acoustics units inventory, which is being worked on within the LARYNGO-Voice project.

Over the years, the ARTIC TTS system has shown its value through its universality, platform independency and per-purpose tunability.

References

1. Hanzlíček, Z.: Czech HMM-based speech synthesis. In: Sojka, P., Horák, A., Kopeček, I., Pala, K. (eds.) TSD 2010. LNCS (LNAI), vol. 6231, pp. 291–298. Springer, Heidelberg (2010). https://doi.org/10.1007/978-3-642-15760-8_37
2. Hanzlíček, Z.: Czech HMM-based speech synthesis: experiments with model adaptation. In: Habernal, I., Matoušek, V. (eds.) TSD 2011. LNCS (LNAI), vol. 6836, pp. 107–114. Springer, Heidelberg (2011). https://doi.org/10.1007/978-3-642-23538-2_14
3. Hanzlíček, Z.: Optimal Number of States in HMM-Based Speech Synthesis. In: Ekštein, K., Matoušek, V. (eds.) TSD 2017. LNCS (LNAI), vol. 10415, pp. 353–361. Springer, Cham (2017). https://doi.org/10.1007/978-3-319-64206-2_40
4. Hanzlíček, Z., Matoušek, J., Tihelka, D.: Experiments on reducing footprint of unit selection TTS system. In: Habernal, I., Matoušek, V. (eds.) TSD 2013. LNCS (LNAI), vol. 8082, pp. 249–256. Springer, Heidelberg (2013). https://doi.org/10.1007/978-3-642-40585-3_32
5. Hanzlíček, Z., Romportl, J., Matoušek, J.: Voice conservation: towards creating a speech-aid system for total laryngectomees. In: Kelemen, J., Romportl, J., Zackova, E. (eds.) Beyond Artificial Intelligence. TIEI, vol. 4, pp. 203–212. Springer, Heidelberg (2012). https://doi.org/10.1007/978-3-642-34422-0_14
6. Hanzlíček, Z., Vít, J., Tihelka, D.: WaveNet-based speech synthesis applied to Czech: a comparison with the traditional synthesis methods. In: Sojka, P., Horák, A., Kopeček, I., Pala, K. (eds.) TSD 2018. LNAI, vol. 11107, pp. 445–452. Springer, Cham (2018)
7. Ircing, P., Romportl, J., Loose, Z.: Audiovisual interface for Czech spoken dialogue system. In: Proceedings of ICSP 2010, pp. 526–529. IEEE, Beijing (2010)
8. ITU Recommendation BS.1534-2: Method for the subjective assessment of intermediate quality level of coding systems. Technical report, International Telecommunication Union (2014)

9. Jůzová, M., Tihelka, D.: Minimum text corpus selection for limited domain speech synthesis. In: Sojka, P., Horák, A., Kopeček, I., Pala, K. (eds.) TSD 2014. LNCS (LNAI), vol. 8655, pp. 398–407. Springer, Cham (2014). https://doi.org/10.1007/978-3-319-10816-2_48

10. Jůzová, M., Tihelka, D.: Tuning limited domain speech synthesis using general text-to-speech system. In: Sojka, P., Horák, A., Kopeček, I., Pala, K. (eds.) TSD 2014. LNCS (LNAI), vol. 8655, pp. 408–415. Springer, Cham (2014). https://doi.org/10.1007/978-3-319-10816-2_49

11. Jůzová, M., Tihelka, D., Matoušek, J.: Designing high-coverage multi-level text corpus for non-professional-voice conservation. In: Ronzhin, A., Potapova, R., Németh, G. (eds.) SPECOM 2016. LNCS (LNAI), vol. 9811, pp. 207–215. Springer, Cham (2016). https://doi.org/10.1007/978-3-319-43958-7_24

12. Jůzová, M., Tihelka, D., Matoušek, J., Hanzlíček, Z.: Voice conservation and TTS system for people facing total laryngectomy. In: Proceedings of Interspeech 2017, pp. 3425–3426. ISCA, Stockholm (2017)

13. Kala, J., Matoušek, J.: Very fast unit selection using Viterbi search with zero-concatenation-cost chains. In: Proceedings of ICASSP 2014, pp. 2569–2573. IEEE, Florence (2014)

14. Krňoul, Z., Železný, M.: A development of Czech talking head. In: Proceedings of Interspeech (ICSLP) 2008, Brisbane, Australia, pp. 2326–2329 (2008)

15. Legát, M., Matoušek, J.: Pitch contours as predictors of audible concatenation artifacts. In: Proceedings of WCECS 2011, San Francisco, USA, pp. 525–529 (2011)

16. Matoušek, J., Hanzlíček, Z., Campr, M., Krňoul, Z., Campr, P., Grůber, M.: Web-based system for automatic reading of technical documents for vision impaired students. In: Habernal, I., Matoušek, V. (eds.) TSD 2011. LNCS (LNAI), vol. 6836, pp. 364–371. Springer, Heidelberg (2011). https://doi.org/10.1007/978-3-642-23538-2_46

17. Matoušek, J., Legát, M.: Is unit selection aware of audible artifacts? In: Proceedings of SSW8, ISCA, Barcelona, pp. 267–271 (2013)

18. Matoušek, J., Romportl, J.: Recording and annotation of speech corpus for Czech unit selection speech synthesis. In: Matoušek, V., Mautner, P. (eds.) TSD 2007. LNCS (LNAI), vol. 4629, pp. 326–333. Springer, Heidelberg (2007). https://doi.org/10.1007/978-3-540-74628-7_43

19. Matoušek, J., Tihelka, D.: Annotation errors detection in TTS corpora. In: Proceedings of Interspeech 2013, pp. 1511–1515. ISCA, Lyon (2013)

20. Matoušek, J., Tihelka, D.: Voting detector: a combination of anomaly detectors to reveal annotation errors in TTS corpora. In: Proceedings of Interspeech 2016, pp. 1560–1564. ISCA, San Francisco (2016)

21. Matoušek, J., Tihelka, D., Romportl, J.: Current state of czech text-to-speech system ARTIC. In: Sojka, P., Kopeček, I., Pala, K. (eds.) TSD 2006. LNCS (LNAI), vol. 4188, pp. 439–446. Springer, Heidelberg (2006). https://doi.org/10.1007/11846406_55

22. Matoušek, J., Tihelka, D., Romportl, J.: Building of a speech corpus optimised for unit selection TTS synthesis. In: Proceedings of LREC 2008, pp. 1296–1299. ELRA, Marrakech (2008)

23. Matoušek, J., Tihelka, D., Šmídl, L.: On the impact of annotation errors on unit-selection speech synthesis. In: Sojka, P., Horák, A., Kopeček, I., Pala, K. (eds.) TSD 2012. LNCS (LNAI), vol. 7499, pp. 456–463. Springer, Heidelberg (2012). https://doi.org/10.1007/978-3-642-32790-2_55

24. van den Oord, A., et al.: WaveNet: a generative model for raw audio. CoRR abs/1609.03499 (2016)

25. van den Oord, A., et al.: Parallel WaveNet: fast high-fidelity speech synthesis. CoRR abs/1711.10433 (2017)
26. Qian, Y., Soong, F.K., Yan, Z.J.: A unified trajectory tiling approach to high quality speech rendering. IEEE Trans. Audio Speech Lang. Process. **21**(2), 280–290 (2013)
27. Romportl, J.: Structural data-driven prosody model for TTS synthesis. In: Proceedings of the Speech Prosody 2006, pp. 549–552. TUDpress, Dresden (2006)
28. Romportl, J., Matoušek, J.: Formal prosodic structures and their application in NLP. In: Matoušek, V., Mautner, P., Pavelka, T. (eds.) TSD 2005. LNCS (LNAI), vol. 3658, pp. 371–378. Springer, Heidelberg (2005). https://doi.org/10.1007/11551874_48
29. Romportl, J., Zovato, E., Santos, R., Ircing, P., Relaño, J.G., Danieli, M.: Application of expressive TTS synthesis in an advanced ECA system. In: Proceedings of SSW7, pp. 120–125. ISCA, Kyoto (2010)
30. Stanislav, P., Šmídl, L., Švec, J.: An automatic training tool for air traffic control training. In: Proceedings of Interspeech 2016, pp. 782–783. ISCA, San Francisco (2016)
31. Taylor, P.: Text-to-Speech Synthesis, 1st edn. Cambridge University Press, New York (2009)
32. Tihelka, D.: Symbolic prosody driven unit selection for highly natural synthetic speech. In: Proceedings of Interspeech 2005 - Eurospeech, pp. 2525–2528. ISCA, Lisboa (2005)
33. Tihelka, D., Grůber, M., Hanzlíček, Z.: Robust methodology for TTS enhancement evaluation. In: Habernal, I., Matoušek, V. (eds.) TSD 2013. LNCS (LNAI), vol. 8082, pp. 442–449. Springer, Heidelberg (2013). https://doi.org/10.1007/978-3-642-40585-3_56
34. Tihelka, D., Hanzlíček, Z., Jůzová, M., Matoušek, J.: First steps towards hybrid speech synthesis in Czech TTS system ARTIC. In: SPECOM 2018 (2018, submitted for review)
35. Tihelka, D., Kala, J., Matoušek, J.: Enhancements of Viterbi search for fast unit selection synthesis. In: Proceedings of Interspeech 2010, pp. 174–177. ISCA, Makuhari (2010)
36. Tihelka, D., Matoušek, J., Kala, J.: Quality deterioration factors in unit selection speech synthesis. In: Matoušek, V., Mautner, P. (eds.) TSD 2007. LNCS (LNAI), vol. 4629, pp. 508–515. Springer, Heidelberg (2007). https://doi.org/10.1007/978-3-540-74628-7_66
37. Tihelka, D., Stanislav, P.: ARTIC for assistive technologies: transformation to resource-limited hardware. In: Proceedings of WCECS 2011, pp. 581–584. IANG, San Francisco (2011)
38. Vít, J., Matoušek, J.: Concatenation artifact detection trained from listeners evaluations. In: Habernal, I., Matoušek, V. (eds.) TSD 2013. LNCS (LNAI), vol. 8082, pp. 169–176. Springer, Heidelberg (2013). https://doi.org/10.1007/978-3-642-40585-3_22
39. Vít, J., Matoušek, J.: On the analysis of training data for WaveNet-based speech synthesis. In: Proceedings of ICASSP 2018, IEEE, Calgary (2018)
40. Zen, H.: Acoustic modeling in statistical parametric speech synthesis - from HMM to LSTM-RNN. In: Proceedings of MLSLP (2015, invited paper)
41. Železný, M., Krňoul, Z., Císař, P., Matoušek, J.: Design, implementation and evaluation of the Czech realistic audio-visual speech synthesis. Sig. Process. **12**, 3657–3673 (2006)

Semantic Role Labeling of Speech Transcripts Without Sentence Boundaries

Niraj Shrestha[✉] and Marie-Francine Moens

Department of Computer Science, KU Leuven, Leuven, Belgium
{niraj.shrestha,marie-francine.moens}@cs.kuleuven.be,
http://liir.cs.kuleuven.be

Abstract. Speech data is an extremely rich and important source of information. However, we lack suitable methods for the semantic annotation of speech data. For instance, semantic role labeling (SRL) of speech that has been transcribed by an automated speech recognition (ASR) system is still an unsolved problem. SRL of ASR data is difficult and complex due to the absence of sentence boundaries, punctuation, grammar errors, words that are wrongly transcribed, and word deletions and insertions. In this paper we propose a novel approach to SRL of ASR data based on the following idea: (1) train the SRL system on data segmented into frames, where each frame consists of a predicate and its semantic roles without considering sentence boundaries; (2) label it with the semantics of PropBank roles; and to assist the above (3) train a part-of-speech (POS) tagger to work on noisy and error prone ASR data. Experiments with the OntoNotes corpus show improvements compared to the state-of-the-art SRL applied on ASR data.

Keywords: Frame semantics · Speech

1 Introduction

In natural language processing semantic role labeling (SRL) regards the task of recognizing the basic event structure of a sentence. More specifically, events and their argument roles are identified, i.e., recognizing the "who" "does what", "to whom" or "to what", "where", "when" and "how". SRL is considered as an essential task in natural language understanding and often is integrated in question answering and human computer interaction systems. SRL is commonly applied on written text, and not on speech. The reason might be that current semantic role labelers require as input well-formed sentences, which are lacking in speech data.

We propose a semantic role labeler which is trained on annotated semantic frames instead of on annotated sentences. A semantic frame is defined as a realization of a predicate and its arguments roles. We show that the frame based semantic role labeler performs equally well on written data as a sentence based

© Springer Nature Switzerland AG 2018
P. Sojka et al. (Eds.): TSD 2018, LNAI 11107, pp. 379–387, 2018.
https://doi.org/10.1007/978-3-030-00794-2_41

SRL and has a better performance than sentence based SRL when evaluated on the speech data of the OntoNotes corpus.

We report the following contributions. First, we define the new task of semantic role labeling of speech data. Second, we propose a semantic frame based SRL which successfully recognizes events and their semantic arguments in language data that lack sentence boundaries. Third, our work is a first step towards conveying a new paradigm for semantic recognitions, which in this case are centered around event structures, instead of well formed sentences. The paper is organized as follows. First, we discuss the related work, which is followed by a methodology section. After that we describe the experimental setup followed by a discussion of the results obtained.

2   Related Work

Although SRL of written text is well-established (e.g., Illinois SRL: [1], Lund University SRL: [2]) reaching F_1 scores higher than 85% [3] on a standard collection such as the CoNLL dataset, SRL of (transcribed) speech data is very limited, perhaps due to the non-availability of benchmarking corpora. Nevertheless, several authors have stressed the importance of semantic role labeling of speech data, for instance, in the frame of question answering speech interfaces (e.g., [4,5]), speech understanding by robots (e.g., [6]), and speech understanding in general [7]. Favre [8] developed a system for joint dependency parsing and SRL of transcribed speech data in order to be able to handle speech recognition output with word errors and sentence segmentation errors. He uses an unpublished classifier for segmenting the sentences trained on sentence-split ASR data taking into account sentence parse information, lexical features and audio features such as pause duration. The performance of semantic role labelers drops significantly when applied on ASR data of broadcast news and broadcast conversation of OntoNotes release 3 [9] due to transcription errors and lack of sentence boundaries. [8] report a F_1 score decrease to 50.76%. A decrease in performance is also reported when performing SRL on non-well formed texts such as tweets [10].

3   Methodology

A main objective of this work is to identify suitable ASR segments that represent a predicate with its PropBank style arguments which together form a semantic frame. This entails generating candidate frame segments and labeling these with their semantic roles.

3.1   Preprocessing

The transcribed speech data[1] is already tokenized. The performance of a state-of-the-art part of speech (POS) tagger trained on written text substantially

---

[1] The transcribed corpus is provided by [8] with the consent of SRI (http://www.sri.com).

```
[ARGM-MNR much  better] [TARGET looking] [ARG0 News  Night] .

[ARG0 I] [ARGM-MOD might] [TARGET add] .

[ARG1 Paula Zahn] [TARGET sits] in [ARG2 for Anderson and Aaron] .
```

Fig. 1. Frame segments of the example sentence: "A much better looking News Night I might add as Paula Zahn sits in for Anderson and Aaron".

degrades in performance due to the noisy nature of speech data and speech transcription errors. We have retrained the Stanford POS tagger on the *WSJ* subset of the OntoNotes data. This data is converted into lowercase characters and also sentence punctuation is deleted. Additionally, we have removed some random number of tokens from the left and right parts of a sentence in order to simulate improper starts and endings of candidate frame segments. The result is a POS tagger adapted to lowercase data and to language fragments, which we refer to as the **POS-Speech**. This model is applied on various data segmentations of the speech data.

3.2   Generating Candidate Frame Segments

The candidate frame segments from the ASR data are generated by taking a window of a fixed size, and moving it word-by-word. Considering each segment as a candidate frame segment, the SRL system identifies predicates as verbs detected by our POS-Speech model. For each predicate, we generate a candidate frame segment by expanding its left and right context by k tokens where k starts from 5 to 20. We call this model **LR-n**. A frame segment is a realization of a predicate and its argument roles. If a sentence contains a single predicate then there will be one frame segment and, if there are more than one predicate, then there will be an equal number of frame segments as there are predicates.

The generated segments often have wrong start and ending words. We use written data that is annotated with ground truth semantic role labels to train a part-of-speech tag language model (POS-LM). A frame segment for each predicate argument structure starts from the first occurrence of its argument role to the last occurrence of its argument role. For example, for the predicate *sits*, its frame segment starts from [*ARG1 Paula Zahn*] and ends at [*ARG2 for Anderson and Aaron*]. The purpose of the POS-LM model is to learn valid start and end POS patterns of frame segments from written text. We apply the Stanford POS tagger on a large written text corpus (in our case OntoNotes). We then split the POS-tagged text into frame segments and learn the POS-LM that represents the start and end POS tags of a frame segment. We learn the probability of trigrams of consecutive POS tags that form the start or ending of the frames using the *SRILM* [11] language model tool.

We then remove potential invalid starts or endings of the candidate frame segments in the test data with the trained part-of-speech language model (POS-LM). If a segment does not start or end with a valid 3-gram POS pattern, we remove a token at the respectively start or end of the segment and again check for a valid start or end POS pattern. This word pruning process is repeated until a valid start and end POS pattern of the segment is obtained. We call this method **LR-n-pruned**. The frame segments for the given sentence are shown in Fig. 1.

3.3   Predicate Recognition and Semantic Role Labeling

We train the *nlpnet* SRL tool [12] on the English *WSJ* corpus from OntoNotes release 3. Before training, we import the word embeddings from the SENNA tool, which is trained on the entire English Wikipedia. We prepare two different dataset formats of the OntoNotes training corpus to train the *nlpnet* tool and generate two corresponding SRL models.

Sentence model (SM): The *nlpnet* tool is trained with sentence split text, where each sentence is annotated with its predicate-argument structure.

Frame segment model (FM): The *nlpnet* tool is trained with frame segmented text, where each segment is annotated with its predicate-argument structure [1].

4   Dataset and Experimental Setup

4.1   Dataset for POS-Speech

Train Data: We have used the *WSJ* subset from the OntoNotes corpus [9] consisting of 1,555 files and 32,956 sentences to train the Stanford POS tagger after removing sentence boundaries and lowercasing its words.

Test Data: To test the **POS-Speech** model, we use 100 sentences from the CoNLL 2000 test set[2] referred to as *100-sent*. We also created a kind of noisy data from these 100 sentences by removing a random number of tokens from the left and right side of the sentences to make noisy data as discussed above. To test the POS tagger model in a domain different from the CoNLL test data, we also make a small test data set. We take the first 200 sentences from the *cnn_0008.txt* file from the */bc/cnn* subset data from OntoNotes. We refer to this data as *bc-cnn-200*. We compare our **POS-Speech** against the Stanford standard POS tagger *100-sent-noisy* which is trained on written text and also the Stanford POS tagger trained on the lowercase data.

4.2   Dataset for SRL Training

Train Data: The *WSJ*[3] subset data from the OntoNotes corpus [9] is used to train the *nlpnet* SRL tool with the two different data models, viz. **SM**, and **FM**. These data are annotated in constituent role form.

Test Data: As test data we use the data set described in [13] which is a subset of the OntoNotes corpus containing English broadcast and conversation news BN/BC^4 and which also contains speech data annotated with ground-truth semantic roles (total of 52181 predicates). There are two variants of the test dataset, one regards the well-formed manual transcription of the speech data and the other regards the output of the automated speech recognition (ASR). For speech data, we use different methods for segmentation as described above (**LR-n** and **LR-n-pruned**) to test the performance of the systems. In our experiments we also use a variant of the ASR dataset that is already segmented into sentences using audio information [8] (**ASR-sent**). We have also tested the performance of the SRL systems on two variants of written data. The first one is in sentence form, which we call **written-sent**, the second is in frame segment format called **written-FrameSeg**.

4.3 Evaluation

We define the following evaluation settings:

- *pEval*: Evaluation of the generated predicates. After the predicate identification step, we use these predicates in the subsequent evaluations.
- *paEval*: Evaluation of a predicate with each argument type.
- *cpaEval*: Evaluation of predicate-argument pairs, but now given a gold labeled correct predicate.
- *pafEval*: Evaluation of predicate-argument full frames.
- *cpafEval*: Evaluation of predicate-argument full frames, but now given a gold labeled correct predicate.

4.4 Evaluation Metrics

We evaluate the system performance in terms of precision (P), recall (R) and F_1 scores in two different settings. One is *strict evaluation* for **written-sent**, **written-FrameSeg**, and **ASR-sent** where we evaluate the exact match between system generated labels and ground truth labels. Another one is *lenient evaluation* for the different variations of the segmented data models, which are **LR-n** and **LR-n-pruned**, where we evaluate the system and ground truth labels based on the similarity of the labeled texts. In both evaluation settings, semantic role arguments can be composed of several words, some of which might be missed or wrongly transcribed in the ASR data. In the lenient evaluation, we compute the cosine string similarity between the two vectors representing the text spans of an argument generated by the system and the ground truth:

$$SimiWeight(t_{si}, t_{gj}) = \frac{\overrightarrow{v}(t_{si}).\overrightarrow{v}(t_{gj})}{|\overrightarrow{v}(t_{si})||\overrightarrow{v}(t_{gj})|}$$

---

[4] The subsets of the corpus that we use in this work are: bc/cnn, bc/msnbc, bn/abc, bn/cnn, bn/mnb, bn/nbc, bn/pri, bn/voa as done in [8].

where t_{si} and t_{gj} are the system generated and ground truth text spans respectively of an argument type. This similarity is used as a weight in computing the counts of correct system generated answers when evaluating recall and precision. The dimensions of the vector is the size of the vocabulary.

5   Results and Discussion

5.1   POS-Speech on Lowercase and Noisy Data

We run our **POS-Speech** model, the standard Stanford POS tagger (**SS-POS**) and the Stanford POS tagger trained on lowercase data (**SL-POS**) on *100-sent, 100-sent-noisy* and *bc-cnn-200* data. As claimed by [14], the accuracy of the standard Stanford POS tagger on written text is 97.3%. When the same POS tagger runs on the *100-sent* data, the performance degrades by nearly 10% as shown in Table 1. But when the same POS tagger trained on lowercase data (**SL-POS**) runs on the *100-sent* data, the performance is similar as on the written data. Our **POS-Speech** model performs as good as the Stanford

SL-POS model on the *100-sent*. When **SL-POS** is applied on a different domain i.e., on *bc-cnn-200*, it performance drops by 4.61%. But when our **POS-Speech** is applied on this new domain, the performance only drops by 2.23% as shown in Table 1. This is because of **POS-Speech** is trained on noisy lowercase data. It is also observed that **POS-Speech** performs marginally better than **SL-POS** on this new domain, making **POS-Speech** a good choice for POS tagging of the speech data segments.

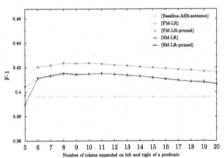

Fig. 2. F_1 of the *paEval* evaluation applied on **LR-n** data obtained by different SRL models.

5.2   SRL on Written Data

The **SM** model trained on written sentences and its **FM** variant trained on frame segments have a similar performance when evaluated on written data which is on par with the performance of state-of-the-art SRL systems (Table 2), and sometimes they outperform these (e.g., when identifying argument types like *argm-neg, r-arg0, r-arg1, r-argm-loc*). The **FM** model trained on frame segments performs quite close to the **SM** model.

From Table 2 it is observed that the **FM** model performs better in identifying the predicates than the **SM** model. As shown in Table 2, using the *pafEval* evaluation, the **FM** model outperforms the other models. It improves the F_1 measure by around 10% compared to the state-of-the-art SRL tool. The **FM** model performs better than the **SM** model with an improvement of 2.27% in F_1

Table 1. Accuracy results of POS-Speech and Stanford models on different lowercase test data.

Models	100-sent	100-sent-noisy	bc-cnn-200
SS-POS	0.8793	0.8816	0.8301
SL-POS	0.9605	0.9541	0.9144
POS-Speech	0.9425	0.9376	0.9202

Table 2. Results of *pEval, paEval, cpaEval, pafEval* and *cpafEval* evaluation settings using different SRL models.

	pEval			paEval			cpaEval			pafEval			cpafEval		
	P	R	F_1	P	R	F_1	P	R	F_1	P	R	F_1	P	R	F_1
Written-sent															
SM	0.8904	0.8333	0.8609	0.2984	0.1871	0.2300	0.3180	0.1871	0.2356	0.2553	0.2389	0.2468	0.2818	0.2349	0.2562
FM	0.8872	0.8358	0.8607	0.2708	0.1763	0.2136	0.2895	0.1766	0.2194	0.2452	0.2310	0.2379	0.2701	0.2265	0.2464
Written-frameSeg															
SM	0.8904	0.8333	0.8609	0.2984	0.1871	0.2300	0.3180	0.1871	0.2356	0.2553	0.2389	0.2468	0.2818	0.2349	0.2562
FM	0.8872	0.8358	0.8607	0.2708	0.1763	0.2136	0.2895	0.1766	0.2194	0.2452	0.2310	0.2379	0.2701	0.2265	0.2464
ASR-sent															
SM	0.8904	0.8333	0.8609	0.2984	0.1871	0.2300	0.3180	0.1871	0.2356	0.2553	0.2389	0.2468	0.2818	0.2349	0.2562
FM	0.8872	0.8358	0.8607	0.2708	0.1763	0.2136	0.2895	0.1766	0.2194	0.2452	0.2310	0.2379	0.2701	0.2265	0.2464
Segmentation-data															
SM	0.8904	0.8333	0.8609	0.2984	0.1871	0.2300	0.3180	0.1871	0.2356	0.2553	0.2389	0.2468	0.2818	0.2349	0.2562
FM	0.8872	0.8358	0.8607	0.2708	0.1763	0.2136	0.2895	0.1766	0.2194	0.2452	0.2310	0.2379	0.2701	0.2265	0.2464

score. From the above two experiments on *written-sent* and *written-frameSeg*, it is observed that the performance of the state-of-the-art SRL system degrades when applied on frame segmented data.

5.3   SRL on Speech Data

SM and FM models on *ASR-sent* **data:** Table 2 shows that the **SM** model outperforms the other models. In *paEval* evaluation the **SM** model marginally improves the F_1 measure by 0.96% compared to the **FM** model. In *pafEval* evaluation, the model marginally improves the F_1 score by 0.45% as compared to the **FM** model.

SM and FM Models Evaluated on Frame Segmented Data: For the **pred-LR-n** methods we set the window size to 15 as determined on a small held-out validation set. Table 2 shows the *pEval*, *paEval* and *cpaEval* evaluations, where frame segmented models usually perform better than sentence based methods. It is also observed that our pruning method using **POS-LM** on *LR-n* data marginally improves the F_1 score (from 17.75% to 18.17%) This also reveals that our POS tagger trained on noisy data with no sentence boundaries works quite satisfactory when applied on speech data. In *pafEval* evaluation, the **FM-LR-15** model outperforms the other models. Figure 2 shows the *paEval* evaluation of different SRL models on *LR-n* data where n is the number of tokens expanded left and right from the given predicate, showing that the **FM-LR-8-pruned** model outperforms the other models. The graph also reveals that the results are not very sensitive to the number of tokens in the expansion, even though an optimal size of the segment is expected.

6   Conclusion

We have proposed a novel approach to identify PropBank-style semantic predicates and their semantic role arguments in speech data that are lowercased and lack sentence boundaries. We identify semantic frame segments in the language data and train the SRL system on frame segmented data. Our models perform better than state-of-the-art SRL systems when identifying predicates and their semantic roles in speech data. We have also successfully trained a POS tagger to work on noisy, lowercase, and error prone ASR data.

References

1. Punyakanok, V., Roth, D., Yih, W.: The importance of syntactic parsing and inference in semantic role labeling. Comput. Linguist. **34**(2), 257–287 (2008)
2. Johansson, R., Nugues, P.: Dependency-based semantic role labeling of PropBank. In: Proceedings of the EMNLP, Stroudsburg, PA, USA. ACL, pp. 69–78 (2008)
3. Zhao, H., Chen, W., Kit, C., Zhou, G.: Multilingual dependency learning: a huge feature engineering method to semantic dependency parsing. In: Proceedings of the Thirteenth CoNLL 2009, Boulder, Colorado, USA, pp. 55–60 (2009)

4. Stenchikova, S., Hakkani-Tür, D., Tür, G.: QASR: question answering using semantic roles for speech interface. In: Proceeding of INTERSPEECH, ISCA (2006)
5. Kolomiyets, O., Moens, M.F.: A survey on question answering technology from an information retrieval perspective. Inf. Sci. **181**(24), 5412–5434 (2011)
6. Hüwel, S., Wrede, B.: Situated speech understanding for robust multi-modal human-robot communication. In: Proceedings of the COLING/ACL 2006 Main Conference Poster Sessions, pp. 391–398. ACL (2006)
7. Huang, X., Baker, J., Reddy, R.: A historical perspective of speech recognition. Commun. ACM **57**(1), 94–103 (2014)
8. Favre, B., Bohnet, B., Hakkani-Tür, D.: Evaluation of semantic role labeling and dependency parsing of automatic speech recognition output. In: Proceedings of ICASSP 2010, pp. 5342–5345, March 2010
9. Hovy, E., Marcus, M., Palmer, M., Ramshaw, L., Weischedel, R.: OntoNotes: the 90% solution. In: Proceedings of NAACL HLT, Stroudsburg, PA, USA, pp. 57–60. ACL (2006)
10. Mohammad, S., Zhu, X., Martin, J.: Semantic role labeling of emotions in Tweets. In: Proceedings of the 5th Workshop on WASSA, Maryland, pp. 32–41. ACL June 2014
11. Stolcke, A.: SRILM - an extensible language modeling toolkit. In: Proceedings of the 7th ICSLP 2002, pp. 901–904 (2002)
12. Fonseca, E., Rosa, J.: A two-step convolutional neural network approach for semantic role labeling. In: Proceedings of IJCNN 2013, pp. 1–7, August 2013
13. Shrestha, N., Moens, M.F.: Semi-automatically alignment of predicates between speech and ontonotes data. In: Proceedings of the 10th edition of LREC 2016 (2016)
14. Manning, C.D.: Part-of-speech tagging from 97% to 100%: is it time for some linguistics? In: Gelbukh, A.F. (ed.) CICLing 2011. LNCS, vol. 6608, pp. 171–189. Springer, Heidelberg (2011). https://doi.org/10.1007/978-3-642-19400-9_14

Voice Control in a Real Flight Deck Environment

Michal Trzos[✉], Martin Dostl, Petra Machkov, and Jana Eitlerov

Honeywell International, Aerospace Advanced Technology Europe,
Tuřanka 1387/100, Brno, Czech Republic
{michal.trzos,martin.dostal2,
petra.machackova2,jana.eitlerova}@honeywell.com

Abstract. In this paper, we present a methodology on how to implement multimodal voice controlled systems by means of automatic speech recognition. The real flight deck environment brings many challenges such as high accuracy requirements, high noise conditions, non-native English-speaking users or limited hardware and software resources. We present the design of an automatic speech recognition system based on a freely available AMI Meeting Corpus and a proprietary corpus provided by Airbus. Then we describe how we trained and evaluated the speech recognition models in a simulated environment using the anechoic chamber laboratory. The tuned speech recognition models were tested in real flight environment on two Honeywell experimental airplanes: Dassault Falcon 900 and Boeing 757.

Keywords: Automatic speech recognition · Multi-modal interaction
Navigation display

1 Introduction

The modern flight decks in business and air transport aircraft are equipped with digital flight instrument displays rather than analog dials and gauges. Such flight decks are termed as glass cockpits. Glass cockpits typically have a Primary Flight Display (PFD), shown on Fig. 1, providing flight information and Navigation Display (ND) (or multi-function display) providing the pilots with navigation route, moving map, flight planning or weather radar to name a few navigation display functions. Navigation displays interaction is based on the graphical user interfaces and point-and-click paradigm. A keyboard and a cursor control devices (CCD) are used as primary interaction means. The current interaction means brings couple of limitations; navigation displays are equipped with a huge number of functionalities, which together with graphical user interfaces and cursor control devices impacts efficiency. Many functions are accessible through time consuming visual navigation and pointing using cursor control device, which are kind of accurate but slow input devices. All the pointing on a graphical user interface requires pilot gaze on the display (head-down interaction) instead of

© Springer Nature Switzerland AG 2018
P. Sojka et al. (Eds.): TSD 2018, LNAI 11107, pp. 388–402, 2018.
https://doi.org/10.1007/978-3-030-00794-2_42

providing the pilot the maximum time to look out the windows (head-up) which is important for the pilot, especially during taxi, take-off, approach and landing phase of flight.

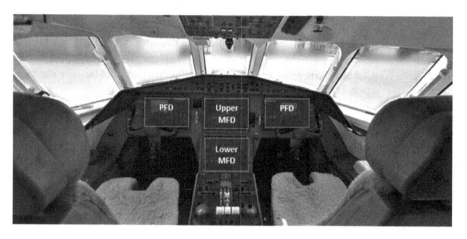

Fig. 1. Position of Primary Functional Display (PFD) and Multi-function Display (MFD) in the cockpit of Falcon F900

In an ideal flight deck all kind of tasks should be applicable with the maximum efficiency possible which cannot be accomplished with the existing glass cockpit user interfaces. All the input devices have benefits and limitations (e.g. fast vs less accurate, slow vs more accurate) and these benefits and limitations have different impact according to the nature of the tasks being controlled. One of the most promising ways how to dramatically increase the efficiency of the flight deck interaction is represented by multi-modal interaction. Multimodal interaction provides the user with multiple modes of interacting with the system. Interaction modes (modalities) can be a cursor control device, touch screen, speech or keyboard, to name a few. Thoroughly and consistently implemented multi-modal interaction provides the pilot to choose and switch between the modalities according to the kind of task, individual pilot preferences and especially operational conditions. For example, during a turbulence pilot may prefer to switch from touchscreen to CCD.

In this paper, we focus on the speech modality in a multimodal Human-Machine Interface (HMI) [1] and its transition from laboratory conditions to a real cockpit environment of the F900 and Boeing 757 experimental airplane. Speech as modality provides an important alternative to graphical user interface and point-and-click paradigm. While graphical user interfaces require a lot of visual attention, pointing and navigation, speech offer non-visual and direct access to functionality. This advantage, however, is balanced by the necessity to understand and memorize speech commands and related rules how to correctly constitute speech utterances.

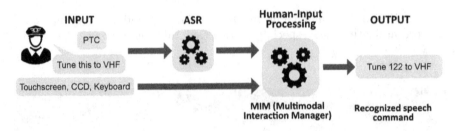

Fig. 2. Overview of the ASR system with Multimodal Interaction Manager showing anaphoric interaction (this is converted to frequency by MIM)

The paper is organized as follows. First, we describe the development of two speech recognition systems, one aimed for use with office headsets in laboratory conditions, and the other for use with aviation headsets in real conditions. Then we present principles of creating grammars for multi-modal interaction with the navigation flight deck. Then we present evaluation of the designed speech recognition system in an anechoic chamber laboratory followed with flight tests on board of the F900 and Boeing 757 experimental airplanes.

2   Automatic Speech Recognition System

The automatic speech recognition (ASR) system is expected to perform automatic speech recognition of non-native spoken English in noisy cockpit environments for selected European accents. Its input is speech in the form of an audio signal and a push-to-command (PTC) signal generated by a PTC button. The PTC button is a preferred form of activating input to voice-controlled system by pilots and is used several voice-controlled avionics systems such as in [2] where it is used to input radio frequencies into an avionics hardware radio. Its output is the recognized text which is to be displayed in the user-interface of the system and is also input to the Multimodal Interaction Manager (MIM), see Fig. 2.

We have employed two approaches in designing the ASR. First speech recognition system was to be used in earlier prototype design stages and for proof-of-concept demonstrations with non-native English speakers. From all the available corpora for training we have chosen AMI Meeting Corpus [3] for this system as it contains approximately 100 h of utterances from mostly non-native English speakers. First, we used a gaussian mixture model [4] AMI-GMM for its ease of use and low latency and moved to a deep neural network-based AMI-TDNN [5] for better accuracy.

Second recognition system was aimed at real use by pilots in an airplane cockpit. Microphones used in aviation mostly use noise-cancelling electret microphones with narrow bandwidth adapted to the use for communication over a VHF channel that has maximum bandwidth of 8.33 kHz though the actual bandwidth varies for different aircraft and control towers [6] and is usually between 6 kHz and 8 kHz. This produces a rather specific audio output that is not easily recognized even when used with ASR systems trained on audio data with sampling frequency 8 kHz that perform well on telephone audio. We have found two viable solutions of this issue. First is done by training one ASR system on a large amount of data from various corpora and various contexts, also called out of domain data and adding a small amount of context in-domain data for adaptation [7,8]. Second is to train an ASR system on large amount of audio from the aviation domain. We have been able to gather and transcribe approximately 47 h of data from Toulouse-Blagnac airport (LFBO) [9] which we used for training the Airbus-TDNN model.

ASR systems typically consist of two main parts, acoustic model and language model. Acoustic model represents the relationship between input audio and some linguistic unit such as phonemes, senones, or letters. Language model usually represents probability distribution of words in a word sequence. Such combination of acoustic model and language model can yield several hypotheses that can be compared with a specified set of commands using dynamic programming [10]. But since the input to our system will be a sentence with predefined structure, we have decided to use a grammar instead of language model. This leads to significantly smaller models and faster execution.

We are using two measures for evaluation of the speech recognition system, word accuracy (WACC) [11] and sentence accuracy (SACC). Word accuracy is computed as a ratio of all correctly recognized words and number of words in reference commands. Sentence accuracy is a ratio between all correctly recognized commands and all commands in reference. While word accuracy describes the performance of the speech recognition engine, sentence accuracy describes the systems ability to correctly recognize commands.

Two headsets were used in the evaluations, one is office headset Microsoft LifeChat LX-3000 with noise cancellation technology for the microphone and the other is Telex Airman 850, aviation headset used mostly in commercial aircrafts. The Airman 850 features a noise-cancelling electret microphone and provides up to 12 dB noise attenuation between 100 Hz and 2000 Hz.

3   Voice Control in a Multi-modal System

This section defines the general principles, syntax patterns and grammar of multimodal command control language. The command control language covers selected functionality of a navigation display which includes moving map interaction, flight planning and radio tuning.

The voice multimodal interaction philosophy follows these principles:

1. Simple, minimalistic and predictable syntax. A few syntax patterns should be used across the whole flight deck to minimize the need for memorization.
2. Flexibility in syntax. Although the language should be simple and minimalistic, there should be certain level of freedom to improve usability of the complete system. The versatility is accomplished via the following features:
   - Generic commands. One of the important concepts to maintain simplicity of the command controls language is genericity of commands. Generic commands operate regardless of object type. For instance, there is one

single command for setting a value regardless it is an altitude value or map layer state, or one single command for displaying a visual object regardless if it is a menu category or a dialog, for example.

- Synonyms. All objects or commands may have multiple names, to improve usability of Command Control Language.
- Implicit commands. In defined cases, a command may be omitted if it is beneficial to the user and does not introduce ambiguity.
- Optional keywords. Some keywords may be omitted to provide as much as possible efficiency in syntax. With omitting an optional keyword the language syntax becomes shorter but also more artificial for the user.
- Implicit states. For some objects, there is default state defined and thus user can omit the state specification. For example, if a user performs SET TERRAIN command, then the Multimodal Interaction Manager (MIM) will automatically assign the default state ON to this command. As a result, the terrain layer on the lateral map is set on.

Simple syntax patterns are defined consistently across all applications to minimize the need for memorization. This follows the same design principles. For example, speech utterance for radio frequency input uses the following syntax:

<command> <target> <destination>

as an example, if the <command> is SET then <target> is the frequency to be set, e.g. 123.5. <destination> specifies where to set the target frequency. As an example of zoom functionality zoom sets view to a pre-specified zoom value and range in and out incrementally move between the values, see Table 1.

Table 1. Sample grammar for controlling zoom functionality

Command	Target
Zoom	Number (e.g. 500)
Range	In
	Out

The navigation display grammar covers moving map, graphical flight planning and radio tuning. The moving map grammar consists of 6 commands with 50 targets/destinations, the graphical flight planning grammar consists of 5 commands with 84 targets, and the radio tuning grammar consists of 3 commands and 238 targets. Numbers specifying a radio frequency are counted as one target despite its format ranging from one number to a six-decimal number with a decimal point. It should be noted that pilots are thoroughly trained to use aircraft systems including speech recognition.

3.1   Multimodal (Cross-Modal) Utterances

Multimodal interaction philosophy outlined in this paper allows also for utterances provided by multiple modalities. This is termed as cross-modal interaction.

For example, one part of the input may be provided by speech while the other part using another modality. Using an underlying technology, the MIM, the system is able to associate these separate inputs into correct, meaningful input. For instance, the <destination> to set a new frequency can be specified by a combination of speech and touch modality. For example, pilot presses the PTC button and says CENTER. At the same time, he touches Toulouse-Blagnac airport (LFBO) on the touchscreen. The result of this multimodal command is that the map is centered around the LFBO airport.

3.2   Anaphoric Interaction

Anaphoric interaction builds on cross-modal interaction. Anaphora are represented by deictic terms such as IT, THIS, THAT or THERE. When pilot uses an anaphora in speech utterance the system will try to associate the pronoun reference to the corresponding object.

For example, user says TUNE 118.1 ON THIS. System tries to associate THIS with inputs from other modalities at relevant time window. Since the user was pointing on VHF text box the system will understand the user input as: TUNE 118.1 ON VHF.

Advanced multimodal interaction features such as cross-modal interaction or anaphoric interaction have an impact on speech recognition grammar. The grammar must accept also partial inputs (cross-modal use-cases) or abstract pronouns (anaphoric interaction).

3.3   JSpeech Grammar Format

Grammar constructed according to the rules in this document is implemented in JSpeech Grammar Format (JSGF), a W3C standard [12] for speech recognition grammar with notation in Backus-Naur Form style. JSGF was chosen, because its well readable for humans and compatible with speech recognition engine Kaldi.

3.4   System Architecture

The system architecture, shown on Fig. 2, consists of human-input devices such as automatic speech recognition or touch screen input, MIM, which receives the inputs from input devices and performs parsing, association, filtering, and disambiguation of the user input. The output from MIM is sent to the user application where it is executed afterwards.

3.5   Aviation-Specific Objects

In aviation, there are specific objects such as waypoints and airports that have to be incorporated into the grammar. Waypoints identify a point in physical space and have five letters or a combination of three letters and two numbers (e.g.

ABITO, BR53V), airports have four letters (e.g. LKPR - Vclav Havel Airport Prague). There are several ways of pronouncing the letters. The most widely used is specified by the ICAO alphabet [13], where the letters are pronounced as words such as bravo for the letter b or victor for the letter v. Waypoints and airports sometimes form pronounceable words (e.g. SABER or MALOT) that work well with native English speakers but are almost unusable with non-native speakers as the pronunciation of such varies greatly from pilot to pilot. The letters can also be pronounced letter-by-letter (e.g. ODUDA as ou-di-ju-di-ei), but pilots generally do not prefer to read them this way.

3.6   Integrating Grammar with ASR

After the grammar was written, we used a tool from CMUSphinx jsgf2fsg to transform the JSGF-based grammar to a finite state automaton. Then we transformed the finite state automaton to a finite state transducer by copying the input string to the output string. In the last step we added weights to each final state based on the number of input states so that every possible word in the grammar has equal probability which forms a weighted finite state transducer (WFST). Then using the WFST we have been able to create a HCLG graph for the use with a Kaldi decoder.

4   Speech Recognition System Evaluation

The purpose of this evaluation was to assess the presented ASR systems in a series of tests that are progressively closer to real use on an aircraft during flight. The first evaluation was performed in a simulated environment with representative cockpit noise. The same system was then used on a Falcon F900 aircraft during flight and later with Boeing 757 experimental airplane.

Noise measurement was a significant part of the evaluations. Our goal was to assess the overall noise profile during flight from taxiing, taking off, climb, cruise, to landing to determine if our ASR system was usable in all phases of flight. Sound pressure level (SPL) meter used in the evaluations was NTi XL2 Sound Level Meter with M2230 Class 1 certified measurement microphone. All noise was recorded in WAV 24-bit linear format with 48 kHz sampling frequency. A-weighting was used in the measurements as it approximates human hearing. All the data was measured using a time constant $t = 5$ s meaning that measurement has been taken each 5 s for the maximum, minimum and average SPL values for the measured interval.

4.1   Evaluation in Anechoic Chamber

To estimate performance of our speech recognition engine in an airplane cockpit, we decided to perform a measurement in as close environment as possible to a Falcon F900 airplane. Since airplane time is expensive and scarcely available, we decided to perform the measurement in a simulation of cockpit environment.

Measurement Method. The measurement was carried out in an anechoic chamber where we put a chair in the center and placed a loudspeaker with flat frequency response around 2 m away from the center in forward direction (see Fig. 3). The loudspeaker was oriented at the chair and was placed in such height that matched head of a person sitting on the chair. While it is true that in an aircraft the noise is coming also from the back and sides, we decided that due to directivity and noise-cancellation nature of the headsets used in airplanes, it would be enough to reproduce the sound only from the front side.

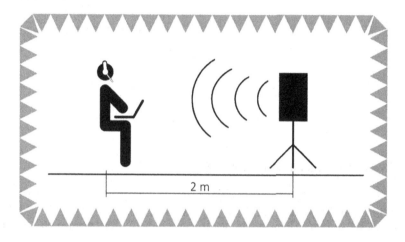

Fig. 3. Setup of evaluation in anechoic chamber

The noise used in the experiment was measured in a Falcon F900 airplane in the descent flight phase as that was when the cockpit was consistently very noisy. The noise was measured using a sound pressure level meter and its average spectrum can be seen on Fig. 4.

After setting up the chair and the loudspeaker, we put a measurement microphone with flat frequency response in position where we expected speakers head. Then we played cockpit noise through the loudspeaker and set it to frequency response shown in Fig. 4 with A-weighted sound pressure level set to 74.5 dBA.

Participants were then seated on the chair and read a list of 111 commands consisting of exhaustive list of commands for radio control. All the commands were recorded and saved for accuracy estimation and for further acoustic model training.

Participants. We tested our system on 4 non-native English speakers speaking 3 different accents Czech, Slovak, and Chinese. Two of the speakers were male and two female.

Evaluation Outcomes. After recording all the speakers using both headsets we compared the output from the speech recognition system and the list of the 111 commands and computed the system accuracy. Results are shown on Table 2. While the accuracy is high for office headset, the results for pilot headset were 27% and 10% worse for sentence accuracy and word accuracy respectively. This has been expected as the AMI corpus, basis for acoustic model training, has been trained on office headsets and recorded at sampling frequency 16 kHz while aviation headsets employ active noise-cancelling electret microphones that have a specific sound characteristic and produce frequencies only up to 6 kHz.

Table 2. Accuracy of our speech recognition system with AMI-GMM model in an anechoic chamber

	WACC	SACC
Office headset	98.19%	88.00%
Aviation headset	88.21%	61.26%

4.2   Evaluation in Falcon F900

The flight test took place on April 17th during a flight from Vienna to Brno. The test pilot conducted a set of given tasks during the cruise phase of the flight.

The added value of this testing lies in the following for the speech modality:

- Real aircraft engine & aerodynamic noise
- Real background ATC communication
- Real physical environment for using Push-To-Command buttons

Speech interaction was triggered by PTC which is the standard operation for ATC communication in the cockpit. We used three Bluetooth buttons as PTC buttons. They were identical in functionality, but only one should be pressed at a time. All three buttons were used in flight test to compare different locations. Microphone used in this testing was a common office headset connected to the Surface via USB.

Two Honeywell flight test pilots participated in this testing. One of them only participated in the ground test. The other one attended both ground and flight test. This one had some previous experience with an older version of the prototype because he participated in the exploration study in 2016.

The flight test was conducted during the cruise phase of a flight from Vienna to Brno. Due to limited time, training was not repeated. The test pilot relied on his training and experiences from the ground test.

The Surface tablet hosting speech recognition system was located on top of the lower multi-functional display, (see Fig. 5), and was providing a multimodal navigation map. Three PTCs were all attached to the cockpit at locations illustrated in Fig. 5. The pilot monitoring was the test pilot. Microsoft headset was placed around his head instead of the standard pilot headset.

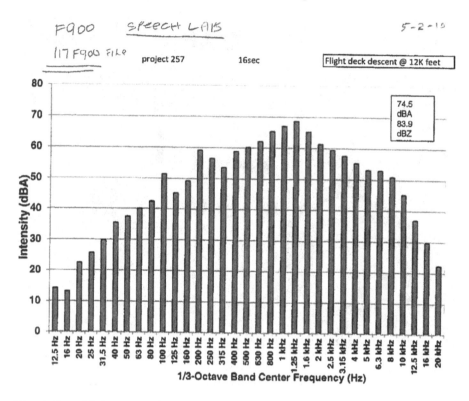

Fig. 4. One-third octave noise intensity in a cockpit of a Falcon F900 airplane during cruise phase used in the anechoic chamber evaluation

During cruise, the test pilot conducted a list of tasks under three noise conditions:

- Low noise (lower cruise speed) free choice between speech and touch
- High noise (higher cruise speed) use only speech
- Airbrake (higher cruise speed + airbrake) use only speech

In addition, cockpit noise was measured and recorded. The aim was to measure the noise in the cockpit during different flight phases and to record it so we have an accurate rendition of the environment for further development and testing. Each flight phase has a distinct noise characteristic (level and spectral content), e.g. during taxi, most of the noise comes from the gear running on the runway; while during descent extension of landing gears adds significant aerodynamic noise, see Fig. 7.

The noise measurement was conducted by a person sitting in a jump seat of the cockpit with the Sound Pressure Level meter in his hands. The recording consisted of two phases, first was from taxi to cruise (before the test session) and second was from top of descend (after the test session) to taxi.

Fig. 5. Hardware used in the flight tests; MS Surface tablet, office headset, and PTC buttons

Table 3. Speech engine accuracy of the AMI-GMM model in the Falcon F900 evaluation

	WACC	SACC
Ground	99.47%	96.97%
Air	99.97%	89.37%

A total of 237 utterances were collected during this testing and 55 of them (23.2%) were not constructed according to the currently implemented phraseology. During ground test, the phraseology design was introduced to the pilots prior to the tasks and a cheat sheet was available. These 55 out-of-phraseology utterances are valuable outcome of the study to understand what level of flexibility pilots would like to have which is, however, out of this paper scope.

Sentence Accuracy and Word Accuracy were calculated from 176 utterances (74.2% of all recorded utterances). Table 3 summarizes the results. Cockpit noise only slightly decreased the recognition accuracy.

Fig. 6. Hardware installation in the cockpit

4.3   Evaluation in Boeing 757

In this evaluation we wanted to test the performance of our speech recognition system we have tested in the Falcon F900 with AMI-GMM, AMI-TDNN, and Airbus-TDNN models in a real environment using an aviation headset. It has been carried out over a three-day flight test campaign in the cockpit of the experimental Honeywell Boeing 757 airplane (B757). On each of the days the airplane performed a flight from LTBK (Brno airport) to EHAM (Schiphol airport) and back. Cockpit and cabin of the airplane is not representative of airline version of B757 as it has most of seats and internal noise suppressing material removed and replaced with test equipment. While this has increased noise in the cockpit, overall signal-to-noise ratio of the recorded voice commands remained the same due to the aviation headsets noise-cancelling microphone (Fig. 6).

The evaluation consisted of 194 utterances spoken by two pilots (97 each) with US and German accent. The 97 utterances consist of randomly generated commands from graphical flight planning and map grammar which are then selected so the set of utterances represents all possible use-cases. The speech recognition system was running on a Microsoft Surface tablet with the aviation headset connected over microphone input. A PTC held by the pilots was used to trigger recording and speech recognition. The Surface tablet was placed on the lower multi-functional display (see Fig. 1). The pilots were instructed to use the PTC as they would normally use it for controlling the aircraft radio. Before the evaluation, the pilots switched all controls to the second pilot, while the first pilot went through the evaluation. After evaluation of the first pilot, the pilots switched control to the second pilot who then continued in the evaluation. All commands were recorded for later evaluation with different models.

Results of the evaluation show Airbus-TDNN as the best performing system with 92.16% and 72.68% word and sentence accuracy respectively (see Table 4). This has been expected as the Airbus-TDNN is trained mainly on narrow band-

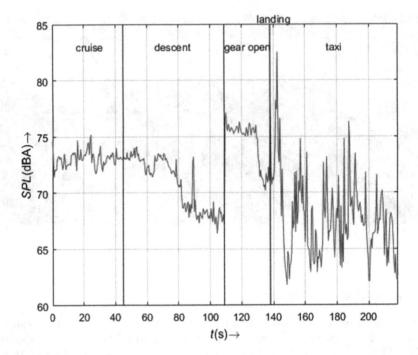

Fig. 7. A-weighted SPL of noise in the cockpit of the Falcon F900 during cruise, descent, landing, and taxi phases of flight

Table 4. Word and sentence accuracy for Airbus-TDNN, AMI-TDNN, and AMI-GMM models tested on board of the B757 with the aviation headset

Model	WACC	SACC
Airbus-TDNN	92.16%	72.68%
AMI-TDNN	66.36%	36.60%
AMI-GMM	63.98%	33.90%

width data where aviation headsets were used whereas the AMI-based models were trained on data recorded with office headsets that are downsampled from sampling frequency 16 kHz to 8 kHz before training.

5 Conclusion

We have presented the development of speech recognition system for our new multi-modal human-machine interface. We created an ASR system with three models: AMI-GMM, AMI-TDNN built on the AMI corpus using gaussian mixture models and time delay deep neural networks respectively, and Airbus-TDNN built on data gathered from ATC communication.

Then we presented principles which we used to create multi-modal command control language for moving map interaction, flight planning, and radio tuning functionality of a navigation display along with some examples and a summary of all the commands we implemented using grammars.

After integrating the grammars into the presented models, we have performed an evaluation of the resulting system in an anechoic chamber with non-native English speakers. Results using the office headset were 98.19% and 88% for WACC and SACC respectively. Accuracy for the aviation headset are 10% and 26% percent worse.

In the evaluation on the Falcon F900 airplane we have performed to tests with the office headset; one on ground, one during flight. WACC of both test was close to 99%, SACC was 96.97% for the ground test and 89.37% for the test in flight.

Last evaluation was performed on the Boeing 757 experimental airplane with the aviation headset. Both AMI-GMM and AMI-TDNN performed rather poorly at 66.36% and 63.98% respectively. Airbus-TDNN, model trained on ATC communication data performed the best with 92.16% WACC and 72.68% SACC.

The noise measurements point to the possibility to use speech recognition in all phases of flight during normal conditions which have been present during the Falcon F900 flight. Noise conditions during the Boeing 757 flight were not representative of commercial airplane due to the experimental nature of the airplane but point to the need for a model specifically tailored to aviation headsets or a universal model usable both by aviation and office headsets as the AMI-GMM and AMI-TDNN models both perform at least 30% WACC worse than the Airbus-TDNN model.

References

1. Dostal, M., Kolcarek, P.: Multimodal navigation display. In: 2015 IEEE/AIAA 34th Digital Avionics Systems Conference (DASC), Prague, pp. 3B1-1–3B1-11 (2015)
2. Swearingen, P.A.: United States Patent No. 8,234,121 B1. U.S. Patent and Trademark Office, Washington, DC (2012)
3. Mccowan, I., et al.: The AMI meeting corpus. In: Proceedings Measuring Behavior 2005, 5th International Conference on Methods and Techniques in Behavioral Research. In: Noldus, L.P.J.J., Grieco, F., Loijens, L.W.S., Zimmerman, P.H. (eds.) Noldus Information Technology, Wageningen (2005)
4. Povey, D., et al.: The Kaldi speech recognition toolkit. In: IEEE 2011 Workshop on Automatic Speech Recognition and Understanding. IEEE Signal Processing Society (2011)
5. Peddinti, V., Povey, D., Khudanpur, S.: A time delay neural network architecture for efficient modeling of long temporal contexts. In: proceedings of INTERSPEECH 2015, Dresden, Germany, pp. 3214–3218 (2015)
6. Airband. https://en.wikipedia.org/wiki/Airband. Accessed 20 Mar 2018
7. Srinivasamurthy, A., Motlicek, P., Himawan, I., Szaszk, G., Oualil, Y., Helmke, H.: Semi-supervised learning with semantic knowledge extraction for improved speech recognition in air traffic control. In: Proceedings of the Interspeech 2017, pp. 2406–2410 (2017). https://doi.org/10.21437/Interspeech.2017-1446

8. Oualil, Y., Klakow, D., Szaszk, G., Srinivasamurthy, A., Helmke, H., Motlicek, P.: A context-aware speech recognition and understanding system for air traffic control domain. In: 2017 IEEE Automatic Speech Recognition and Understanding Workshop (ASRU), Okinawa, pp. 404–408 (2017)

9. Delpech, E., et al.: A real-life, french-accented corpus of air traffic control communications. In: Proceedings of the 11th Language Resources and Evaluation Conference (LREC 2018), Miyazaki, Japan (2018)

10. Ranzenberger, T., Hacker, Ch., Gallwitz, F.: Integration of a Kaldi speech recognizer into a speech dialog system for automotive infotainment applications. In: Conference on Electronic Speech Signal Processing (ESSV 2018), Ulm (2018)

11. Word Error Rate. https://en.wikipedia.org/wiki/Word_error_rate. Accessed 20 Mar 2018

12. JSpeech Grammar Format. http://www.w3.org/TR/jsgf. Accessed 20 Mar 2018

13. ICAO. Manual of Radiotelephony. Document 9432-AN/925, 4th edn (2007)

Data Augmentation and Teacher-Student Training for LF-MMI Based Robust Speech Recognition

Asadullah[(⊠)] and Tanel Alumäe

Laboratory of Language Technology, Tallinn University of Technology,
Tallinn, Estonia
{asad.ullah,tanel.alumae}@ttu.ee

Abstract. Deep neural networks (DNN) have played a key role in the development of state-of-the-art speech recognition systems. In recent years, lattice-free MMI objective (LF-MMI) has become a popular method for training DNN acoustic models. However, domain adaptation of DNNs from clean to noisy data still remains a challenging problem. In this paper, we compare and combine two methods for adapting LF-MMI-based models to a noisy domain that do not require transcribed noisy data: multi-condition training and teacher-student style domain adaptation. For teacher-student training, we use lattices obtained via decoding untranscribed clean speech as supervision for adapting the model to noisy domain. We use in-domain noise extracted from a large untranscribed speech corpus using voice activity detection for noise-augmentation in multi-condition training and teacher-student training. We show that combining multi-condition training and lattice-based teacher-student training gives better results than either of the methods alone. Furthermore, we show the benefits of using in-domain noise instead of general noise profiles for noise augmentation. Overall, we obtain 7.4% relative improvement in word error rate over a standard multi-condition baseline.

Keywords: Speech activity detection · Noise augmentation
Domain adaptation · Weighted prediction error · Deep neural networks

1 Introduction

The emergence of deep neural networks (DNN) have played key role in the development of the state of the art acoustic model for large vocabulary continuous speech recognition (ASR) [4]. The main reason is that DNN learns internal hierarchical structure in the training data. The hierarchical structure is relatively invariant to speaker variability, speaking style and environmental backgrounds. However, DNN is unable to extrapolate from clean speech training data to noisy speech test data. Multi-condition training using noise-augmentation

© Springer Nature Switzerland AG 2018
P. Sojka et al. (Eds.): TSD 2018, LNAI 11107, pp. 403–410, 2018.
https://doi.org/10.1007/978-3-030-00794-2_43

and domain adaptation using teacher-student style techniques are used to handle the mismatch between clean speech training and noisy speech test data. Noise-augmentation [7] is used to combine clean speech data with in-domain extracted noises. There have been some challenges of building a robust speech recognition such as automatic speech recognition in reverberant environments (ASpIRE) [3], speech separation and recognition challenge (CHiME) [2] and reverberant voice enhancement and recognition benchmark (REVERB) [6]. The winners of ASpIRE have used noise-augmentation technique for training robust speech recognition system [5, 12].

Teacher-student style domain adaptation is used to train a new model (target) from an already well-trained model (source) [18]. Domain adaptation helps in bootstrapping a new system from an already well-trained supervised system. Teacher-student domain adaptation approach has been proposed in [8] where the posterior probabilities of teacher network were used as output labels for student network. Another approach in domain adaptation is knowledge distillation [16] where soft labels generated by a teacher model are interpolated with conventional hard target labels so as to provide regularization. Domain adaptation has been also used for adapting to new dialects [9]. However, the performance gap between supervised source model and unsupervised target model is large.

In this paper, we experiment with two methods for adapting an acoustic model for noisy speech: multi-condition training and teacher-student style domain adaptation. In order to better handle the actual target domain, we first extract a set of background noises from our large untranscribed in-domain training data using speech activity detection (SAD), instead of using background noise recordings from various external resources. We then use the extracted noises for standard multi-condition training. Secondly, we experiment with teacher-student style domain adaptation using models trained with lattice free maximum mutual information (LF-MMI) [15]. Instead of using posteriors as soft labels for domain adaptation, we use lattices generated by the teacher model as supervision. The lattices are produced for a large clean unlabeled speech corpus, and the target model is adapted using the same data that is reverberated and mixed with in-domain background noise. The proposed method does not require any manual transcriptions for target data.

The results of our experiments were further improved by cleaning the noisy test data with a pre-processing step called weighted prediction error (WPE) [11], which is a de-reverberation technique widely used in ASR systems dealing with distant and noisy speech (e.g., [1]).

The contents of the paper is organized as follows: Sect. 2 presents the methods including in-domain noise augmentation, teacher-student style domain adaptation and weighted prediction error. Section 3 describes our experiments which consists of dataset, implementation and results. Section 4 shows the conclusion part.

2   Methods

2.1   In-Domain Noise Augmentation

In this work SAD is used for extracting long segments of non-speech (i.e., in-domain background noise) from our unsupervised in-domain training data. We use those segments for noise-augmenting clean training data, instead of using noises from different external resources, such as the Freesound database.

We use a relatively simple feed-forward neural network for performing SAD. Input features are 40-dimensional MFCCs extracted every 10 ms within a 25 ms sliding window. Cepstral mean normalization is applied to the features of all recordings. 20 frames from both sides are used for frame splicing, forming an input to the network that covers 210 ms. The model has three hidden layers, each with 1024 units, using the "leaky" ReLU non-linearities. Dropout layers (using a dropout probability of 0.5) and batch normalization layers are used after each hidden layer. The final softmax layer has two outputs, corresponding to speech and non-speech. Training targets are extracted from phone-aligned clean ASR training data, with all phones except silence (including spoken noise) mapped to speech. The neural network is trained using data augmentation: one copy of reverberated and artificially noised (with Freesound noises [17]) is added to the clean data. In order to smooth the predictions emitted by the SAD model, we average the frame-level speech/non-speech probabilities using a sliding window of 20 frames. All frames with a resulting speech probability of 0.15 or larger are marked as speech. We found that using such low speech detection threshold is important for avoiding having short and low energy speech sounds (such as hesitations and backchannels) in the extracted non-speech segments. Finally, all resulting speech segments are padded by 500 ms from both sides. The non-speech segments of each recording in the unsupervised training data are concatenated, using a cross-fade length of 500 ms. Finally, we only keep the concatenated extracted noises that are at least one minute long.

Data augmentation is performed by reverberating clean training recordings with various small and medium room impulse responses [7] and mixing the extracted in-domain non-speech segments with the resulting signal. The non-speech segments are extended, by repetition, to cover the entire clean speech recording. A block diagram of the whole process is shown in Fig. 1.

2.2   Teacher-Student Style Domain Adaptation

In teacher-student style domain adaptation, unlabeled data from the source domain is processed by the source-domain (teacher) model to generate senone posterior probabilities. Those posterior probabilities are used as labels to train a student model, with parallel unlabeled data from the target domain. In the case of adapting a clean speech model to noisy speech, parallel data from the target domain can be obtained by mixing clean data with noise [8]. Training the student model on posteriors rather than automatically generated one-hot hard labels provides more informations to the student model, so that the student

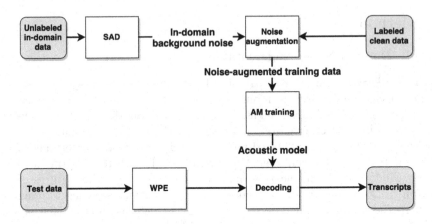

Fig. 1. Noise augmentation using in-domain noise.

network in the target domain behaves very similarly to the well-trained teacher network in the source domain.

This approach cannot be straightforwardly applied to acoustic models trained using the LF-MMI criterion, since such models are usually trained using utterance labels, rather than frame labels. Therefore, in order to propagate the uncertainty of the teacher model on unlabeled source data to the student model, we train the student model using lattices generated by the teacher model, and the corresponding audio data that is reverberated and mixed with in-domain noise. We employ the recently proposed method of using lattice-based supervision for semi-supervised training of LF-MMI based acoustic models [10] for this. We use a beam of 2.0 and a language model weight of 0 to prune the lattices before training the student model. The student model is trained on both noisy unlabeled data and noise-augmented labeled data, using equal weights. The whole process is depicted in Fig. 2.

2.3   Weighted Prediction Error (WPE)

In this work, WPE is used as a pre-processing step for cleaning of test data [11]. This approach cancel the effect of late reverberation in noisy speech data without prior knowledge of room impulse response (RIR). The captured speech signal composed of a time-varying variance Gaussian source and a delayed linear prediction observation process. The objective function of de-reverberation is the sum of the squared prediction errors normalized by the source variance. WPE can be represented by the following equation:

$$x_t^{(m)} = \sum_{k=0}^{L_h-1} h_k^{(m)} S_{(t-k)} + b_t^{(m)} \tag{1}$$

where $x_t^{(m)}, S_{(t-k)}$ and $b_t^{(m)}$ are observed signal, source signal and noise signals respectively; m and t represent speech samples and microphone indices; and $h_k^{(m)}$

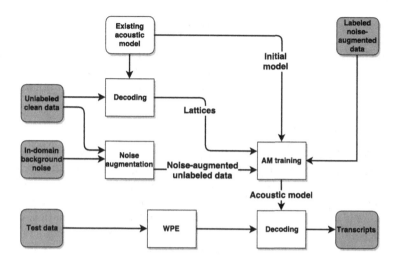

Fig. 2. Teacher-student domain adaptation using lattices for supervision.

is the room impulse response. The goal is to build a de-reverberation filter that can cancel the effect of reverberation as:

$$y_t = \sum_{m=1}^{L_m} \sum_{k=0}^{L_w-1} W_k^{(m)} X_{t-k}^{(m)} \tag{2}$$

where y_t is an enhanced output speech signal and $W_k^{(m)}$ is a de-reverberation filter of length L_w.

3   Experiments

3.1   Data

The out-of-domain transcribed clean training data contains 216 h of Estonian speech data from various domains (broadcast news, broadcast conversations, lectures and conference speeches, studio-recorded spontaneous dialogues), mostly collected at Tallinn University of Technology in the last 10 years. Note that this data is not entirely noise-free, as it also contains telephone interviews, news reports recorded in noisy environments, etc. In addition, we use about 500 h of untranscribed broadcast conversations as unlabeled out-of-domain speech data.

The in-domain speech data originates from the real usage data of the public Estonian speech transcription service provided by Tallinn University of Technology. The service allows users to upload audio files which are then processed by a speech recognition system and the transcripts is sent to the user. The service is provided for free and has been running for more than five years. The user-made audio recordings are however often recorded in noisy environments, using a microphone positioned relatively far from the user (e.g., a smart phone lying on the table). Therefore, this kind of data has often significant reverberation and background noise. About five hours of randomly selected user data was manually transcribed, resulting in a development set of about two hours and a test set of about three hours in length. In addition, we use 585 h of untranscribed user data as unlabeled in-domain data for extracting in-domain background noises.

3.2   Implementation

We use Kaldi [14] for experiments. All acoustic models are TDNN-BLSTM neural networks [13] trained using the LF-MMI objective function [15]. 40-dimensional MFCCs and 100-dimensional i-vectors are used as an input to our neural networks after some pre-processing steps. Three-fold speed perturbation and volume perturbation is used to augment the training data. Clean speech data is used for context dependent triphone state clustering and for generating numerator lattices for LF-MMI training. All models are trained using 2 epochs as increasing above 2 epochs causes overfitting problem. A proportional-shrink value of 35 is used for regularization.

3.3   Results

We measure the word error rates (WER) of transcribing noisy user-uploaded data based on the manually transcribed development and test sets. All word error rate results of our experiments are listed in Table 1.

The first row (A) of the table shows the performance of the system that is trained only on out-of-domain (i.e., clean) data. Model B is trained on augmented training data, using one copy of noisy and reverberated training data in addition to the clean data. Background noise recordings downloaded from *Freesound* were used for mixing clean data with noise. As expected, such multi-condition training improves the performance of the system on our in-domain data by a large margin. In the following experiments, we used this system as the baseline.

The following four models (C, D, E, F) demonstrate that the system benefits from more aggressive augmentation: using two copies of noisy and reverberated data results in lower WER for in-domain data. Also, using background noise extracted automatically from 585 h of unsupervised in-domain data results in slightly lower WERs, although the best result is obtained when Freesound and in-domain noise is used intermixed.

Finally, models G and H are trained using teacher-student style domain adaptation. First, we experimented with using a system trained only on clean speech as the teacher for obtaining the lattices for 500 h of out-of-domain clean unlabeled data. We then mixed the unlabeled clean data with automatically extracted in-domain noise and used this data, together with the corresponding lattices, for training a new model, using the teacher model as a starting point. Labeled clean data was mixed with unlabeled data when training the student model. As can be seen, the performance of the resulting model is not better than the multi-condition baseline. Model H is trained using our best multi-condition model (F) as a starting point in teacher-student style domain adaptation. Such combination of multi-condition training and teacher-student style domain adaptation resulted in the best WER, 5.6% relative improvement over the multi-condition baseline.

Table 1 also demonstrates that WPE consistently improves the WER on in-domain data, suggesting that it is useful even when the acoustic model is adapted to noisy and reverberated speech. The final relative improvement of WER on in-domain test data over the simple multi-condition baseline was 7.4%. The 95% confidence interval of the baseline system (B) is (35.0, 36.9). The confidence interval for the best system (H) on test data with WPE is (32.3, 34.2). We also experimented with applying WPE to training data but it degraded performance.

Table 1. Word error rate results of domain adaptation and noise-augmentation.

ID	Supervised training		Unsup. training	Without WPE			With WPE		
	Noise aug- mentation	#copies		Dev	Test	Relat. improvem.	Dev	Test	Relat. improvem.
A	None		No	52.5	45.8	−27.5%	50.7	43.7	−21.5%
B	Freesound	1	No	42.9	35.9	**Baseline**	41.9	34.9	2.9%
C	In-domain	1	No	42.9	35.6	0.9%	42.1	34.9	2.9%
D	Freesound	2	No	41.8	35.1	2.4%	41.2	34.6	3.8%
E	In-domain	2	No	41.3	34.6	3.8%	40.5	33.8	6.0%
F	Mixed	2	No	42.1	34.5	4.1%	41.5	33.7	6.3%
G	None		Yes	45.8	36.3	−0.9%	44.8	35.8	0.4%
H	**In-domain**	**2**	**Yes**	**40.5**	**33.9**	**5.6%**	**39.6**	**33.3**	**7.4%**

4   Conclusion

In this paper, we evaluated the results of combining teacher-student domain adaptation and multi-condition training using noise-augmentation. We extracted noise from our in-domain untranscribed audio data using speech activity detection, instead of using a wide range of various ambient noises. We showed that results obtained with in-domain noise-augmentation are better than those obtained with wide-range noise augmentation. For teacher-student training, we used lattices generated from clean data as supervision, instead of using frame posteriors, as our acoustic models are trained using the LF-MMI criterion. From our results we conclude that combining teacher-student style domain adaptation with multi-condition training performs better than either of the methods alone.

Furthermore, we conclude that using WPE for pre-processing test data is beneficial even when multi-condition training and domain adaptation are used for artificially reverberating training data.

In the future, we plan to add more advanced methods (e.g., GAN or VAE based data augmentation) to the combination in order to further improve speech recognition under noisy conditions.

References

1. Alumäe, T., et al.: The 2016 BBN Georgian telephone speech keyword spotting system. In: ICASSP, pp. 5755–5759 (2017)
2. Barker, J., Marxer, R., Vincent, E., Watanabe, S.: The third CHiME speech separation and recognition challenge dataset, task and baselines. In: ASRU (2015)
3. Harper, M.: The automatic speech recognition in reverberant environments (ASpIRE) challenge. In: ASRU (2015)
4. Hinton, G.: Deep neural networks for acoustic modeling in speech recognition: the shared views of four research groups. IEEE Signal Process. Mag. **29**, 82–97 (2012)
5. Hsiao, R., et al.: Robust speech recognition in unknown reverberant and noisy conditions. In: ASRU (2015)
6. Kinoshita, K., et al.: The REVERB challenge: a common evaluation framework for dereverberation and recognition of reverberant speech. In: Applications of Signal Processing to Audio and Acoustics (2013)
7. Ko, T., Peddinti, V., Povey, D., Seltzer, M.L., Khudanpur, S.: A study on data augmentation of reverberant speech for robust speech recognition. In: ICASSP (2017)
8. Li, J., Seltzer, M., Wang, X., Zhao, R., Gong, Y.: Large-scale domain adaptation via teacher-student learning. In: INTERSPEECH (2017)
9. Lippmann, R., Martin, E., Paul, D.: Multi-style training for robust isolated-word speech recognition. In: ICASSP (1987)
10. Manohar, V., Hadian, H., Povey, D., Khudanpur, S.: Semi-supervised training of acoustic models using lattice-free MMI. In: ICASSP (2018)
11. Nakatani, T., Yoshioka, T., Kinoshita, K., Miyoshi, M., Juang, B.H.: Speech dereverberation based on variance normalized delayed linear prediction. IEEE Trans. Audio, Speech Lang. Process. **18**, 1717–1731 (2010)
12. Peddinti, V., Chen, G., Manohar, V., Ko, T., Povey, D., Khudanpur, S.: JHU ASpIRE system: robust LVCSR with TDNNS, iVector adaptation and RNN-LMS. In: ASRU (2015)
13. Peddinti, V., Povey, D., Khudanpur, S.: A time delay neural network architecture for efficient modeling of long temporal contexts. In: INTERSPEECH (2015)
14. Povey, D., et al.: The Kaldi speech recognition toolkit. In: IEEE Workshop on ASRU (2011)
15. Povey, D., et al.: Purely sequence-trained neural networks for ASR based on lattice-free MMI. In: INTERSPEECH (2016)
16. Sak, H., Senior, A., Rao, K., Beaufays, F.: Fast and accurate recurrent neural network acoustic models for speech recognition. In: INTERSPEECH (2015)
17. Synder, D., Chen, G., Povey, D.: MUSAN: a music, speech, and noise corpus. arXiv (2015)
18. Yu, D., Yao, K., Su, H., Li, G., Seide, F.: KL-divergence regularized deep neural network adapation for improved large vocabulary speech recognition. In: ICASSP (2013)

Using Anomaly Detection for Fine Tuning of Formal Prosodic Structures in Speech Synthesis

Martin Matura[(⊠)] and Markéta Jůzová

Department of Cybernetics and New Technologies for the Information Society,
Faculty of Applied Sciences, University of West Bohemia, Pilsen, Czech Republic
{mate221,juzova}@kky.zcu.cz

Abstract. Consistent prosody description of speech corpora is a fundamental requirement for a high quality speech synthesis generated by current TTS systems. In this preliminary study, we are using One-class SVM anomaly detection approach to predict formal prosodic structure outliers (a prosodic mismatch) in recorded utterances, that can negatively influence the overall quality of synthesized speech, especially in unit selection. To evaluate the outcome of our detection system, we performed a listening test with encouraging results.

Keywords: Anomaly detection · One-class SVM
Formal prosodic grammar · Prosodemes
Unit selection speech synthesis · Legendre polynomials

1 Introduction

In general, in the unit selection speech synthesis [20], the contours of fundamental frequency F_0 are not generated and subsequently followed in the unit selection process (compared to e.g. HMM [8], hybrid [16] or DNN [25] speech synthesis). However, to keep the required communication functions on the phrase level, the symbolic prosody features (called *prosodemes* [18], Sect. 2) are used in the synthesis. There are usually no other limits concerning F_0 behaviour since the symbolic prosody should ensure the natural increase/decrease of fundamental frequency, e.g. to distinguish declarative sentences from questions.

The speech corpora, recorded in a "newspapers style", should, in general, follow the formal prosody grammar generating the symbolic prosody description. However, despite being recorded by professional speakers, the corpus may

The work has been supported by the grant of the University of West Bohemia, project No. SGS-2016-039 and by Ministry of Education, Youth and Sports of the Czech Republic project No. LO1506. Access to computing and storage facilities owned by parties and projects contributing to the National Grid Infrastructure MetaCentrum provided under the programme "Projects of Large Research, Development, and Innovations Infrastructures" (CESNET LM2015042), is greatly appreciated.

P. Sojka et al. (Eds.): TSD 2018, LNAI 11107, pp. 411–418, 2018.
https://doi.org/10.1007/978-3-030-00794-2_44

contain some inconsistencies, especially in the *null* prosodeme which should not be manifested by any special F_0 behaviour compared to other prosodeme types. When used in the synthesis, these samples of speech sometimes cause a speech artefact appearance. In other words, the correct prosody labelling of the speech corpora is crucial for the high-quality speech synthesis.

For that reason, some studies and experiments, usually using Gaussian mixture models or HMM models, were performed to reveal wrongly labelled speech data in the corpora and to correct their symbolic prosody description [6,7,9]. The presented paper focuses mainly on the *null* prosodemes and presents another technique of the fine tuning of formal prosodic structures – the use of anomaly detection, based on the *Legendre polynomials* (Sect. 3.1) for the data representation.

2   Symbolic Prosody Features

The formal prosody grammar, introduced and described in details in [17–19], parses the given text sentence in a derivation tree. As a result of this parsing, each prosodic word (*PW*, i.e. a group of words with only one words stress) is assigned with an abstract prosodic unit, a *prosodeme*, marked as P_X. Since each prosodeme type represents a specific F_0 behaviour, it is used in the speech synthesis to ensure the required communication function on the phrase level of synthesized sentences [23] – the usage of correct prosodeme type is controlled by a *target cost* computation in the unit selection method.

However, the inaccurate description of speech corpora (and thus the wrong usage of some speech units in the synthesis itself) may lead to an unnatural excessive increase or decrease of F_0 contour in a non-phrase-final prosodic word with *null* prosodeme which could be manifested by an inappropriate stress or an unnatural melody or, eventually, it may result in a misunderstanding due to not keeping the required communication function.

In our TTS *ARTIC* [21], we distinguish the following prosodeme types:

- P_1 – prosodeme terminating satisfactorily (last *PW*s of declar. sentences)
- P_2 – prosodeme terminating unsatisfactorily (last *PW*s of questions)
- P_3 – prosodeme non-terminating (last *PW*s in intra-sentence phrases)
- P_0 – *null* prosodeme (assigned otherwise)
- $P_{0.1}$[1] – special type of *null* prosodeme assigned to first PWs in phrases [11]

2.1   Analysis of Prosodemes in Speech Corpora

In the presented paper, the experiments are carried out on large speech corpora recorded by professional speakers for the purposes of speech synthesis – a male voice AJ and a female voice MR [14,21]. For this study, the authors intentionally chose these two voices because of their difference. The male synthetic voice,

---

[1] This prosodeme type has been established recently and tested in [11].

built from AJ corpus, is widely used in commercial products for its high naturalness. On the other hand, the female synthetic voice MR is not very consistent in prosody (her prosody is very dynamic) – given the original prosodic description baseline, synthesized sentences quite often contain an unnatural intonation pattern.

Both corpora contain about 12,000 sentences (about 15 hours of speech) and more than 80,000 prosodic words. The complete statistics concerning the prosodeme types in the corpora are listed in Table 1.

Table 1. Number of prosodemes in corpora.

Corpus	Number of sentences	Number of prosodemes	P_0	P_1	P_2	P_3	$P_{0.1}$
AJ	12,277	84,733	35,781	9,850	922	12,141	26,039
MR	12,308	83,486	41,728	11,017	905	7,953	21,883

3   Anomaly Detection

Anomaly (or novelty) detection [3] is a well-known approach which is used to find items that do not have the same or similar properties as other items in a dataset, e.g. [12,13]. We are working with One-class Support Vector Machine (OCSVM), an unsupervised algorithm that learns a decision function and classifies new data as similar or different to a training set. We are using the implementation of OCSVM from *scikit-learn* [15] which is based on *libsvm* [4] with radial basis function as a kernel and $\gamma = 0.1$. We also set the ν parameter which influences the upper bound on the fraction of training errors, in our case to 10%. Apart from OCSVM, other anomaly detection algorithms available in *scikit-learn* (e.g. isolation forest) have also been tested, however, they did not perform so well when used on a different prosodeme type data which should be (almost) all classified as outliers. The results of OCSVM are presented in Sect. 4.

Since we are looking for anomalies only in our closed dataset, we can afford to train OCSVM on all data to get the best decision function possible. To train the OCSVM model, a choice of a suitable data description is a principal issue. Despite prosodemes being the only symbolic prosody features, each prosodeme type corresponds to specific changes in fundamental frequency contour – these could be modelled by *Legendre polynomials* (see Sect. 3.1) whose first four coefficients are used as the only representation of our data in the presented experiment.

3.1   Legendre Polynomials

Legendre polynomials are defined by Eq. 1,

$$L_n(x) = 2^n \sum_{k=0}^{n} x^k \binom{n}{k} \binom{\frac{n+k-1}{2}}{n},$$

(1)

and they form an orthogonal basis (i.e., non-correlated) suitable for modeling of F_0 contours [5,24]. An F_0 contour is described by coefficients as a linear combination of these polynomials. Because of the orthogonality, the coefficients can be estimated using cross-correlation at a time lag of 0 (i.e., a mutual energy of F_0 contour and Legendre polynomial). The first four polynomials $L_0(x)$, $L_1(x)$, $L_2(x)$ and $L_3(x)$ (see Fig. 1a) match linguistic interpretation as $L_0(x)$ responds to mean value of the pitch, $L_1(x)$ to rise or fall depending on the positive or negative sign of the coefficient (the slope is determined by its absolute value), $L_2(x)$ to peak or valley and $L_3(x)$ to the wave shape of F_0 contour.

(a)   **(b)**

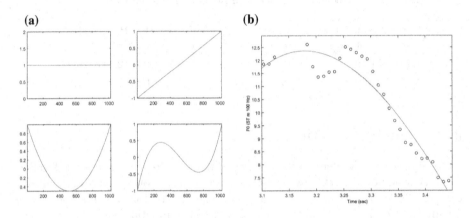

Fig. 1. Legendre polynomials. (a) The first four polynomials - mean value, slope, valley shape and curvature. (b) Interpolation of F_0 contour to estimate Legendre coefficients.

Using mPraat toolbox for Matlab [1,2], for each F_0 contour, we have transferred frequency values to semitone scale, interpolated the contour in 1,000 equidistant points and estimated the first four Legendre coefficients (for example, see Fig. 1b, coefficients are 10.7407 (*mean value*), −2.6880 (*falling slope*), −1.5522 (*valley shape*), 0.1685 (*only a slight wave curvature*)).

4   Evaluation of OCSVM Detection

We trained two OCSVM models. The first one was trained by using 35,781 P_0 prosodemes from AJ corpus and the second one by using 41,728 P_0 prosodemes from MR corpus. After training the models to be able to satisfactorily detect 10% of outliers in P_0 prosodemes, we tested how models react to the different type of prosodemes and also to the training data.

For each corpus we used appropriate OCSVM model to detect anomalies in each group of prosodemes (P_1, P_2, P_3, $P_{0.1}$ and P_0). Since each model was trained with P_0 prosodemes, where we supposed 10% of anomalies, we expected the number of outliers to be about 10% for the P_0 group and significantly higher for the other prosodeme groups.

Table 2. Number of anomalies detected by OCSVM for each type of prosodeme.

Corpus	P_0	P_1	P_2	P_3	$P_{0.1}$
AJ	3,578 (10.0%)	8,508 (86.4%)	299 (32.4%)	2,094 (17.2%)	10,774 (41.4%)
MR	4,174 (10.0%)	10,317 (93.6%)	51 (5.6%)	2,182 (27.4%)	12,412 (56.7%)

The results shown in Table 2 confirm our assumption, especially for the prosodeme P_1 where the number of outliers is 86.4% for AJ corpus and 93.6% for MR corpus. Nevertheless, in the case of P_2 in MR corpus, the number of outliers is only 5.6%. In the authors' opinion, the main problem is a very dynamic prosody of that corpus and the reason for this could be also hidden in the way how the female speaker finished the questions. She may not have raised her voice properly thus there could be only a small change in F_0 in P_2 prosodemes versus P_0 prosodemes. Therefore the OCSVM model could consider most of the P_2 to be P_0 and not mark them as outliers. Unfortunately, further exploration of this issue needs more elaboration and it is beyond the scope of this work.

5   Outlier Units Penalization in Unit Selection TTS

Since there is no clear evaluation of OCSVM, the results of anomaly detection will be evaluated directly in the speech synthesis itself – in the unit selection TTS [21]. In the baseline TTS system, the speech units originated from the anomalous prosodic words, i.e. outlier units, are more likely to cause a disturbing artefact due to an inappropriate F_0 contour, as described in Sect. 2. In a modified TTS, the outlier units are banned during the Viterbi search [22] for the optimal sequence of units. This may, hopefully, increase the naturalness of speech synthesis. Unfortunately, about 10% of all P_0 units are dropped by this approach – this should, however, not be a big problem since the corpora are quite large and they were carefully designed [14] to cover all different units sufficiently.

To analyse the frequency of outliers occurrences in the synthesized speech, we synthesized 6,000 sentences (containing 3–6 prosodic words). It emerged that at least one such a unit appeared approximately in every fourth synthesized sentence, only 6% of sentences contained minimally 5 anomalous units and at least 8 units appeared in only 1% of sentences (the percentages are almost the same for both AJ corpus and MR corpus). Afterwards, twenty sentences containing at least 8 anomaly units were randomly selected for the listening test.

5.1   Listening Test

To verify the contribution of our approach, i.e. penalization of units classified as outliers, we prepared a 3-scale preference listening test. The listening test consisted of 40 pairs of synthesized sentences (20 per voice), each pair included two variants of the same sentence – one generated by the original (baseline) system *TTS-base* and one generated by the modified system *TTS-new* where the outlier units were highly penalized in the *target cost* computation. The huge penalization should ensure that these units are not selected into the optimal units sequence found by Viterbi search [22], unless another (non-anomalous) unit exists. Naturally, this approach results in a different sequence of units compared to that generated by *TTS-base*.

Sixteen listeners participated in the listening test, five of them being speech experts. They were instructed to use earphones and to compare the overall quality of samples *A* and *B* in each pair by selecting one from these options:

– *Sentence A sounds better.*
– *I cannot decide.*
– *Sentence B sounds better.*

The order of the synthesized samples in the pairs was randomized so that the listeners did not know which sentence was generated by *TTS-base* or *TTS-new*.

Afterwards, the listeners' answers where normalized for each pair p to $p = 1$ where the *TTS-new* output was preferred, to $p = -1$ where the *TTS-base* output was preferred and $p = 0$ otherwise. These values are used for the final computation of the listening test score s, defined by Eq. 2, whose positive value indicates the improvement of the overall quality when using *TTS-new*.

$$s = \frac{\sum_{p \in T} p}{\sum_{p \in T} 1} \tag{2}$$

5.2   Results

The answers of the listening test participants were firstly counted separately for each particular speech corpus and then they were counted up together to determine the contribution of outlier units penalization in general. *TTS-new* was preferred for both voice corpora, the numbers of listeners' answers in *TTS-new* column are quite high and the score values s are always positive. The results of the listening test are clearly summarized in Table 3.

Table 3. The results of the listening test – the numbers and percentages of the answers from 16 listeners and the *sign test* results.

Corpus	TTS-base better	Same quality	TTS-new better	score s	p-value	sign test result
AJ	62 (19.4%)	76 (23.7%)	182 (56.9%)	**0.375**	<0.0001	*TTS-new* better
MR	104 (32.5%)	79 (24.7%)	137 (42.8%)	**0.103**	0.0391	*TTS-new* better
total	166 (25.9%)	155 (24.2%)	319 (49.9%)	**0.239**	<0.0001	*TTS-new* better

Despite the positive score value s for MR corpus, the results are not so convincing. Thus, the authors of this study decided to carried out a *sign test* (similar to e.g. [10]) to verify the statistical significance of the obtained results. The null and alternative hypothesis were defined as follows:

$H0$: The outputs of both systems are of the same quality.
$H1$: The outputs of one system sound better.

The computed p-values and the results of the *sign test* are also listed in Table 3. It was proved that the results are statistically significant on the significance level $\alpha = 0.05$, however, the p-value is very close to α for MR corpus.

6 Conclusion and Future Work

In this paper, we described the usage of anomaly detection to reduce the prosody artefacts in the speech synthesis. We used OCSVM detectors, based on Legendre polynomials features representing F_0, to find anomalies among our P_0 prosodemes in two corpora. Afterwards, the outcomes of OCSVM were used in the unit selection – the authors investigated the contribution of the "ban" of these units in the listening test. The results, presented in Sect. 5.2, indicate an improvement in overall quality of speech synthesis (even despite not using 10 % of P_0 prosodic words).

The authors plan to perform the presented anomaly detection of P_0 prosodemes on more speech corpora and to verify its contribution in a large listening test. Moreover, as a future work, we are considering classification of P_0 outliers to other prosodeme types by using a multi-class classifier. The usage of these re-labelled units, i.e. P_0 outliers with a new prosodeme assigned, could be more beneficial than reducing data by excluding these outliers totally.

References

1. Boersma, P., Weenink, D.: PRAAT: doing phonetics by computer [computer program]. http://www.praat.org/ (2018)
2. Bořil, T., Skarnitzl, R.: Tools rPraat and mPraat. In: Sojka, P., Horák, A., Kopeček, I., Pala, K. (eds.) TSD 2016. LNCS (LNAI), vol. 9924, pp. 367–374. Springer, Cham (2016). https://doi.org/10.1007/978-3-319-45510-5_42
3. Chandola, V., Banerjee, A., Kumar, V.: Anomaly detection: a survey. ACM Comput. Surv. **41**(3), 1–58 (2009)
4. Chang, C.C., Lin, C.J.: LIBSVM: a library for support vector machines. ACM Trans. Intell. Syst. Technol. **2**, 27:1–27:27 (2011). http://www.csie.ntu.edu.tw/~cjlin/libsvm
5. Grabe, E., Kochanski, G., Coleman, J.: Connecting intonation labels to mathematical descriptions of fundamental frequency. Lang. Speech **50**(Pt 3), 281–310 (2007)
6. Hanzlíček, Z.: Correction of prosodic phrases in large speech corpora. In: Sojka, P., Horák, A., Kopeček, I., Pala, K. (eds.) TSD 2016. LNCS (LNAI), vol. 9924, pp. 408–417. Springer, Cham (2016). https://doi.org/10.1007/978-3-319-45510-5_47
7. Hanzlíček, Z., Grůber, M.: Initial experiments on automatic correction of prosodic annotation of large speech corpora. In: Sojka, P., Horák, A., Kopeček, I., Pala, K. (eds.) TSD 2014. LNCS (LNAI), vol. 8655, pp. 481–488. Springer, Cham (2014). https://doi.org/10.1007/978-3-319-10816-2_58

8. Hanzlíček, Z.: Czech HMM-based speech synthesis. In: Sojka, P., Horák, A., Kopeček, I., Pala, K. (eds.) TSD 2010. LNCS (LNAI), vol. 6231, pp. 291–298. Springer, Heidelberg (2010). https://doi.org/10.1007/978-3-642-15760-8_37

9. Hanzlíček, Z.: Classification of prosodic phrases by using HMMs. In: Král, P., Matoušek, V. (eds.) TSD 2015. LNCS (LNAI), vol. 9302, pp. 497–505. Springer, Cham (2015). https://doi.org/10.1007/978-3-319-24033-6_56

10. Jůzová, M., Tihelka, D., Skarnitzl, R.: Last syllable unit penalization in unit selection TTS. In: Ekštein, K., Matoušek, V. (eds.) TSD 2017. LNCS (LNAI), vol. 10415, pp. 317–325. Springer, Cham (2017). https://doi.org/10.1007/978-3-319-64206-2_36

11. Jůzová, M., Tihelka, D., Volín, J.: On the extension of the formal prosody model for TTS. In: Sojka, P., Horák, A., Kopeček, I., Pala, K. (eds.) TSD 2018. LNCS, vol. 11107, pp. 351–359. Springer, Cham (2018)

12. Matoušek, J., Tihelka, D.: Annotation errors detection in TTS corpora. In: INTERSPEECH 2013, Lyon, France, pp. 1511–1515 (2013)

13. Matoušek, J., Tihelka, D.: Anomaly-based annotation error detection in speech-synthesis corpora. Comput. Speech Lang. **46**(C), 1–35 (2017)

14. Matoušek, J., Tihelka, D., Romportl, J.: Building of a speech corpus optimised for unit selection TTS synthesis. In: LREC 2008, Proceedings of 6th International Conference on Language Resources and Evaluation, pp. 1296–1299. ELRA, Marrakech (2008)

15. Pedregosa, F., et al.: Scikit-learn: machine learning in Python. J. Mach. Learn. Res. **12**, 2825–2830 (2011)

16. Qian, Y., Soong, F.K., Yan, Z.J.: A unified trajectory tiling approach to high quality speech rendering. IEEE Trans. Audio Speech Lang. Process. **21**(2), 280–290 (2013)

17. Romportl, J.: Structural data-driven prosody model for TTS synthesis. In: Proceedings of the Speech Prosody 2006, pp. 549–552. TUD Press, Dresden (2006)

18. Romportl, J., Matoušek, J.: Formal prosodic structures and their application in NLP. In: Matoušek, V., Mautner, P., Pavelka, T. (eds.) TSD 2005. LNCS (LNAI), vol. 3658, pp. 371–378. Springer, Heidelberg (2005). https://doi.org/10.1007/11551874_48

19. Romportl, J., Matoušek, J., Tihelka, D.: Advanced prosody modelling. In: Sojka, P., Kopeček, I., Pala, K. (eds.) TSD 2004. LNCS (LNAI), vol. 3206, pp. 441–447. Springer, Heidelberg (2004). https://doi.org/10.1007/978-3-540-30120-2_56

20. Tihelka, D.: Symbolic prosody driven unit selection for highly natural synthetic speech. In: INTERSPEECH 2005, pp. 2525–2528. ISCA, Bonn (2005)

21. Tihelka, D., Hanzlíček, Z., Jůzová, M., Vít, J., Matoušek, J., Grůber, M.: Current state of text-to-speech system ARTIC: a decade of research on the field of speech technologies. In: Sojka, P., Horák, A., Kopeček, I., Pala, K. (eds.) TSD 2018. LNCS, vol. 11107, pp. 369–378. Springer, Cham (2018)

22. Tihelka, D., Kala, J., Matoušek, J.: Enhancements of Viterbi search for fast unit selection synthesis. In: INTERSPEECH 2010, pp. 174–177. ISCA, Makuhari (2010)

23. Tihelka, D., Matoušek, J.: Unit selection and its relation to symbolic prosody: a new approach. In: INTERSPEECH 2006, vol. 1, pp. 2042–2045. ISCA, Bonn (2006)

24. Volín, J., Tykalová, T., Bořil, T.: Stability of prosodic characteristics across age and gender groups. In: INTERSPEECH 2017, pp. 3902–3906 (2017)

25. Vít, J., Matoušek, J.: On the analysis of training data for WaveNet-based speech synthesis. In: Proceedings of ICASSP 2018 (2018)

The Influence of Errors in Phonetic Annotations on Performance of Speech Recognition System

Radek Šafařík[1]([⊠]), Lukáš Matějů[1], and Lenka Weingartová[2]

[1] Institute of Information Technology and Electronics,
Technical University of Liberec, Liberec, Czech Republic
{radek.safarik,lukas.mateju}@tul.cz
[2] NEWTON Technologies, Prague, Czech Republic
Lenka.Weingartova@newtontech.cz

Abstract. This paper deals with errors in acoustic training data and the influence on speech recognition performance. The training data can be prepared manually, automatically or by combination of these two. In all cases, some mislabeled phonemes can appear in phonetic annotations. We conducted series of experiments which simulate some common errors. The experiments deal with various amount of changes in phonetic annotations such as different types of changes in voicing of obstruents, random substitution of consonants or vowels and random deleting of phonemes. All experiments were done for Czech language using GlobalPhone speech data set and both Gaussian mixture models and deep neural networks were used for acoustic modeling. The results show that some amount of such errors in training data does not influence speech recognition accuracy. The accuracy is significantly influenced only by large amount of errors (more than 50%).

Keywords: Speech recognition · Gaussian mixture models
Deep neural networks · Phonetic annotations · Phoneme substitution

1 Introduction

Modern Automatic Speech Recognition (ASR) systems are compiled from different modules which can be divided into language dependent and language independent groups. Whereas the language dependent modules have to be altered for every given language, the language independent modules remain unchanged. The main language dependent modules are the acoustic model (AM) and the language model (LM). These are created using state-of-the-art machine learning methods. For that, a sufficient amount of training data is needed. Training data for AM consist of speech recordings and their phonetic annotations. Such training data can be obtained in different ways: (a) to buy already created training data (b) to download some free data from the Internet, (c) to create your own data. The first option is the most comfortable. There are several companies offering

© Springer Nature Switzerland AG 2018
P. Sojka et al. (Eds.): TSD 2018, LNAI 11107, pp. 419–427, 2018.
https://doi.org/10.1007/978-3-030-00794-2_45

different speech data for many languages (e.g. GlobalPhone [11]). However, the data can cost a lot of money depending on their amount and quality and they are mostly available for major languages only. If we have limited funding, another option is to use free sources available on the Internet. It is free but mostly in low quality and it is not certain whether phonetic transcriptions perfectly correspond to the speech in recordings, if it contains non-speech elements, etc. Last option is to create our own training data from available sources. If we work on ASR system for some minor language, generally no training data are available for the language and create own data is the only option. We have developed our own automatic approach for creation of training data. It downloads text and audio data from free available sources on the Internet and process them in iterative retraining approach. All details are described in [5].

The amount of related works about influence of errors in training data is very scarce. We found only one work in field of speech recognition [12]. It shows effect of errors using Gaussian Mixture Models. In the work, errors were simulated by substituting of similar and dissimilar phonemes and results show that even if the training data contains 20% of errors it does not significantly influence the performance. Similar works but from different fields focused on dealing with errors in training data in image processing [4], class imbalances in training data [1] and identifying and filtering mislabeled training data [2].

2   ASR System Settings

Our ASR system used in our experiments has been developed since 90s and it is capable of real-time transcription using a lexicon of size over a half million words. Originally, it was developed for Czech language and used for many applications such as voice dictation, transcription of historical audio archive [10] or broadcast monitoring [7]. Thanks to its modular basis, it was easily ported for another Slavic languages [8,9].

The system works with two types of acoustic models: Gaussian Mixture Models (GMM) and Deep Neural Networks (DNN). The GMMs trained within the scope of this work were triphone multi-gaussian models with 32 mixtures. The models were speaker-independent and context-dependent. 39-dimensional Mel-Frequency Cepstral Coefficients (MFCC) were used for parametrization of the input signal. Additionally, Cepstral Mean Subtraction (CMS) and HLDA transformation were applied. The frame length was 25 ms and the time shift was 10 ms. The training was done using HTK speech recognition toolkit.

The DNNs trained within the scope of this work were based on a deep neural network – hidden Markov model (DNN-HMM) hybrid architecture [3]. They used the same acoustic data as the baseline GMM models. The hyper-parameters of DNNs were set as described in detailin [6]. Each network had five hidden layers with a decreasing number of neurons per layer (tree structure - 1024-1024-768-768-512). The hidden layers used ReLU activation function with the softmax function being applied in the output layer. The networks were trained within 15 epochs with a learning rate set to 0.08. The features were 39-dimensional log filter banks, and five previous frames, the current frame and five following frames formed the input feature vector (5-1-5). The inputs were also locally normalized within a one-second-long window. Note that Torch framework (http://torch.ch/) was used to train the DNNs.

In all experiments, the same LM and vocabulary were used. The system works with statistical n-gram language model with Kneser–Ney smoothing. The vocabulary contains 621k words with 651k pronunciations and LM was trained on 5 GB of text data.

3   Experimental Work

The main goal of our experiments was to find how the annotation errors in training data influence the training of acoustic model and the final recognition score of our ASR system. We conducted several experiments where we simulated different kind of errors on Czech training set. The experiments can be divided into experiments with substitution of vowels and consonants, experiments with change of voicing of obstruents and obstruent clusters and experiments with deletion of phonemes. All experiments were done with both GMM and DNN models.

3.1   Train and Test Data

Czech language was chosen for experiments as we have suitable data and it is our native language. Czech part of GlobalPhone database [11] was used for experiments, and thus, the experiments can be easily replicable as the GlobalPhone database is accessible. It consist of recordings of read speech from 102 hundred native speakers each reading 100 hundred sentences. The database was divided for test and training sets. The first 10 speakers (1–10) were used for testing and the rest (11–102) for experimenting and training. Details about train and test sets are in Table 1.

Table 1. Train and test data sets

Speech set	Hours	Words
GlobalPhone – Test set	3.8	26,535
GlobalPhone – Training set	27	222,215

The phoneme set we use for Czech language contains 40 phonemes and 5 type of noises. Phonemes are divided into 11 vowels (6 short and 5 long) and 30 consonants (10 voiced obstruents, 10 unvoiced obstruents and 9 other consonants). In our experiments, we focused on groups of consonants, long and short vowels and obstruents. We also did experiments with word-final obstruents and obstruent clusters. About 95% of clusters are composed from two phonemes, the rest is from three phonemes. Distribution of these different phoneme groups in the training set is shown in Table 2.

Table 2. Distribution of phonemes in the training set

Phoneme group	Count
Vowels	432,639
Short vowels	343,628
Long vowels	89,011
Consonants	598,067
Voiced obstruents	138,704
Unvoiced obstruents	238,500
Ending obstruents	32,317
Voiced obst. clusters	8,208
Unvoiced obst. clusters	40,336
Ending obst. clusters	1,952

3.2   Experimental Setup

In our experiments, we wanted to find out how different kind of errors in phonetic annotations influence the recognition score. The experiments can be divided into three groups.

– substitution of phonemes
– voice change of obstruents
– random deleting of phonemes

Every group of experiments was applied on different kinds of phonemes such as vowels (long and short), consonants, obstruents and obstruent clusters. To study the effect on speech recognition performance, we altered phoneme annotations in the GlobalPhone training set. In each experiment, a probability was rolled for every phoneme (or a whole cluster) from examined group and if the number was lower than a given threshold, the phoneme (or a whole cluster) was changed. Different thresholds in range from 1% to 100% were used. Moreover, both GMM and DNN were used in all experiments to compare their behavior.

Random substitutions of phonemes were done for three groups: consonants, all vowels together and vowels distinguished by their length. In each experiment, a given amount of random phonemes in phonetic annotations was replaced by different random phonemes from their group (i.e. consonant for another consonant, vowel for another vowel, short vowel for another short vowel). It can simulate the situation when an annotator makes a typing error or speaker does not pronounce the word properly, but our automatic approach for creation of annotation recognizes it in its orthographically correct form and adds it to the annotation.

Second group of experiments was focused on a random changing of obstruent voicing. The experiments were done in different groups: for single obstruents, for whole obstruent clusters and for word-final single obstruents and obstruent clusters. In each experiment, a given amount of random obstruents was replaced by their voicing counterparts (e.g. 'k' for 'g', 'd' for 't', etc.). In case of single obstruents, a random obstruent was replaced, even as a part of a cluster while the rest of the cluster remained unchanged. In case of clusters, the whole consonant cluster changed voicing. These experiments again simulate incorrect annotation done by annotator or an error in automatic annotations where can appear some pronunciation error in vocabulary caused by manual edition or incorrect G2P conversion. The same experiments were done for word-final obstruents which is a very common error in phonetic annotations.

The last group of experiments was focused on random deletion of phonemes. It was done for three different group of phonemes: consonants, vowels and all phonemes. It simulates either annotation errors or G2P conversion errors.

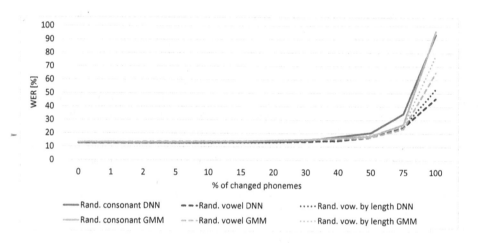

Fig. 1. WER for random substitution of phonemes in different groups.

4   Results

All experiments were evaluated using standard Word Error Rate (WER) metric:

$$\text{WER}[\%] = \frac{I + S + D}{N} \cdot 100, \tag{1}$$

where I, D and S are the numbers of insertions, deletions and substitutions, respectively. N is the total number of words in the reference text.

Figure 1 shows the results of how random phoneme substitution in different groups influences the performance. It is evident that WER is not affected until about 40% of phonemes from a given group are changed. Random substitutions in the consonant group affect performance more than substitutions in vowel groups because there are more phonemes in the consonant group which are phonetically more diverse than vowels. However, the difference within vowel groups is more interesting. It can bee seen from the results that the performance is a slightly

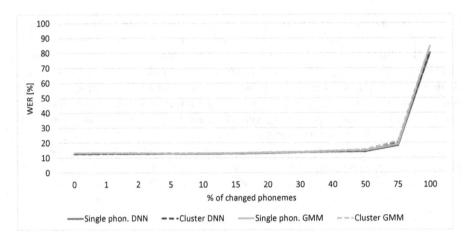

Fig. 2. WER for voice change of obstruents.

more affected when the vowels are substituted by other vowels with the same length than when the substitution does not take the length into account. From the comparison between GMM and DNN models can be seen that both models have relatively similar WER until 50% of substitution. Then, DNN models are more affected by consonant substitutions whilst GMM are more affected by vowel substitutions.

Figures 2 and 3 show results from experiments with voicing change of obstruents. The performance is not significantly affected until 50% of changes. Then, the performance is significantly worse. However, the performance is not affected at all when only word-final obstruents are changed (neither single obstruents nor whole clusters). The difference between changing single obstruents and whole obstruent clusters is negligible. But note that 36% of all phonemes are obstruents and only 3% are word-final obstruents. The difference between GMM and DNN models was negligible in these experiments.

Figure 4 shows the last experiments with deletion of phoneme. The deletion was done only up to 90% because the training process crashed with higher thresholds due to small amount of phonemes in the training set. In case of random phonemes deletion, even up to 30% of deleted phonemes the performance is not largely affected so much. The WER grows significantly from 50%. Afterwards, when only consonants or vowels are deleted, the performance is not affected as much. Even when 90% of consonants or vowels are deleted, the WER is still not as high as expected. Note that 90% of consonants or vowels are 52% and 37% of all phonemes, respectively. It can be seen that DNN models are more affected by deletion than GMMs. It is obvious in case of all phonemes deletion where the performance of DNN is getting significantly worse already with 10% of deleted phonemes.

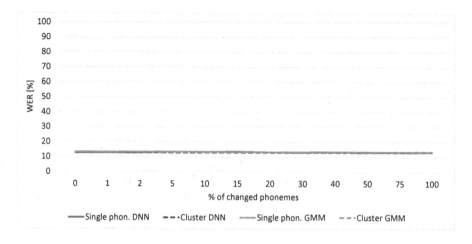

Fig. 3. WER for voice change of word-final obstruents.

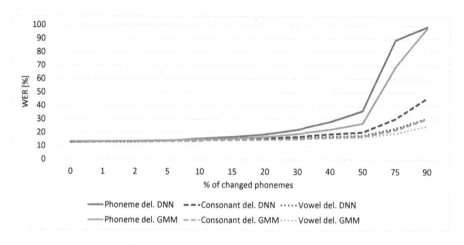

Fig. 4. WER for deletion of phonemes.

5   Conclusions

In this paper, we presented several experiments focused on how errors in training data influence the performance of our speech recognition system. We conducted three types of experiments: (1) substitution of phonemes, (2) voicing change of obstruents and (3) deletion of phonemes. Each type of experiment was applied on different groups of phonemes such as vowels (also distinguished by length), consonants, obstruents and all phonemes.

All experiments were done for Czech language using GlobalPhone speech data set. The results showed that both used acoustic models (GMM and DNN) are very robust and the performance is not influenced even with relatively high amount of errors, but DNN models are more vulnerable to deleted phonemes in phonetic annotations than GMM. In other types of experiments, the behavior of both types of models was relatively similar. From the point of view of training data creation, the results showed us that it is not necessary to spend too much effort to create perfectly precise phonetic annotations. Even when up to a quarter of whole training set is mislabeled, it does not significantly affect the performance of the system. It also highly depends on the type of error. If a phoneme is mislabeled as some other phonetically similar phoneme, it does not affect the model as much as when it is mislabeled for some random phoneme.

It should be noted that our experiments were done on a bigger amount of training data. It follows that with well-resourced languages (such as Czech in our case) it is possible to substitute manual labor with data, at least to some point. Especially our results with voicing substitutions are applicable to real-world situations: in Slavic languages, the voicing of word-final obstruents (or clusters) is governed by the following word. Thus, vocabulary creators are often faced with the decision whether to preserve the difference and have two variants for every word that ends with an obstruent, or whether to keep only one – and which? Our results indicate that any alternative might be valid and annotators can use whatever speeds up their work.

Acknowledgements. This work was supported by the Technology Agency of the Czech Republic in project no. TH03010018 and by the Student Grant Scheme 2018 of the Technical University in Liberec.

References

1. Batista, G.E., Prati, R.C., Monard, M.C.: A study of the behavior of several methods for balancing machine learning training data. ACM SIGKDD Explor. Newsl. **6**, 20–29 (2004)
2. Brodley, C.E., Friedl, M.A.: Identifying mislabeled training data. J. Artif. Intell. Res. **11**, 131–167 (1999)
3. Dahl, G.E., Yu, D., Deng, L., Acero, A.: Context-dependent pre-trained deep neural networks for large-vocabulary speech recognition. IEEE Trans. Audio Speech Lang. Process. **20**, 30–42 (2012)
4. Hansen, M.S., Kozerke, S., Pruessmann, K.P., Boesiger, P., Pedersen, E.M., Tsao, J.: On the influence of training data quality in k-t BLAST reconstruction. Magn. Reson. Med. **52**(5), 1175–1183 (2004)
5. Nouza, J., Cerva, P., Safarik, R.: Cross-lingual adaptation of broadcast transcription system to polish language using public data sources. In: Vetulani, Z., Mariani, J., Kubis, M. (eds.) LTC 2015. LNCS (LNAI), vol. 10930, pp. 31–41. Springer, Cham (2018). https://doi.org/10.1007/978-3-319-93782-3_3
6. Matějů, L., P.C., Ždánský, J.: Investigation into the use of deep neural networks for LVCSR of Czech. In: IEEE International Workshop of Electronics, Control, Measurement, Signals and Their Application to Mechatronics (ECMSM), pp. 1–4 (2015)
7. Nouza, J., Ždánský, J., Červa, P.: System for automatic collection, annotation and indexing of Czech broadcast speech with full-text search. In: Melecon 2010–2010 15th IEEE Mediterranean Electrotechnical Conference, pp. 202–205, April 2010. https://doi.org/10.1109/MELCON.2010.5476306

8. Safarik, R., Nouza, J.: Unified approach to development of ASR systems for East Slavic languages. In: Camelin, N., Estève, Y., Martín-Vide, C. (eds.) SLSP 2017. LNCS (LNAI), vol. 10583, pp. 193–203. Springer, Cham (2017). https://doi.org/10.1007/978-3-319-68456-7_16

9. Nouza, J., Šafaří k, R., Červa, P.: ASR for South Slavic languages developed in almost automated way. In: INTERSPEECH 2016, pp. 3868–3872, September 2016

10. Nouza, J., et al.: Speech-to-text technology to transcribe and disclose 100,000+ hours of bilingual documents from historical Czech and Czechoslovak radio archive. In: INTERSPEECH, pp. 964–968. ISCA (2014)

11. Schultz, T.: Globalphone: a multilingual speech and text database developed at Karlsruhe University. In: Proceedings of the ICSLP, pp. 345–348 (2002)

12. Sundaram, R., Picone, J.: Effects on transcription errors on supervised learning in speech recognition. In: Proceedings of IEEE International Conference on Acoustics, Speech, and Signal Processing, ICASSP 2004, vol. 1, p. I–169. IEEE (2004)

Deep Learning and Online Speech Activity Detection for Czech Radio Broadcasting

Jan Zelinka[✉]

Faculty of Applied Sciences, New Technologies for the Information Society,
University of West Bohemia, Univerzitní 8, 306 14 Pilsen, Czech Republic
zelinka@kky.zcu.cz

Abstract. In this paper, enhancements of online speech activity detection (SAD) is presented. Our proposed approach combines standard signal processing methods and modern deep-learning methods which allows simultaneous training of the detector's parts that are usually trained or designed separately. In our SAD, an NN-based early score computation system, an NN-based score smoothing system and proposed online decoding system were incorporated in a training process. Besides the CNN and DNN, spectral flux and spectral variance features are also investigated. The proposed approach was tested on a Czech Radio broadcasting corpus. The corpus was used for investigation supervised and also semi-supervised machine learning.

Keywords: Speech activity detector · Differentiable decoding
Semi-supervised learning

1 Introduction

A large percentage of a usual radio broadcasting is irrelevant music mostly with a singing voice. Even acapella solos aren't rare. Standard voice activity detection might detect these irrelevant vocal parts or neglect a relevant speech activity with a music on its background. Thus, a specialized SAD is necessary.

Applying delta coefficients and a variance computation as a spectral flux measure is briefly investigated in this paper. Nevertheless, we mostly focused on using Convolutional Neural Networks (CNN) [2] and Deep Neural Network (DNN). There are three different parts of our proposed SAD: The first part computes early speech activity scores from a relatively short context of input observation (i.e. spectrum). The second part uses a large context (about 2 s) of the early scores and smoothes the early scores. The third part uses the smoothed early scores and makes the final decision by means of a decoder.

J. Zelinka—This work was supported by the European Regional Development Fund under the project AI&Reasoning (reg. no. CZ.02.1.01/0.0/0.0/15 003/0000466).

P. Sojka et al. (Eds.): TSD 2018, LNAI 11107, pp. 428–435, 2018.
https://doi.org/10.1007/978-3-030-00794-2_46

In our work, we try to avoid to train any part of a SAD separately because separate design could lead to a suboptimal solution. In order to accomplish this goal, we incorporated the score smoothing part and the final online decoding into a training process. Usually, an HMM decoder is used for making a final decision. While the smoothing part incorporation is truly unproblematic because it is a simple average computation or an NN, an incorporation of an HMM decoder seems almost unfeasible. Hence, we replaced a standard HMM decoding with a simpler and differentiable online decoding method that keeps important qualities of the HMM detector and that can be applied online without any modification.

Because our intention was to allow some fine tuning of the SAD in a speech recognition system, we avoided to make hard 0/1 decision and we used score computation. Hence, a minor modification of standard HMM decoder is also presented. The modification allows us to compare not only Frame Error Rate (FER) but also Equal Error Rate (EER).

The proposed approach was tested on the Czech Radio broadcasting corpus that consists of circa 50% music with predominant singing voice. The corpus was used for a supervised machine learning and a semi-supervised machine learning.

2   Related Work

A common SAD utilizes the fact that a voiced and an unvoiced sound changing much more rapidly than in generic noise or a music (even so-called fast music pieces). Thus, even a simple SAD could reach a high-quality detection by means of measuring a spectral flux. Some successful methods compute so called vocal variance [4]. Another measure is based on time domain features and/or measures a distance between two subsequent feature vectors [6].

Nowadays, a CNN replaces such features except delta (and delta-delta) coefficient [9] that indicate some perceptual spectral flux, too. Since delta coefficients computation and absolute values computation could be seen as an application of standard CNN with max pooling, using only CNN is a clear choice [8] in case of supervised learning. In this paper, we combined standard feed-forward NNs and a long context processing.

Using recurrent NNs is also a promising possibility [3]. In this paper, we focused on a differentiable online decoding that could be incorporated in a training process. Other papers deal with complex transducers [5] or sophisticated boosting methods such as Boosted DNN [11].

In our paper, we used ordinary spectrogram as an input of our SAD. There are several spectrum-based feature vectors such as the Perceptual Spectral Flux [6], the Multi-Resolution Cochleagram [1] or the Gammatonegram [10]. These features are reported as highly beneficial in case of an unsupervised SAD. Feature vectors for speech recognition such as the PLP could be applied [3] at least as an extension of specialized features [7].

3   Softmax and a Score Computation

In this paper, we suppose that a SAD produces a score for each input feature vector. As usual, all SADs were trained by means of cross-entropy criterion applied on a layer with softmax activation function. These SADs estimate two posteriors: conditional probability of speech and non-speech. We found that the posteriors are usually too sharp (even after applying an inverse sigmoid function) to be used for some following fine tuning in a speech recognition system. Therefore, we have applied another approach where last hidden layers (or another signal processing module that immediately precedes an output layer) have one single output that serves as a score computation and output layers have fixed weights and biases. Both fixed biases are equal to zero and the fixed weights are -1 or $+1$ respectively. Hence, a negative score produces posteriors that indicate non-speech segment and a positive score produces posteriors indicating speech segment.

This approach doesn't limit a potency in comparison with an NN that doesn't include such ultimate bottleneck and such fixed weights in its output layer since we could add a linear bottleneck producing a score s and the fixed output layer without changing input/output behavior. The proof is simple: After short reasoning, one can see that nonexistence of s implies contradiction $y_1 + y_2 \neq 1$ where y_1, y_2 are softmax outputs. Hence, s exists and it could be computed as $s = (\xi_2 - \xi_1)/2$ where ξ_1 and ξ_2 are inputs on the softmax function. Naturally, there is a possible but implausible risk that this approach could train an inferior NN. On the other hand, this approach certainly produces scores that have useful bimodal distribution.

4   HMM-Based and Max-Pooling-Based Decoder

The first part of the proposed SAD computes only an early score. Either this score nor used smoothing does guarantee to produce segments suitable for speech recognition. Hence, applying some decoder which could even decrease an error is necessary.

In our experiments, we used an HMM decoder with two states. Since we intended to compare EERs, we slightly modified the algorithm in order to produce a score for each time point. The modification lies in a utilization of conditional probabilities $\phi_{\omega_t} = P(\omega_t | o_t, \omega_{t-1})$ for observations o_t and states ω_t computed by Viterbi's algorithm. We compute score s for an HMM as

$$s = \begin{cases} +sp(\Delta) & \text{for } \omega_t \in SP \\ -sp(\Delta) & \text{for } \omega_t \in NSP \end{cases}, \quad \Delta = Ln\left(\sum_{\omega \in SP} \phi_\omega\right) - Ln\left(\sum_{\omega \in NSP} \phi_\omega\right), \quad (1)$$

where v_t is a state in the optimal path at a time point t, SP is a set of speech production states, NSP is a set of other states and sp is softplus function, i.e. a smooth approximation to ReLU. Obviously, positive scores indicate speech activity and negative scores indicate non-speech activity.

Incorporation of the HMM decoder into a gradient descent seems to be impracticable. Thus, we proposed a simpler final decision-making process that fulfills these two requirements: (1) The process is differentiable by means of a standard machine learning toolbox such as Tensorflow. (2) The process makes speech segments long enough. These requirements led us to design decoder that uses a window of scores and finds the maximal score. Obviously, this max-pooling based decoding produces segments that have lengths at least equal to the length of the window. Although minimum computation fulfills the same requirements, we prefer maximum because a false positive is usually less harmful than a false negative. Figure 1 shows a real example of the proposed score decoding.

The described HMM decoder and the training process described in Sect. 3 should lead to a threshold which equals to zero. However, the statement about speech segments lengths holds for any threshold.

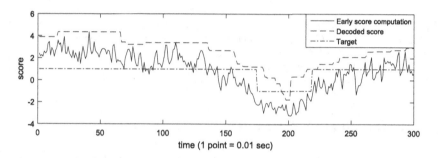

Fig. 1. A real example of the proposed score decoding.

5   Simultaneously Trained NNs

Our proposed SAD has three parts. All these three parts are showed in details in Fig. 2 part f. The first part processes a spectrogram. This part combines CNNs with max pooling and full-connected layers and produces an early score. All CNN use a 3×3 window, the number of kernels is always six and ReLU is always an activation function. Standard full-connected layers use ReLU too and relevant layers are followed by dropout layers in a training process.

The second part smoothes the early score by means of a window where timeshifts are from -100 to $+100$. One single neuron with linear (diagonal) activation function produces a smoothed early score. Instead of the neuron, a simple average could be applied. But the neuron could also perform a filter such as hamming widow application. Unfortunately, visualization of the neuron weights didn't show any meaningful pattern.

The last part is the differentiable decoding process described in Sect. 4. This process doesn't include any NN layer but we add the fixed layer in the training process as we mentioned in Sect. 3.

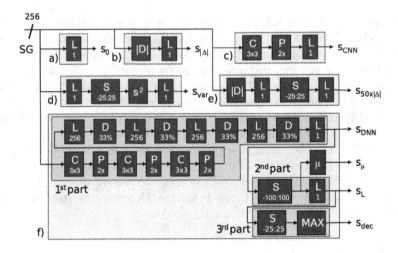

Fig. 2. Schemas of investigated SADs without any decoder (a–e) the SAD with the proposed decoder (f). SG stands for spectrogram, C for convolution layer, P for (max) pooling, L for standard full-connected layer, D for dropout, S for splicing, $|\Delta|$ for absolute values of delta coefficients, μ for an average computation, σ^2 for a variance computation and MAX for the maximum finding.

6   Experiments and Results

Since the proposed SAD process large context twice (see Fig. 2, part f), used SGD algorithm have to use at least 2.56 s. long segments. We used 20 s. long segments during the training process. Each batch in the training process contained 64 of these long segments. The segments have been selected randomly but the training process kept a percentage of speech activity as close to 50% as possible. This training process was used in every presented experiment.

The sampling rate was 16 kHz. Our SAD processes a logarithm of amplitude spectrum that is computed from 512 samples long windows with 160 samples step (hop). Thus, the spectrum has 256 features.

All results are shown in Table 1 and some selected Detection Error Tradeoff (DET) curves are shown in Fig. 3. The way how the SADs are trained and the fact that the amount of music and speech almost equal make FER and EER also almost equal in many cases.

6.1   Czech Radio Broadcasting Corpus

The train set contains approximately 42 h of an annotated signal from one selected radio station and approximately 405 h of an unannotated signal for five selected radio stations for semi-supervised learning. These unannotated recordings were supplemented with 189 h of selected popular music and 71 h of archived mp3 files with cut out music. The test set contains 24 h from one selected radio station. The annotated part of the train set doesn't contain the same radio

Table 1. Observed FERs and EERs. No HMM means that no HMM decoding was applied and HMM means that the described modified HMM decoder was applied. See Fig. 2 for understanding the notation.

s	FER (no HMM)	EER (no HMM)	FER (HMM)	EER (HMM)
0	23.0%	23.2%	15.9%	16.8%
$\|\Delta\|$	19.6%	19.9%	10.6%	10.4%
CNN	16.5%	16.9%	10.3%	12.2%
var	8.3%	8.3%	10.6%	7.9%
$50 \times \|\Delta\|$	6.7%	6.7%	**6.2%**	**6.1%**
DNN	10.9%	11.3%	8.3%	9.9%
μ	6.0%	6.5%	5.9%	6.5%
L	4.5%	4.0%	**3.8%**	**4.0%**
dec	3.6%	3.6%	3.4%	3.5%
$semisup$	**2.8%**	**2.8%**	2.8%	2.8%

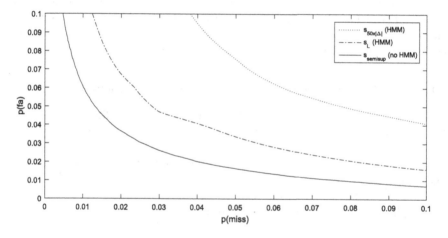

Fig. 3. DET curves for selected SADs. See Fig. 2 for understanding the notation.

stations that are included in the test set. Thus, the test set contains different voices, different jingles and slightly different music style. The test set contains approximately 50% of a speech activity.

6.2 Spectral Flux

In order to investigate the quality of spectral flux measures, we used only single layers to compute scores in the first experiments (see Fig. 2, parts a–e). We investigated raw spectrogram (s_0), absolute values of delta coefficients ($s_{|\Delta|}$), a single CNN layer with 2 kernels (s_{CNN}) and variance computation (s_{var}). Results for s_0 shows that applying a simple SAD based on an energy threshold is highly inappropriate. Variance computation is much more successful in comparison with the other simple method. But this success might lie in a window which is much longer than for example used context for delta coefficient computation. Thus, we used a new window that makes a total amount of data which is used in decision making equal ($s_{50 \times |\Delta|}$). Accuracies and EERs are in the first part of Table 1. The results show that the CNN outperforms the delta coefficients computation and the variance computation isn't beneficial in comparison with the delta coefficient computation. The results also show that process spectral flux could give decent results even when utterly simple NNs are used.

6.3   Score Smoothing

In the second experiment, score smoothing methods are investigated. All used NNs have two parts: (1) 3 CNN layers and 5 full-connected NN layers that compute an early score and (2) a score smoothing. We trained two different NNs with the same structure of the first part. At first, we didn't smooth the scores at all (s_{DNN}). Then, we computed an average from several windows (s_μ). In our experiments, the optimal window was $-167{:}167$. Finally, we trained one single layer as the average substitution (s_L). The results (see Table 1) show that smoothing significantly decreases the errors.

6.4   Max Pooling Decoder and Semi-supervised Machine Learning

In the last experiment, the proposed system with max pooling decoder introduced in Sect. 5 (see Fig. 2, part f) was investigated. We trained a system by means of by means of fully supervised learning (s_{dec}). The unannotated part of the training set was annotated using s_{dec} and another SAD was trained (s_{dec}). The results are in the third part of Table 1. Technically, the HMM decoder could be also applied on the decoded scores. As we anticipated, the results show that such application doesn't decrease the errors.

7   Conclusion

In this paper, our proposed NN-based SAD was presented. We focused on the simultaneously trained early score smoothing and the proposed online decoding process. Czech radio broadcast corpus was used for training and testing. Our experiments show that a relatively low error could be achieved by means of absolute values of delta coefficients in case of a simple SAD.

Since the mechanism of our decoder is a simple max pooling and since only a prevention of an undesirable cutting off speech segments was the intended contribution, we didn't expect that our decoder must significantly outperform the systems without any decoder which have a sufficient length of context for smoothing. However, the max-pooling based decoding allows the SAD to focus on reliable symptoms of a speech activity and prevents from wasting its performance on frames including a confusing information. This might be the reason for the reported error decreasing. Naturally, this is possible only when the decoder is involved in the training process as we described.

Unfortunately, this approach doesn't allow to use reliable symptoms of non-speech activity in the same way as reliable symptoms of speech activity. Because false alarms weren't as serious as missed speech activities, we ignored this disadvantage in this paper. In the future, we plan to eliminate this issue and to employ some musical pitch or rhythm detectors.

References

1. Chen, J., Wang, Y., Wang, D.: A feature study for classification-based speech separation at very low signal-to-noise ratio. In: 2014 IEEE International Conference on Acoustics, Speech and Signal Processing (ICASSP), pp. 7039–7043, May 2014
2. Goodfellow, I., Bengio, Y., Courville, A.: Deep Learning. MIT Press, Cambridge (2016)
3. Hughes, T., Mierle, K.: Recurrent neural networks for voice activity detection. In: ICASSP, pp. 7378–7382 (2013)
4. Lehner, B., Widmer, G., Sonnleitner, R.: On the reduction of false positives in singing voice detection. In: 2014 IEEE International Conference on Acoustics, Speech and Signal Processing (ICASSP), pp. 7480–7484 (2014)
5. Mateju, L., Cerva, P., Zdansky, J., Malek, J.: Speech activity detection in online broadcast transcription using deep neural networks and weighted finite state transducers. In: 2017 IEEE International Conference on Acoustics, Speech and Signal Processing (ICASSP), pp. 5460–5464, March 2017
6. Sadjadi, S.O., Hansen, J.H.L.: Unsupervised speech activity detection using voicing measures and perceptual spectral flux. IEEE Signal Process. Lett. **20**, 197–200 (2013)
7. Saon, G., Thomas, S., Soltau, H., Ganapathy, S., Kingsbury, B.: The IBM speech activity detection system for the DARPA RATS program, pp. 3497–3501, January 2013
8. Sehgal, A., Kehtarnavaz, N.: A convolutional neural network smartphone app for real-time voice activity detection. IEEE Access **6**, 9017–9026 (2018)
9. Thomas, S., Ganapathy, S., Saon, G., Soltau, H.: Analyzing convolutional neural networks for speech activity detection in mismatched acoustic conditions. In: 2014 IEEE International Conference on Acoustics, Speech and Signal Processing (ICASSP), pp. 2519–2523, May 2014
10. Thomas, S., Saon, G., Segbroeck, M.V., Narayanan, S.S.: Improvements to the IBM speech activity detection system for the DARPA RATS program. In: 2015 IEEE International Conference on Acoustics, Speech and Signal Processing (ICASSP), pp. 4500–4504 (2015)
11. Zhang, X.L., Wang, D.: Boosting contextual information for deep neural network based voice activity detection. IEEE/ACM Trans. Audio, Speech, Lang. Process. **24**, 252–264 (2016)

A Survey of Recent DNN Architectures on the TIMIT Phone Recognition Task

Josef Michálek[(✉)] and Jan Vaněk

University of West Bohemia, Univerzitní 8, 301 00 Pilsen, Czech Republic
{orcus,vanekyj}@kky.zcu.cz

Abstract. In this survey paper, we have evaluated several recent deep neural network (DNN) architectures on a TIMIT phone recognition task. We chose the TIMIT corpus due to its popularity and broad availability in the community. It also simulates a low-resource scenario that is helpful in minor languages. Also, we prefer the phone recognition task because it is much more sensitive to an acoustic model quality than a large vocabulary continuous speech recognition (LVCSR) task. In recent years, many DNN published papers reported results on TIMIT. However, the reported phone error rates (PERs) were often much higher than a PER of a simple feed-forward (FF) DNN. That was the main motivation of this paper: To provide a baseline DNNs with open-source scripts to easily replicate the baseline results for future papers with lowest possible PERs. According to our knowledge, the best-achieved PER of this survey is better than the best-published PER to date.

Keywords: Neural networks · Acoustic model · Survey · Review
TIMIT · LSTM · Phone recognition

1 Introduction

The Texas Instruments/Massachusetts Institute of Technology (TIMIT) corpus of read speech [2] is available since 1993 in LDC as an LDC93S1 corpus. It has been designed for the development and evaluation of automatic speech recognition systems. TIMIT contains speech from 630 speakers representing 8 major dialect divisions of American English, each speaking 10 phonetically-rich sentences. The TIMIT corpus includes time-aligned orthographic, phonetic, and word transcriptions, as well as speech waveform data for each spoken sentence.

This paper was supported by the project no. P103/12/G084 of the Grant Agency of the Czech Republic and by the grant of the University of West Bohemia, project No. SGS-2016-039. Access to computing and storage facilities owned by parties and projects contributing to the National Grid Infrastructure MetaCentrum provided under the programme "Projects of Large Research, Development, and Innovations Infrastructures" (CESNET LM2015042), is greatly appreciated.

© Springer Nature Switzerland AG 2018
P. Sojka et al. (Eds.): TSD 2018, LNAI 11107, pp. 436–444, 2018.
https://doi.org/10.1007/978-3-030-00794-2_47

Very valuable part is a definition of training, development, and test sets. It helped to develop the TIMIT corpus to be a very popular phone recognition benchmark task.

Very detailed overview until 2011 was published by Lopes and Fernando in [7]. It mapped the pre-DNN era and the start of the deep learning represented by a Mohamed et al. monophone deep belief network (DBN) [8] with PER 20.7% on the core test set. A triphone version of the DBN with a speaker adaptive training and a fMLLR adaptation was developed by Bagher BabaAli and Karel Vesely in the TIMIT Kaldi example s5 [5]. The Kaldi example achieved PER 18.5% on the core test set. Better results were then obtained by DNNs with rectified linear units (ReLU). The ReLU DNNs do not need the DBN pretraining and, if dropout is applied, they perform well on held out data. Laszlo Toth reported PER 17.76% on core test set with a convolutional bottle neck ReLU DNN in [12] and a year later he reported PER 16.5% with a 2D convolutional bottleneck maxout DNN in [13]. We also reported PER 16.5% with an ensemble of DBN DNNs augmented by regularization post-layer [14]. Taesup Moon then achieved stable PER 16.9% with a droupout bi-directional long-short term memory recurrent DNN (DBLSTM) and a peak PER with a larger net up to 16.29% [9].

In this paper, we have evaluated several recent deep neural network architectures. We also published our scripts to easily repeat our work and results. We followed the Kaldi s5 example and limit the experiments to a triphone model obtained by the Kaldi example together with the fMLLR speaker adapted training, development, and test data. First, we evaluated a simple feed-forward ReLU DNNs. Then, we tested time delay neural networks (TDNNs). Finally, we evaluated long-short term memory (LSTM) recurrent DNNs which gave the lowest PER. Because of the common feature processing stage, we did not try any 2D convolutional DNNs. We plan to investigate them as a future work.

2   Neural Network Architectures

2.1   Feed Forward DNNs

First DNNs used a sigmoid activation function, which suffers from the vanishing gradient problem. Hinton et al. in [3] proposed a greedy layer-wise unsupervised pre-training learning procedure. This procedure relies on the training algorithm of restricted Boltzmann machines (RBM) and initializes the parameters of a deep belief network (DBN), a generative model with many layers of hidden causal variables. The greedy layer-wise unsupervised training strategy helps the optimization by initializing weights in a region near a good local minimum, but also implicitly acts as a sort of regularization that brings better generalization and encourages internal distributed representations that are high-level abstractions of the input [6].

Later DNNs used ReLU, that do not suffer from the vanishing gradient problem. Therefore, pre-training is not necessary. On the other hand, the ReLU

DNNs are more prone to overfitting. The most effective regularization technique is dropout [11].

2.2 Time Delay Neural Network

The time delay neural network (TDNN) is a network designed to classify patterns shift-invariantly. It was first proposed to classify phonemes in speech recognition systems [15].

In standard DNN, initial layers learn an affine transform of the entire temporal context. However, in TDNN, the initial transforms are learnt on narrow contexts and the deeper layers process the hidden activations from a wider temporal context. Therefore, the higher layers have the ability to learn wider temporal dependencies. Usually, each layer operates on different temporal resolution, which increases as we go higher in the layers. The transforms in TDNN are tied across time steps and this is why TDNNs are seen as precursors to convolutional networks.

The hyperparameters defining a TDNN network are input context size for each layer and a number of filters in each layer.

In our work, we used ReLU as an activation function for TDNN. Other authors such as Peddinti use p-norm nonlinearity, although in [10] he proposes switching to ReLU due to the better results of his preliminary experiments.

2.3 Long Short-Term Memory

Long short-term memory (LSTM) is a widely used type of recurrent neural network (RNN). Standard RNNs suffer from both exploding and vanishing gradient problems.

The exploding gradient problem can be solved simply by truncating the gradient. On the other hand, the vanishing gradient problem is harder to overcome. It does not simply cause the gradient to be small; the gradient components corresponding to long-term dependencies are small while the components corresponding to short-term dependencies are large.

The LSTM was proposed in 1997 by Hochreiter and Scmidhuber [4] as a solution to the vanishing gradient problem. Let c_t denote a hidden state of a LSTM. The main idea is that instead of computing c_t directly from c_{t-1} with matrix-vector product followed by an activation function, the LSTM computes Δc_t and adds it to c_{t-1} to get c_t. The addition operation is what eliminates the vanishing gradient problem.

Each LSTM cell is composed of smaller units called gates, which control the flow of information through the cell. The forget gate controls what information will be discarded from the cell state, input gate controls what new information will be stored in the cell state and output gate controls what information from the cell state will be used in the output.

The LSTM has two hidden states, c_t and h_t. The state c_t fights the gradient vanishing problem while h_t allows the network to make complex decisions over short periods of time. There are several slightly different LSTM variants. The architecture used in this paper is specified by the following equations:

$$i_t = \sigma(W_x i x_t + W_h i h_{t-1} + b_i)$$
$$f_t = \sigma(W_x f x_t + W_h f h_{t-1} + b_f)$$
$$o_t = \sigma(W_x o x_t + W_h o h_{t-1} + b_o)$$
$$c_t = f_t * c_{t-1} + i_t * \tanh(W_x c x_t + W_h c h_{t-1} + b_c)$$
$$h_t = o_t * \tanh(c_t)$$

3 Experiments

The TIMIT corpus contains recordings of phonetically-balanced prompted English speech. It was recorded using a Sennheiser close-talking microphone at 16 kHz rate with 16 bit sample resolution. TIMIT contains a total of 6300 sentences (5.4 h), consisting of 10 sentences spoken by each of 630 speakers from 8 major dialect regions of the United States. All sentences were manually segmented at the phone level.

The prompts for the 6300 utterances consist of 2 dialect sentences (SA), 450 phonetically compact sentences (SX) and 1890 phonetically-diverse sentences (SI).

The training set contains 3696 utterances from 462 speakers. The core test set consists of 192 utterances, 8 from each of 24 speakers (2 males and 1 female from each dialect region). The training and test sets do not overlap.

3.1 Speech Data, Processing, and Test Description

As mentioned above, we used TIMIT data available from LDC as a corpus LDC93S1. Then, we ran the Kaldi TIMIT example script s5, which trained various NN-based phone recognition systems with a common HMM-GMM tied-triphone model and alignments. The common baseline system consisted of the following methods: It started from MFCC features which were augmented by Δ and $\Delta\Delta$ coefficients and then processed by LDA. Final feature vector dimension was 40. We obtained final alignments by HMM-GMM tied-triphone model with 1909 tied-states (may vary slightly if rerun the script). We trained the model with MLLT and SAT methods, and we used fMLLR for the SAT training and a test phase adaptation. We dumped all training, development and test fMLLR processed data, and alignments to disk. Therefore, it was easy to do compatible experiments from the same common starting point. We employed a bigram language/phone model for final phone recognition. A bigram model is a very weak model for phone recognition; however, it forced focus to the acoustic part of the system, and it boosted benchmark sensitivity. The training, as well as the recognition, was done for 48 phones. We mapped the final results on TIMIT core test set to 39 phones (as it is usual by processing TIMIT corpus), and phone error rate (PER) was evaluated by the provided NIST script to be compatible with previously published works. In contrast to the Kaldi recipe, we used a different phone decoder. It is a standard Viterbi-based triphone decoder. It gives better results than the Kaldi standard WFST decoder on the TIMIT phone recognition task. We have used an open-source Chainer 3.0 DNNs Python tranining tool that supports NVidia GPUs [1]. It is multiplatform and easy to use.

3.2 Feed-Forward DNNs

First, we re-trained feed-forward (FF) DNN with sigmoid activation function from the Kaldi example. We used the identical topology and the DBN pre-trained parameters.

In the other experiments with FF DNNs, we used a simple DNNs with ReLU without any pre-training. We used lower dropout $p = 0.2$, we have obtained better results than $p = 0.5$ like in [12]. We stacked 11 input fMLLR feature frames to 440 NN input dimension, like in Kaldi example s5. All the input vectors were transformed by an affine transform to normalize input distribution. We have tested a range from 6 to 9 hidden layers with 512, 1024, and 2048 ReLU neurons. The final softmax layer had 1909 neurons. We used SGD with momentum 0.9. The learning rate was three-times reduced according to development data training criterion change. Together with the learning rate reduction, the batch size was gradually increased from initial 256 to 512, 1024, and final 2048.

Besides ReLU, we tried also other activation functions. Leaky-ReLU gave almost identical error rates and criterion like ReLU. We have tested also maxout units (1/2, 1/4, and 1/8) that gave worse results than ReLU.

3.3 Time Delay Neural Network

First we used a network with 4 layers and context sizes 5, 5, 9, 9 as in [10] without sub-sampling. We used dropout $p = 0.2$. Input data were stacked to the size required by the context sizes and normalized. We also evaluated context sizes 5, 5, 5, 5 and 9, 9; 9, 9. We tested hidden layers with 256, 512 and 1024 filters with ReLU activation function. The final layer was again softmax with 1909 neurons.

The networks were trained first using Adam optimization algorithm and then using SGD with momentum 0.9. SGD training stage was used three times, each with lower learning rate. Batch size was the same as in FF DNN case.

3.4 Long Short-Term Memory

We used standard LSTM architecture as specified in Sect. 2.3. We tried the number of hidden layers in range from 2 to 6 and 256, 512 and 1024 LSTM cells in each hidden layer. Input data were transformed to normalize the input distribution. We used output time delay equal to 5 time steps. Dropout used was again $p = 0.2$.

The network was trained first using Adam and then momentum SGD as in TDNN case. Batch size used was 512 for Adam and 128 for SGD training stages.

3.5 Results

After we trained all the networks, we have evaluated their performance on the development and test dataset. Each of the networks was trained several times, 4 times in case of TDNN and 10 times for other network types. The Table 1 contains the minimum, maximum and average phone error rates for each type of used feed-forward networks. The FF network with sigmoid activation function and DBN pre-training has average 17.04% PER, but it was outperformed by several simple FF networks with ReLU without pretraining. The best average PER

Table 1. Feed-Forward DNN phone error rate

Network	Development PER			Test PER		
	Min	Max	Avg	Min	Max	Avg
FF sigmoid 7x1024	16.17	16.48	16.31	16.76	17.24	17.04
FF ReLU 6x512	15.97	16.64	16.40	17.34	18.03	17.63
FF ReLU 6x1024	16.06	16.37	16.23	16.90	17.34	17.09
FF ReLU 6x2048	15.94	16.48	16.27	16.88	17.39	17.17
FF ReLU 7x512	16.05	16.49	16.26	17.20	18.09	17.50
FF ReLU 7x1024	15.70	16.26	15.95	16.62	17.23	16.93
FF ReLU 7x2048	15.92	16.49	16.17	16.83	17.41	17.06
FF ReLU 8x512	16.23	16.65	16.43	17.33	18.30	17.72
FF ReLU 8x1024	15.79	16.21	16.02	16.80	17.27	17.03
FF ReLU 8x2048	15.78	16.24	15.99	16.66	17.13	16.91
FF ReLU 9x512	16.34	16.57	16.50	17.38	17.71	17.57
FF ReLU 9x1024	15.73	16.01	15.91	16.49	17.41	16.99
FF ReLU 9x2048	15.61	16.08	15.91	16.69	17.23	16.96

Table 2. TDNN phone error rate

Context size	# Filters	Development PER			Test PER		
		Min	Max	Avg	Min	Max	Avg
5, 5, 5, 5	256	17.55	17.65	17.61	18.27	18.75	18.57
5, 5, 5, 5	512	17.19	17.38	17.30	18.02	18.30	18.20
5, 5, 5, 5	1024	17.33	17.72	17.46	17.77	18.66	18.23
5, 5, 9, 9	256	17.71	17.96	17.82	18.49	18.84	18.67
5, 5, 9, 9	512	17.32	17.59	17.50	18.30	18.82	18.52
5, 5, 9, 9	1024	17.67	17.83	17.79	18.50	19.24	18.85
9, 9, 9, 9	256	17.95	18.28	18.14	18.71	19.03	18.90
9, 9, 9, 9	512	17.61	18.04	17.86	18.66	19.07	18.81
9, 9, 9, 9	1024	18.26	20.34	18.91	18.82	21.07	19.72

we have achieved is 16.91% PER in the case of 8x2048 FF network. However, all the deeper networks with at least 1024 neurons in the hidden layers have similar performance.

The Table 2 contains the results for TDNN networks. All the results are worse than the results for our FF networks. Interestingly enough, we obtained the best results for context sizes 5, 5, 5, 5. The larger context sizes resulted in worse network performance. Also, the networks with higher number of filters have better performance. The best average PER we have received was 18.20% for the network with the context size 5, 5, 5, 5 and 512 filters in each layer.

The results for our LSTM experiments are in the Table 3. Our LSTM networks have better performance than other network types used in this work. We have received the best average PER, 15.58%, for the network with 4 hidden layers each with 1024 LSTM units. The best PER from all experiments was 15.02%, also belonging to the network with 4 hidden layers with 1024 LSTM units. However, we have also trained other networks with similar PER to the best one. The network with 3 layers with 1024 units and the network with 5 layers with 1024 units have the second and third best PER, 15.69% and 15.71% respectively. The networks with 256 units in each layer have comparable or worse PER than FF networks. The network with 6 layers with 1024 units has worse performance than less deep layers due to overfitting.

Table 3. LSTM phone error rate

Network	Development PER			Test PER		
	Min	Max	Avg	Min	Max	Avg
2x256	16.21	16.99	16.62	16.96	18.13	17.51
3x256	15.63	16.40	16.04	16.37	17.08	16.76
4x256	15.39	16.19	15.80	16.09	16.99	16.58
5x256	15.39	15.95	15.79	15.98	17.03	16.49
6x256	15.73	16.21	16.00	15.97	17.16	16.73
2x512	15.14	16.07	15.59	16.11	16.91	16.41
3x512	14.82	15.39	15.11	15.77	16.55	16.05
4x512	14.75	15.32	15.08	15.69	16.19	15.96
5x512	14.65	15.31	14.97	15.36	16.27	15.83
6x512	14.91	15.59	15.25	15.80	16.29	16.02
2x1024	14.92	15.36	15.08	15.62	16.17	15.97
3x1024	14.37	15.02	14.68	15.38	15.95	15.69
4x1024	14.43	15.16	14.67	15.02	15.84	15.58
5x1024	14.49	14.85	14.66	15.34	16.04	15.71
6x1024	14.82	15.20	15.04	15.34	16.48	15.87

4   Conclusion

We have trained several neural network types on the TIMIT phone recognition task. Each of the networks was trained several times, 4 times in case of TDNN and 10 times for other network types. We have evaluated the phone error rate (PER) for each trained model and showed the minimum, maximum and average PER in tables.

First we trained FF networks with ReLU and obtained 16.49% PER as our best result. We have also tried FF network with sigmoid activation function and DBN pre-training, but the best model has worse PER than some of our FF ReLU networks, 16.76%. Then we trained TDNN networks with several settings, but we couldn't get better results than our simple FF networks. Our best TDNN network has 17.77% PER. The last network type we trained was recurrent LSTM network. We have found that the best performing networks have 4 layers each with 1024 LSTM units. Their average PER was 15.58%, although networks with 3 or 5 layers have similar results. The best result of all of our experiments was 15.02% PER for LSTM network with 4 layers each with 1024 units. These results are the best results published to date according to our knowledge.

The scripts used in our experiments are freely available at https://github.com/OrcusCZ/NNAcousticModeling.

References

1. A flexible framework of neural networks for deep learning. https://chainer.org
2. Garofolo, J.S., et al.: TIMIT Acoustic-Phonetic Continuous Speech Corpus. Linguistic Data Consortium LDC93S1 (1993)
3. Hinton, G.E., Osindero, S., Teh, Y.W.: A fast learning algorithm for deep belief nets. Neural Comput. **18**(7), 1527–1554 (2006)
4. Hochreiter, S., Schmidhuber, J.: Long short-term memory. Neural Comput. **9**(8), 1735–1780 (1997)
5. Kaldi speech recognition toolkit. https://github.com/kaldi-asr/kaldi
6. Larochelle, H., Bengio, Y., Louradour, J., Lamblin, P.: Exploring strategies for training deep neural networks. J. Mach. Learn. Res. **10**(Jan), 1–40 (2009)
7. Lopes, C., Perdigao, F.: Phoneme Recognition on the TIMIT Database. Speech Technologies (2011). https://doi.org/10.5772/17600, http://www.intechopen.com/books/speech-technologies/phoneme-recognition-on-the-timit-database
8. Mohamed, A., Dahl, G.E., Hinton, G.: Acoustic modeling using deep belief networks. IEEE Trans. Audio Speech Lang. Process. **20**(1), 14–22 (2012). https://doi.org/10.1109/TASL.2011.2109382
9. Moon, T., Choi, H., Lee, H., Song, I.: RNNDROP: a novel dropout for RNNs in ASR. In: Proceedings of the ASRU (2015)
10. Peddinti, V., Povey, D., Khudanpur, S.: A time delay neural network architecture for efficient modeling of long temporal contexts. In: Sixteenth Annual Conference of the International Speech Communication Association (2015)
11. Srivastava, N., Hinton, G., Krizhevsky, A., Sutskever, I., Salakhutdinov, R.: Dropout: a simple way to prevent neural networks from overfitting. J. Mach. Learn. Res. **15**(1), 1929–1958 (2014)

12. Tóth, L.: Convolutional deep rectifier neural nets for phone recognition. In: Proceedings of the Annual Conference of the International Speech Communication Association, INTERSPEECH, pp. 1722–1726, August 2013
13. Tóth, L.: Convolutional deep maxout networks for phone recognition. In: Proceedings of the INTERSPEECH, pp. 1078–1082 (2014). https://doi.org/10.1186/s13636-015-0068-3
14. Vaněk, J., Zelinka, J., Soutner, D., Psutka, J.: A regularization post layer: an additional way how to make deep neural networks robust. In: Camelin, N., Estève, Y., Martín-Vide, C. (eds.) SLSP 2017. LNCS (LNAI), vol. 10583, pp. 204–214. Springer, Cham (2017). https://doi.org/10.1007/978-3-319-68456-7_17
15. Waibel, A., Hanazawa, T., Hinton, G., Shikano, K., Lang, K.J.: Phoneme recognition using time-delay neural networks. In: Readings in Speech Recognition, pp. 393–404. Elsevier (1990)

WaveNet-Based Speech Synthesis Applied to Czech
A Comparison with the Traditional Synthesis Methods

Zdeněk Hanzlíček(✉), Jakub Vít, and Daniel Tihelka

NTIS - New Technology for the Information Society, Faculty of Applied Sciences, University of West Bohemia, Univerzitní 22, 306 14 Pilsen, Czech Republic
{zhanzlic,jvit,dtihelka}@ntis.zcu.cz
http://www.ntis.zcu.cz/en

Abstract. WaveNet is a recently-developed deep neural network for generating high-quality synthetic speech. It produces directly raw audio samples. This paper describes the first application of WaveNet-based speech synthesis for the Czech language. We used the basic WaveNet architecture. The duration of particular phones and the required fundamental frequency used for local conditioning were estimated by additional LSTM networks. We conducted a MUSHRA listening test to compare WaveNet with 2 traditional synthesis methods: unit selection and HMM-based synthesis. Experiments were performed on 4 large speech corpora. Though our implementation of WaveNet did not outperform the unit selection method as reported in other studies, there is still a lot of scope for improvement, while the unit selection TTS have probably reached its quality limit.

Keywords: Speech synthesis · WaveNet · Deep neural network
Unit selection · HMM-based speech synthesis

1 Introduction

Recently, deep neural networks (DNNs) have been successfully applied for acoustic modeling in statistical parametric speech synthesis [1]. At first, DNNs were used for mapping from linguistic features (derived from an input text) into acoustic features, e.g. spectral parameters and fundamental frequency. The resulting speech waveform was created by an additional vocoder. DNNs have naturally replaced the formerly used hidden Markov models [2,3].

WaveNet is a recently introduced convolutional deep neural network [4] for generating high-quality synthetic speech. It works directly with the raw audio signal and no speech parameters are needed in the training or synthesis stage. WaveNet apparently outperforms other synthesis methods, both concatenative and statistical parametric. For its ability to produce high-quality speech, WaveNet has been also utilized in other tasks, e.g. in statistical voice conversion [5] or as a statistical vocoder [6,7].

© Springer Nature Switzerland AG 2018
P. Sojka et al. (Eds.): TSD 2018, LNAI 11107, pp. 445–452, 2018.
https://doi.org/10.1007/978-3-030-00794-2_48

In this paper, we describe the first application of WaveNet-based speech synthesis for the Czech language.

2   General WaveNet Description

Our implementation of WaveNet is based on the basic architecture described in [4,8]. The speech waveform is given as a sequence of samples $\mathbf{x} = \{x_1, x_2, \ldots, x_T\}$ and each sample x_t is supposed to be conditioned by all previous samples. Then, the probability of the whole waveform $\mathbf{x}$ is given as

$$p(\mathbf{x}) = \prod_{t=1}^{T} p(x_t | x_1, \ldots, x_{t-1}) \tag{1}$$

The distribution $p(x)$ describes a general evolution of speech samples. To be able to generate required utterances, the model must be conditioned by an additional input $\mathbf{h}$ corresponding to all required characteristics of produced speech

$$p(\mathbf{x}|\mathbf{h}) = \prod_{t=1}^{T} p(x_t | x_1, \ldots, x_{t-1}, \mathbf{h}) \tag{2}$$

The model can be conditioned globally and locally. Global conditions are constant for the whole waveform, e.g. the speaker identity in a multi-speaker system. Local conditions correspond to characteristics that are evolving throughout the utterance, e.g. phonetic or linguistic information and prosody. In case the network is locally conditioned by acoustic parameters, the network is working as a vocoder [6,7].

The conditional probability distribution (2) is modeled by a stack of convolutional layers with a gated activation function

$$\mathbf{z} = \tanh\left(W_f * \mathbf{x} + V_f * \mathbf{h}\right) \odot \sigma\left(W_g * \mathbf{x} + V_f * \mathbf{h}\right) \tag{3}$$

where σ is a sigmoid function, f and g denote filter and gate, W and V are learnable weights, $\mathbf{h}$ is a time series containing the local conditioning for particular audio samples; operator $*$ stands for the convolution and $\odot$ is the element-wise multiplication.

3   Implementation Details

Our network contains 20 dilated convolution layers with the same dilatation pattern as in the original paper [4]. Each layer has 128 residual connections to the following layer and 128 skip connections. Skip outputs of all layers are concatenated and passed through two ReLU postprocessing layers into a softmax layer. The model has two ReLU layers for preprocessing local conditions and one more in each stack layer. Similarly to [4], waveform samples were quantized with the μ-law algorithm into 256 discrete values.

For a consistent comparison with other TTS systems (see Sect. 4), we used 16 kHz sample rate. The network was therefore conditioned with 120 ms of previous speech samples (so-called receptive field). We also performed experiments with different sample rates; the higher is the sampling frequency, the shorter is the receptive field, e.g. 85 ms for 24 kHz. However, the receptive field can be easily extended by using more dilated convolution layers or by changing the dilatation pattern.

Neural networks were trained by using the TensorFlow[1] framework. The training of one network takes approximately two days on one GTX 1080 Ti graphic card. Since each network was trained for one individual voice, the global conditioning was not used.

3.1   Local Conditioning

The local conditioning corresponds to the required characteristics of produced speech and is applied to guide WaveNet during the speech generation. In our experiments, we used the following local conditions:

- phonetic information: identity of the current phone and its left and right context (represented as one-hot vectors)
- sample position within the current phone (coarse coded vector with dimension experimentally set to 100)
- $\log F_0$ (interpolated in unvoiced segments)
- binary flag for voicedness.

However, the local conditions and the actual properties of produced speech are not guaranteed to match exactly. It depends on WaveNet how the input conditions are transformed into the resulting waveform. In our experiments, we detected a difference between the required and the resulting values of fundamental frequency. It was probably caused by a complex relation between the audio signal and corresponding F_0 that is not easy to model precisely. However, in most cases, the resulting fundamental frequency had a proper range and shape and we did not perceive any disruptive effects.

To be able to assign the local conditions to the particular speech samples, additional models for prediction of phone duration and fundamental frequency from the input text (or derived linguistic features) are needed. In our experiments, we used additional long short-term memory recurrent neural networks LSTM-RNN [3].

The network for phone duration consists of 2 bidirectional LSTM layers with 32 neurons and 2 preprocessing layers with 64 neurons with the ReLU activation function. Pitch and voicedness were modeled by a network composed of 2 bidirectional LSTM layers with 256 neurons and 2 preprocessing layers with 512 neurons with the ReLU activation function. This network actually outputs all vocoder features for parametric speech synthesis but only the pitch and voicing

---

[1] TensorFlow is an open-source machine learning framework available at http://www. tensorflow.org.

outputs were used as local conditions for WaveNet. The LSTM network works with a 5 ms frame shift; for using as local conditions, the predicted values are interpolated to the particular samples.

The complete diagram of our experimental WaveNet-based TTS system is depicted in Fig. 1.

4 Experiments and Results

This section describes a comparison of the experimental WaveNet-based TTS system with 2 other TTS systems:

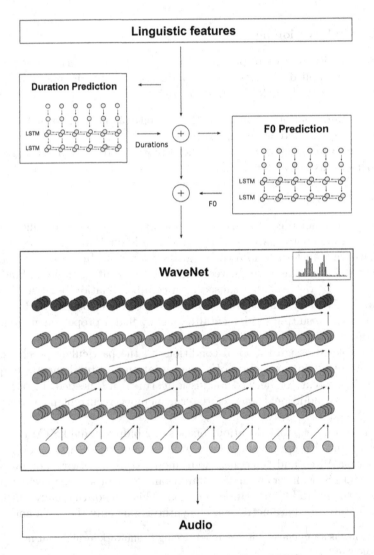

Fig. 1. A basic diagram of our WaveNet-based TTS system.

- unit selection TTS system [9,10]
- HMM-based TTS system [11,12].

Thanks to the long-term research and development in the field of speech synthesis, both systems are precisely tuned and utilize the potential of the particular methods. Thus, WaveNet is apparently disadvantaged in this comparison.

4.1 Experimental Speech Data

For our experiments, we used 4 large Czech speech corpora recorded by professional speakers for the purposes of unit selection speech synthesis [13]: 2 male voices (M1 and M2) and 2 female voices (F1 and F2). All corpora contain the same sentences. About 10,000 sentences (approximately 14 h of speech) were used for training, 20 utterances were excluded from the training set and used for the following evaluation. The selected sentences were about 5–6 words long. Such sentences are convenient for a listening test.

4.2 Listening Test

We conducted a MUSHRA (multiple stimuli with hidden reference and anchor) listening test [14] to compare the overall speech quality generated by the particular synthesis methods. The test was composed of 20 queries for each speaker (i.e. 80 together). Each query contained a reference natural utterance and 4 utterances for evaluation (in random order): 3 synthesized by the compared systems and again the natural utterance. The overall quality of particular utterances was rated from 0 (very poor) to 100 (sounds like natural). Since the definition of the lower quality anchor is not clear in the case of synthetic speech [15], it was not used in our test.

4.3 Results and Discussion

11 listeners participated in the test, everybody evaluated all utterances. The overall results are presented in Table 1 and Fig. 2. The unit selection was rated as the best, the WaveNet-based synthesizer was placed second and, as expected, the HMM-based synthesizer took the last place. This is not in accordance with the other studies where WaveNet is reported to outperform both methods [4,16]. However, as mentioned above, WaveNet was at a disadvantage in this comparison since both competing systems have been developed and tuned for a long time.

The results are consistent even for particular voices as shown in Fig. 3. The stable quality of the unit selection is worth noticing, whereas the quality of statistical parametric speech synthesis is rather speaker-dependent.

A considerable variability in evaluation by individual listeners is evident in Fig. 4. Some listeners used almost the whole evaluation range while others utilized only its smaller part. That is a consequence of the missing lower quality anchor in the MUSHRA test. Moreover, individual listeners were probably not equally sensitive to various types of speech degradation typical for particular

Table 1. Results of MUSHRA listening test.

System	Mean score $\pm$ std	Median
Natural voice	98.6 ± 5.2	100
Unit selection	82.9 ± 19.3	90
HMM synthesis	47.8 ± 23.8	51
WaveNet	66.1 ± 23.0	73

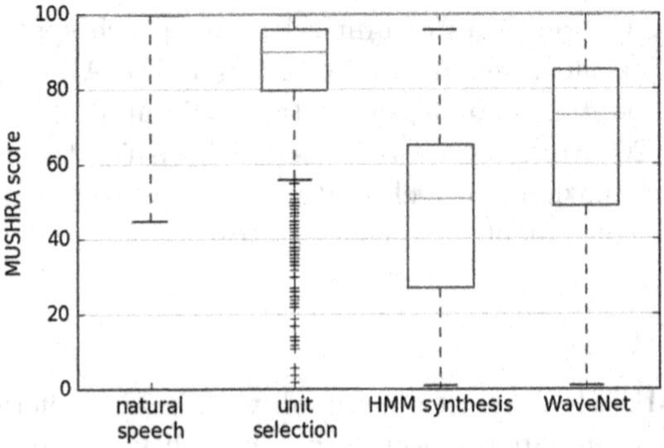

Fig. 2. MUSHRA listening test: the overall results.

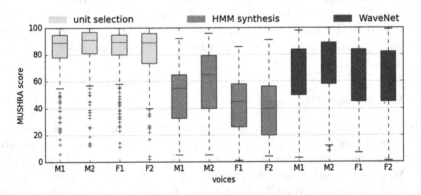

Fig. 3. MUSHRA listening test: results for particular voices. The evaluation of natural speech is not included in the figure.

methods, e.g. some listeners might prefer the average constant quality and penalized the sudden local fluctuation and vice versa. The large variability of listeners explains also the large number of outliers in Figs. 2 and 3.

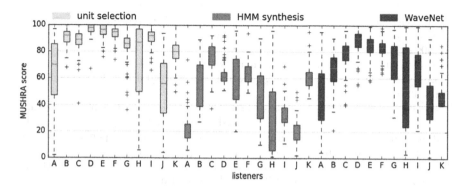

Fig. 4. MUSHRA listening test: boxplots for particular listeners, each boxplot was created from 80 values. Again, natural speech is not included.

5   Conclusion

This paper has presented the first experiments with WaveNet-based speech synthesis for the Czech language. Though our experimental system did not outperform the unit selection method, the results are very promising. Produced speech sounds very naturally, comparable to the unit selection. However, the overall good impression was affected by sudden local quality drops or in some cases also by gradual quality degradation.

Our future work will be focused on links between the training setup and the resulting speech quality. We would like to find optimal setting for an average voice and for various size and quality of training data.

We also intend to use data from more speakers together and experiment with multi-speaker systems. It could open the possibility to produce voices for which only less training data or low-quality training data is available.

Acknowledgment. This research was supported by the Czech Science Foundation (GACR), project No. GA16-04420S and by the grant of the University of West Bohemia, project No. SGS-2016-039. Access to computing and storage facilities owned by parties and projects contributing to the National Grid Infrastructure MetaCentrum provided under the program "Projects of Large Research, Development, and Innovations Infrastructures" (CESNET LM2015042), is greatly appreciated.

References

1. Ling, Z.H., Kang, S.Y., Zen, H., et al.: Deep learning for acoustic modeling in parametric speech generation: a systematic review of existing techniques and future trends. IEEE Signal Process. Mag. **32**(3), 35–52 (2015)
2. Zen, H., Tokuda, K., Black, A.W.: Statistical parametric speech synthesis. Speech Commun. **51**(11), 1039–1064 (2009)
3. Zen, H.: Acoustic modeling in statistical parametric speech synthesis - from HMM to LSTM-RNN. In: Proceedings of MLSLP (2015)

4. van den Oord, A., Dieleman, S., Zen, H., Simonyan, K., et al.: WaveNet: a generative model for raw audio. CoRR abs/1609.03499 (2016). http://arxiv.org/abs/1609.03499

5. Kobayashi, K., Hayashi, T., Tamamori, A., Toda, T.: Statistical voice conversion with WaveNet-based waveform generation. In: Proceedings of Interspeech 2017, pp. 1138–1142 (2017)

6. Hayashi, T., Tamamori, A., Kobayashi, K., et al.: An investigation of multi-speaker training for WaveNet vocoder. In: Proceedings of ASRU 2017, pp. 712–718 (2017)

7. Tamamori, A., Hayashi, T., Kobayashi, K., et al.: Speaker-dependent WaveNet vocoder. In: Proceedings of Interspeech 2017, pp. 1118–1122 (2017)

8. Arik, S.O., Chrzanowski, M., Coates, A., et al.: Deep voice: real-time neural text-to-speech. CoRR abs/1702.07825 (2017). https://arxiv.org/abs/1702.07825

9. Matoušek, J., Tihelka, D., Romportl, J.: Current state of Czech text-to-speech system ARTIC. In: Sojka, P., Kopeček, I., Pala, K. (eds.) TSD 2006. LNCS, vol. 4188, pp. 439–446. Springer, Heidelberg (2006). https://doi.org/10.1007/11846406_55

10. Matoussek, J., Legát, M., Tihelka, D.: Is unit selection aware of audible artifacts? In: Proceedings of SSW8, pp. 267–271. ISCA (2013)

11. Hanzlíček, Z.: Czech HMM-Based Speech Synthesis. In: Sojka, P., Horák, A., Kopeček, I., Pala, K. (eds.) TSD 2010. LNCS, vol. 6231, pp. 291–298. Springer, Heidelberg (2010). https://doi.org/10.1007/978-3-642-15760-8_37

12. Hanzlíček, Z.: Optimal number of states in HMM-based speech synthesis. In: Ekštein, K., Matoušek, V. (eds.) TSD 2017. LNCS, vol. 10415, pp. 353–361. Springer, Cham (2017). https://doi.org/10.1007/978-3-319-64206-2_40

13. Matoussek, J., Tihelka, D., Romportl, J.: Building of a speech corpus optimised for unit selection TTS synthesis. In: Proceedings of LREC (2008)

14. Method for the subjective assessment of intermediate quality level of coding systems. ITU Recommendation ITU-R BS.1534-2 (2014)

15. Henter, G.E., Merritt, T., Shannon, M., et al.: Measuring the perceptual effects of modelling assumptions in speech synthesis using stimuli constructed from repeated natural speech. In: Proceedings of Interspeech 2014, pp. 1504–1508 (2014)

16. van den Oord, A., Li, Y., Babuschkin, I., et al.: Parallel WaveNet: fast high-fidelity speech synthesis. CoRR abs/1711.10433 (2017). https://arxiv.org/abs/1711.10433

Phonological Posteriors and GRU Recurrent Units to Assess Speech Impairments of Patients with Parkinson's Disease

Juan Camilo Vásquez-Correa[1,2]([✉]), Nicanor Garcia-Ospina[1],
Juan Rafael Orozco-Arroyave[1,2], Milos Cernak[3], and Elmar Nöth[2]

[1] Faculty of Engineering, University of Antioquia UdeA, Medellín, Colombia
[2] University of Erlangen-Nüremberg, Erlangen, Germany
juan.vasquez@fau.de
[3] Logitech, Lausanne, Switzerland

Abstract. Parkinson's disease is a neurodegenerative disorder characterized by a variety of motor symptoms, including several impairments in the speech production process. Recent studies show that deep learning models are highly accurate to assess the speech deficits of the patients; however most of the architectures consider static features computed from a complete utterance. Such an approach is not suitable to model the dynamics of the speech signal when the patients pronounce different sounds. Phonological features can be used to characterize the voice quality of the speech, which is highly impaired in patients suffering from Parkinson's disease. This study proposes a deep architecture based on recurrent neural networks with gated recurrent units combined with phonological posteriors to assess the speech deficits of Parkinson's patients. The aim is to model the time-dependence of consecutive phonological posteriors, which follow the sound patterns of English phonological model. The results show that the proposed approach is more accurate than a baseline based on standard acoustic features to assess the speech deficits of the patients.

Keywords: Parkinson's disease · Dysarthria assessment
Phonological posteriors · Gated recurrent units
Recurrent neural network

1 Introduction

Parkinson's disease (PD) is a neurological disorder characterized by the progressive loss of dopaminergic neurons in the mid-brain, producing several motor and non-motor impairments [8]. The motor symptoms include different speech deficits including reduced loudness, monopitch, monoloudness, reduced stress, breathy, hoarse voice quality, and imprecise articulation. These impairments are

© Springer Nature Switzerland AG 2018
P. Sojka et al. (Eds.): TSD 2018, LNAI 11107, pp. 453–461, 2018.
https://doi.org/10.1007/978-3-030-00794-2_49

grouped together and called *hypokinetic dysarthria* [11]. The disease progression in motor activities is currently evaluated with the third section of the movement disorder society, unified Parkinson's disease rating scale (MDS-UPDRS-III) [2].

Several studies in the literature have described the speech impairments developed by PD patients in terms of four different dimensions: phonation, articulation, prosody, and intelligibility [7,12]. These feature extraction strategies have shown to be suitable to support the diagnosis process and to assess the neurological state of the patients. Although the success of these classical feature extraction approaches, in the recent years deep learning methods have shown to be highly accurate to assess the speech of PD patients [15,16]. In [15] the authors proposed a deep learning model to assess the severity of dysarthria in speech. The model considers an intermediate interpretable hidden layer to assess four perceptual dimensions: nasality, vocal quality, articulatory precision, and prosody. The authors reported a Spearman's correlation of up to 0.82 between the output of the deep learning model and a perceptual score of the severity of dysarthria provided by speech and language therapists. In [16] the authors considered a convolutional neural network and time-frequency representations to model articulation impairments in the speech of PD patients [16]. The model classified PD patients vs. healthy control subjects with accuracies of up to 89%. To the best of our knowledge, most of the related studies that consider deep learning methods to assess pathological speech only consider static features computed from a complete utterance. Those methods are not able to model the dynamics of the speech signal properly when the patients pronounce different sounds in continuous speech such as sentences or monologues. From the different deep learning architectures, the recurrent neural networks (RNN) have been designed to process the time-dependence of sequential inputs such as text or speech, which makes it suitable to model the dynamics of speech features computed for different frames. On the other hand, voice quality of the speech can be characterized using phonological features [4], by computing phonological posterior features for modal and non-modal phonations on consecutive speech frames.

This study combines the phonological analysis and RNNs to model the dynamics of the features computed on consecutive frames to assess the speech deficits of PD patients. RNNs are formed with gate recurrent units (GRUs) [5], which have shown similar results than the standard long short-term memory (LSTM) units, but with less parameters to learn. The results indicate that the dynamics phonological posterior features are better to model the speech impairments of PD patients than the standard acoustic features based on Mel frequency Cepstral coefficients (MFCCs). In addition, the phonological posteriors provide interpretable results for the medical examiner to evaluate the speech state of the patients.

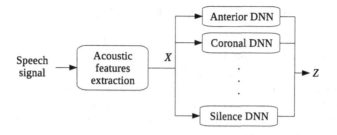

Fig. 1. Phonological feature extraction process

2 Methods

2.1 Phonological Features Extraction

Phonological features were extracted using the deep learning approach from [3]. This process involves the following steps: (1) the speech signal is segmented in short-time frames, (2) 13 MFCCs and their derivatives are computed for every frame of the speech signal, and (3) a set of 15 pre-trained DNNs infers the phonological posteriors from the acoustic feature vector. These posteriors are concatenated in a phonological feature vector z_t. The process is summarized in Fig. 1, where X is the set of acoustic features and Z is the set of phonological features. A total of 15 phonological features are computed. Table 1 shows details of each phonological feature.

Table 1. List of phonological features

Feature	Brief description
Vocalic	Refers to the vocal folds vibration without constriction in the vocal tract
Consonantal	Models sounds where there is obstruction of the vocal tract
High	The body of the tongue is above its neutral position
Back	The body of the tongue is retracted from its neutral position
Low	The body of the tongue is below its neutral position
Anterior	Indicates an obstruction located in front of the palato-alveolar region of the mouth
Coronal	The blade of the tongue is raised from its neutral position
Round	Refers to narrowed lips
Rising	Differentiates diphthongs from monophthongs
Tense	Models stressed vowels
Voice	Indicates voiced sounds
Continuant	Differentiates plosives from non-plosives
Nasal	Models a lowered velum, where the air to escape through the nose
Strident	Refers to sounds with more energy in high frequency components

2.2 Recurrent Neural Network and GRU Units

The RNNs process the sequence one element at a time, a state vector in their hidden units, which contains information about the history of all the past elements of the sequence [10]. The RNNs can be formed with different recurrent units including the conventional recurrent units, the long short-term memory (LSTM) units, or the gated recurrent units (GRUs). The conventional recurrent units can be seen as very deep networks where all the layers share the same weights. Although their main purpose is to learn long-term dependencies, there is evidence that shows difficulties to learn very long sequences [1]. This problem may be fixed with the LSTM units, which have a memory cell to model the long-term time-dependency. The GRU were proposed as a modification of the LSTM replacing the separate input and forget gates with a reset gait to control the input information to the network. GRUs and LSTMs have provided similar results for several tasks including speech and language modeling [9]; however, the GRUs are faster to train and require less parameters [6], which make these units more suitable to be used when less train data is available.

Figure 2 shows the architecture used in this study to process the phonological posteriors sequences and the MFCC feature vectors. The phonological features are processed individually by two GRU layers. On the other path, classical MFCC features are modeled by other two GRU units. The output of the two paths are merged with two fully connected layers (h1, and h2), followed by the output layer to make the final decision. Three architectures are considered in this study: (1) a network to process only the phonological posteriors, (2) a network to process only the acoustic features, and (3) a network to combine the phonological and acoustic features (see Fig. 2).

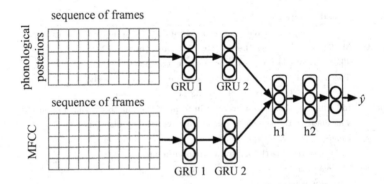

Fig. 2. Deep architecture to assess speech impairments of PD patients using phonological posteriors and GRU units

2.3   Validation

The experiments are validated using 80% of the data for training, 10% to optimize the hyper-parameters, i.e., development set, and the remaining 10% to test. The process is repeated 10 times with different partitions to produce different and independent test sets. The hyper-parameter tuning is performed with a Bayesian optimization approach [14] due to the large number of hyper-parameters to optimize. The tuning is performed based on an optimization problem, where the hyper-parameters that maximize the performance of the model on the development set are found. The range of the hyper-parameters to be optimized is shown in Table 2. A batch-size of 128 samples and a total of 100 epochs are considered with an early stopping strategy.

Table 2. Range of the hyper-parameters used to train the RNN.

Hyper-parameter	Values
GRU units in all layers	$\{8, 16, 32, 64\}$
Hidden units in fully connected layers	$\{16, 32, 64, 128\}$
Learning rate	$\{0.0001, \cdots, 0.01\}$
Dropout rate	$\{0.1, 0.2 \cdots 0.7\}$
Recurrent dropout rate	$\{0.1, 0.2 \cdots 0.7\}$

3   Data

3.1   m-FDA Scale

The evaluation of PD patients according to the MDS-UPDRS-III scale has shown to be suitable to assess general motor impairments of PD patients; however, the deterioration of the communication skills of the PD patients is not properly evaluated because such a scale only considers speech impairments in one of its items. A modified version of the Frenchay dysarthria assessment scale (m-FDA), which can be administered based on speech recordings was recently developed [4, 12]. The scale includes several aspects of speech: respiration, lips movement, palate/velum movement, larynx, tongue, monotonicity, and intelligibility. The scale has a total of 13 items and each of them ranges from 0 (normal or completely healthy) to 4 (very impaired), thus the total score of the scale ranges from 0 to 52. The labeling process of the recordings was performed by three phoniatricians who agreed in the first ten speakers. Afterwards, each phoniatrician evaluated the remaining recordings independently. The inter-rater reliability among the labelers is 0.75.

3.2 Participants

We consider the PC-GITA database [13]. The data contain speech utterances from 50 PD and 50 HC Colombian Spanish native speakers balanced in age and gender. The participants pronounce several utterances including the rapid repetition of the syllables /pa-ta-ka/, /pa-ka-ta/, /pe-ta-ka/, /pa/, /ta/, /ka/, isolated sentences, a read text, and a monologue. All patients were recorded in ON state, i.e., no more than three hours after their morning medication, and were evaluated by a neurologist expert. Additional information from the participants is shown in Table 3. In addition, Fig. 3 shows the distribution of the clinical scores for the patients. We divided the patients in three groups according to their level of the total MDS-UPDRS-III and to the speech item of the MDS-UPDRS-III scores. For the m-FDA score, the subjects are divided in four groups because that scale was applied also to HC subjects (white bars). The division consider the same number of subjects in each group.

Table 3. Demographic information of the participants from this study

	PD patients		HC subjects	
	Male	Female	Male	Female
Number of subjects	25	25	25	25
Age ($\mu \pm \sigma$)	61.3 ± 11.4	60.7 ± 7.3	60.5 ± 11.6	61.4 ± 7.0
Range of age	33–81	49–75	31–86	49–76
Duration of the disease ($\mu \pm \sigma$)	8.7 ± 5.8	12.6 ± 11.6	-	-
MDS-UDRS-III ($\mu \pm \sigma$)	37.8 ± 22.1	37.6 ± 14.1	-	-
MDS-UDRS-III speech ($\mu \pm \sigma$)	1.4 ± 0.9	1.3 ± 0.7	-	-
Total m-FDA ($\mu \pm \sigma$)	29.8 ± 8.6	28.2 ± 9.0	7.6 ± 9.2	5.1 ± 7.3

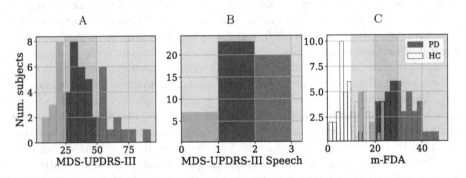

Fig. 3. Distribution of the clinical scores for the participants of this study. Figure includes the distribution of the total MDS-UPDRS-III score (A), the speech item of the MDS-UPDRS-III score (B), and the m-FDA score (C). The scores for the PD patients are grouped into three classes: low (green), intermediate (blue), and severe (red) according to the severity of the disease. The scores for the m-FDA scale also include HC subjects, represented with the white bars. (Color figure online)

4   Experiments and Results

Three experiments are performed: (1) classification of PD vs. HC subjects, (2) classification of HC vs. PD patients in three stages of the disease divided according to the speech item of the MDS-UPDRS-III score (see Fig. 3B), and (3) classification of HC and PD patients divided into four groups according to the total m-FDA scale (see Fig. 3C). The results are shown in Table 4.

The phonological features provide higher accuracies than those obtained with MFCCs to discriminate between PD patients and HC subjects. On the other hand, note that when we consider the multi-class experiments e.g., the classification of the UPDRS-speech item and the m-FDA scores, the highest accuracies are obtained with the fusion of MFCC and phonological features, which indicate that these two feature sets provide complementary information to assess the speech of PD patients in several stages of the disease. Further experiments with other deep architectures are required to improve the results for multi-class assessment of the patients.

Table 4. Results of the proposed approach to classify PD patients vs. HC subjects, and to assess the speech deficits of the patients following the speech item of the MDS-UPDRS-III and the m-FDA scores. **ACC:** accuracy in the test set, **AUC:** Area under receiving operating characteristic curve for the two-class experiments.

Features	Classification task	Num. classes	ACC.	AUC
Phonological	PD vs HC	2	76.0 ± 5.8	0.78
Phonological	UPDRS-speech	3	57.0 ± 4.0	-
Phonological	m-FDA	4	30.8 ± 1.4	-
MFCC	PD vs HC	2	65.0 ± 4.7	0.66
MFCC	UPDRS-speech	3	59.4 ± 6.9	-
MFCC	m-FDA	4	33.7 ± 2.0	-
Phonological + MFCC	PD vs HC	2	64.0 ± 5.6	0.69
Phonological + MFCC	UPDRS-speech	3	59.4 ± 6.9	-
Phonological + MFCC	m-FDA	4	39.5 ± 11.3	-

5   Conclusion

This study considers phonological posterior features and recurrent neural networks based on GRU units to assess speech impairments of PD patients. A total of 15 phonological features are computed based on the sound pattern of English to model several aspects of the speech production system. The phonological posteriors used in this study can be interpreted by medical experts, which may support the evaluation of the speech state of the patients.

The results obtained with phonological features are compared with those obtained with standard acoustic features based on MFCCs. The phonological features are better to model the speech impairments of PD patients than the standard acoustic features, specially to discriminate between PD patients and HC subjects, however, the combination of acoustic features with the phonological posteriors shows to be suitable to assess the speech deficits of the patients in several stages of the disease. Further experiments are required with other deep architectures to assess the neurological state and the dysarthria level of the patients. Additionally, other feature sets based on phonation, articulation, or prosody analyses could be considered.

Acknowledgments. The work reported here was financed by CODI from University of Antioquia by grants Number 2015–7683. This project has received funding from the European Union's Horizon 2020 research and innovation programme under the Marie Skłodowska-Curie Grant Agreement No. 766287.

References

1. Bengio, Y., Simard, P., Frasconi, P.: Learning long-term dependencies with gradient descent is difficult. IEEE Trans. Neural Netw. **5**(2), 157–166 (1994)
2. Goetz, C.G., et al.: Movement Disorder Society-sponsored revision of the Unified Parkinson's Disease Rating Scale (MDS-UPDRS): scale presentation and clinimetric testing results. Mov. Disord. **23**(15), 2129–2170 (2008)
3. Cernak, M., Potard, B., Garner, P.N.: Phonological vocoding using artificial neural networks. In: IEEE International Conference on Acoustics, Speech and Signal Processing, ICASSP, pp. 4844–4848. IEEE (2015)
4. Cernak, M., et al.: Characterisation of voice quality of Parkinsons disease using differential phonological posterior features. Comput. Speech Lang. **46**, 96–208 (2017)
5. Cho, K., et al.: Learning phrase representations using RNN encoder-decoder for statistical machine translation. In: Conference on Empirical Methods in Natural Language Processing, EMNLP, pp. 1724–1734 (2014)
6. Chung, J., Gulcehre, C., Cho, K., Bengio, Y.: Empirical evaluation of gated recurrent neural networks on sequence modeling. In: NIPS 2014 Deep Learning and Representation Learning Workshop (2014)
7. Hlavnicka, J., Cmejla, R., Tykalova, T., Sonka, K., Ruzicka, E., Rusz, J.: Automated analysis of connected speech reveals early biomarkers of Parkinson's disease in patients with rapid eye movement sleep behaviour disorder. Nat. Sci. Rep. **7**(12), 1–13 (2017)
8. Hornykiewicz, O.: Biochemical aspects of Parkinson's disease. Neurology **51**(2 Suppl. 2), S2–S9 (1998)
9. Irie, K., Tüske, Z., Alkhouli, T., Schlüter, R., Ney, H.: LSTM, GRU, highway and a bit of attention: an empirical overview for language modeling in speech recognition. In: Proceedings of INTERSPEECH, pp. 3519–3523 (2016)
10. LeCun, Y., Bengio, Y., Hinton, G.: Deep learning. Nature **521**(7553), 436–444 (2015)
11. Logemann, J.A., Fisher, H.B., Boshes, B., Blonsky, E.R.: Frequency and cooccurrence of vocal tract dysfunctions in the speech of a large sample of Parkinson patients. J. Speech Hear. Disord. **43**(1), 47–57 (1978)
12. Orozco-Arroyave, J.R., Vásquez-Correa, J.C., et al.: NeuroSpeech: an open-source software for Parkinson's speech analysis. Dig. Signal Process. (2017, in press)

13. Orozco-Arroyave, J.R., et al.: New Spanish speech corpus database for the analysis of people suffering from Parkinson's disease. In: Language Resources and Evaluation Conference, LREC, pp. 342–347 (2014)
14. Snoek, J., Larochelle, H., Adams, R.P.: Practical Bayesian optimization of machine learning algorithms. In: Advances in Neural Information Processing Systems, NIPS, pp. 2951–2959 (2012)
15. Tu, M., Berisha, V., Liss, J.: Interpretable objective assessment of dysarthric speech based on deep neural networks. In: Proceedings of INTERSPEECH, pp. 1849–1853 (2017)
16. Vásquez-Correa, J.C., Orozco-Arroyave, J.R., Nöth, E.: Convolutional neural network to model articulation impairments in patients with Parkinson's disease. In: Proceedings of INTERSPEECH, pp. 314–318 (2017)

Phonological i-Vectors to Detect Parkinson's Disease

N. Garcia-Ospina[1]([✉]), T. Arias-Vergara[1,2], J. C. Vásquez-Correa[1,2],
J. R. Orozco-Arroyave[1,2], M. Cernak[3], and E. Nöth[2]

[1] Faculty of Engineering, University of Antioquia UdeA, Medellín, Colombia
nicanor.garcia@udea.edu.co
[2] Pattern Recognition Lab, University of Erlangen-Nürnberg, Erlangen, Germany
[3] Logitech Europe S.A., Lausanne, Switzerland

Abstract. Speech disorders are common symptoms among Parkinson's disease patients and affect the speech of patients in different aspects. Currently, there are few studies that consider the phonological dimension of Parkinson's speech. In this work, we use a recently developed method to extract phonological features from speech signals. These features are based on the Sound Patterns of English phonological model. The extraction is performed using pre-trained Deep Neural Networks to infer the probabilities of phonological features from short-time acoustic features. An i-vector extractor is trained with the phonological features. The extracted i-vectors are used to classify patients and healthy speakers and assess their neurological state and dysarthria level. This approach could be helpful to assess new specific speech aspects such as the movement of different articulators involved in the speech production process.

Keywords: Parkinson's disease · Phonological features · i-vectors

1 Introduction

Parkinson's disease (PD) is the second most common neuro-degenerative disorder worldwide [21]. PD patients suffer several motor symptoms including tremor, rigidity, slowed movement, postural instability, lack of coordination, and speech impairments. These symptoms limit the mobility and communication skills of patients, making it hard for them to attend appointments and therapy and to adequately convey their symptoms to their physicians and caregivers [20]. Most PD patients develop hypokinetic dysarthria, which includes a group of speech disorders such as reduced loudness, monopitch, monoloudness, reduced stress, breathy, hoarse voice quality, and imprecise articulation. The disease severity is evaluated by neurologists following several tests. One of them is the Movement Disorder Society-Unified Parkinson's Disease Rating Scale (MDS-UPDRS) [9], a perceptual scale that assesses motor and non-motor abilities of PD patients.

© Springer Nature Switzerland AG 2018
P. Sojka et al. (Eds.): TSD 2018, LNAI 11107, pp. 462–470, 2018.
https://doi.org/10.1007/978-3-030-00794-2_50

As PD affects several aspects of speech [17], it makes sense to model motor capabilities from speech considering different dimensions such as phonation, articulation, prosody, and intelligibility [13,16]. In recent years, the scientific community has been developing computer-based aids to help physicians with the detection and evaluation of the disease. Different features extracted from the speech signal have been proposed to analyze different dimensions of the affected speech: Phonation impairments in PD patients include stability and periodicity problems in vocal fold vibration due to an inadequate closing of the vocal fold and vocal fold bowing [10]. Phonation in PD was automatically analyzed in [22], using features related to perturbation, noise content, and non-linear dynamics to evaluate whether the response of 14 PD patients to the Lee Silverman voice treatment is "acceptable" or "unacceptable". The authors considered only sustained vowels and reported an accuracy close to 90% when discriminating between "acceptable" vs. "unacceptable" utterances. However, speaker independence between train and test sets was not guaranteed, which may lead to optimistic results. Articulation problems include a reduced articulatory capability in PD patients to produce vowels [19] and continuous speech. These are related to reduced amplitude and velocity of the articulator movements [1]. In [13] the authors modeled six different articulatory deficits in PD analyzing a diadochokinetic speech task uttered by 24 Czech native speakers and reported an accuracy of 88% classifying between PD patients and healthy controls (HC). Prosody refers to intonation, loudness, and rhythm during continuous speech. Prosodic problems in PD patients includes a decrease in loudness and low variations of pitch, which is related to the frequency of vocal fold vibration (F0) [6,11]. Prosody features were computed in [2]. The authors consider voiced segments as speech unit to compute features based on the fundamental frequency F0 contour, energy contour, duration, and pitch periods to classify PD patients and HC speakers, and to classify the patients according to their neurological state in a 3-class approach (low, middle, and severe) state. The authors report an accuracy of up to 74 classifying PD patients and HC speakers, and of 37% for the 3-class problem. Phonology studies how the sounds of speech are represented in the human mind [12]. A phonological representation refers to an abstract system of rules that converts words of a language into phonemes [5]. Few studies have analyzed the speech of PD patients in phonological terms, and they focus on evaluating phonology from a neurological point of view usually with phonological fluency tests [18]. This is in part due to the difficulty to estimate phonological features. Recently, a method to reliably estimate phonological features was proposed in [3]. These phonological features could be used in the analysis of dysarthric speech to assess specific articulators and parts of the speech production system. This method was used in [4] to evaluate the voice quality of PD patients.

In this work, we extract the phonological features with the method in [3] and model them using the i-vector approach. The extracted i-vectors will be referred to as phonological i-vectors. The proposed model is tested in three scenarios: (1) the classification of PD patients vs. HC, (2) the assessment of the neurological state of the patients following the MDS-UPDRS-III scale, and (3) the assessment of the dysarthria level of the patients following a modified version of the Frenchay Dysarthria assessment (m-FDA) scale. This approach is compared to a baseline which models the phonological features using four statistical functionals.

2 Methods

2.1 Phonological Features

Phonological features were extracted using the deep learning approach from [3]. This process involves the following steps: (1) the speech signal is segmented into short-time frames, (2) 13 Mel Frequency Cepstral Coefficients (MFCCs) and their derivatives are computed for every frame of the speech signal, and (3) a set of 15 pre-trained Deep Neural Networks (DNNs) infers the phonological posteriors from the acoustic feature vector. These posteriors are concatenated into a phonological feature vector z_t. A total of 15 phonological features based on the Sound Patterns of English [5] are computed. Table 1 indicates a brief description of each feature and the Pearson's correlation to the scores used to evaluate the neurological state and dysarthria level of patients.

Table 1. List of phonological features

Feature	Brief description
Vocalic	The vocal folds vibrating and no constriction in the vocal tract
Consonantal	Indicates there is an obstruction of the vocal tract
High	The body of the tongue is above its neutral position
Back	The tongue is retracted from its neutral position
Low	A position lower than neutral for the body of the tongue
Anterior	Refers to an obstruction in front of the palato-alveolar region of the mouth
Coronal	The blade of the tongue is raised from its neutral position
Round	Refers to narrowed lips
Rising	Differentiates diphthongs from monophthongs
Tense	Indicates stressed vowels
Voice	The vocal folds are vibrating
Continuant	Differentiates plosives from non-plosives
Nasal	Indicates a lowered velum, where the air to escape through the nose
Strident	Refers to sounds with more energy in high frequency components
Silence	Indicates that there is no speech in the frame

2.2 i-Vector Extraction

In the i-vector approach, factor analysis is used to define a new low-dimensional space known as the total variability space. Initially, for speaker verification applications, this space had the aim of modeling the speaker and the channel variability [7]. In pathological speech analysis applications, the speaker variability carries the information about the disorders in speech due to the disease. In this work, the i-vector extractor was trained using recordings from a speech corpus collected for Speaker Verification (SV) experiments. The configuration with the lowest EER on the SV experiments was selected for this experiment: an Universal Background Model (UBM) with 64 Gaussians and 100-dimensional i-vectors. The extracted i-vectors are processed with the following steps: (1) the i-vectors computed from utterances of the same speaker are averaged to obtain one i-vector per speaker, (2) a whitening is process is applied by subtracting the mean of the training i-ivectors and performing a Principal Component Analysis [8]. No further processing is applied to the i-vectors as all the speech signals used in this study were recorded in similar acoustic conditions.

2.3   Classification Methods

Cosine Distance Threshold. The score computed from a test signal is compared with respect to a threshold θ. The score used in the i-vector approach to compare two i-vectors is the cosine distance. In this case, the average cosine distance to a reference set of i-vectors is computed according to Eq. 1.

$$\text{score}(w_{\text{test},j}) = \frac{1}{N}\sum_{i=1}^{N} C_i \frac{w_{\text{test},j} \cdot w_{\text{ref},i}}{||w_{\text{test},j}||||w_{\text{ref},i}||} \tag{1}$$

where C_i is the condition label of the reference i-vector: 1 for HC and -1 for PD patients. Our hypothesis is that the larger the distance to the HC, the more affected the speech. Considering this argument, the condition is: if $\text{score}(w_{\text{test},j}) > \theta$ then $w_{\text{test},j}$ is considered from a PD patient and the threshold was set at $\theta = 0$. No parameters needed to be optimized for this method.

Support Vector Machines (SVMs). This classifier discriminates data points by using a separating hyperplane which maximizes the margin between two classes. A soft margin Support Vector Machine (SVM) with Gaussian kernel is used to classify PD vs. HC subjects. This classifier was chosen as it has shown good results in this kind of tasks [15]. Two hyper-parameters need to be optimized: the margin cost C and the bandwidth of the Gaussian kernel γ. The optimization process is described in Sect. 4.

2.4   Evaluation of the Neurological State

The prediction of the neurological state and the dysarthria level are evaluated with the Spearman's correlation between the real label and the score given by Eq. 1. In this case, two different reference sets are used: the first only includes i-vectors from HC and the second is formed with i-vectors from PD patients only. The condition label is $C_i = 1$ in both reference sets.

3 Data

3.1 Speakers Considered for the i-Vector Extractor

The recordings used to train the i-vector extractor come from a speech corpus collected for Speaker Verification experiments. The corpus contains recordings from 103 young healthy native Colombian Spanish speakers. The speakers were asked to read aloud ten short utterances ten times each.

3.2 PC-GITA Speech Corpus

The PC-GITA speech corpus contains recordings of 50 PD patients (25 male and 25 female) and age balanced 50 healthy controls (HC), all of them native Colombian Spanish speakers. The recordings were captured in a sound-proof booth using professional audio equipment. The recordings were down-sampled to 16 kHz for this study. During the recordings, the participants were asked to perform different speech tasks including ten read short sentences. All the patients were diagnosed by a neurologist expert and their neurological state was assessed according to the MDS-UPDRS [9]. None of the HC had a history of symptoms related to any kind of neurological disorder. Additional information of this corpus can be found in [14].

3.3 m-FDA Scale

The evaluation of the neurological state PD patients according to the MDS-UPDRS-III scale is suitable to assess general motor impairments of PD patients; however, the speech deficits are only evaluated in one of its 33 items. A modified version of the FDA scale (m-FDA) based only on speech recordings was developed in [4,16,23]. This modified scale includes several aspects of speech: respiration, lips movement, palate/velum movement, larynx, tongue, monotonicity, and intelligibility. The scale has a total of 13 items and each of them ranges from 0 (normal or completely healthy) to 4 (very impaired), thus the total score of the scale ranges from 0 to 52. The labeling process of the recordings of the PC-GITA database was performed by three phoniatricians who agreed in the first ten speakers. Afterwards, each phoniatrician evaluated the remaining recordings independently. The inter-rater reliability among the labelers is 0.75.

4 Experiments and Results

The experimental methodology used in this work comprises the following steps: (1) The phonological features from the speech signals are extracted, (2) the phonological features from training signals are used to train the i-vector extractor, (3) i-vectors are extracted from the features of the signals in PC-GITA corpus and are processed, and (4) the i-vectors are analyzed as will be described next. This methodology is summarized in Fig. 1.

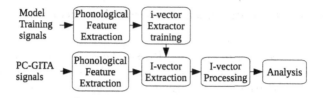

Fig. 1. General experimental methodology

4.1   Experimental Setup and Validation

For this study the PC-GITA i-vectors are split into training and test as follows: 60% is used to perform training and development and the remaining 40% is used to test. This means that the i-vectors of 60 of the speakers (30 PD patients and 30 HC) are used to train the system, and the remaining 40 i-vectors (20 PD patients and 20 HC) are used to test. Speaker independence is guaranteed between training and test sets. The optimization of the hyper-parameters of the classifiers is done by performing a 3-fold cross validation over the 60% training data. The optimal hyper-parameters found in this process are then used to perform the test. To validate the results, this process is repeated ten times. The train and test sets are randomly chosen for each repetition of the experiment. The mean and standard deviation over the ten iterations are reported. As a baseline to compare the results the phonological features are modeled with four functionals (mean, standard deviation, skewness, and kurtosis) computed from all the phonological features of a given speaker to form a 60-dimensional feature vector. SVM classifier and regressor are trained on these features vectors.

4.2   Classification Results

The results in Table 2 show that the threshold classification method has a better accuracy. Note that the threshold-based classifier exhibits similar results of accuracy, specificity and sensitivity, which indicates that the system is not biased to any of the two classes (HC and PD). Conversely, the SVM shows a high specificity and a low sensitivity, indicating that it is good to detect healthy speakers but not as good to detect Parkinson's patients. The baseline shows poor results.

Table 2. PD vs. HC classification results

Method	Accuracy [%]	Sensitivity [%]	Specificity [%]	F1-score
Baseline	55.2 ± 3.1	49.0 ± 14.8	61.5 ± 14.3	0.51 ± 0.10
Threshold	77.5 ± 7.3	77.0 ± 12.1	78.0 ± 8.4	0.77 ± 0.08
i-vectors SVM	73.5 ± 8.2	64.0 ± 16.7	83.0 ± 8.7	0.70 ± 0.11

4.3 Estimation of Neurological State and Dysarthria Level

Two different reference i-vector sets are considered for this experiment, each consisting of the i-vectors from HC and PD speakers of the training set, respectively. The assessment of the neurological state and dysarthria level of a patient is evaluated by computing the Spearman's correlation coefficient between the MDS-UPDRS-III or m-FDA label and the average cosine distance. This is the same distance that was taken into account by the threshold classifier.

Table 3. Spearman's correlation

Label	Baseline	HC reference	PD reference
MDS-UPDRS-III	0.13 ± 0.1	0.52 ± 0.2	-0.52 ± 0.2
m-FDA	0.11 ± 0.1	0.30 ± 0.1	-0.28 ± 0.1

The results in Table 3 show that the phonological i-vectors average cosine distance is more correlated to the neurological state of the patient than to the phonological evaluation. The negative correlation found when using the PD reference indicates that a shorter distance to PD i-vectors means a more affected speech. Again, the baseline shows poor results.

5 Conclusion

In this work we introduced the use of phonological i-vectors extracted from the speech of PD patients and age balanced HC to perform the classification of PD vs. HC and to estimate their neurological state and their dysarthria level. These i-vectors are extracted from phonological posteriors obtained using DNNs.

The average cosine distance had better classification results than the SVM and its sensitivity and specificity indicate that is not biased to any of the classes. One of the main advantages of this approach is that no parameters need to be optimized, unlike classifiers like SVMs or those based on neural networks. Additionally, the same cosine distance can be used to assess the neurological state and dysarthria level.

Future work includes modeling subsets of the phonological features with i-vectors to assess specific items in the m-FDA scale. This can help in obtaining interpretable results such that are suitable to guide the phoniatrician or clinician when defining the patient's therapy. Also, we want to test the language independence assertion of the phonological features and perform similar analyses in different languages.

Acknowledgments. The work reported here was financed by CODI from University of Antioquia by grants Number 2015–7683. This project has received funding from the European Union's Horizon 2020 research and innovation programme under the Marie Sklodowska-Curie Grant Agreement No. 766287.

References

1. Ackermann, H., Ziegler, W.: Articulatory deficits in Parkinsonian dysarthria: an acoustic analysis. J. Neurol., Neurosurg. Psychiatry **54**(12), 1093–1098 (1991)
2. Bocklet, T., et al.: Automatic evaluation of Parkinson's speech - acoustic, prosodic and voice related cues. In: Annual Conference of the International Speech Communication Association, pp. 1149–1153 (2013)
3. Cernak, M., et al.: Phonological vocoding using artificial neural networks. In: 2015 IEEE International Conference on Acoustics, Speech and Signal Processing (ICASSP), pp. 4844–4848, April 2015. https://doi.org/10.1109/ICASSP.2015. 7178891
4. Cernak, M., et al.: Characterisation of voice quality of Parkinson's disease using differential phonological posterior features. Comput. Speech Lang. **46**, 96–208 (2017)
5. Chomsky, N., Halle, M.: The Sound Pattern of English. Studies in Language. Harper and Row (1968). https://books.google.com.co/books?id=PbtZAAAAM AAJ
6. Darley, F.L., et al.: Differential diagnostic patterns of dysarthria. J. Speech Lang. Hear. Res. **12**(2), 246–269 (1969)
7. Dehak, N., et al.: Front-end factor analysis for speaker verification. IEEE Trans. Audio Speech Lang. Process. **19**(4), 788–798 (2011). https://doi.org/10.1109/ TASL.2010.2064307
8. Garcia-Romero, D., Espy-Wilson, C.: Analysis of i-vector length normalization in speaker recognition systems. In: Proceedings of the 12th INTERSPEECH, September 2011
9. Goetz, C.G., et al.: Movement disorder society-sponsored revision of the unified Parkinson's disease rating scale (MDS-UPDRS): scale presentation and clinimetric testing results. Mov. Disord. **23**(15), 2129–2170 (2008). https://doi.org/10.1002/ mds.22340
10. Hanson, D.G., et al.: Cinegraphic observations of laryngeal function in Parkinson's disease. Laryngoscope **94**(3), 348–353 (1984)
11. Ho, A.K., et al.: Speech impairment in a large sample of patients with Parkinson's disease. Behav. Neurol. **11**(3), 131–137 (1999)
12. de Lacy, P.: The Cambridge Handbook of Phonology. Cambridge Handbooks in Language and Linguistics. Cambridge University Press (2007). https://books. google.com.co/books?id=7sxLaeZAhOAC
13. Novotný, M., et al.: Automatic evaluation of articulatory disorders in Parkinson's disease. IEEE/ACM Trans. Audio Speech Lang. Process. **22**(9), 1366–1378 (2014)
14. Orozco, J.R., et al.: New Spanish speech corpus database for the analysis of people suffering from Parkinson's disease. In: Proceedings of the 9th LREC, pp. 342–347 (2014)
15. Orozco-Arroyave, J.R., et al.: Characterization methods for the detection of multiple voice disorders: neurological, functional, and laryngeal diseases. IEEE J. Biomed. Health Inform. **19**(6), 1820–1828 (2015). https://doi.org/10.1109/JBHI. 2015.2467375
16. Orozco-Arroyave, J.R., et al.: Neurospeech: An open-source software for Parkinson's speechanalysis. Dig. Sign. Process. (2017). (In press)
17. Orozco-Arroyave, J.: Analysis of Speech of People with Parkinson's Disease, 1st edn. Logos-Verlag, Berlin (2016)
18. Santangelo, G., et al.: A neuropsychological longitudinal study in Parkinson's patients with and without hallucinations. Mov. Disord. **22**(16), 2418–2425 (2007). https://doi.org/10.1002/mds.21746

19. Skodda, S., et al.: Vowel articulation in Parkinson's disease. J. Voice **25**(4), 467–472 (2011)
20. Stamford, J.A., et al.: What engineering technology could do for quality of life in Parkinson's disease: a review of current needs and opportunities. IEEE J. Biomed. Health Inform. **19**(6), 1862–1872 (2015). https://doi.org/10.1109/JBHI. 2015.2464354
21. Sveinbjornsdottir, S.: The clinical symptoms of Parkinson's disease. J. Neurochem. **139**(S1), 318–324 (2016). https://doi.org/10.1111/jnc.13691
22. Tsanas, A., et al.: Objective automatic assessment of rehabilitative speech treatment in Parkinson's disease. IEEE Trans. Neural Syst. Rehabil. Eng. **22**(1), 181–190 (2014)
23. Vásquez-Correa, J., et al.: Towards an automatic evaluation of the dysarthria level of patients with parkinson's disease. J. Commun. Disord. (under review) (2018)

Dialogue

Subtext Word Accuracy and Prosodic Features for Automatic Intelligibility Assessment

Tino Haderlein[1]([⊠])(iD), Anne Schützenberger[2], Michael Döllinger[2], and Elmar Nöth[1]

[1] Friedrich-Alexander-Universität Erlangen-Nürnberg (FAU), Lehrstuhl für Informatik 5 (Mustererkennung), Martensstraße 3, 91058 Erlangen, Germany
Tino.Haderlein@fau.de
https://www5.cs.fau.de
[2] Universitätsklinikum Erlangen, Phoniatrische und Pädaudiologische Abteilung in der HNO-Klinik, Waldstraße 1, 91054 Erlangen, Germany

Abstract. Speech intelligibility for voice rehabilitation can successfully be evaluated by automatic prosodic analysis. In this paper, the influence of reading errors and the selection of certain words (nouns only, nouns and verbs, beginning of each sentence, beginnings of sentences and subclauses) for the computation of the word accuracy (WA) and prosodic features are examined. 73 hoarse patients read the German version of the text "The North Wind and the Sun". Their intelligibility was evaluated perceptually by 5 trained experts according to a 5-point scale. Combining prosodic features and WA by Support Vector Regression showed human-machine correlations of up to $r = 0.86$. They drop for files with few reading errors, however, but this can largely be evened out by feature set adjustment. WA should be computed on the whole text, but for some prosodic features, a subset of words may be sufficient.

Keywords: Intelligibility · Automatic assessment · Prosody
Reading errors

1 Introduction

Established methods for objective voice and speech evaluation in therapy analyze only sustained vowels. In our approach for the assessment of speech intelligibility, the test persons read a given standard text that undergoes prosodic analysis. Usually, each prosodic feature has been averaged over all words in the text. However, it is widely known that intelligibility varies among different word classes which is mostly caused by prosodic properties [1–5]. Hence, putting all content and function words, long and short words, and words at different positions in sentences, together bears the risk of losing information. In previous work [6], the influence of the position and type of words, which are selected from a read-out text, on the reliability of the automatic analysis has already been addressed.

© Springer Nature Switzerland AG 2018
P. Sojka et al. (Eds.): TSD 2018, LNAI 11107, pp. 473–481, 2018.
https://doi.org/10.1007/978-3-030-00794-2_51

However, this was restricted to single prosodic features. The word accuracy (WA) of a speech recognizer has also been used as basic measure for intelligibility [7,8]. It has also always been computed for an entire text sample. In this follow-up study, the suitability of the WA computed on subunits of a text will be examined, and the combination of these features and prosodic features by Support Vector Regression will be presented for the first time. It has further been shown that the automatic analysis is influenced by the number of reading errors in the sample [6]. This will be tested for the new feature sets. These main questions are addressed in this paper:

- How can prosodic features and the word accuracy together model the human intelligibility rating when they are computed on different subparts of a standard text?
- Does the number of reading errors in a speech sample have an influence on the human and automatic intelligibility rating?

This work is organized as follows: Sect. 2 introduces the test data and the perceptual evaluation reference. The computation of the features is described in Sect. 3. The results of the experiments (Sect. 4) will be discussed in Sect. 5.

2   Test Data and Subjective Evaluation

73 German subjects with chronic hoarseness participated in this study (Table 1). Patients suffering from cancer were excluded. Each person read the text "Der Nordwind und die Sonne" ("The North Wind and the Sun", [9]), a phonetically rich standard text which is frequently used in clinical speech evaluation in German-speaking countries. It contains 108 words (71 distinct) with 172 syllables. The data were recorded with a sampling frequency of 16 kHz and 16 bit amplitude resolution using an AKG C 420 microphone (AKG Acoustics, Vienna, Austria). They were recorded in a quiet room in our university and digitally stored on a server by a client/server-based system [10, Chap. 4]. The study respected the principles of the World Medical Association (WMA) Declaration of Helsinki on ethical principles for medical research involving human subjects and has been approved by the ethics committee of our clinics.

Five voice professionals (one ear-nose-throat doctor, four speech therapists) evaluated the intelligibility of each original recording perceptually. The samples were played to the experts once via loudspeakers in a quiet seminar room without disturbing noise or echoes. Rating was performed on a five-point Likert scale. For computation of average scores for each patient, the grades were converted to integer values (1 = 'very high', 2 = 'rather high', 3 = 'medium', 4 = 'rather low', 5 = 'very low'). For each patient, an intelligibility mark, expressed as a floating point value, was calculated as the arithmetic mean of the single scores. These marks served as ground truth in our experiments.

Due to reading errors, repetitions, and additional remarks, such as "read now?", the recordings did not only contain words appearing in the text reference but also additional words and word fragments. In order to describe the

Table 1. The test speakers (entire set, group with few and group with many reading errors)

group	persons			age				reading errors			
	all	men	women	μ	σ	min	max	μ	σ	min	max
overall	73	24	49	48.3	16.8	19	85	3.10	3.50	0	17
low-error	32	9	23	48.5	13.7	26	76	0.34	0.47	0	1
high-error	41	15	26	48.1	18.9	19	85	5.24	3.34	2	17

errors, a manual word-based counting of errors was adopted (see details in [6,7]). In order to study the effect of errors on the evaluation on subsets of reasonable size, the overall data set was divided into a 'low-error' group with at most one reading error per speaker and a 'high-error' group with 2 to 17 errors per speaker (Table 1). The left side of Fig. 1 shows that there are also low-error readers with a bad perceptual ranking. The human intelligibility rating and the number of reading errors are just weakly correlated ($r = 0.35$).

3 Prosodic Features

The speech recognizer used for the experiments [11] is based on semi-continuous Hidden Markov Models (HMM). For each 16 ms frame, a 24-dimensional feature vector is computed. It contains short-time energy, 11 Mel-frequency cepstral coefficients, and the first-order derivatives of these 12 static features. The recognition vocabulary of the recognizer was changed to the 71 words of the standard text. Only a unigram language model was used so that the results mainly depend on the acoustic models.

In order to find counterparts for intelligibility, a 'prosody module' was used to compute features based upon frequency, duration, and speech energy (intensity) measures. The prosody module processes the output of the word recognition module and the speech signal itself. 'Local' prosodic features are computed for each word position. Originally, there were 95 of them. After several studies on voice and speech assessment, however, a relevant core set of 33 features has been defined for further processing [12]. The components of their abbreviated names are given in parentheses:

- Length of pauses (Pause): length of silent pause before (–before) and after (–after), and filled pause before (Fill-before) and after (Fill-after) the respective word
- Energy features (En): regression coefficient (RegCoeff) and the mean square error (MseReg) of the energy curve with respect to the regression curve; mean (Mean) and maximum energy (Max) with its position on the time axis (Max-Pos); absolute (Abs) and normalized (Norm) energy values
- Duration features (Dur): absolute (Abs) and normalized (Norm) duration

- F_0 features (F0): regression coefficient (RegCoeff) and mean square error (MseReg) of the F_0 curve with respect to its regression curve; mean (Mean), maximum (Max), minimum (Min), voice onset (On), and offset (Off) values as well as the position of Max (MaxPos), Min (MinPos), On (OnPos), and Off (OffPos) on the time axis; all F_0 values are normalized.

The last part of the feature name denotes the context size, i.e. the interval of words on which the features are computed (see Table 2). They can be computed on the current word (W) or in the interval that contains the second and first word before the current word and the pause between them (WPW). A full description of the features used is beyond the scope of this paper; details and further references are given in [11,13].

Besides the 33 local features per word, 15 'global' features were computed for intervals of 15 words length each. They were derived from jitter, shimmer, and the number of detected voiced and unvoiced sections in the speech signal [13]. They covered the means and standard deviations of jitter and shimmer, the number, length, and maximum length of voiced and unvoiced sections, the ratio of the numbers of voiced and unvoiced sections, the ratio of the length of the voiced sections to the length of the signal, and the same for unvoiced sections. The last feature was the standard deviation of the F_0.

The human listeners gave ratings for the entire text. In order to receive also one single value for each feature that could be compared to the human ratings, the average of each prosodic feature over all selected words served as final feature value.

Table 2. Local prosodic features; the context size denotes the interval of words on which the features are computed (W: one word, WPW: word-pause-word interval).

features	context size	
	WPW	W
Pause: before, Fill-before, after, Fill-after		•
En: RegCoeff, MseReg, Abs, Norm, Mean	•	•
En: Max, MaxPos		•
Dur: Abs, Norm	•	•
F0: RegCoeff, MseReg	•	•
F0: Mean, Max, MaxPos, Min, MinPos, Off, OffPos, On, OnPos		•

4   Experiments

Earlier experiments averaged each prosodic feature over the entire read-out text. For this study, we examined whether the restriction to certain subsets might be beneficial:

- averaging over *all words* (the baseline; 108 words, denoted by the suffix '_all')
- *nouns only* (24 words, '_nouns')
- *nouns and verbs* (44 words, '_n+v')
- *beginnings of sentences* (first 3 words of each of the 6 sentences; 18 words, '_sent_i')
- *beginnings of sentences and subclauses* (first 3 words of each of the 6 sentences and 10 subclauses; 48 words, '_s+s_i').

Nouns and verbs were chosen because content words generally show less predictability and hence intelligibility than function words, such as articles, prepositions, and conjunctions [14]. Adjectives were not taken into account because there are very few in the text. These words contribute to intelligibility mainly because of their stress patterns, one of the main aspects the prosodic features were designed for. The beginnings of sentences and subclauses, without the regard of the word classes, were chosen with respect to the medical application. Many voice and speech patients show higher speaking effort and shorter phonation time, so they will have to pause more often and fragment the paragraph to be read. Breaks usually occur at syntactic boundaries.

In former studies [7,8], the word accuracy (WA) of a speech recognizer was an important feature to model intelligibility. It is computed from the comparison of the recognized word sequence and the reference text consisting of the $n_{\text{all}} = 108$ words of the read text. With the number of words that were wrongly substituted (n_{sub}), deleted (n_{del}) and inserted (n_{ins}) by the recognizer, the word accuracy in percent is given as

$$\text{WA} = [1 - (n_{\text{sub}} + n_{\text{del}} + n_{\text{ins}})/n_{\text{all}}] \cdot 100.$$

In this study, it is also added to the feature set, but just like the prosodic features, several versions of it will be used for the first time: besides WA_all, computed on all words of the text, there is also WA_nouns, WA_n+v, WA_sent_i, and WA_s+s_i.

In order to find the best subset of WA and the prosodic features to model the subjective intelligibility ratings, Support Vector Regression (SVR) was applied. For this study, the sequential minimal optimization algorithm [15] of Weka [16] was applied. Due to the small amount of available data, a 10-fold cross-validation was used. For the regression, the automatically computed measures (WA and all prosodic features) served as the training set. The test set consisted of the subjective, perceptual intelligibility scores. Due to the small amount of data, we also refrained from using deep learning technology.

5   Results and Discussion

The first experiment was performed using the different word accuracy types as a single feature. The human-machine correlation showed the best results for WA_all ($r = -0.74$, first part of Table 3). The other values are – in some cases substantially – lower.

The next part of the table shows a comparison of the correlations for the best single prosodic features, as determined in [6], and the predicted intelligibility of the SVR when all prosodic features were put together. The single features could compete with the WA results, the combination of all prosodic features achieves substantially better values. The best result was $r = 0.84$ for all features being averaged over all words.

The next lines of Table 3 show the use of prosodic features from a subtext scenario x supplemented either by WA_all or WA_x. These results are even better than for the prosodic features alone. In general, WA_all contributes better to the human-machine agreement than the WA from the other scenarios (with two non-significant exceptions). WA_all and prosodic features obtained on the whole text achieved $r = 0.86$. The right side of Fig. 1 shows this case. The 'real' ratings can only be between 1 and 5. If the predicted values outside that range are replaced by the possible minimum and maximum values (i. e. in our data 0.909 by 1, 5.714 and 6.023 by 5), the correlation does not improve, however. Without the outlier at position $(0.909, 2.400)$, the correlation for the low-error files would rise from 0.74 to 0.80.

Computing the prosodic features on a lower number of words keeps the human-machine agreement at a high level in general, with the best result for sentence and subunit beginnings (_s+s_i) where only a drop in correlation of $\Delta r = 0.01$ is measured.

Next, the influence of the reading errors on the combined feature set was examined. The low-error files show significantly lower correlations, just like for single features [6], while for the high-error files, the results stay at about $r = 0.80$ if WA_all is used.

Table 4 contains the regression weights for the case that all 73 speech samples are used in the SVR. The baseline is given in the first data column, i. e. using WA_all and the prosodic features computed on all words. WA_all shows a consistently high weight in all setups. The other WA_x for the respective scenarios x show lower weights on the average. WA_nouns is not even part of the best set for the whole database. The same holds for WA_sent_i and WA_s+s_i on the low-error files. Like in a similar study with a slightly different SVR setup on the _all scenario [8], also the duration feature DurNormWPW, representing the speaking rate, becomes important in some cases. Further, the F_0 value at voice offset (F0OffW) appears in some sets. It likely resembles voice quality and stability. Similar information may be inherent in #+Voiced, i. e. the number of voiced segments, that appears in the sets for the _sent_i and _s+s_i cases (when using WA_all).

The new results mostly confirm the earlier study [6] that identified single prosodic features with a high human-machine correlation. Pause–before and the regression coefficient of the energy in a word-pause-word interval (EnRegCoef-fWPW) were in the new tests not among the best sets, however. The normalized duration of a word-pause-word interval (DurNormWPW) has also been a good indicator for intelligibility in earlier studies and could mostly replace the energy

EnNormWPW [7]. Here, it plays only a minor role. Both DurNormWPW and Pause–before reveal the overall speaking rate.

EnNormWPW is a good indicator for intelligibility [6,7,11]. Especially for low-error reading, a selection of words from the text lowers the correlation to the perceptual scores, however. MeanJitter showed the highest correlation of all in [6], namely $r = 0.73$ for low-error reading and the _n+v case. In this new study, it is present in all scenarios.

In a final experiment, all features from all scenarios were combined. The best subset comprised EnNormWPW_all, F0OffW_all, MeanJitter_n+v, and WA_all, achieving $r = 0.86$ for the whole data set. In this case only the MeanJitter_n+v had to be computed for only 44 words and the other features for all 108 words. So the benefit of restricting the word set for computation seems too small to be useful. However, with these features, the correlation was $r = 0.73$ for the low-error and $r = 0.87$ for the high-error files. With MeanJitter_all, only $r = 0.59$ had been measured on the low-error files, so the use of mixed scenarios (_all, _n+v) shows on the average the most stable results for all tested cases.

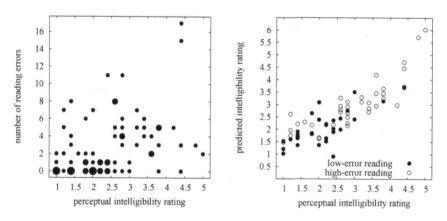

Fig. 1. Left side: average human intelligibility rating vs. number of reading errors in the speech files; the point size indicates single, double, or triple occurrence. Right side: average human vs. predicted intelligibility rating for WA_all and the _all prosodic features

We are aware of the problem arising when standard texts are used for measuring intelligibility. However, it was shown that text-based evaluation performed by trained listeners is as reliable as an inverse intelligibility test, where naïve raters write down a previously unknown word sequence read by the test person. For more details, see [6].

As a conclusion, it can be stated that the word accuracy (WA) should always be used in a feature set for intelligibility assessment, and it should be computed on the full text while it is sufficient to compute some of the prosodic features only on the first three words of a sentence and subclause. Only WA_n+v and the _n+v prosodic features give a slightly better result than WA_all and _n+v

Table 3. Human-machine correlation r for SVR values obtained from prosodic features and WA on all 73 speech samples, depending on the scenario x used for computation; bold-face: best results of each line; italics: WA_x not part of best feature set

feature set	scenario x files	all	nouns	n+v	sent_i	s+s_i
WA_x alone	all files	**−0.74**	−0.65	−0.70	−0.62	−0.66
best single pros. feature [6]	all files	**0.69**	0.66	0.67	0.61	0.66
pros. features on x	all files	**0.84**	0.78	0.79	0.72	0.79
pros. features on x + WA_all	all files	**0.86**	0.83	0.82	0.80	0.85
pros. features on x + WA_x	all files	—	0.78	**0.83**	0.77	0.81
pros. features on x + WA_all	low-error files	0.59	0.56	**0.74**	0.50	0.55
pros. features on x + WA_x	low-error files	—	0.44	**0.66**	*0.63*	*0.55*
pros. features on x + WA_all	high-error files	**0.86**	0.79	0.80	0.78	0.83
pros. features on x + WA_x	high-error files	—	*0.79*	0.68	0.79	**0.81**

Table 4. Regression weights for single features in the best feature sets for all 73 speech samples, depending on the scenario x and the additional use of WA_all or WA_x, respectively

scenario x	all	nouns	n+v	sent_i	s+s_i	nouns	n+v	sent_i	s+s_i
feature name		using WA_all					using WA_x		
EnNormWPW_x (local)	0.368	0.197	0.170	0.349	0.458	—	0.286	0.464	0.748
DurNormWPW_x (local)	—	—	—	—	−0.149	0.556	—	—	−0.336
F0OffW_x (local)	−0.198	—	−0.172	−0.141	—	—	−0.193	—	—
MeanJitter_x (global)	0.375	0.453	0.380	0.163	0.294	0.455	0.493	0.322	0.328
#+Voiced_x (global)	—	—	—	0.196	0.135	—	—	—	—
WA_all	−0.331	−0.486	−0.463	−0.445	−0.524	—	—	—	—
WA_x	—	—	—	—	—	—	−0.243	−0.326	−0.387
Human-machine corr. r	0.86	0.83	0.82	0.80	0.85	0.78	0.83	0.77	0.81

features for low-error files. The _n+v case already showed the best results for single prosodic features [6]. The influence of many reading errors is a positive one at first sight since the human-machine correlation is better for samples with many errors. However, Fig. 1 (left) shows that the low-error files are concentrated on a much smaller perceptual range than the high-error files. Hence, it is more difficult to find a feature set mapping these small differences. More effort has to be put on this in the next experiments. Future work also includes tuning of the SVR parameters. Preliminary tests changing the kernel parameter C in a range of 0.01 to 1000 have shown no better results, however.

Acknowledgments. Dr. Döllinger's contribution was supported by the German Research Foundation (DFG), grant no. DO1247/8-1 (no. 323308998).

References

1. Hustad, K., Dardis, C., McCourt, K.: Effects of visual information on intelligibility of open and closed class words in predictable sentences produced by speakers with dysarthria. Clin. Linguist. Phon **21**, 353–367 (2007)
2. Cutler, A.: Phonological cues to open- and closed-class words in the processing of spoken sentences. J. Psycholinguist Res. **22**, 109–131 (1993)
3. Grosjean, F., Gee, J.: Prosodic structure and spoken word recognition. Cognition **25**, 135–155 (1987)
4. Pichney, M., Durlach, N., Braida, L.: Speaking clearly for the hard of hearing. II: acoustic characteristics of clear and conversational speech. J. Speech Hear. Res. **29**, 434–446 (1986)
5. Turner, G., Tjaden, K.: Acoustic differences between content and function words in amyotrophic lateral sclerosis. J. Speech Lang. Hear. Res. **43**, 769–781 (2000)
6. Haderlein, T., Schützenberger, A., Döllinger, M., Nöth, E.: Robust automatic evaluation of intelligibility in voice rehabilitation using prosodic analysis. In: Ekštein, K., Matoušek, V. (eds.) TSD 2017. LNCS (LNAI), vol. 10415, pp. 11–19. Springer, Cham (2017). https://doi.org/10.1007/978-3-319-64206-2_2
7. Haderlein, T., Nöth, E., Maier, A., Schuster, M., Rosanowski, F.: Influence of reading errors on the text-based automatic evaluation of pathologic voices. In: Sojka, P., Horák, A., Kopeček, I., Pala, K. (eds.) TSD 2008. LNCS (LNAI), vol. 5246, pp. 325–332. Springer, Heidelberg (2008). https://doi.org/10.1007/978-3-540-87391-4_42
8. Haderlein, T., Döllinger, M., Matoušek, V., Nöth, E.: Objective voice and speech analysis of persons with chronic hoarseness by prosodic analysis of speech samples. Logop. Phoniatr Vocol **41**, 106–116 (2016)
9. International Phonetic Association (IPA): Handbook of the International Phonetic Association. Cambridge University Press, Cambridge (1999)
10. Maier, A.: Speech of Children with Cleft Lip and Palate: Automatic Assessment. Studien zur Mustererkennung, vol. 29. Logos Verlag, Berlin (2009)
11. Haderlein, T., Moers, C., Möbius, B., Rosanowski, F., Nöth, E.: Intelligibility rating with automatic speech recognition, prosodic, and cepstral evaluation. In: Habernal, I., Matoušek, V. (eds.) TSD 2011. LNCS (LNAI), vol. 6836, pp. 195–202. Springer, Heidelberg (2011). https://doi.org/10.1007/978-3-642-23538-2_25
12. Haderlein, T., Schwemmle, C., Döllinger, M., Matoušek, V., Ptok, M., Nöth, E.: Automatic evaluation of voice quality using text-based laryngograph measurements and prosodic analysis. Comput. Math. Methods Med. **2015**, 11 (2015)
13. Batliner, A., Buckow, J., Niemann, H., Nöth, E., Warnke, V.: The Prosody Module. In: Wahlster, W. (ed.) Verbmobil: Foundations of Speech-to-Speech Translation, pp. 106–121. Springer, Berlin (2000). https://doi.org/10.1007/978-3-662-04230-4_8
14. Rubenstein, H., Pickett, J.: Intelligibility of words in sentences. J. Acoust. Soc. Am. **30**, 670 (1958)
15. Smola, A., Schölkopf, B.: A tutorial on support vector regression. Stat. Comput. **14**, 199–222 (2004)
16. Witten, I., Frank, E.: Data Mining: Practical Machine Learning Tools and Techniques, 2nd edn. Morgan Kaufmann, San Francisco (2005)

Prosodic Features' Criterion for Hebrew

Ben Fishman[1], Itshak Lapidot[2], and Irit Opher[2]([⊠])

[1] Tel Aviv University, Tel Aviv, Israel
benf22@gmail.com
[2] Afeka Academic College of Engineering, Tel Aviv, Israel
{itshakl,irito}@afeka.ac.il

Abstract. Prosody provides important information about intention and meaning, and carries clues regarding dialogue turns, phrase emphasis and even the physiological or emotional condition of the speaker. Prosody has been researched extensively by linguists and speech scientists; However, little attention has been given to formulating and ranking the acoustic features that represent prosodic information. This paper aims at defining a simple methodology that allows us to test whether a feature conveys prosodic information. This way, we can compare different features and rate them as prosodic or content related (In this paper the word "content" refers to the verbal information of the utterance.). We explore many features using a Hebrew dataset especially designed for validating prosodic features, and as the first step of our research we chose two prosody classes: neutral and question. We apply our methodology successfully and find that prosodic features indeed are invariant to the content of the utterance, while correlating with prosodic manifestations. We validate our methodology by showing that our ranking of prosodic features yields similar results to classification based feature selection.

Keywords: Prosody · Prosodic features · Hebrew database

1 Introduction

Prosody can be defined as the study that relates to non-contextual aspects of speech. Prosody provides valuable information that can be perceived by the listener and plays an important role in everyday life – it helps maintaining dialogue structure [6], deciphering higher level utterances (e.g. sarcasm) and assessing the speaker's emotional state, attitude or intentions [13]. It also contributes to the medical field, especially in Neurology [4,12]. Prosody is also essential for many speech based systems, such as Text to Speech (TTS) [2], Speech Morphing [14] or Speech based Analysis [11].

In the past years there has been extensive work towards standardization of prosody transcription, for example ToBI [15] that is used for annotating intonation or the IPrA Prosodic Alphabet [7]. Still, little attention has been given so far to generalizing and formulating the acoustic features that represent the perceived

© Springer Nature Switzerland AG 2018
P. Sojka et al. (Eds.): TSD 2018, LNAI 11107, pp. 482–491, 2018.
https://doi.org/10.1007/978-3-030-00794-2_52

intonation and other prosodic building blocks, especially in under resourced languages such as some of the Semitic languages.

In this work we take a few steps towards such a formulation and define a simple methodology for determining whether an acoustic or spectral feature represents prosodic information and to what degree.

2   Features

Prosodic features are widely used for various tasks – emotion detection [1], language identification [16], TTS [2], etc'. There has been extensive work on extracting various acoustic and spectral features for prosodic research, e.g. [17]. The openSMILE project [5] lists a few hundred features for emotion recognition. Other works, such as [3] limit themselves to features that can be derived from F0, duration and energy only, that are the most commonly used prosodic features. So, there are many features one can use, but is there a way to decide whether a feature indeed carries prosodic information? Our suggested methodology tries to address this issue and is presented in Sect. 3. To demonstrate and test this methodology, we use a 48 feature set, most of which are considered standard in prosodic research, e.g. F0 and its derivatives, while some are hand-crafted features such as amplitude-tilt $(A = \frac{|A_r|-|A_f|}{|A_r|+|A_f|}$, when A_r/A_f is the amount of F0 rise/fall respectively) or duration-tilt $(D = \frac{|D_r|-|D_f|}{|D_r|+|D_f|}$, where D_r/D_f is the F0 rise/fall duration respectively) [11]. All features are listed in Table 1.

Naturally, these features can be scalars, e.g. max value of F0 or vectors e.g. average energy per syllable, or MFCC entries per frame. To obtain syllable boundaries, we used word level forced-alignment using Hebrew acoustic models trained with the Kaldi engine.

Table 1. Feature set list

Directly calculated	Derived features	Segments types	Num
F0, dF0, energy	Max, min, mean, var	Per-syllable, accumulated	24
F0, dF0	Max-range		4
F0	Peak-position, ampTilt, durTilt		6
MFCC		Per-frame	13
Duration		Per-syllable	1

Per-syllable: evaluated over a single syllable. Accumulated: evaluated over a segment starting at the beginning of the syllable and ending at the end of the utterance.

3   Methodology for Evaluating a Prosodic Feature

We wish to define a criterion for measuring how well does a feature represent prosodic information conveyed in speech utterances. The proposed methodology is simple and requires the following: (1) prosodic features should be correlated with some prosodic manifestations and (2) should not be correlated with significant changes in other "dimensions", when the same prosody is used. Note that requirement (2) relates in our paper only to the content of the utterance, as we use an over simplification where only two attributes characterize a feature – prosodic or content related. Naturally this does not hold for some languages such as tonal languages, where different tonal patterns convey content [8] hence these languages will have to be considered separately.

3.1 Formulation

Let us formulate the above requirements: Suppose we have a set of utterances U_{pc}^k, where $p = 1, 2, \ldots, N_P$ is an index representing the different prosodies in our dataset, $c = 1, 2, \ldots, N_C$ represents the different content types, i.e. different phrases, and $k = 1, 2, \ldots, K_{pc}$ runs through all utterances of type pc, i.e. all the utterances in our dataset with prosody p and phrase type c. Feature F is denoted prosodic if the following requirements hold:

Requirement 1: The dissimilarity between the extracted features is sufficiently small, for most utterance pairs with the same prosody p: $d_F\left(F_{pc}^k, F_{pr}^l\right) < T_1$ for more than $x_1\%$ of the pairs, where $d_F\left(\cdot, \cdot\right)$ is the dissimilarity between two features, and T_1 is some threshold we use to define "sufficiently small". x_1 and T_1 can be tuned, and d_F is defined for all features taking into account feature type – scalar or vector.

Requirement 2: The dissimilarity between extracted features for utterances with different prosodies is higher than T_2 for most such utterance pairs, $(q \neq p)$: $d_F\left(F_{pc}^k, F_{qc}^l\right) > T_2$ for more than $x_2\%$ of the pairs. This requirement should hold for both cases of same or different content type, but it is naturally stronger when the content is unchanged.

T_1, T_2, x_1, x_2 can be tuned for each feature and for each dataset. Instead of tuning these parameters, we propose to combine the two requirements in the following way – we require that the PMFs of the dissimilarities of the two sets (same prosody and different prosody) will be well separated reflecting the different behavior of the feature's values for the two sets; This means that we require low values of dissimilarities for the same prosody set, and high values of dissimilarity for the different prosody set.

3.2 Proposed Methodology

Our proposed methodology is depicted in Fig. 1 using a block diagram to determine the nature of the feature: (i) Calculate dissimilarities between feature values over all possible utterance pairs (ii) Group all pairs into two sets – "same prosody" set (S_{same}^P) and "different prosody" set (S_{diff}^P) (iii) For each set, evaluate the Probability Mass Function (PMF) using the normalized histograms of the

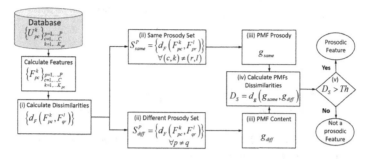

Fig. 1. Flow chart describing the proposed methodology for evaluating the prosodic nature of a feature

dissimilarities (g_{same} and g_{diff}) (iv) Calculate dissimilarity score, denoted D_s, between the two PMFs (v) Use a threshold Th to decide whether the feature can be considered prosodic, i.e. conveys prosodic information. The threshold Th reflects the requirement stated above regarding large enough separability between dissimilarities PMFs of the two sets – same prosody set and different prosody set.

In this work we use the Euclidean distance as the dissimilarity function $d_F(\cdot)$ for step (i) and the symmetrized KL-divergence as the function $d_g(\cdot)$ for step (iv) when calculating D_s. If the feature is indeed prosodic, we expect to see a high degree of separability between the two dissimilarities PMFs evaluated for the two sets – "same prosody" and "different prosody". If, on the other hand, we find that there is a high similarity between the two dissimilarities PMFs (g_{same} and g_{diff}), we conclude that this feature does not carry any prosodic information, at least for the prosodies that were used.

4 Datasets

We have used two different datasets:

4.1 Hebrew Dataset

Our main dataset[1] is in the Hebrew language, and was designed specifically for prosody research. As this was a preliminary stage, we used only two prosody types: question and neutral. 36 speakers were recorded (males: 47%, females: 53%) of various ages (20–30: 22%, 30–40: 33%, 40–50: 8%, 50–60: 20%, 60–70: 17%). Each speaker uttered the same three short phrases, that consisted of four syllables each. All phrases were syntactically correct, and contained mostly voiced phonemes. Each phrase was recorded in two different prosodies (neutral: 46%, question: 54%). The data was recorded using personal cellular phones, in

---

[1] Hebrew dataset is freely available for research purposes only, by contacting the authors.

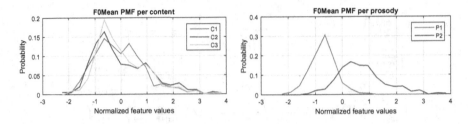

Fig. 2. Normalized F0Mean PMFs. Right – good separation between prosody classes. Left – no separation between content classes

a quiet room environment. In total there are 252 short phrases. To validate the data, two experienced listeners tagged all utterances in a random blind test. The manual tagging was 97% correct, therefore we consider the prosody labeling to be accurate.

4.2   Validation Dataset

In order to validate our results obtained with the Hebrew dataset, we used a small subset of LDC2002S28 – "Emotional Prosody Speech and Transcripts" corpus [9]. This English dataset contains recordings of professional actors reading a series of semantically neutral utterances using different emotional categories. We used a single speaker and five emotional categories only (anxiety, boredom, sadness, panic, and elation). This dataset is relevant for our research as prosody is often used to expresses emotional states.

5   Data Analysis

Based on our highly simplified approach, where acoustic features can be associated with either prosody or content, the first clue regarding the nature of the feature is evident when we look at the features' PMFs for different dataset partitionings – according to prosodies and according to content. Figure 2 shows an example for F0Mean, evaluated for each syllable in the Hebrew dataset. This feature seems to be prosodic – the feature values PMFs are significantly different for the prosody tagging, while content tagging yields similar PMFs. Indeed, F0Mean is considered to convey prosodic information. In Fig. 3 we can see the opposite behavior for one of the MFCC features (which are indeed more suitable for representing content).

Next we look at averages of features over time to explore the separation ability of a feature. In Fig. 4 we can see the behavior of two features over time. First we comapre between different prosodies and different contents for the Herbrew set. It is evident that DurTilt is prosodic while it does not separate well between our different phrases. We also look at F0Max for the validation dataset that includes five different prosody classes, and see that when using this single feature, it is possible to distinguish between some of these classes but not between all of them. This behavior indicates that naturally, F0Max carries prosodic information.

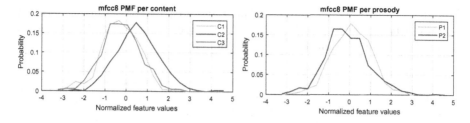

Fig. 3. Normalized MFCC8 PMFs. Right – no separation between prosody classes. Left – some separation between content classes

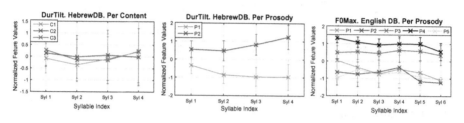

Fig. 4. Averaged feature values over syllables. Left & center: Hebrew dataset tagged per-content and per-prosody. Right: English dataset tagged per-prosody

5.1 Dissimilarities

Following steps (ii) and (iii) in our methodology, Fig. 5 shows the dissimilarity PMFs for the different sets – "same prosody" and "different prosody" for the duration-tilt feature for the Hebrew dataset. We can see good separation between the sets of same and different prosody, while there is no separation between the sets of same and different content. According to our criterion, this feature is definitely a "prosodic feature".

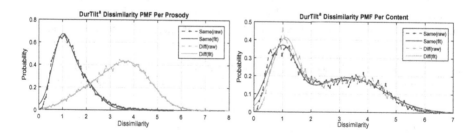

Fig. 5. Duration-tilt dissimilarity PMFs for different conditions: same and different prosodies, and same and different content

In Fig. 6 we can see another example of the dissimilarity PMFs between same and different prosody sets, for F0Max over the validation dataset. This feature separates well between prosody P4 (Panic) and P1 (Anxiety), while it does not separate between P4 and P3 (Elation). This feature also separates well between P4 and P2 (Boredom) and P5 (Sadness) (not shown).

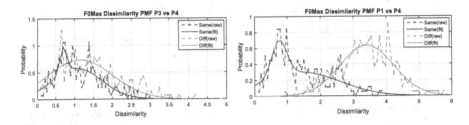

Fig. 6. PMFs dissimilarity for same and different prosody classes

5.2   Classification

One of our goals in grading and estimating feature quality is choosing the best features to be used in a classification task. Hence, to validate our proposed methodology we compare it with classifier based feature analysis. Since we currently defined only a single feature's ranking, we compare our results to ranking based on 1D classification results, i.e. classification using a single feature. We use a simple classifier, so for each feature separately we trained a logistic regression classifier using 66% of our data for training, while making sure train and test sets did not contain the same speakers. For each feature, we used a threshold that yields the best F_1 measure[2] over the train set. Applying this score to the test set, we obtained classification accuracy for the test set. These accuracy scores were used to rank the features for the classification task.

Next, we chose the 14 highest ranked feature according to our methodology, i.e. with highest D_s score, and compared them with the 14 best classification features i.e. the features that yielded the highest F_1 scores. We found that 13 features appeared in both lists (see Table 2), some of them with similar ranks.

Table 2. Comparison between ranking produced by the proposed methodology (D_s) and by classification (F_1). X[a] denotes accumulated features as explained in Table 1

Feature	AmpTilt[a]	DurTilt[a]	AmpTilt	DurTilt	F0Mean[a]	dF0Mean[a]	F0Max[a]
D_s	11.24	10.28	6.13	5.21	4.33	3.69	2.23
F_1	0.89	0.88	0.78	0.78	0.81	0.89	0.82
Feature	F0Mean	F0Max	dF0Max[a]	dF0Mean	dF0Max	F0Range[a]	F0Var[a]
D_s	1.98	1.59	1.3	0.85	0.73	0.71	0.52
F_1	0.72	–	0.83	0.73	0.74	0.77	0.82

---

[2] F_1 measure is the harmonic mean of recall and precision: $F_1 = \frac{2}{1/recall + 1/precision}$.

5.3   Dimensionality Reduction

When using more than one feature, we need to visualize this high-dimensional data and check separability between different classes. This can be done by applying dimension reduction schemes. We chose the t-SNE algorithm [10] and applied it to the best 15 prosodic features obtained using our D_s score (as explained in Sect. 3). Figure 7 shows very good separation between the two prosodic classes, while there is no separation at all between the content classes. This shows that the 15 best features do not represent the phrases content, in addition to conveying prosodic information. When repeating this process for the best content related features, we do not get any separation between different prosodies.

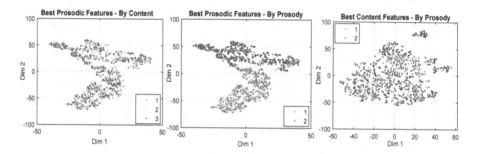

Fig. 7. 2D representation. Left & center – best prosodic features colored by content and by prosody. Right – best content features colored by prosody

6   Conclusions and Future Work

In this paper we introduced a methodology for validating the prosodic relevance of an acoustic or spectral feature. We refer to a feature as prosodic if its values differ significantly for utterances spoken with different prosodies, and show little or no change for utterances spoken with the same prosody. This methodology can be further extended to provide an estimation as to the "prosodic ranking" of a feature, taking into account its separation ability for different prosodic classes, and its insensitivity to changes in content and other non-prosodic information. We believe that creating a formal standard ranking mechanism for prosodic features can assist in finding representations for known perceptual notations such as IrPA or ToBI, as well as in revealing new prosodic features. This methodology can also be used for analyzing prosodic features and manifestations in different languages. For our Hebrew dataset, we have successfully shown that features that were ranked high based on our methodology, are indeed relevant for conveying prosodic information. This was done by ranking the single features according to their classification accuracy scores and comapring this ranking to the one induced by our proposed prosodic score.

Future work should address a few issues that were not covered: (1) Extending the methodology to deal with: a. tonal languages b. more than two prosody classes c. additional non-prosodic dimensions other than content (2) Providing a full mathematical formalization for D_s (3) Validating the proposed methodology using larger datasets in additional languages, as well as for other prosody classes, using known feature sets such as openSMILE [5] (4) Dealing with multi-feature classification results.

Acknowledgments. The authors thank Ella Erlich, Ruth Aloni-Lavi and Noga Hellman for their help with the Hebrew dataset.

References

1. Ang, J., Dhillon, R., Krupski, A., Shriberg, E., Stolcke, A.: Prosody-based automatic detection of annoyance and frustration in human-computer dialog. In: Seventh International Conference on Spoken Language Processing (2002)
2. Chen, S.H., Hwang, S.H., Wang, Y.R.: An RNN-based prosodic information synthesizer for mandarin text-to-speech. IEEE Trans. Speech Audio Process. **6**(3), 226–239 (1998)
3. Rose, R.C.: Prosody recognition from speech utterances using acoustic and linguistic based models of prosodic events. In: Sixth European Conference on Speech Communication and Technology (1999)
4. Diehl, J.J., Paul, R.: The assessment and treatment of prosodic disorders and neurological theories of prosody. Int. J. Speech-Lang. Pathol. **11**(4), 287–292 (2009)
5. Eyben, F., Wöllmer, M., Schuller, B.: OpenSMILE: the Munich versatile and fast open-source audio feature extractor. In: Proceedings of the 18th ACM International Conference on Multimedia, pp. 1459–1462. ACM (2010)
6. Hastie, W.H., Poesio, M., Isard, S.: Automatically predicting dialogue structure using prosodic features. Speech Commun. **36**, 63–79 (2002)
7. Hualde, J., Prieto, P.: Towards an international prosodic alphabet (IPrA). Lab. Phonol. **7** (2016)
8. Li, S., Wang, Y., Sun, L., Lee, L.: Improved tonal language speech recognition by integrating spectro-temporal evidence and pitch information with properly chosen tonal acoustic units. In: INTERSPEECH (2011)
9. Liberman, M.: Emotional Prosody Speech and Transcripts LDC2002S28 (2002). https://catalog.ldc.upenn.edu/LDC2002S28
10. Maaten, L., Hinton, G.: Visualizing data using t-sne. J. Mach. Learn. Res. **9**, 2579–2605 (2008)
11. Mary, L., Yegnanarayana, B.: Extraction and representation of prosodic features for language and speaker recognition. Speech Commun. **50**(10), 782–796 (2008)
12. McCann, J., Peppé, S.: Prosody in autism spectrum disorders: a critical review. Int. J. Lang. & Commun. Disord. **38**(4), 325–350 (2003)
13. Pierre-Yves, O.: The production and recognition of emotions in speech: features and algorithms. Int. J. Hum.-Comput. Stud. **59**(1–2), 157–183 (2003)
14. Qavi, A., Khan, S.A., Basir, K.: Voice morphing based on spectral features and prosodic modification. In: Multi-Topic Conference (INMIC), pp. 401–405. IEEE (2014)
15. Silverman, K., et al.: ToBI: a standard for labeling English prosody. In: Second International Conference on Spoken Language Processing (1992)

16. Tong, R., Ma, B., Zhu, D., Li, H., Chng, E.S.: Integrating acoustic, prosodic and phonotactic features for spoken language identification. In: Acoustics, Speech and Signal Processing, vol. 1, p. I. IEEE (2006)
17. Vaissière, J.: Language-independent prosodic features. In: Cutler, A., Ladd, D.R. (eds.) Prosody: Models and Measurements, pp. 53–66. Springer, Heidelberg (1983). https://doi.org/10.1007/978-3-642-69103-4_5

The Retention Effect of Learning Grammatical Patterns Implicitly Using Joining-in-Type Robot-Assisted Language-Learning System

AlBara Khalifa[(✉)], Tsuneo Kato, and Seiichi Yamamoto

Graduate School of Science and Engineering, Doshisha University,
1-3 Tatara Miyakodani, Kyotanabe-shi, Kyoto 610-0394, Japan
{tsukato,seyamamo}@mail.doshisha.ac.jp, ajkhalifa@taibahu.edu.sa

Abstract. Conducting a multiparty conversation among two robots and a human learner for the purpose of language learning is a novel idea. It can help in conveying grammatical information to the human learner in an implicit manner. The main focus in this paper is the quantification of the level of retention of what was learned implicitly over a period of four weeks. We had evaluated the utterances of the human learners on the level of similarity of n-grams with a reference answer, and on the basis of grammatical correctness of use. The experiments revealed effect of repletion of implicit learning for learning corrective use of grammatical patterns.

Keywords: Computer Assisted Language Learning (CALL)
Robot Assisted Language Learning (RALL) · Implicit learning

1 Introduction

Learning a second language (L2) is becoming more important than ever. In order to ease the task of learning and overcome the issues of time and money, computer assisted language learning (CALL) systems offered a proper environment and a good solution.

Dialogue based CALL systems offer a more practical learning style. They help the learner to be closer to a real life environment to use the language. The motivation factor plays a major role in such systems. They also have the merit of introducing the effect of implicit learning during the conversation instead of only having explicit instructions of how to use the language. It was found that in a conversation, the participants tend to align their utterances with each other in an unconscious manner, which is referred to as the *interactive-alignment* [1]. Borrowing each other's linguistic features is a sign of *implicit learning*, which

---

A. Khalifa—Lecturer at Computer Science and Engineering College in Taibah University, Medina, Saudi Arabia. Currently studying at Doshisha University.

P. Sojka et al. (Eds.): TSD 2018, LNAI 11107, pp. 492–499, 2018.
https://doi.org/10.1007/978-3-030-00794-2_53

is "how one develops intuitive knowledge about the underlying structure of a complex stimulus environment" [2], which is the language in this case.

During the interaction with the system for the purpose of learning a L2, the automatic speech recognition (ASR) engine is essential for providing proper responses by the system. Recognizing a L2 is a challenge for ASR, and it would be helpful to design the scenario in a way to enhance the level of predictability for the ASR engine. Utilizing the effect of implicit learning, which could appear in the form of alignment between the participants in the conversation, can be a supporting factor to regulate the utterances of the learner and enhance the performance of the system accordingly [3].

As a step toward improving the dialogue based CALL systems, humanoid robots were introduced. The effect of having an embodiment for the system, and the multimodality in the interaction that exists in robots, are helpful to have a more realistic environment. Using robots was shown to improve the motivation of the participants to use the system [5].

The early usage of the robot to support language learning was in 2004 [4], in which the "Robovie" robot was used to conduct conversation with elementary school children. Other examples of robot assisted language learning (RALL) systems included the English course designed by [5], and the "L2TOR" project [6]. An effective trial of using robot as an assistive teacher was held by Alemi et al. in 2014 [7].

One-on-one learning style offers the learner with more chances to use the language by communicating with the teacher. It is believed that one-on-one tutoring with a skilled instructor is the best way to learn L2 [8,9]. However, peer learning from other learners and the corrective feedback to other learners' mistakes, are considered benefits of the classroom style that may not be found in one-on-one learning style. Language learning using CALL or RALL systems incorporate the benefits of one-on-one learning style. In order to combine the benefits of both styles, three party conversation can be a good trade-off in this case. Tutoring from a teacher robot, and peer learning from a peer robot, can be a good structure to combine the benefits of both styles, while keeping the participation chances to the maximum possible.

Khalifa et al. proposed a novel structure of a system using two robots to conduct a three-party conversation with a human learner for the purpose of language learning to offer a combined benefits of both tutoring and peer learning styles [10]. The main learning method in their system was focused on implicit learning, where the human learner is implicitly learning from the interaction between the two robots.

They revealed effect of implicit learning but did not reveal retention effect of learning grammatical patterns implicitly. We intended in this step to measure the effect of retention of using specific grammatical patterns that were learned implicitly over a period of four weeks. This work confirms their findings on the effect of the implicit learning, and added a quantification measure of the retention.

2   Joining-in-Type Robot-Assisted Language Learning System

2.1   The System

We used two humanoid NAO robots standing on a table in front of the human learner in a triangular form in order for them to rotate their heads to face the participant they are talking to, whether facing each other or facing the human learner. The robots had to speak up the pre-designed scenario using their TTS engine and generate specific gestures according to their utterances. They were distinguishable with their voice and their colors. One robot plays the role of a teacher or a high proficiency participant and the other robot played the role of an advanced learner.

The robots were controlled remotely in Wizard-of-Oz method because the system is under development and our main focus was to measure the effect of implicit learning as a step toward a fully automated system. The control program offered a limited control to the experimenter to decide some actions. The experimenter had to decide for the robot according to the answer of the learner whether to repeat the question, utter a sample answer and repeat the question, or move on to the next question.

2.2   The Scenarios

We created four different fixed scenarios that were used over four consecutive weeks and on the retention test. They were designed in a way to have a repetition of the usage of specific grammatical patterns, which can help the human learner learn these patterns in an implicit way. When the teacher robot asks a question to the learner robot, and then the learner robot answers using a chosen grammatical patterns, the human learner can implicitly learn how to use this pattern by aligning his/her utterance to the robot's utterance. As two typical difficult grammatical patterns for Japanese learners of English, we chose causative verbs and inanimate subjects to be used in the questions and in the answers of the robots.

The scenarios were designed in a question-answer style. They had a seamless structure containing several general topics, which helped to have the focus of the learner on the dialogue and implicitly conveyed the grammatical information. In other words, the system conducted conversations with the human learners and keep using specific grammatical patterns throughout the conversations in order for the learner to learn how to use these patterns without explicitly instructing them to do so.

In every week, a session of the conversation was held. In every session, there were two 10-min turns with a 5-min break between them. For the four weeks dedicated for training the learners on the grammatical patterns, the questions in the second turn used the same grammar and contextual flow as in the first turn, however different wording were applied. Applying the same grammatical pattern using different sentences is thought to offer the learner with the opportunity to

Table 1. Sample of the scenario between the two robots and the human learner. *R1*: the teacher robot. *R2*: the learner robot.

Speaker	Listener	Utterance
R2	R1	Oh, our sweets and drinks are coming now
R1	R2	Wow, everything looks so delicious
R1	R2	Besides these sweets, what gives you an appetite?
R2	R1	Cooking or gourmet TV programs give me an appetite
R1	R2	I suppose so
R1	Learner	What makes you feel excited?
Learner	R1	Action dramas make me feel excited
R1	Learner	I see
R2	R1	I am going to post some photos on Instagram
R1	R2	What makes you post on Instagram so often?
R2	R1	Getting more likes makes me post photos so often
R1	R2	I think it is addictive
R1	Learner	What makes you use your smartphone so often?
Learner	R1	Exchanging messages makes me use my smartphone so often

practice and learn how to use that pattern correctly. In the case of the retention test, the scenario of the first turn of week1 was used as the first turn of the test, and scenario of the first turn of week3 was used as the second turn of the test. Each learner were prompted with 10 questions in every turn, that were repeatedly using one or both of the grammatical patterns selected, expecting the learner to use them in his/her answer. The data collected in this experiment were analyzed in order to measure the effect of the implicit learning, and to measure the effect of retention. Retaining what was learned implicitly over several weeks is the main goal of this work. Table 1 shows a sample of the conversation between the two robots and the human learner.

2.3   Design of Retention Evaluation Test with JI-RALL System

The training of the learners to use the selected grammatical patterns were held over a period of four consecutive weeks. They had a chance to be exposed to use the selected grammatical patterns repeatedly. After the training period, the learners had a retention test session, which happen to be three to five weeks after the training period.

The learners were divided into two groups, the control group and the experimental group. The control group is used to measure the baseline performance, while the difference between the two groups represent the effect of implicit learning. Each learner in both groups were asked 10 questions in every turn in every session (i.e., a total of 20 questions). They had the chance to listen to the answers of all the questions expressed in the selected grammatical patterns. For all the

10 questions that were asked to the learner the teacher robot asked a question to the human learner first then asked the same question to the learner robot for the control group. In the case of the experimental group, six questions out of ten were asked first to the learner robot, then to the human learner as shown in Table 2. The idea was to give an implicit information to the human learner in the experimental group so that he/she can use it in answering the question. This chance was not offered to the control group, whereas they had to answer the question before listening to the learner robot's answer. In order to have a fair comparison between the two groups, we have chosen the answers of the four questions that were having the same order of asking for both groups for the analysis.

Table 2. Orders of asking the questions in both groups. The experimental group had six questions that were asked to the robot first then to the human learner.

	R2 then learner	Learner then R2
Control group	Non	10 questions
Experimental group	6 questions	4 questions

Used Measures. We have used two measures to measure the effect of implicit learning in different perspectives.

1. BLEU score: It stands for BiLingual Evaluation Understudy, and it is originally an automatic translation evaluation measure. We have used the smoothed version of BLEU using one reference answer. The evaluation using BLEU is done on a text format of the data, and it results in a number ranging from zero to one. The smoothed BLEU formula is given by Eq. (1).

$$\text{Smoothed BLEU} = BP \cdot \exp\left(\sum_{n=1}^{N} \frac{1}{N} \log p_n\right) \tag{1}$$

where p_n is the precision of n-grams in a learner utterance that is determined through comparison with reference sentences. N is usually set at 4, and BP is a brevity penalty, a coefficient for correction. A sample answer presented by the learner robot is used as the reference in this case.

2. Correctness of use: It is a manual evaluation of the utterances of the learners. We have focused on evaluating the two grammatical patterns we have used in the scenarios. The criteria used in the evaluation are:
   - Evaluating the use of inanimate subject pattern:
     (a) The use of inanimate subject (e.g. Eating the local food)
     (b) The use of appropriate verb (e.g. attracts them)
     (c) Having a correct order (e.g. Eating the local food attracts them)
     Each point results in a one or a zero and the total is divided by three.

– Evaluating the use causative verb pattern:
  (a) The use of a causative verb (i.e. Let, Make, Have, Get, Help)
  (b) Using the verb correctly (e.g. New exciting experiences makes him go abroad)

  Each point results in a one or a zero and the total is divided by two.
Then, the maximum of the two evaluations was used for every utterance. This was done since the scenarios provided a mixed exposure of the two patterns, and it may not be valid to evaluate using only one pattern.

3 Experiments

3.1 Experimental Setup

We had invited 23 participants to participate in the experiment. Among them, 16 could participate in all sessions, eight were in the control group and eight were in the experimental group. They were Japanese university students, between the age of 18 to 24. They had acquired Japanese as their L1 and had learned English as their L2. About half of the participants provided their TOEIC (Test of English for International Communication) scores, and their average is 685 (990 being the highest attainable score).

In order to have a multimodal corpus, we have used two cameras to record the audio/video of the conversation. A headset microphone was also used to have a clearer recording for the audio.

After wearing the headset microphone and receiving a brief instruction on how to respond to the robots, every participant performed the 10-min long scenario and then asked to repeat the conversation using a modified scenario after a 5-min break. Hence, every participant was asked to respond to the system naturally ten times in every conversation, expecting them to use the chosen grammatical patterns, and the performance was measured using four questions in the two rounds.

3.2 Objective Measures of Repetitive Training and Retention

The data collected in the experiment were annotated in order to extract a transcription of the utterances of the learners. The corpus was collected from the conversation with 16 participants who were expected to utter 1600 sentences. Total duration of conversations is 26.6 h.

The left part of Fig. 1 shows the average BLEU scores of the initial performance in week1, the performance after the training sessions, and the average scores in the retention test. From the perspective of n-gram similarity, the implicit learning effect is higher in the experimental group with an average difference of about 38% after the training sessions. The retention was higher in the experimental group with an average difference of 135% compared to the initial performance.

The right part of Fig. 1 shows the average scores of using the grammatical patterns correctly of the initial performance in week1, the performance after the

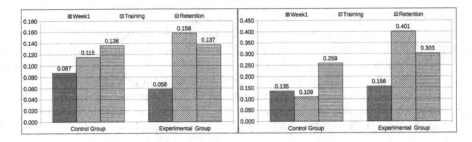

Fig. 1. *Left*: BLEU scores for both groups showing the initial performance, the performance after training, and in the retention test. *Right*: correctness of use scores for both groups showing the initial performance, the performance after training, and in the retention test.

training sessions, and the average scores in the retention test. The implicit learning effect seems to be much higher in the experimental group with an average difference of about 267% after the training sessions. The retention using this measure was higher in the experimental group with an average difference of 94% compared to the initial performance.

4 Discussion

As clearly shown, both BLEU scores and correctness of use scores in retention test were improved comparing with those of the average during the training weeks. This revealed effect of repletion of implicit learning for learning corrective use of grammatical patterns.

Both groups gained a level of implicit learning, however, the effect were much clearer for the experimental group. The difference between their level in the initial state in week1 and after the training weeks, exceeds the difference in the case of the control group, which indicate a higher effect of implicit learning. That effect were kept even couple of weeks after the training, which was shown in the results of the retention test when compared to the initial state of week1.

The BLEU score was helpful in detecting the partial alignment between the utterances of the human learner and the reference answers which were uttered by the learner robot. The correctness of use measure helped in detecting the proper usage of the grammatical patterns by the human learner. Difference of correctness of use scores between both groups was larger than that of BLEU scores in training session. This result suggests that implicit learning adopted in JI-RALL system improves competence of participants in composing utterances and let them to be able to compose grammatically more correct utterances.

Some participants did not improve their performance during four training weeks. The scenario used in the first week was a difficult one and they were looked like lose their motivation at the first week. Having an easy start could add better motivation to the human learner and could improve total performance.

5   Conclusion

We have conducted an experiment of joining-in-type robot assisted language learning system to analyze effectiveness of teaching grammatical usage in an implicit style. The answers by learners in experimental and control groups were evaluated using BLEU score for similarity test, and grammar correctness of use to evaluate learning of their grammatically correct usage. The effect of implicit learning were found in both groups, but it was higher in the case of the experimental group. Likewise, the effct of retaining what was learned implicitly was higher for the experiment group. We have a plan to introduce an adaptive discourse control to change the flow of the conversational scenario depending on the attitude of the learner to cope with their variety of language proficiency.

Acknowledgement. This research was supported in part by a grant from the Japan Society for the Promotion of Science (JSPS) (No. 15K02738).

References

1. Pickering, M.J., Garrod, S.: Toward a mechanistic psychology of dialogue. Behav. Brain Sci. **27**(2), 169–190 (2004)
2. Reber, A.S.: Implicit learning of artificial grammars. J. Verbal Learn. Verbal Behav. **77**, 317–327 (1967)
3. Fandrianto, A., Eskenazi, M.: Prosodic entrainment in an information-driven dialog system. In INTERSPEECH, pp. 342–345 (2012)
4. Kanda, T., Hirano, T., Eaton, D., Ishiguro, H.: Interactive robots as social partners and peer tutors for children: a field trial. Hum.-Comput. Interact. **19**(1), 61–84 (2004)
5. Lee, S.: On the effectiveness of robot-assisted language learning. ReCALL **23**(01), 25–58 (2011)
6. Belpaeme, T., et al.: L2TOR-second language tutoring using social robots. In: Proceedings of International Workshop on Educational Robots (2015)
7. Alemi, M., Meghdari, A., Ghazisaedy, M.: Employing humanoid robots for teaching English language in Iranian junior high-schools. Int. J. Hum. Robot. **11**(3), 4–16 (2014)
8. Long, M.H.: Input, interaction, and second-language acquisition. Ann. N. Y. Acad. Sci. **379**(1), 259–278 (1981)
9. Bloom, B.S.: The 2 sigma problem: the search for methods of group instruction as effective as one-to-one tutoring. Educ. Res. **13**(6), 4–16 (1984)
10. Khalifa, A., Kato, T., Yamamoto, S.: Measuring effect of repetitive queries and implicit learning with joining-in type robot assisted language learning system. In: Proceedings of 7th ISCA Workshop on Speech and Language Technology in Education, pp. 13–17 (2017)

Learning to Interrupt the User at the Right Time in Incremental Dialogue Systems

Adam Chýlek$^{(\boxtimes)}$, Jan Švec, and Luboš Šmídl

NTIS – New Technologies for Information Society, Faculty of Applied Sciences,
University of West Bohemia, Pilsen, Czech Republic
{chylek,honzas,smidl}@ntis.zcu.cz

Abstract. Continuous processing of input in incremental dialogue systems might result in the need of interrupting a user's utterance when clarification or rapport is needed. Being able to predict the right time when to interrupt the utterance can be another step to a more human-like dialogue. On the other hand, annotation of corpora with different types of possible interruptions requires additional human resources. In this paper, we discuss how to process a corpus that does not have interruptions specifically annotated. We also present initial experiments on two corpora and show that it is possible to model the desired behaviour from these corpora.

Keywords: Incremental dialogue system · Model of interruptions
Corpora preparation

1 Introduction

Incremental dialogue systems are an important evolution of human-machine interaction [1]. While typical dialogue systems process user's input as a sequence of utterances that are separated by a silence of certain duration, the incremental systems have the ability to process shorter segments [2]. This can improve the user's experience and it can result in a behaviour that is closer to human conversation.

As a toy example and a motivation for this work, we can imagine a spoken dialogue system that has to take a phone number as an input from the user [3].

This work was supported by the European Regional Development Fund under the project Robotics for Industry 4.0 (reg. no. CZ.02.1.01/0.0/0.0/15_003/0000470) and by the grant of the University of West Bohemia, project No. SGS-2016-039. Access to computing and storage facilities owned by parties and projects co ntributing to the National Grid Infrastructure MetaCentrum provided under the programme "Projects of Large Research, Development, and Innovations Infrastructures" (CESNET LM2015042), is greatly appreciated.

© Springer Nature Switzerland AG 2018
P. Sojka et al. (Eds.): TSD 2018, LNAI 11107, pp. 500–508, 2018.
https://doi.org/10.1007/978-3-030-00794-2_54

Let's assume that due to errors in automatic speech recognition (ASR), a noisy environment or other sources of errors, the system recognizes a different number than what the user has said.

In a non-incremental dialogue system, the user has to wait for the end of the system's utterance. Let's say that the system reads the number back to the user and now she wants to correct the incorrectly recognized number. Depending on the design of a dialogue manager, the user may be able to correct only a part of the number (e.g. "The second digit was nine.") or in the worst case try to say the whole number again.

On the other hand, the incremental system should allow the user to barge in at any time during its turn. The system should remember what has been already said and change only the part of the number that the user is referring to (e.g. the user, after a second digit is read back, says "No, that was nine.").

In an incremental dialogue system, we can also easily imagine the following situation: the user dictates his number, the confidence score from the ASR for the last digit goes below a certain threshold and the system wants to ask the user to repeat that number. In this situation, it is the system that should plan whether and when to interrupt the user in order to get the number right. For example, the system could ask to clarify the second digit of a phone number when the user hesitates after first three digits (e.g. "Was that three one?", the user could respond "No, three nine one"). Interrupting the user can be useful and it can also be another challenging part of the incremental dialogue systems research.

One could argue that being interrupted by a machine may be annoying, rude or hostile. The authors of [4] show that interruptions (from a linguistic and psychological point of view during human conversations) do not necessarily need to have a negative connotation (deemed as competitive). They can also be used to request clarification, convey rapport or help the other party finish their utterance. These can be classified as collaborative interruptions. A user can interrupt a dialogue system both in the competitive and collaborative fashion. It's up to a dialogue manager's strategy whether its interruptions will stay only collaborative. The definition of such strategy that would not be perceived as competitive is beyond the scope of this paper, as our method will only provide information whether it is the right time to interrupt.

As our research focuses mainly on unimodal spoken dialogue with a single user, we leveraged the availability of large corpora of speech data. We then processed them in a way that allowed us to obtain the training data for supervised learning methods using deep neural networks. Our addition to the existing research is also the description of the corpora preparation. The results of our experiments look promising, we can show that even though the used corpora were not created specifically for the task, it is possible to leverage the available data and predict the interruptions based on a short history of an audio signal.

2   Related Work

Although research that explores interruptions and turn-taking cues already exists, it was done either on a different kind of dialogue or with different intentions than our research (using corpora without explicit annotation of interruptions). For example, turn-taking phenomena were examined in multi-user multimodal systems [5], while our research focuses specifically on a spoken dialogue between two participants. An incremental dialogue system with interruptions presented in [6] focused on the dialogue as a whole and not so much on the mid-utterance interruptions. In [7] a simulated system with interruptions was also evaluated as a whole.

Many researchers also focus on an end-of-turn detection [8–10]. Their methods would allow the dialogue manager to know when the user stopped her thought (e.g. whether a pause in a speech meant an end of a sentence or a hesitation). This would, of course, be a meaningful time for the system to take the ground and start speaking. On the other hand, it may be possible to interrupt the user earlier and our approach should provide this information. In their work, we can also see the prominence of processing either raw audio signals or low-level features and the use of deep learning methods, which we also chose to adapt.

For a theory of interruptions and statistical corpus analysis, we'd like to mention [11] as a related work. Unfortunately, their interruption taxonomy was created with more detailed annotations in mind and their granularity could not be reached in our datasets, where we were missing such annotations. We had to resort to a simpler description of interruptions.

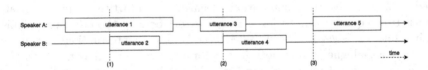

Fig. 1. Example of possible interruption types. The interruption (1) results in an internal overlap (INT), because utterance 2 ends before utterance 1. (2) marks beginning of an overlap that results into switching of speakers (OSW) and (3) marks the time of a clean switch without any overlap (CSW).

3   Choice of Corpora

We have reviewed several speech corpora that we have obtained in the course of several years of ASR development and analyzed their suitability for our current task. Our main requirement was to have a separate channel for each speaker so we can clearly distinguish which speaker is speaking. Additional information about overlapping segments of speech will also allow us to distinguish several interruption types. These requirements resulted in experiments on two corpora: USC-SFI MALACH [12] and BH [13].

The BH corpus (Bezplatné hovory, standing for "free calls" in Czech) contains recordings of spontaneous telephone conversations between pairs of speakers in the Czech language. Each speaker was recorded on a separate channel. In contrast with similar datasets, there were no restrictions on the topics of the calls. The only restriction was the length of the call (10 min maximum). A large portion of the dataset has human-created transcriptions of speech, the annotations are time-aligned and assigned to the respective channels. This makes the data ideal for our task, as these conversations were rich in interruptions and overlapping speech. Although neither the interruptions nor the overlaps were specifically annotated, the ground-truth data could be automatically derived from the aligned transcription. The mixture of different topics and speakers could also help us create a more general model of interruptions.

The MALACH corpus contains interviews with holocaust survivors in Czech. The recordings have two channels - one for the interviewer and the other for the interviewee. Several hours of speech from the recordings were transcribed by human annotators. Timestamps of a start and of an end of overlapping segments were available.

We've also made an effort to find other corpora that were not only meeting our criteria, but that were freely available to other researchers. The closest match to our requirements was the CALLFriend corpus [14]. This corpus has timestamps of overlaps, but channel-to-speaker assignment is missing.

The approach for the automatic assignment that was used on MALACH archive could not be used here, because one channel could have more than one speaker assigned, breaking the already fragile automation. Therefore, our approach has not been tested on this corpora, but future work could allow us to test and compare our efforts with others on this dataset in the future.

4   Interruptions from Overlaps

For the purpose of our studies, we derive interruption from overlapping speech segments. We defined three types of overlaps: internal overlap, overlap resulting in a new turn and a clean switch of turns, as illustrated in Fig. 1.

If speaker B starts speaking during speaker A's utterance, but speaker A continues her turn even after the end of the overlapping part, we mark the overlapping event as internal (INT). It is possible that some of the INT overlaps may be just a backchannel (e.g. "ehm"), but the annotated data did not have a consistent format for these events, so we consider them as interruptions. We can reason that even the dialogue system might want to provide a backchannel in its response and this would allow it to know when to do that.

If speaker A ends her utterance during the overlapping segment, we call that event an overlap resulting in a switch of turns (OSW). The last type of event is a clean switch (CSW) when there is no overlap and the other speaker starts the turn.

Only the OSW and INT events will be considered as interruptions. Even though these events may have a different meaning in a conversation [11], we have decided we will also evaluate the system without differentiating between the INT and OCW events. We will refer to these simply as overlap (OVR) events. This had to be done because the MALACH corpus does not provide enough information to distinguish the type of an overlap.

From the point of a dialogue manager, we think that the OSW event may be more useful to the system than the INT. When the system decides it would like to interrupt the user, it would be more reasonable to use the moment when the user is more likely to end his turn after the interruption (the OSW type). As for the usefulness of INT, we can consider it as a good moment for a backchannel information.

5   Corpora Preparation

The time-aligned annotations of each channel in the BH corpus allow us to clearly detect any overlap of the speakers. We simply take the time of the beginning of one speaker's utterance and check whether it started during the other speaker's utterance. If it did start, we mark it as the beginning of an overlap. When either of the speakers stops speaking, we mark that time as the end of an overlap.

Although the MALACH had overlap annotations for each channel, they were assigned to a recording and not to its channels. This meant that we had to automatically assign who was responsible for the overlap (which channel initiated the interruption). Furthermore, complete textual transcriptions were not available, only an output of phoneme recognizer that did not prove to be useful for this task. The assignment of an overlap to a speaker has been done for each overlap automatically based on average energy levels before the beginning of an overlap. The channel with lower average energy was marked as the initiator of the overlap. We've tried to use the sequences of phonemes for the decision, but from the nature of the recording set-up, one speaker could often be heard on both channels and that introduced errors into the automatic assignment.

The statistics of the datasets created from the corpora are in Table 1.

Table 1. Statistics of the datasets. For MALACH, the information on who continues the utterance was missing, so OSW and INT events could not be distinguished.

	MALACH	BH
# of speakers	94	8150
# of overlaps (train/test)	665/312	91123/4658
# of OSW overlaps	-	56750/2910
# of INT overlaps	-	34373/1748

6   Experiment Setup

In previous sections, we have shown how we infer the time of interruption from overlaps of different types. Now we can use speech data preceding these timestamps to predict whether it is the right time to interrupt the user.

We have focused our attention on the speech signal itself, as this offers us the chance to process the interruption detection in parallel with automatic speech recognition and it is also a common practice as mentioned in the related work. We always take t seconds of audio preceding each moment of interruption (for both the positive and the negative example) and extract several features that are described in the following paragraphs.

For a feature set that we will call MFCC, we have extracted 12 Mel-Frequency cepstral coefficients from the t seconds of an audio signal with a window length of 50 ms and also the frame's energy.

We have also extracted features using openSMILE and their Interspeech 2009 Emotion Challenge (IS09) feature set [15]. This contains not only the 12 Mel-Frequency cepstral coefficients but also root-mean-square signal frame energy, a frame-based zero-crossing rate of a time signal, a voicing probability from autocorrelation function and a fundamental frequency from the cepstrum. For these low-level descriptors, their moving averages with a window length of 3 were appended as well as first-order delta coefficients of the smoothed descriptors.

Although we had plenty of positive examples of interruptions, obtaining negative examples was a challenge, as these were not annotated.

It is clear that we can't mark everything that wasn't a positive example as a negative example. There could have been many opportunities when the speaker could have been interrupted, but the other party simply chose not to interrupt or did not have a reason to interrupt the speaker. The best way would be to let human annotators mark such examples, but that would defeat the purpose of using the already available datasets without any additional manual work.

To work around this issue, we have made an assumption that the current speaker was purposefully not interrupted during the t seconds preceding the actual interruption, making this our negative example. To reduce the space for the parameter search, we have also used t as the length of the audio segment preceding the interruption, from which the features were extracted. This means that when we see an interruption at time t_p, we generate features from time $t_p - t$ to t_p and mark them as our positive example and the features from $t_p - 2t$ to $t_p - t$ form the negative example.

Table 2. Results for both feature sets.

Corpus	Type	IS09				MFCC			
		Accuracy	Precision	Recall	f-measure	Accuracy	Precision	Recall	f-measure
BH	OVR	0.654	0.748	0.464	0.573	0.645	0.611	0.795	0.691
	OSW	0.693	0.737	0.601	0.662	0.692	0.697	0.682	0.689
	INT	0.639	0.633	0.662	0.647	0.629	0.625	0.649	0.636
MALACH	OVR	0.614	0.628	0.561	0.593	0.587	0.602	0.511	0.553

Another assumption we've made was about the exact moment of an interruption. To compensate for possible annotation error in the range of milliseconds, we have added an offset parameter m. We augmented our data by offsetting the actual moment of interruption by 1 to m samples and the same principle has been used for the negative examples to keep the classes balanced.

We have conducted initial experiments on the development part of the BH dataset (2656 samples) using several ML methods, including support vector machines, decision trees, and neural networks. We have decided to use deep neural networks for their performance. Specifically, deep residual learning network (ResNet-152, [16]) was used, because it was shown to be performing well not only on image classification tasks but also in automatic speech recognition [17].

The input was normalized, the output was a softmax layer with 2 neurons. We have used categorical cross-entropy as a loss function and Adam as an optimizer. We have used a grid search to find parameters t and m using a development part of the BH dataset.

7   Results

The best performing setup used history of $t = 0.7$ s. The training data were augmented with an offset of up to $m = 3$ frames and the negative examples were generated 0.7 s before the actual interruption. The accuracy achieved on this setup and other monitored metrics can be seen in the Table 2. Although this performance might not be suitable for production systems, it is significantly better than chance (binomial test, $p < 0.005$).

Moreover, we've reasoned in previous sections that we perceive the task of predicting OSW interruption as more valuable to the system. This type was the best performing in terms of accuracy and f-measure. It may be reasonable to pursue only this type of interruptions in further research.

Detecting only the general OVR type of interruption proved to be less useful than anticipated. This means that using datasets similar to MALACH (where we weren't able to automatically distinguish between the more specific OSW and INT interruptions) may not be possible without additional manual work.

The IS09 feature set did not achieve significantly better results (in accuracy). This means that we can use computationally much less expensive MFCC features without risking significantly worse performance.

8   Conclusion

We can conclude that it is possible to predict the right time to interrupt the user even when the source of the data was not intended for this task and segments where speakers overlap were used instead of specific interruption annotations.

The results of initial experiments indicate that making an effort to differentiate between the OSW and INT types of overlaps might result in an improved performance. This differentiation could also be used by different strategies in a dialogue management.

One of the directions of future research to improve the performance is the incorporation of features from an output of a phonetic recognizer or a spoken language understanding system. Another direction is to train the classifier on one dataset and test it on a different one to see how well can be the interruption prediction generalized.

In a more distant future, analysis of more possible metrics and most importantly their relation to a factual improvement of the dialogue from the user's perspective may be needed.

References

1. Ward, N.G., Devault, D.: Ten challenges in highly-interactive dialog systems. In: AAAI Spring Symposium on Turn-Taking and Coordination in Human-Machine Interaction, pp. 104–107 (2015)
2. Schlangen, D., Skantze, G.: A general, abstract model of incremental dialogue processing. In: Proceedings of the 12th Conference of the European Chapter of the Association for Computational Linguistics, pp. 710–718 (2009)
3. Walker, M., Langkilde, I., Wright, J., Gorin, A., Litman, D.: Learning to predict problematic situations in a spoken dialogue system: experiments with how may I help you? In: Proceedings of the 1st North American Chapter of the Association for Computational Linguistics Conference, pp. 210–217 (2000)
4. Li, H.Z.: Cooperative and intrusive interruptions in inter- and intracultural dyadic discourse. J. Lang. Soc. Psychol. **20**(3), 259–284 (2001)
5. Skantze, G., Johansson, M., Beskow, J.: Exploring turn-taking cues in multi-party human-robot discussions about objects. In: Proceedings of the 2015 ACM on International Conference on Multimodal Interaction, pp. 67–74 (2014)
6. Zhao, T., Black, A.W., Eskenazi, M.: An incremental turn-taking model with active system barge-in for spoken dialog systems. In: 16th Annual Meeting of the Special Interest Group on Discourse and Dialogue, pp. 42–50, September 2015
7. Khouzaimi, H., Laroche, R., Evre, F.: Turn-taking phenomena in incremental dialogue systems. In: Proceedings of the 2015 Conference on Empirical Methods in Natural Language Processing, pp. 1890–1895, September 2015
8. Heeman, P.A., Lunsford, R.: Turn-taking offsets and dialogue context. Interspeech **2017**, 1671–1675 (2017)
9. Masumura, R., Asami, T., Masataki, H., Ishii, R., Higashinaka, R.: Online end-of-turn detection from speech based on stacked time-asynchronous sequential networks. INTERSPEECH **2017**, 1661–1665 (2017)
10. Maier, A., Hough, J., Schlangen, D.: Towards deep end-of-turn prediction for situated spoken dialogue systems. INTERSPEECH **2017**, 1676–1680 (2017)
11. Gravano, A., Hirschberg, J.: A corpus-based study of interruptions in spoken dialogue. In: INTERSPEECH-2012 (2012)
12. Psutka, J., Radová, V., Ircing, P., Matoušek, J., Müller, L.: USC-SFI MALACH Interviews and Transcripts Czech LDC2014S04 (2014)
13. Valenta, T., Šmídl, L., Švec, J., Soutner, D.: Inter-annotator agreement on spontaneous Czech language. In: Sojka, P., Horák, A., Kopeček, I., Pala, K. (eds.) TSD 2014. LNCS (LNAI), vol. 8655, pp. 390–397. Springer, Cham (2014). https://doi.org/10.1007/978-3-319-10816-2_47
14. Canavan, A., Zipperlen, G.: CALLFRIEND American English-non-southern dialect. Linguist. Data Consort. Phila. **10**, 1 (1996)

15. Schuller, B., Steidl, S., Batliner, A.: The INTERSPEECH 2009 emotion challenge. In: Interspeech, pp. 312-315 (2009)
16. He, K., Zhang, X., Ren, S., Sun, J.: Deep residual learning for image recognition. Multimed. Tools Appl. 1–17 (2015)
17. Xiong,W., et al.: The Microsoft 2016 conversational speech recognition system. In: ICASSP, pp. 5255-5259 (2017)

Towards a French Smart-Home Voice Command Corpus: Design and NLU Experiments

Thierry Desot[1]($\boxtimes$), Stefania Raimondo[1,2], Anastasia Mishakova[1],
François Portet[1], and Michel Vacher[1]

[1] Univ. Grenoble Alpes, CNRS, Grenoble INP, LIG, 38000 Grenoble, France
{thierry.desot,francois.portet,michel.vacher}@univ-grenoble-alpes.fr,
anastasia.mishakova@gmail.com
[2] University of Toronto, Toronto, ON M5S 3H7, Canada
stefania.raimondo@mail.utoronto.ca

Abstract. Despite growing interest in smart-homes, semantically annotated *large* voice command corpora for Natural Language development
(NLU) are scarce, especially for languages other than English. In this
paper, we present an approach to generate customizable *synthetic* corpora of semantically-annotated French commands for a smart-home. This
corpus was used to train three NLU models – a triangular CRF, an
attention-based RNN and the Rasa framework – evaluated using a small
corpus of real users interacting with a smart home. While the attention
model performs best on another large French dataset, on the small smart
home corpus the models vary performance across to intent, slot and slot
value classification. To the best of our knowledge, no other French corpus
of semantically annotated voice commands is currently publicly available.

Keywords: Natural Language Understanding
Corpora and language resources · Ambient intelligence
Voice-user interface

1 Introduction

Smart-homes with integrated voice-user interfaces (VUI) can provide in-home
assistance to aging individuals, allowing them to retain autonomy [13]. However,
speech can only provide effective interaction with a home automation system if
its semantics are properly understood. Since users tend to deviate from predefined sets of voice commands [12,14,16], placing restrictions on their vocabulary
and syntax is unrealistic and prohibitive. Instead, we must train robust Natural
Language Understanding (NLU) models on well-balanced voice command corpora. But, the removal of such constraints is a huge bottleneck for NLU and
would necessitate a massive dataset.

In this paper, we present a customizable domain-specific corpus generator as
an alternative to a large manually annotated data set. It can be developed quickly

© Springer Nature Switzerland AG 2018
P. Sojka et al. (Eds.): TSD 2018, LNAI 11107, pp. 509–517, 2018.
https://doi.org/10.1007/978-3-030-00794-2_55

without the cost of manual semantic annotations, and is easily adaptable to new smart-home settings. For performance evaluation, a real smart-home corpus has been acquired from a limited set of users. This part is presented in Sect. 4. To validate the approach, three state-of-the-art NLU models were trained on the synthetic dataset and evaluated on the real smart-home dataset to show how the trained models perform in realistic conditions. This part is presented in Sect. 5. The paper ends with a conclusion and an outlook on future work.

2   Related Work

While early slot-filling systems were rule-based [18], modern methods are data-driven. Conditional random fields [6], have recently been superseded by deep neural networks, including basic RNNs [11], Bi-directional LSTM RNN encoder-decoders [1], Attention-based RNNs [9] and Attention based CNNs [5]. Most approaches treat slot-filling as sequence labeling, attaching a slot to each word in the input utterance. However, other approaches are possible, such as treating it as a dependency parsing task [5], template matching, used by the Sweet-Home system [17], or string matching as in [3]. While intent detection has traditionally been seen as a separate task from slot-filling [15], since both tasks are highly correlated, much recent work performs slot-filling (sequence labeling) and intent detection (sequence classification) simultaneously. Such work includes Tri-CRF [6], which extends the linear sequence labeling CRF with a node to represent the dialogue act, and Att-RNN [9], which extends the slot-filling encoder-decoder RNN with an extra intent decoder. The Cassandra system [4] performs NLU via neural networks, using an LSTM for intent prediction and deep networks to identify slot locations and slot types. These simultaneous approaches are the most relevant to the work described here.

Since slots and intents are typically domain-dependent, new domains cannot benefit from models trained on massive, well-studied corpora. One common approach is to instead extend limited domain-specific corpora with synonyms and syntactic replacement [10]. [3] focuses on statistical decision-making using contextual information. Other work has instead targeted cross-domain prediction [2,7], including the Tri-CRF model [7] mentioned above. In this work, we take a third approach: without a large domotic corpus as a starting point, we develop an artificial, automatically generated corpus to bootstrap our models as outlined in the following sections.

3   Method

3.1   Task, Intent and Slot Definition

Two main challenges for NLU in the smart-home environment are syntactic and linguistic variability; and underspecified commands. For the ambiguous command "turn on the fan", the NLU must identify the correct fan in the home

Table 1. Examples of NLU annotated voice commands.

Sentence	Intent and slots, in format INTENT(SLOT-LABEL=slot-value="text")
"are the lights upstairs on?"	CHECK_DEVICE_GROUP(DEVICE=light="lights", LOCATION-FLOOR=1="upstairs", device-setting=on="on")
"call the doctor"	CONTACT(PERSON-OCCUPATION=doctor="doctor")
"what time is it?"	GET_WORLD_PROPERTY(WORLD_PROPERTY=time="time")
"hey, can you hit the light?"	SET_DEVICE(ACTION=change="hit", DEVICE=light="the light")

based on the user's current location and activity. The NLU must also identify the same intent from a more syntactically complex utterance such as "can you turn on the fan". Similarly, "a bit more" following the command "raise the blinds a bit" must be *inferred* to be a request to repeat the previous action. This syntactic variability and underspecification make NLU development a daunting task. For the current version of our artificial corpus, we focused on understanding commands *without context* and with one intent per utterance, while still tackling the issue of syntactic and linguistic variability.

The semantics of our artificial corpus were defined and developed around an existing smart-home Amiqual4Home (https://amiqual4home.inria.fr) as described in more detail in Sect. 4. The resulting artificial corpus contains seventeen slot categories and eight intent classes. Intents are divided into four main categories: contact which allows a user to place a call; set to make changes to the state of objects in the smart-home; get to query the state of objects as well as properties of the world at large; and check to check the state of an object.

The slot labels are divided into eight categories: the action to perform, the device to act on, the location of the device or action, the person or organization to be contacted, a device component, a device setting and the property of a location, device, or world. Table 1 provides representative examples of the annotated voice commands, used in a flat slot-filling approach. Differing from previous work, *slot-label* prediction is augmented with *slot-value* prediction to pass sufficient required information to the decision making unit.

4   Data

The semantics for intents and slots defined in Sect. 3.1 were used to automatically generate artificial data as well as to annotate a real dataset.

4.1   Artificial Corpus Generation

The core of the corpus generator is a feature-based generative grammar, built around an open-source NLTK python library to which feature-respecting top-down grammar generation was added. Unification functions limit feature propagation between rules to only those features which are explicitly specified in order to avoid conflicting features. The grammar defines intents (Sect. 3.1) as a composition of their possible constituents, with fine-grained constraints on generation.

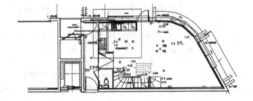

Fig. 1. Instrumented kitchen.   **Fig. 2.** Ground floor: kitchen and living room.

Table 2. NLU datasets

Dataset	# intents	# slots	# values	Avg. # values/slot	# utterances
Port-media*	4	32	450	13.6 ± 21.3	20260
Synthetic data	8	17	57	3.35 ± 4.6	28000
VocADom	7	12	46	3.91 ± 3.7	4610

*4 intent classes were extracted based on the manually labeled slots in Port-Media

For a rule that defines the slots of the intent set_device and can generate the command "open the door", the Slot_action has the feature ACTION whereas Slot_device has the feature ALLOWABLE_ACTION. Both those features are set to the same variable value which makes sure we only generate phrases with an allowable action to a particular device. Subsequent rules, contain other linguistic features such as gender and number agreement. Furthermore, domain constraints are defined for object location in the smart-home. Unification of features disallows nonsensical utterances such as "turn on the dishwasher in the bedroom".

Furthermore, syntactical variation was added to the grammar rules, such as, French interrogative constructions with the particle ("est-ce que"). The resulting vocabulary comprises 207 word types. Counting only lemmas, it contains 23 nouns denoting devices and 23 verbs denoting actions. The grammar generates about 28,000 phrases, each annotated with an intent and slots.

4.2 VocADom **Real Dataset Acquisition**

The real dataset was recorded with users interacting in realistic conditions in Amiqual4Home. This $87\,m^2$ smart-home with a kitchen, living room, bedroom and bathroom, is equipped with home automation systems, multimedia devices, and microphone arrays (Figs. 1 and 2). A control room centralizes remote monitoring, recording of sensors, and control of the home devices. Eleven participants uttered voice commands while performing scripted activities of daily living for about one hour of recording per participant. In the first half of the experiment, participants uttered unrestricted voice commands; in the second half, voice commands were restricted by a pre-defined grammar. Using a wizard-of-Oz approach, out-of-sight experimenters enacted user commands, acting as a 'perfect' NLU system. The VocADom corpus includes about twelve hours of audio signal and logs from the automation system. Speech was manually transcribed and 1,650 utterances, annotated with intents and slots, were used as the validation dataset for our experiments. Sentences without intent class were excluded.

For comparison purposes, we make use of the Port-Media dataset [8] of French-language tourist information and ticket reservations for the 2010 Avignon music festival. It is of the same size as the artificial data and rich in terms of slot and value labels. The dataset contains natural utterances of 140 speakers in a simulated telephone booking task with slot and value label annotations. A comparison between the Port-Media corpus, the synthetic and VOCADOM datasets is provided in Table 2.

5 Experiment

To evaluate the synthetic smart-home corpus, we examine performance of state-of-the-art NLU models trained on the artificial corpus and tested on the real corpus. We chose a Triangular Conditional Random Field model (Tri-CRF), a neural network with attention (Att-RNN), and one open-source commercial tool, Rasa, as a baseline. For comparison, we also evaluated performance of the models on the Port-Media dataset.

5.1 Tri-CRF, Att-RNN and Rasa-NLU Models

The Tri-CRF model from [6,7] is an extension of a linear chain Conditional Random Field (CRF). Linear CRFs model the conditional probability distribution of the output label sequence, given the input sequences (sentences): each observed word x_t in a sequence is conditionally dependent on its corresponding *unobserved* label y_t. The label y_t is also conditionally dependent on the previous label y_{t-1}. The Tri-CRF extends this model by adding an intent z for which each slot y_t (and also potentially each word x_t) is dependent on the overall sentence intent z. To reduce training time, we pruned low-probability intents (<0.1%) and initialized the weights using the pseudo-likelihood (for 30 training iterations). Training proceeded for 200 iterations.

The Attention RNN (Att-RNN) model from [9], is a recurrent encoder decoder architecture for simultaneous intent detection and slot labeling. In our implementation of Att-RNN, the input words are first passed to a 128-unit embedding layer. The bi-directional LSTM encoder and decoder are each a single layer of 128 units. Training is performed using stochastic gradient descent (SGD) with a batch size of 16, using gradient clipping at a norm of 5.0, dropout with a keep-probability of 0.5 and training was allowed to continue for 10,000 training steps. We selected the trained model with the highest F1 score on the slot labeling task on the validation set. For Tri-CRF and Att-RNN, two models are trained, one to predict intent and slot-labels (Att-RNN-Labels) and one for slot-values (Att-RNN-Values).

Rasa NLU (https://rasa.ai/products/rasa-nlu/), an open-source tool for building NLU pipelines, is used as a baseline. Unlike Tri-CRF and the Att-RNN, Rasa does not predict a sequence of slots for each input word, but rather a set of slot-labels and slot-values associated with different segments of the input. The selected Rasa configuration, 'spacy_sklearn', uses a linear chain CRF to classify slot-labels and a lookup table to determine slot-values. Separately, the model uses a linear SVM based on pre-trained word-embeddings to classify intents. The embeddings are drawn from the spacy language model 'fr_depvec_web_lg', trained using word2vec on text data from Wikipedia, OpenSubtitles and Wikinews. The final vocabulary contains 1,184,651 words and the embeddings are vectors of length 300.

5.2   Results

Standard metrics were used for both intent classification and slot-labeling: precision, recall and F1-score. For slot-label and slot-value classification, metrics are calculated by comparing labels for words across all examples. The Port-Media and the artificial datasets are randomly divided into a training set of 90% and a development set of 10%.

Table 3. Learning results on Port-Media and VOCADOM datasets

Model	Precision	Recall	F1
Att-RNN-Intent	**97.56**	**97.56**	**97.56**
Tri-CRF-Intent	96.42	96.43	96.36
Rasa-Intent	92.20	92.52	92.26
Att-RNN-Labels	**95.96**	**96.36**	**96.11**
Tri-CRF-Labels	95.31	95.74	95.39
Rasa-Labels	95.17	94.22	94.16
Att-RNN-Values	**94.85**	**95.73**	**95.08**
Tri-CRF-Values	92.01	93.49	92.32
Rasa-Values	93.94	93.73	93.34

(a) NLU performances on Port-Media development dataset.

Model	Precision	Recall	F1
Att-RNN-Intent	**93.77**	**90.28**	**91.30**
Tri-CRF-Intent	84.11	79.47	76.36
Rasa-Intent	90.48	71.39	76.57
Att-RNN-Labels	69.19	66.24	66.09
Tri-CRF-Labels	77.28	52.65	60.64
Rasa-Labels	**85.72**	**73.54**	**79.03**
Att-RNN-Values	43.02	30.51	35.00
Tri-CRF-Values	51.33	25.52	33.51
Rasa-Values	**68.56**	**56.73**	**61.95**

(b) NLU performances on VOCADOM dataset.

Table 3a reports results on the Port-Media dataset. The performance is only reported for the development set. Att-RNN provides the highest performances for the three tasks, with Tri-CRF competitive in all but the slot-value tasks. Both outperform Rasa. These results demonstrate the level of performance that can be achieved on real data, similar to real smart-home data. Accuracies on the artificial data development set are quasi-perfect, due to the very homogeneous nature of the synthetic corpus. Results of testing these models trained on the artificial data on the real VOCADOM validation dataset of 1,650 utterances are provided in Table 3b. Overall performances on VOCADOM are worse than on Port-Media and particularly bad for slot-label and slot-value prediction. However, the high intent prediction accuracies on Port-Media are biased due to the high frequency of 'none' intents in the corpus, the low number of intent classes (4) and overlap between slots and intents. The results of slot-filling on VOCADOM are unsatisfactory. Errors are more randomly distributed over several categories and therefore more difficult to analyze. This is probably due to the significantly higher syntactic and lexical variation in the VOCADOM real dataset. Repetitions, disfluencies

Table 4. Detailed results for Att-RNN

Intent	F1
check_device	76.47
check_device_group	71.69
set_device	97.09
set_device_group	88.65
set_room_property	70.59
(a) intents	

Slot	F1
action	62.03
device	84.06
device-setting	10.42
location-room	66.90
room-property	70.00
(b) slots	

Value	F1
close	72.81
light	80.83
lower	17.53
open	67.51
turn off	41.51
turn on	68.07
(c) values	

and interjections (ex. "euh") result in utterances that are syntactically different from the artificial dataset. The 3-gram artificial language model perplexity on the real corpus is of 58 (without the <s> tag). The number of OOV is also high, with 142 words not occurring in the artificial dataset.

Contrary to TRi-CRF and Att-RNN, Rasa performs well on slot labeling as it uses a word embedding layer which allows it to deal with the high number of OOV words indicating that artificial data generation benefits from external resources. Compared to results on the manually annotated Port-Media corpus, the poorer slot-filling results indicate that the automatic slot-label generation algorithm of the synthetic corpus can still be improved. Detailed results for Att-RNN predictions are given in Table 4.

As hypothesized by [6,9], the joint approaches of the Tri-CRF and Att-RNN outperform Rasa's SVM-based intent classification on the VocADom dataset. This shows that the synthetic corpus contains enough information to train isolated intent models to be applied to real data.

6 Conclusion

In this paper, we address the lack of smart-home NLU training corpora by building a customizable automatic corpus generator for the smart-home domain. The corpus was evaluated by training two state-of-the-art models which were tested on a small but real smart-home dataset and compared to our baseline. Comparison of the models allowed us to pinpoint the artificial corpus or the models as main source of prediction errors. Both the Tri-CRF and the Att-RNN performed well on the large real Port-Media dataset and on the artificial voice command dataset. However lower performance on the small real smart-home dataset demonstrates difficulty handling its increased naturalness, vocabulary and syntactic variation. Both corpora are intended to be made available to the community. Future research aims to increase the naturalness of the generated corpus, by jointly inserting generic language models into the task-specific learning phase and taking into account command context and history (see [1]), and aims to simultaneously predict intents, slots and values. Such compound models have, at the time of writing, not been previously explored in the slot-filling literature and would be a useful and novel contribution to the field.

Acknowledgments. This work is part of the VOCADOM project founded by the French National Research Agency (Agence Nationale de la Recherche)/ANR-16-CE33-0006.

References

1. Bapna, A., Tur, G., Hakkani-Tur, D., Heck, L.: Sequential dialogue context modeling for spoken language understanding. In: Proceedings of the 18th Annual SIGdial Meeting on Discourse and Dialogue (2017)
2. Bapna, A., Tur, G., Hakkani-Tur, D., Heck, L.: Towards zero-shot frame semantic parsing for domain scaling. arXiv:1707.02363 [cs] (2017)
3. Chahuara, P., Portet, F., Vacher, M.: Context-aware decision making under uncertainty for voice-based control of smart home. Expert. Syst. Appl. **75**, 63–79 (2017). https://doi.org/10.1016/j.eswa.2017.01.014. http://www.sciencedirect.com/science/article/pii/S0957417417300234
4. Dumitrescu, S.D.: Cassandra smart-home system description. In: 2017 International Conference on Speech Technology and Human-Computer Dialogue (SpeD), pp. 1–6 (2017). https://doi.org/10.1109/SPED.2017.7990440
5. Huang, L., Sil, A., Ji, H., Florian, R.: Improving slot filling performance with attentive neural networks on dependency structures. arXiv:1707.01075 [cs] (2017)
6. Jeong, M., Lee, G.G.: Triangular-chain conditional random fields. IEEE Trans. Audio Speech Lang. Process. **16**(7), 1287–1302 (2008). https://doi.org/10.1109/TASL.2008.925143
7. Jeong, M., Lee, G.G.: Multi-domain spoken language understanding with transfer learning. Speech Commun. **51**(5), 412–424 (2009). https://doi.org/10.1016/j.specom.2009.01.001. http://www.sciencedirect.com/science/article/pii/S0167639309000028
8. Lefèvre, F., et al.: Leveraging study of robustness and portability of spoken language understanding systems across languages and domains: the PORTMEDIA corpora. In: LREC, pp. 1436–1442 (2012)
9. Liu, B., Lane, I.: Attention-based recurrent neural network models for joint intent detection and slot filling. In: Interspeech, pp. 685–689 (2016). https://doi.org/10.21437/Interspeech.2016-1352. http://www.isca-speech.org/archive/Interspeech_2016/abstracts/1352.html
10. Manishina, E., Jabaian, B., Huet, S., Lefèvre, F.: Automatic corpus extension for data-driven natural language generation. In: Proceedings of the Tenth International Conference on Language Resources and Evaluation LREC 2016, Portorož, Slovenia, 23–28 May 2016 (2016)
11. Mesnil, G., et al.: Using recurrent neural networks for slot filling in spoken language understanding. IEEE/ACM Trans. Audio Speech Lang. Process. (TASLP) **23**(3), 530–539 (2015). http://dl.acm.org/citation.cfm?id=2876380
12. Möller, S., Gödde, F., Wolters, M.: Corpus analysis of spoken smart-home interactions with older users. In: Proceedings of the 6th International Conference on Language Resources and Evaluation (2008)
13. Portet, F., Vacher, M., Golanski, C., Roux, C., Meillon, B.: Design and evaluation of a smart home voice interface for the elderly – acceptability and objection aspects. Pers. Ubiquitous Comput. **17**(1), 127–144 (2013). https://doi.org/10.1007/s00779-011-0470-5

14. Takahashi, S., Morimoto, T., Maeda, S., Tsuruta, N.: Dialogue experiment for elderly people in home health care system. In: Matoušek, V., Mautner, P. (eds.) TSD 2003. LNCS (LNAI), vol. 2807, pp. 418–423. Springer, Heidelberg (2003). https://doi.org/10.1007/978-3-540-39398-6_60

15. Tran, Q., Zukerman, I., Haffari, G.: A hierarchical neural model for learning sequences of dialogue acts, pp. 428–437 (2017). https://doi.org/10.18653/v1/E17-1041

16. Vacher, M.: Evaluation of a context-aware voice interface for ambient assisted living: qualitative user study vs. quantitative system evaluation. ACM Trans. Access. Comput. **7**(2), 5:1–5:36 (2015). https://doi.org/10.1145/2738047

17. Vacher, M., Lecouteux, B., Chahuara, P., Portet, F., Meillon, B., Bonnefond, N.: The sweet-home speech and multimodal corpus for home automation interaction. In: The 9th Edition of the Language Resources and Evaluation Conference (LREC), pp. 4499–4506 (2014). https://hal.archives-ouvertes.fr/hal-00953006/

18. Wang, Y., Deng, L., Acero, A.: Semantic frame-based spoken language understanding. In: Spoken Language Understanding: Systems for Extracting Semantic Information from Speech. Wiley (2011)

Classification of Formal and Informal Dialogues Based on Emotion Recognition Features

György Kovács[1,2(✉)]

[1] MTA Research Institute for Linguistics, Budapest, Hungary
[2] MTA-SZTE Research Group on Artificial Intelligence, Szeged, Hungary
gykovacs@inf.u-szeged.hu

Abstract. Social context is an important part of human communication, hence it is also important for improved human computer interaction. One aspect of social context is the level of formality. Here, motivated by the difference observed between the emotional annotation of formal and informal dialogues in the HuComTech corpus, we introduce a content-free classification scheme based on feature sets designed for emotion recognition. With this method we attain an error rate of 8.8% in the classification of formal and informal dialogues, which means a relative error rate reduction of more than 40% compared to earlier results. By combining our proposed method with earlier models, we were able to further reduce the error rate to below 7%.

Keywords: HuComTech · SVM · MultiBoost · Affective computing

1 Introduction

In human dialogue, facets of social context such as the formality of a speech situation, are important aspects of human communication [13] and by extension human-computer interaction (HCI). To understand how the social context affects HCI, let us consider the case of speech recognition. Human speakers adjust their speaking style automatically to the social context, which in turn obliges the listener to make the same adjustment [11], regardless of whether the listener is another human or a computer. Another reason for considering social context is that by doing so, we can achieve a more natural user experience in dialogue systems [1]. For these reasons, our goal here is to classify dialogue segments into two categories based on the formality of the speaking situation.

For the classification of dialogue segments into the formal and informal categories, we examine three different feature sets taken from the OpenSMILE toolkit [6] designed for emotion recognition. We do so based on the assumption that the formal context not only prompts more careful speech [12], but also a more reserved speaking style, displaying emotion in a different distribution to that displayed in an informal setting. We will examine this assumption in Sect. 2.

© Springer Nature Switzerland AG 2018
P. Sojka et al. (Eds.): TSD 2018, LNAI 11107, pp. 518–526, 2018.
https://doi.org/10.1007/978-3-030-00794-2_56

Afterwards, we will describe the feature sets and the classification methods used in Sect. 3. Then we will discuss our experiments and the resulting error rates in Sect. 4, and finally round off with some key conclusions and possible directions for future work in Sect. 5.

2 Research Material

The experiments reported in this study were carried out on the dialogues of the multimodal corpus designed within the framework of the HuComTech project [10].

2.1 HuComTech Corpus

The corpus was recorded using conversations taken from 111 native Hungarian speakers with the aim of analysing the underlying structure of human-human communication [9], and it contains approximately 50 h of spontaneous speech. With each speaker two dialogues were recorded; that is, a simulated job interview, and an informal conversation discussing various predetermined topics ranging from jokes to their saddest memories. In each dialogue there were only two participants, namely the speaker (interviewee), and the agent (interviewer), who always conducted formal and the informal dialogues.

In our experiments for the classification of segments taken from these dialogues, we used the train/development/test partitioning established earlier for the task [19]. The reason for this decision was to make our results comparable with those reported by Szekrényes and Kovács [19]. In their study (representing the only content-free solution we found for this problem in the literature), they attempted to distinguish between dialogue segments from these scenarios based on prosodic information, and turn-taking (attaining a classification error rate of 14.8%). Their method, however, relied on the availability of manual annotation of turn-taking, or an effective automatic speaker diarisation system. Here, we tackle this problem based on the emotional content of dialogues, and apply feature sets that can be calculated in an entirely automatic fashion.

2.2 Annotation of Emotion

The emotional annotation of the audio data was carried out as follows. First, an automatic phrase boundary detector program was applied . Then the resulting boundaries were manually checked, and the proper labels were manually assigned to the phrases at five levels. At the emotional level nine different labels were used. These labels were silence (silences longer than 250 ms), overlapping speech, and seven emotional labels (neutral, sad, happy, surprised, recalling, tense, and other), based on the expressed emotion of the interviewee [14]. We examined the prevalence of these seven emotions in the formal and informal dialogues.

We worked under the supposition that different emotional classes would dominate the informal and the formal dialogues. To test this supposition, we examined the ratio of each of the seven labels over the total in the interviews. Using student's paired t-test, we found that the difference in the proportion of emotional labels between formal and informal dialogues is significant in five out of the seven emotional categories: sad (with $p < 3 \cdot 10^{-13}$), happy (with $p < 3 \cdot 10^{-19}$), surprised (with $p < 2 \cdot 10^{-3}$), recalling (with $p < 2 \cdot 10^{-12}$), and other (with $p < 9 \cdot 10^{-6}$). The ratio of four of these emotions over the total in the formal and informal dialogues can be seen in the box plots of Fig. 1.

As can be seen in Fig. 1, the recalling emotion is significantly more prevalent in the formal dialogues than in the informal ones, while the other three emotions are relatively more common in the informal conversations.

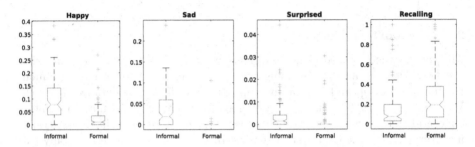

Fig. 1. Comparison of the ratio of happiness, sadness, surprise, and recall over the total in the informal and formal dialogues.

3   Methodology

3.1   Feature Sets for Emotion Recognition

Given the significant difference in the proportion of various emotional labels between the formal and informal dialogues, one possibility for automatic classification would be to train a machine learning model for the recognition of emotions from the audio signal, and then use the output of this model as the basis for features to make the informal/formal distinction. We decided against this course of action for two reasons, however. First, emotional annotation was carried out by only one person on each file, and while his work was checked, we still have no way of calculating useful measures for the reliability of the annotation, such as the inter-rater agreement [17]. Furthermore, the expressed emotion of only one party was annotated, which means that we have no information on the expressed emotions of the interviewer. Instead of using the feature sets proposed for the recognition of emotion on the problem of distinguishing between informal and formal dialogue segments indirectly, we shall try to apply the emotion recognition features directly on our task. We will examine three different feature sets taken from the OpenSMILE toolkit [6]. These feature sets are as follows:

1. IS09: the feature set used in the Classifier sub-challenge of the Interspeech 2009 Emotion challenge, containing 384 features, which were at the time "the most common and [...] promising" [15].

2. Emobase: the OpenSMILE emobase feature set, containing 988 acoustic features [5]
3. Emobase2010: the 1582 dimensional emobase2010 feature set, based on the feature set used in the Interspeech 2010 Paralinguistic challenge [16].

Before further processing, features calculated with each feature set were standardised so as to have a zero mean and unit variance. In the process, features whose values had a standard deviation of zero (i.e. constants) were eliminated.

3.2   Machine Learning Techniques

Here, unlike Szekrényes and Kovács [19], we did not cut dialogue segments where the interviewer and the interviewee discuss one topic on their agenda into smaller, overlapping segments, but used each dialogue segment as a training, development or test instance. This means that we had 1410 examples altogether (1058, 136, and 216 examples in the train, development and test set, respectively). In the experiments, we selected the machine learning techniques accordingly.

Support Vector Machines. Support Vector Machines (SVMs) in classification tasks use hyperplanes for the separation of classes. Here, we applied the libSVM implementation of this algorithm [4], with a linear kernel. For parameter optimisation we used the development set, and ran the complexity parameter between 2^{-13} and 2^{13} (with a step size of 2 in the exponent).

Boosting. For the sake of comparison, we experimented with various boosting techniques using the MultiBoost package [2]. Based on the results of our initial experiments with the Emobase2010 feature set using AdaBoost [7], Lazy-Boost [8], and FilterBoost [3], we decided to carry out our experiments with the FilterBoost method, using a decision tree with eight leaves. To optimise the number of iterations, we trained 100 models for each feature set, and used the average performance of these models after every 10 iterations on the development set as a stopping criterion. Besides the average performance of our models, we also report the performance attained when each instance is classified based on the majority vote of the 100 models.

Combination Methods. In addition to examining the three feature sets described above, we also examined the combination of these feature sets. Here, we employed two methods of fusion for this:

- **Early fusion:** Here, we combined the three feature sets, and repeated our experiments using the resulting feature set. The dimensionality of this feature set was 2,378, as there were several overlaps in the original sets.
- **Late fusion:** Here, we took the weighted sum of the estimated posterior probability scores created using the models trained on the individual feature sets, and classified each instance based on the resulting score. In the case of

the FilterBoost models, we derived the posterior probability estimates from the results of voting (i.e. the estimated posterior value for each class was the number of models voting for the given class divided by the number of models altogether – 100). The final weights were selected based on the results of applying different weighting schemes on the development set.

We also attempted to combine the posterior probability estimates obtained from the Deep Rectifier Neural Net (DRN) models trained by Szekrényes and Kovács [19] on features got from the ProsoTool [18] algorithm. For this combination, we also used the method of late fusion.

Table 1. Recognition error rates attained using FilterBoost on the individual feature sets of the OpenSMILE toolkit (the best results are shown in bold)

Feature set	Meta-parameter	Error rates (average)		Error rates (voting)	
		Dev. set	Test set	Dev. set	Test set
IS09	$\#iteration = 1850$	14.5%	12.8%	14.8%	13.0%
Emobase	$\#iteration = 800$	11.0%	11.2%	11.0%	11.1%
Emobase2010	$\#iteration = 1850$	**9.2%**	**8.8%**	**8.8%**	**8.8%**

4   Experiments and Results

4.1   Experiments Using FilterBoost

Here, we first report our classification results obtained using FilterBoosting. Table 1 lists the results obtained when FilterBoosting was applied on individual feature sets of the OpenSMILE toolkit. When comparing the results obtained using different feature sets, we observe that the lowest error rates (regardless of whether we averaged the results of different models, or based on our decision on their majority vote) were obtained using the Emobase2010 feature set, while the highest error rates were invariably got using the IS09 feature set. We also observe that results obtained by averaging and results obtained by voting are very similar in each case. Most of the differences are likely due to the limited size of the development and test set. Lastly, we can see that the results reported on the test set are very similar, and sometimes even better than those reported on the development set, which suggests that despite the many iterations made, overfitting was probably not an issue.

The results obtained when using the method of FilterBoosting with early fusion and late fusion are listed in Table 2. From the meta-parameter column, we can see here that the best results were obtained when the posterior estimates from the FilterBoost model using the IS09 feature set and the Emobase feature set were assigned a weighting factor of 0.25, while the posterior estimates from the FilterBoost model using the Emobase2010 feature set were assigned a weighting factor of 0.5. However, when comparing these results with those in Table 1 we can see that neither method of combination actually improved the resulting error rates.

Table 2. Recognition error rates attained using FilterBoost and various methods of combination (the best results are shown in bold)

Method of combination	Meta-parameter	Error rates (average)		Error rates (voting)	
		Dev. set	Test set	Dev. set	Test set
Early fusion	$\#iteration = 3000$	11.00%	9.8%	10.3%	9.7%
Late fusion	$weights = [0.25, 0.25, 0.5]$	–	–	**8.8%**	**9.3%**

Table 3. Recognition error rates attained using SVMs on the individual feature sets of the OpenSMILE toolkit (the best results are shown in bold)

Feature set	Meta-parameter	Error rates	
		Dev. set	Test set
IS09	$C = 2^{-9}$	13.2%	13.0%
Emobase	$C = 2^{-9}$	**11.0%**	12.0%
Emobase2010	$C = 2^{-9}$	**11.0%**	**10.2%**

4.2 Experiments Using SVMs

Next, we report our classification results obtained using SVMs. Table 3 lists the results we got using individual emotion recognition feature sets. Here, similar to the FilterBoost results, the error rates are the highest when using the IS09 feature set, and lowest when using the Emobase2010 feature set. We also notice, however, that the performance of the Emobase and Emobase2010 feature sets on the development set are identical. And as we found earlier, the difference between error rates got on the development set and the test set is quite small.

The results obtained by combining the three individual feature sets in various ways are given in Table 4. Here, we observe that when using support vector machines, the performance of early fusion and performance of late fusion do not differ markedly. When comparing the results achieved with SVMs to those achieved with FilterBoost, we find that – unlike in the case of FilterBoost – when using SVMs both fusion methods led to lower error rates. It is also interesting to note that despite the similar individual performance of the Emobase and Emobase2010 feature sets, in the combination found to be optimal on the development set, the weighting factor assigned to the latter is about twice as big as the weighting factor assigned to the former.

4.3   Combination of Different Classifiers Using Late Fusion

In our last set of experiments, we experimented with the late fusion of models derived from applying different machine learning algorithms. The results of these experiments are shown in Table 5. Here, the weighting meta-parameters are reported in such a way that the first element of the vector represents the weighting factor assigned to the method listed in the first column of the table, and the second element of the vector represents the weighting factor assigned to the method listed in the second column of the table.

Table 4. Recognition error rates attained using SVMs and various methods of combination (the best results are shown in bold)

Method of combination	Meta-parameter	Error rates	
		Dev. set	Test set
Early fusion	$C = 2^{-11}$	**10.3%**	9.7%
Late fusion	$weights = [0.20, 0.25, 0.55]$	**10.3%**	**9.3%**

Table 5. Recognition error rates attained with late fusion of our current models, and the DRN models of Szekrényes and Kovács [19] (the best results are shown in bold)

Methods		Meta-parameter	Error rates	
$Method_1$	$Method_2$		Dev. set	Test set
FilterBoost on Emobase2010	SVM early fusion	$weights = [0.25, 0.75]$	8.8%	8.3%
FilterBoost on Emobase2010	DRN	$weights = [0.25, 0.75]$	8.1%	8.8%
SVM early fusion	DRN	$weights = [0.25, 0.75]$	8.8%	8.3%
SVM late fusion	DRN	$weights = [0.20, 0.80]$	**6.6%**	**6.9%**
DRN		*	14.0%	14.8%
Human performance		–	–	43.5%

* A detailed discussion of the meta-parameters of the DRN applied by Szekrényes and Kovács is beyond the scope of this paper. For more details, see [19].

First, we examined the combination of the best performing methods in the current paper. As can be seen in Table 5, when combining the probability estimates obtained from the FilterBoost and SVM models, the resulting error rates on the development set are not lower than those obtained when using just the FilterBoost model. We get similar results when combining the output of the FilterBoost model with those of the DRN model of Szekrényes and Kovács [19]. However, by combining the posterior probability estimates derived from our SVM models with those derived from the above-mentioned DRN models using late fusion, we can greatly lower the classification error rates. Using this technique we can achieve a relative error rate reduction of about 84.1%, 53.4%, and 21.6% compared to the human performance, the DRN results, and the results of our best individual model, respectively. Note that the human performance is surprisingly low, which we attribute to the fact that the human subjects also had to solve the problem in a content-free manner (i.e. without lexical information).

5 Conclusions and Future Work

In this study, we examined different emotion recognition feature sets for the task of the classification of formal and informal dialogue segments. Although we obtained the lowest error rates when combining different models, the use of individual models also dramatically lowered the error rates compared to those reported earlier. This fact, and the investigation of emotion labels in the corpus seem to support our hypothesis regarding the emotional content of formal and informal dialogues. However, as the feature sets utilised do not exclusively measure the emotional content, and the fact that only one corpus was examined, a conclusive confirmation of our hypothesis would require doing more experiments.

Further experiments might also be useful for improving the classification rates attained. One possibility would be to apply feature selection on the feature sets. Another might be to increase the number of instances by either including the dialogue segments shorter than 30 s (that had been excluded from the train/development/test partitioning by Szekrényes and Kovács) in our investigation, or by partitioning dialogues into smaller segments. Lastly, more corpora and feature sets could be included in the study.

Acknowledgments. The research reported in the paper was conducted with the support of the Hungarian Scientific Research Fund (OTKA) grant #K116938.

References

1. André, E., Rehm, M., Minker, W., Bühler, D.: Endowing spoken language dialogue systems with emotional intelligence. In: André, E., Dybkjær, L., Minker, W., Heisterkamp, P. (eds.) ADS 2004. LNCS (LNAI), vol. 3068, pp. 178–187. Springer, Heidelberg (2004). https://doi.org/10.1007/978-3-540-24842-2_17
2. Benbouzid, D., Busa-Fekete, R., Casagrande, N., Collin, F.D., Kégl, B.: MULTI-BOOST: a multi-purpose boosting package. J. Mach. Learn. Res. **13**, 549–553 (2012)
3. Bradley, J., Schapire, R.: FilterBoost: regression and classification on large datasets. In: Advances in Neural Information Processing Systems, vol. 20, pp. 185–192. The MIT Press (2008)
4. Chang, C.C., Lin, C.J.: LIBSVM: a library for support vector machines. ACM Trans. Intell. Syst. Technol. **2**, 27:1–27:27 (2011)
5. Eyben, F., Wöllmer, M., Schuller, B.: The Munich open speech and music interpretation by large space extraction toolkit (2010)

6. Eyben, F., Wöllmer, M., Schuller, B.: openSMILE: the Munich versatile and fast open-source audio feature extractor. In: Proceedings of ACM (MM), pp. 1459–1462 (2010)

7. Freund, Y., Schapire, R.E.: A decision-theoretic generalization of on-line learning and an application to boosting. J. Comput. Syst. Sci **55**, 119–139 (1997)

8. G. Escudero, L.M., Rigau, G.: Boosting applied to word sense disambiguation. In: Proceedings of ECML, pp. 129–141 (2000)

9. Hunyadi, L.: Multimodal human-computer interaction technologies. Theoretical modeling and application in speech processing. Argumentum, pp. 240–260 (2011)

10. Hunyadi, L., Váradi, T., Szekrényes, I.: Language technology tools and resources for the analysis of multimodal communication. In: Proceedings of LT4DH, pp. 117–124. University of Tübingen, Tübingen (2016)

11. Ingram, J.C.L.: Neurolinguistics. Cambridge University Press, Cambridge (2007)

12. Kristiansen, T.: Attitudes, ideology and awareness. In: Wodak, R., Johnstone, B., Kerswill, P. (eds.) The SAGE Handbook of Sociolinguistics, pp. 265–278. SAGE Publishing, Thousand Oaks (2011)

13. Labov, W.: The Social Stratification of English in New York City. Cambridge University Press, Cambridge (1996)

14. Pápay, K., Szeghalmy, S., Szekrényes, I.: HuComTech multimodal corpus annotation. Argumentum **7**, 330–347 (2011)

15. Schuller, B., Steidl, S., Batliner, A.: The INTERSPEECH 2009 emotion challenge. In: Proceedings of INTERSPEECH, pp. 312–315 (2009)

16. Schuller, B., et al.: The INTERSPEECH 2010 paralinguistic challenge. In: Proceedings of INTERSPEECH, pp. 2822–2825 (2010)

17. Siegert, I., Böck, R., Wendmeuth, A.: Inter-rater reliability for emotion annotation in human-computer interaction: comparison and methodological improvements. Multimodal User Interfaces **8**, 17–28 (2014)

18. Szekrényes, I.: ProsoTool, a method for automatic annotation of fundamental frequency. In: Proceedings of CogInfoCom, pp. 291–296 (2015)

19. Szekrényes, I., Kovács, G.: Classification of formal and informal dialogues based on turn-taking and intonation using deep neural networks. In: Proceedings of SPECOM, pp. 233–243 (2017)

Author Index

Printed in the United States
By Bookmasters